Engineering with Excel

Fourth Edition

Ronald W. Larsen
Montana State University

PEARSON

Boston • Columbus • Indianapolis • New York
San Francisco • Upper Saddle River • Amsterdam
Cape Town • Dubai • London • Madrid • Milan
Munich • Paris • Montreal • Toronto • Delhi
Mexico City • Sao Paulo • Sydney • Hong Kong
Seoul • Singapore • Taipei • Tokyo

Vice President/Editorial Director, Engineering/Computer Science: *Marcia J. Horton*
Executive Editor: *Holly Stark*
Senior Marketing Manager: *Tim Galligan*
Marketing Assistant: *Jon Bryant*
Project Manager: *Pat Brown*
Creative Director: *Jayne Conte*
Art Director: *Kenny Beck*
Cover Designer: *Bruce Kenselaar*
Media Editor: *Daniel Sandin*
Full-Service Project Management: *Vijayakumar Sekar*
Composition: *Jouve India*
Printer/Binder: *Edwards Brothers*
Cover Printer: *Lehigh-Phoenix*

All Microsoft screenshots reprinted with permission from Microsoft. All rights reserved.

Credits and acknowledgments borrowed from other sources and reproduced, with permission, in this textbook appear on appropriate page within text.

Many of the designations by manufacturers and seller to distinguish their products are claimed as trademarks. Where those designations appear in this book, and the publisher was aware of a trademark claim, the designations have been printed in initial caps or all caps.

Library of Congress Cataloging-in-Publication Data

Larsen, Ronald W.
  Engineering with Excel / Ronald W. Larsen. — 4th ed.
      p. cm.
  Includes index.
  ISBN 0-13-278865-9
  1. Engineering–Data processing.   2. Microsoft Excel (Computer file)   I. Title.
  TA345.L37 2013
  620.00285'554—dc23

                                   2011044249

PEARSON

10 9 8 7 6 5 4 3 2 1

ISBN 10: 0-13-278865-9
ISBN 13: 978-0-13-278865-6

Contents

About This Book

The *Engineering with Excel* text has been updated to reflect the latest versions of Excel and Windows (Excel 2010 operating in Windows 7). Changes from Excel 2007 to Excel 2010 can be described as refinement rather than major changes, but certain changes, such as the elimination of the Office button, have changed the way you use the program and the text has been updated to reflect the new approach. In general, the changes in Excel 2010 streamline the user interface. With the fourth edition of *Engineering with Excel*:

- All screen captures have been updated for Excel 2010.
- Menu operations have been updated to reflect Ribbon changes in Excel 2010 (but commands for previous versions are also listed for those using older versions).
- New methods for accessing chart formatting dialogs are described.
- The Paste menus in Excel 2010 show icons rather than text. The use of these menus is described in the text.
- The (iterative) Solver has been significantly upgraded with a new dialog. The use of the new Solver is covered in the text.
- There is a new Print dialog that combines features previously available on using Print and Print Preview. This is a significant enhancement that Excel users will appreciate.

1

Introduction to Excel

Objectives

After reading this chapter, you will know

- What an Excel worksheet is
- How to start using Excel
- How the Excel screen is laid out
- The fundamentals of using Excel
- Some options for organizing your worksheets
- How to print your worksheets
- How to save and reopen Excel files

1.1 INTRODUCTION

Spreadsheets were originally paper grids used by accountants and business people to track incomes and expenditures. When *electronic spreadsheets* first became available on personal computers, engineers immediately found uses for them. They discovered that many engineering tasks can be solved quickly and easily within the framework of a spreadsheet. Figure 1.1 shows how a spreadsheet can be used to calculate fuel efficiency.

Figure 1.1
Calculating fuel efficiency using a spreadsheet.

	A	B	C	D	E
1	Vehicle Mileage Calculation				
2					
3		Miles Traveled	Gallons of Fuel		Miles Per Gallon
4					
5		435	16.4		26.5
6					

Powerful modern spreadsheet programs like Microsoft Excel®[1] allow very complex problems to be solved right on an engineer's desktop. This text focuses on using Microsoft Excel, currently the most popular spreadsheet program in the world, to perform common engineering problems.

Excel is now much more than just an electronic implementation of a spreadsheet, and this is reflected in a change of nomenclature; what used to be called a spreadsheet is now more commonly referred to as a *worksheet*, and a collection of worksheets is called a *workbook*.

1.1.1 Nomenclature

Some conventions are used throughout this text to highlight the types of information:

- Key terms, such as the term "active cell" in the example below, are shown in italics the first time they are used.

Press [F2] to edit the *active cell.*

- Variables, formulas, and functions are shown in Courier font.

`=B3*C4`

- Individual keystrokes are enclosed in brackets.

Press the [Enter] key.

- Key combinations are enclosed in brackets.

Press [Ctrl-c] to copy the contents of the cell.

- Buttons that are clicked with the mouse are shown in a bold font.

Click **OK** to exit the dialog.

- Ribbon selections are indicated by listing the Tab, group, and button (or text box) separated by slashes. A boldface font is used for Ribbon selections to help them stand out in the text.

Use **Home/Font/Underline** to underline the selected text.

- Menu options for prior versions of Excel are indicated by listing the menu name and submenu options separated by slashes.

[Excel 2003: File/Open]

- Excel built-in function names are shown as follows:

Excel can automatically calculate the arithmetic mean using the
AVERAGE function.

1.1.2 Examples and Application Problems

We will use sample problems throughout the text to illustrate how Excel can be used to solve engineering problems. There are three levels of problems included in the text:

- *Demonstration Examples:* These are typically very simple examples designed to demonstrate specific features of Excel. They are usually single-step examples.

---

[1] Excel is a trademark of Microsoft Corporation, Inc.

- *Sample Problems:* These problems are slightly more involved and typically involve multiple steps. They are designed to illustrate how to apply specific Excel functions or capabilities to engineering problems.
- *Application Problems:* These are larger problems, more closely resembling the type of problem engineering students will see as homework problems.

The scope of the text attempts to include all engineering disciplines at the undergraduate level, with emphasis, especially in the first two-thirds of the text, on topics appropriate to freshman engineering students. There has been an attempt to cover a broad range of subjects in the examples and application problems and to stay away from problems that require significant discipline-specific knowledge.

1.1.3 What Is a Spreadsheet?

A *spreadsheet* is a piece of paper containing a grid designed to hold values. The values are written into the *cells* formed by the grid and arranged into vertical columns and horizontal rows. Spreadsheets have been used for many years by people in the business community to present financial statements in an orderly way. With the advent of the personal computer in the 1970s, the paper spreadsheet was migrated to the computer and became an *electronic spreadsheet.* The rows and columns of values are still there, and the values are still housed in cells. The layout of an electronic spreadsheet is simple, making it very easy to learn to use. People can start up a program such as Excel for the first time and start solving problems within minutes.

The primary virtue of the early spreadsheets was *automatic recalculation*: Any change in a value or formula in the spreadsheet caused the rest of the spreadsheet to be recalculated. This meant that errors found in a spreadsheet could be fixed easily without having to recalculate the rest of the spreadsheet by hand. Even more importantly, the electronic spreadsheet could be used to investigate the effect of a change in one value on the rest of the spreadsheet. Engineers who wanted to know what would happen if, for example, the load on a bridge was increased by 2, 3, or 4% quickly found electronic spreadsheets very useful.

Since the 1970s, the computing power offered by electronic spreadsheets on personal computers has increased dramatically. *Graphing capabilities* were added early on and have improved over time. *Built-in functions* were added to speed up common calculations. Microsoft added a *programming language* to Excel that can be accessed from the spreadsheet when needed. Also, the computing speed and storage capacity of personal computers has increased to such an extent that a single personal computer with some good software (including, but not limited to, a spreadsheet such as Excel) can handle most of the day-to-day tasks encountered by most engineers.

1.1.4 Why Use a Spreadsheet?

Spreadsheets are great for some tasks, but not all; some problems fit the grid structure of a spreadsheet better than others. When your task requires data consisting of columns of numbers, such as data sets recorded from instruments, it fits in a spreadsheet very well; the analysis of tabular data fits the grid structure of the spreadsheet. But if your problem requires the symbolic manipulation of complex mathematical equations, a spreadsheet is not the best place to solve that problem.

The spreadsheet's heritage as a business tool becomes apparent as you use it. For example, there are any number of ways to display data in pie and bar charts, but the available X–Y chart options, generally more applicable to science and engineering, are more limited. There are many built-in functions to accomplish tasks such as calculating rates of return on investments (and engineers can find those useful), but there is no built-in function that calculates torque.

Spreadsheets are easy to use and can handle a wide range of problems. Many, if not most, of the problems for which engineers used to write computer programs are now solved by using electronic spreadsheets or other programs on their personal computers. A supercomputer might be able to "crunch the numbers" faster, but when the "crunch" time is tiny compared with the time required to write the program and create a report based on its results, the spreadsheet's ease of use and ability to print results in finished form (or easily move results to a word processor) can make the total time required to solve a problem by using a spreadsheet much shorter than that with conventional programing methods.

Spreadsheets are great for

- performing the same calculations repeatedly (e.g., analyzing data from multiple experimental runs),
- working with tabular information (e.g., finding enthalpies in a steam table—once you've entered the steam table into the spreadsheet),
- producing graphs—spreadsheets provide an easy way to get a plot of your data,
- performing parametric analyses, or "what if" studies—for example, "What would happen if the flow rate were doubled?", and
- presenting results in readable form.

There was a time when spreadsheets were not the best way to handle computationally intense calculations such as iterative solutions to complex problems, but dramatic improvements in the computational speed of personal computers has eliminated a large part of this shortcoming, and improvements in the solution methods used by Excel have also helped. Excel can now handle many very large problems that just a few years ago would not have been considered suitable for implementation on a spreadsheet.

But there are still a couple of things that spreadsheets do not do well. Programs such as Mathematica[2] and Maple[3] are designed to handle symbolic manipulation of mathematical equations; Excel is not. Electronic spreadsheets also display only the results of calculations (just as their paper ancestors did), rather than the equations used to calculate the results. You must take special care when developing spreadsheets to indicate how the solution was found. Other computational software programs, such as Mathcad,[4] more directly show the solution process as well as the result.

1.2 WHAT'S NEW IN EXCEL 2010?

While Excel 2007 included a radical change from the menu system to the Ribbon (a combination of menu commands and some dialog box content), Excel 2010 has much less dramatic changes, but there are some new and improved features:

- The **Office** button, introduced with Excel 2007 and used to access file and print features, has been replaced with a **File** tab on the Ribbon.
- You can add your own tabs or groups to the Ribbon in Excel 2010.
- The Print dialog has been redesigned in Excel 2010 and now includes print preview features.

---

[2] Wolfram Research, Inc., Champaign, IL, USA.
[3] Waterloo Maple Inc., Ontario, Canada.
[4] Mathsodt Inc., Cambridge, MA, USA.

- Accessing Format dialog for chart elements (e.g., changing the appearance of an axis on a graph) has been streamlined; you can now open a Format dialog by double-clicking on a chart element. (This is possible in Excel 2003 and Excel 2010, but did not work in Excel 2007.)
- Pivot tables have been improved.
- There are now more conditional formatting options.
- Statistical functions are more accurate.
- The Solver (iterative solver) has been updated with a new interface and a new solution algorithm.
- A 64-bit version of Excel is available that allows even bigger Excel workbooks to be created (requires a 64-bit operating system).

Many of these new features will be mentioned later in the text.

1.2.1 Starting Excel

Excel can be purchased by itself, but it is usually installed as part of the Microsoft Office® family of products. During installation, an Excel option is added to the Start menu. To start Excel use the following menu options, as illustrated in Figure 1.2.

Figure 1.2
Start Excel using menu options: **Start/All Programs/Microsoft Office/Microsoft Excel 2010.**

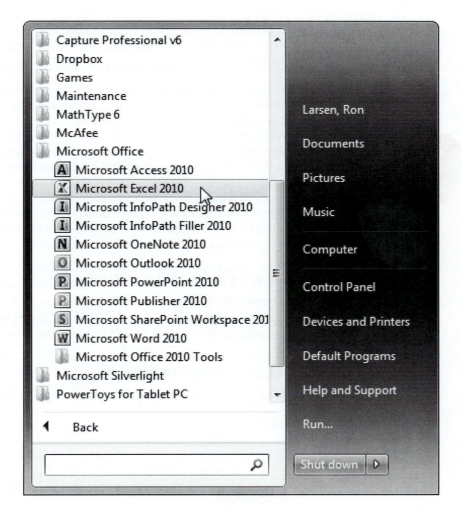

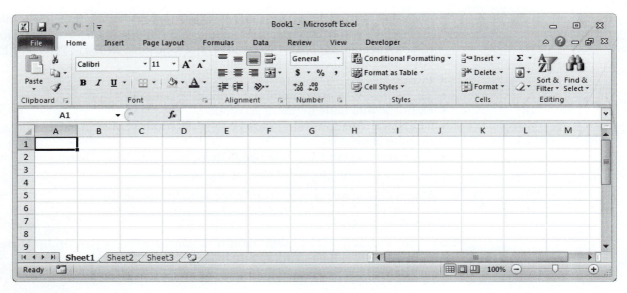

Figure 1.3
The Excel screen at start-up.

Excel 2010: **Start/All Programs/Microsoft Office/Microsoft Excel 2010**

Excel 2007: **Start/All Programs/Microsoft Office/Microsoft Office Excel 2007**

Recently used programs are listed on the left panel of the start menu in Windows XP. If the Excel icon appears on the start menu (as in Figure 1.2), then you can simply click the icon to start Excel.

When Excel starts, the Excel window should look a lot like Figure 1.3. Excel can be configured to suit individual preferences, so your installation of Excel could look a little different. Also, the number of buttons and boxes displayed at the top of the window changes depending on the width of the window; if you use a wide window you will see more information presented.

When you start Excel, a blank workbook is presented that contains, by default, three worksheets named Sheet1, Sheet2, and Sheet3. You can change the default number of sheets; this is covered in Section 1.3.7.

Note: The terms *worksheet* and *spreadsheet* can be interchanged, but Excel's help system uses the term worksheet and that term will be used in this text. A *workbook* is a collection of worksheets.

Since Excel runs within a Windows® environment, let's first review how to work with Windows.

1.3 A LITTLE WINDOWS®

The look of the windows depends on the version of Windows® that you are running. Images in this text are from Excel 2010 running in Windows 7. Other versions of Windows will not change the appearance greatly, but versions of Excel before 2007 will look quite different from the images shown here.

In Figure 1.3, the Excel workbook (called Book1 until it is saved with a different name) is shown maximized in the Excel Window. You can have multiple workbooks open at one time, and they may be easier to access if you do not maximize the workbooks, as illustrated in Figure 1.4.

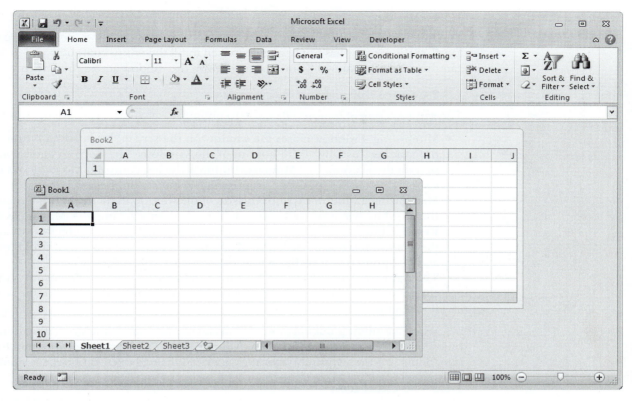

Figure 1.4
The Excel screen with two workbooks displayed.

Notice that the workbook names (Book1 and Book2 in Figure 1.4) are displayed at the top of the individual workbook rather than at the top of the Excel window as when the workbooks are not maximized.

There is a lot of command and control information at the top of an Excel window. Prominent features of Excel are indicated in Figure 1.5.

The top line of the Excel window contains the title bar (labeled 1 in Figure 1.5), the window control buttons (2), the Microsoft Excel Icon (3), the Quick Access toolbar (4), and the Ribbon (5). Each of these items is described in more detail below.

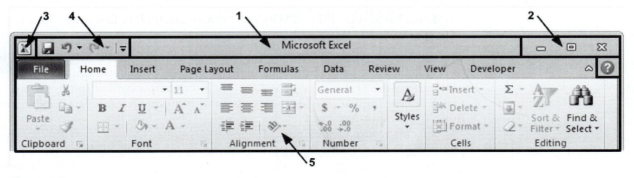

Figure 1.5
Command and control features of the Excel window.

1.3.1 Title Bar

The title bar contains the name of the program running in the window—Microsoft Excel. If the window has not been maximized to fill the entire computer screen, you can drag the title bar with your mouse to move the window across the screen.

1.3.2 Control Buttons

The window control buttons allow you to minimize, maximize, or close the window.

- Minimizing a window closes the window on the screen, but leaves a button on the task bar (usually at the bottom of the Windows® desktop) to allow you to reopen the window. The **Minimize** button is the left control button; it looks like a short line but is supposed to represent the button on the task bar.
- Maximizing a window expands the size of the window to fill the entire screen. If the window has already been maximized, then clicking the **Maximize** (or **Restore**) button will cause the window to shrink back to the original size, before the window was maximized. The **Maximize** button is the middle control button and is supposed to look like a window.
- The right control button (the **X**) is used to **Close** the window. If you attempt to close Excel without saving a workbook, you will be prompted to save the workbook before closing the window.

When a window is not maximized, you can usually change the size of the window by grabbing the border of the window with the mouse and dragging it to a new location. Some windows, typically error messages and dialog boxes, cannot be resized.

1.3.3 The Excel Icon

Microsoft Office 2007 had a very important button in the top-left corner of all Office products, the **Microsoft Office Button**. In Excel 2010, the functionality of that button has been replaced with the **File** tab on the Ribbon, and there is simply an Excel icon in the top-left corner of the Excel window. The only purpose of the Excel icon is to indicate that you are using Excel.

The usage of the File tab will be described later in the chapter, after the Ribbon is introduced.

1.3.4 The Quick Access Toolbar

All of the Office 2010 products feature a *Quick Access Toolbar* which is usually located at the left side of the title bar (see item 4 in Figure 1.5). It can also be located just below the ribbon to provide more space for buttons, if needed. To change the location of the Quick Access Toolbar, right-click on the toolbar and select **Show Quick Access Toolbar Below the Ribbon** from the pop-up menu.

Customizing the Quick Access Toolbar

The purpose of the Quick Access Toolbar is to provide a place for you to put buttons that allow you to access the features that you use frequently. You add buttons by customizing the Quick Access Toolbar. The down arrow symbol to the right of the Quick Access Toolbar provides access to the Customize Quick Access Toolbar drop-down menu, shown in Figure 1.6.

Selecting **More Commands...** from the Customize Quick Access Toolbar menu opens the Excel Options dialog shown in Figure 1.7.

You can also access this dialog using the Office button as **File Tab/Options** to open this dialog, then selecting the **Quick Access Toolbar** option on the dialog. [Excel 2007: Office/Excel Options]

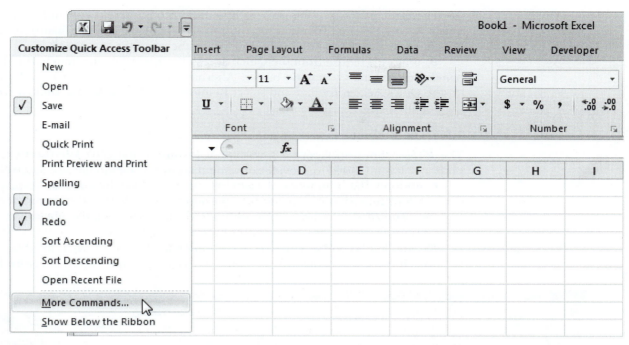

Figure 1.6
Customizing the Quick Access Toolbar.

Figure 1.7
The Quick Access Toolbar panel on the Excel Options dialog is used to customize the Quick Access Toolbar.

The right panel (see Figure 1.7) lists the buttons currently included on the Quick Access Toolbar. To add a new button, highlight the desired feature on the left panel then click the **Add** >> button located between the two panels.

To remove a button from the Quick Access Toolbar, select the item in the right panel and click the <<**Remove** button. Alternatively, you can right-click any button on the Quick Access Toolbar and select **Remove from Quick Access Toolbar** from the pop-up menu.

1.3.5 The Ribbon

The *Ribbon,* shown in Figure 1.8, was a new feature in the 2007 Microsoft Office Products. It is intended to provide convenient access to commonly used features.

The Ribbon has a number of *tabs* across the top. Clicking each tab displays a collection of related *groups* of buttons. (Groups are labeled at the bottom of the Ribbon.) The *Home tab*, shown in Figure 1.8, provides access to the clipboard operations (cut, copy, and paste) and a variety of formatting features.

The Ribbon is context sensitive, and additional tabs appear when needed. For example, if you are working with a graph, such as the example shown in Figure 1.9,

Figure 1.8
The Ribbon, showing the contents of the **Home** tab.

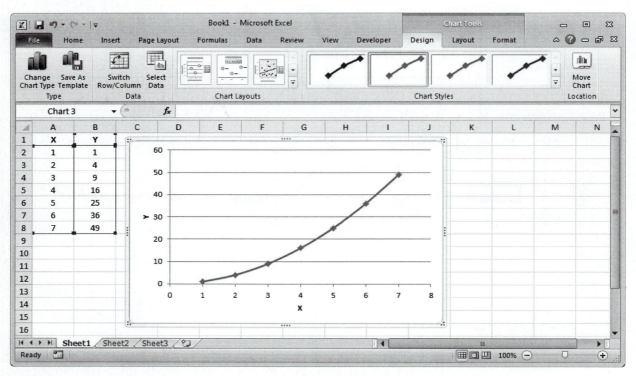

Figure 1.9
The Ribbon showing **Chart Tools** tabs appropriate for modifying a graph.

Design, **Layout**, and **Format** tabs appear on the Ribbon to allow you to customize the graph. The title **Chart Tools** appears on the title bar to let you know that these tabs are used to modify the appearance of the graph. If you click outside of the graph (somewhere on the worksheet grid), the **Chart Tools** tabs will disappear and the standard Ribbon will be displayed. To gain access to the **Chart Tools** tabs, simply click on the graph to select it.

Minimizing the Ribbon

The Ribbon is very useful, but takes up quite a bit of space in the Excel window. When necessary, you can minimize the Ribbon to show only the major tabs (**Home**, **Insert**, etc.). To minimize the Ribbon, right-click on the Ribbon's tab bar and select **Minimize the Ribbon** from the pop-up menu. When the Ribbon is minimized, clicking on any tab causes the groups for that tab to be displayed as a pop-up just below the tab line.

To display the full Ribbon again, right-click on the Ribbon's tab bar and click **Minimize the Ribbon** from the pop-up menu to de-select (i.e., uncheck) the **Minimize** option.

> **Note:** In Excel 2010, there is a **Minimize Ribbon Toggle** button just to the left of the **Help** button (i.e., to the left of the question mark).

1.3.6 Name Box and Formula Bar

Just below the Ribbon are the *Name box* and *Formula bar*, as illustrated in Figure 1.10. The Name box identifies the currently active cell (B3 in Figure 1.10) and the Formula bar displays the contents of the cell (text or formula). Equations can get quite long and a nice feature in Excel is the ability to quickly expand the size of the Formula bar by clicking on the down arrow symbol at the right side of the Formula bar (indicated in Figure 1.10).

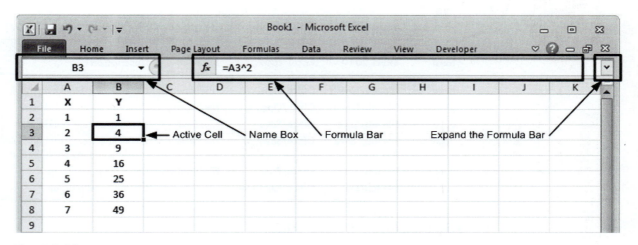

Figure 1.10
The Name box and Formula bar.

The Formula bar is displayed by default, but it can be turned off using the **View** tab on the Ribbon (Figure 1.11.) There is a **Formula Bar** checkbox on the **View** tab to activate or deactivate the display of the Formula bar (and Name box).

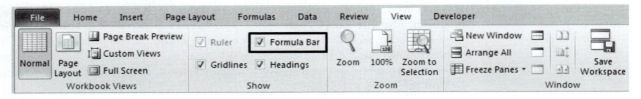

Figure 1.11
Use the **Formula Bar** checkbox on the **View** tab to activate and deactivate the display of the Formula bar.

1.3.7 The File tab

In Excel 2007, many of the tasks common to most programs (e.g., opening and closing files, printing) were collected in a menu under a new feature called the **Office** button, shown in the right panel in Figure 1.12. The **Office** button is gone in Excel 2010, replaced by the **File** tab on the Ribbon.

Figure 1.12
Excel 2007's **Office** button (right panel) has been replaced by the **File** tab in Excel 2010 (left panel).

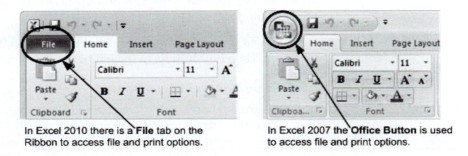

In Excel 2010 there is a **File** tab on the Ribbon to access file and print options.

In Excel 2007 the **Office Button** is used to access file and print options.

The new **File** tab contains many of the features that used to be on the **File** menu versions of Excel prior to 2007. Table 1.1 summarizes the features available through the **File** tab.

Table 1.1 Common tasks available via the File tab (or Office button in Excel 2007)

Menu Item	Excel 2010	Excel 2007	Excel 2003	Description
New	**File tab/New**	**Office/New**	**File/New**	Open a new, blank workbook
Open	**File tab/Open**	**Office/Open**	**File/Open**	Open an existing workbook
Save	**File tab/Save**	**Office/Save**	**File/Save**	Save the current workbook
Save As	**File tab/ Save As...**	**Office/Save As...**	**File/Save As...**	Save the current workbook with a different name or file format
Print	**File tab/Print**	**Office/Print...**	**File/Print...**	Open the print dialog
Exit	**File tab/Exit**	**Office/Exit Excel**	**File/Exit**	Exit the Excel program

1.3.7b Changing Excel Options

You can change the default options for Excel using the Excel Options dialog. The method for accessing the dialog varies depending on the version of Excel that you are using:

- Excel 2010: **File tab/Options**
- Excel 2007: **Office/Excel Options**
- Excel 2003: **File/Options**

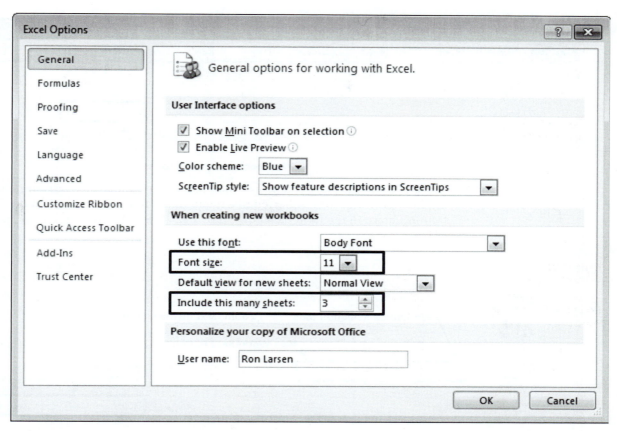

Figure 1.13
The Excel Options dialog, **General** panel.

In Excel 2010, the **File** tab displays a menu of commonly used features, but near the bottom of the menu is a button labeled **Options**. Click the **Options** button to open the Excel Options dialog (Figure 1.13).

The Excel Options dialog allows you to customize your installation of Excel to better fit your needs. There are nine panels in the Excel Options dialog to provide access to various features. The **General** panel (shown in Figure 1.11) (called the **Popular** panel in Excel 2007) shows some commonly changed Excel options. For example, the default font size can be increased to make the spreadsheet easier to read on the screen or decreased to display more information. Also, the default number of worksheets to include in a new workbook can be changed from the default value of 3.

1.3.8 Workbooks and Worksheets

By default, when a new workbook is opened it contains three worksheets, as shown in Figure 1.14.

When the workbook is not maximized (as in Figure 1.14), the workbook control buttons are at the top-right corner of the workbook window, and the worksheet selection tabs are at the bottom-left corner of the worksheet window. The right-most worksheet selection tab is actually a button that is used to create a new worksheet in the workbook.

When the workbook is maximized (as in Figure 1.15), the workbook control buttons appear just below the Excel window control buttons.

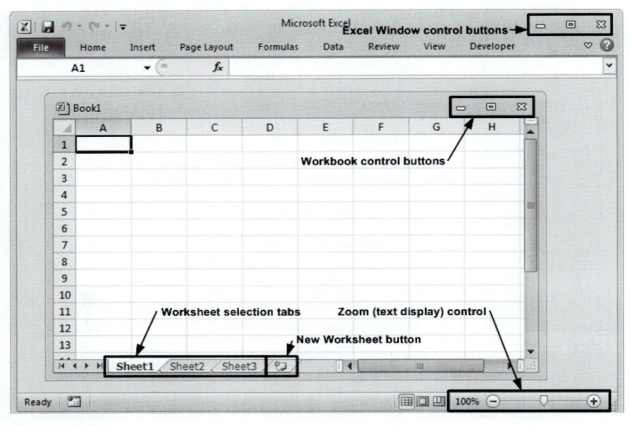

Figure 1.14
A (nonmaximized) workbook.

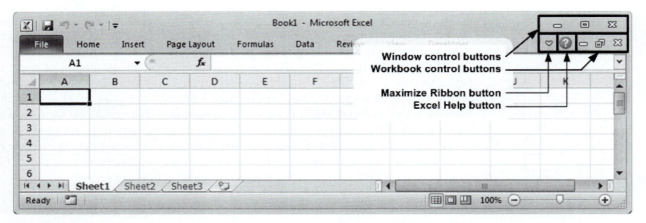

Figure 1.15
A maximized workbook.

While most small problems are solved using only a single worksheet, the option of using multiple worksheets is very helpful in organizing complex solutions. For example (see Figure 1.16), you might keep raw data on one worksheet, calculations on a second, and present results on a third.

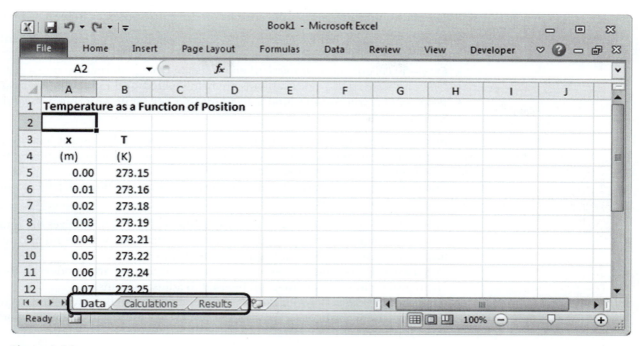

Figure 1.16
Using worksheets to organize a workbook.

To change the name that is displayed on a worksheet's tab, either double-click the tab or right-click on the tab and select **Rename** from the pop-up menu. The tab's pop-up menu can also be used to insert a new worksheet into the workbook, move a worksheet to a different location in the workbook, or create a copy of a worksheet in the workbook.

1.3.9 Customizing the Status Bar

The *Status bar* resides at bottom of the Excel window, as shown in Figure 1.17.

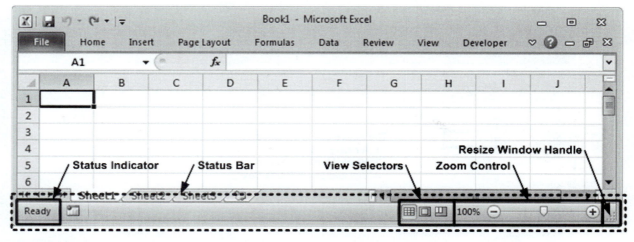

Figure 1.17
The Status bar.

The status indicator reveals the current mode of operation, as follows:

- *Ready*— indicates that Excel is ready (waiting) for you to type information into a cell.
- *Enter*—Excel goes into *enter mode* when you begin typing information into a cell.
- *Edit*—indicates that a cell's contents are being edited, and Excel is in *edit mode.*
- *Point*—when you use the mouse to point to a cell while entering a formula, Excel jumps to *point mode.*

By default, the right side of the status bar includes a slide control that allows you to quickly zoom in on the active cell. Drag the slide indicator or click on the [+] or [−] buttons to change the zoom level.

The view selectors provide an easy way to change between normal, page layout, and page break preview views.

- *Normal View*—the standard view of the worksheet grid that maximizes the number of cells that can be displayed in the Excel window, but provides no information on how the worksheet will appear when printed.
- *Page Layout View*—this view shows the page margins and any headers or footers. It also includes rulers that can be used to adjust margins.
- *Page Break Preview*—no margins or rulers are shown, but page breaks are shown on the worksheet to show where the breaks will occur when the worksheet is printed.

1.4 EXCEL BASICS

1.4.1 The Active Cell

When you start Excel, you see an empty grid on the screen. Each rectangle of the grid is called a *cell*, and you can enter information into any cell in the grid. Each cell is identified by its *cell address*, made up of a column letter and a row number. For example, the cell in the second column from the left and the third row from the top would be called cell B3 (see Figure 1.18).

When you select a cell, either by clicking with the mouse or by moving the active cell indicator (the cell border shown on cell B3 in Figure 1.18) using the arrow keys on the keyboard, the selected cell becomes the *active cell.*

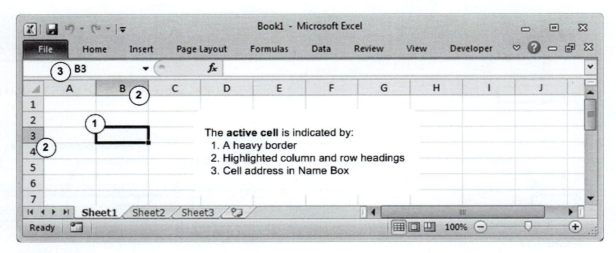

Figure 1.18
The active cell (B3) is indicated in several ways.

The active cell is indicated in several ways:

1. The active cell is surrounded by a heavy border.
2. The row and column headings of the active cell are highlighted.
3. The cell address of the active cell is shown in the Name box at the left side of the Formula bar.

You always type into the active cell, so any time you want to enter information into Excel, you first select a cell (to make it the active cell) and then enter the information.

1.4.2 Labels, Numbers, Formulas

A cell can contain one of three things:

- *Label*—one or more text characters or words.
- *Value*—a number.
- *Formula*—an equation.

Excel attempts to classify the cell contents as you type.

- If you enter a number, Excel treats the cell contents as a value, and the numeric value appears in the cell.
- If the first character you type is an equal sign, Excel will try to interpret the cell's contents as a formula (equation).
- If the first character is not a number or an equal sign, Excel will treat the cell contents as a label. You can also use a single quote ['] as the first character to tell Excel to treat the cell contents as a label.

Notes on entering formulas:

1. While you are entering a formula the characters that you type in are displayed (Excel is in entry mode). But, as soon as you press [Enter] to complete the entry, Excel returns to ready mode and displays the result of the calculation in the cell, not the equation.

    Example:

    If you enter the characters $= 2 + 4$, the formula $= 2 + 4$ would be stored as the contents of the active cell; however the result, 6, would be displayed. This is illustrated in Figure 1.19.

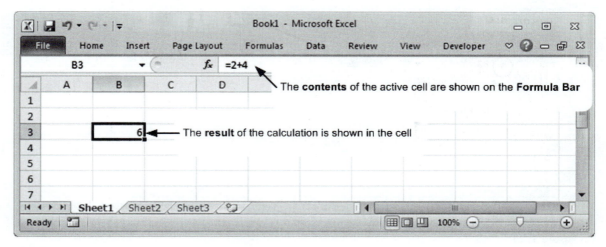

Figure 1.19
Only the results of a calculation appear in the cell, but the contents of the active cell are displayed in the Formula bar.

2. The formula you entered is still stored in the cell and can be edited. Double-click the cell or press [F2] to edit the contents of the active cell.
3. To inspect a formula in a cell, select the cell to make it the active cell. The contents of the active cell are displayed on the Formula bar.

Formulas are an essential part of using Excel to solve engineering problems. What makes formulas especially useful is the ability to use cell addresses (e.g., B3) as variables in formulas. The following example uses a worksheet to compute a velocity from a distance and a time interval.

EXAMPLE 1.1

A car drives 90 miles in 1.5 hours. What is the average velocity in miles per hour?
   This is a pretty trivial example and illustrates another consideration in the use of an Excel worksheet: single, simple calculations are what calculators are for. A spreadsheet program like Excel is overkill for this example, but we'll use it as the starting point for a more detailed problem.

1. Enter labels as follows:
   • "Distance" in cell B3 and the units "miles" in cell B4.
   • "Time" in cell C3 and the units "hours" in cell C4.
   • "Velocity" in cell D3, with units "mph" in cell D4.
2. Enter values as follows:
   • 90 in cell B5 (i.e., in the Distance column (column B)).
   • 1.5 in cell C5 (i.e., in the Time column (column C)).
3. Enter the formula =B5/C5 in cell D5.

Note that cell D5 in Figure 1.20:

   • contains a formula =B5/C5, as indicated on the formula bar, but
   • displays the result of the formula (60 miles per hour).

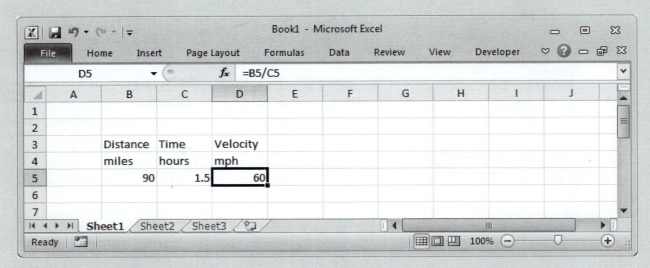

Figure 1.20
Using cell addresses as variables in a formula.

Note: The formula bar will be included in many screen images in this text to show the contents of the active cell.

1.4.3 Using the Mouse to Select Cells While Entering Formulas

Typically, when entering formulas you wouldn't actually type cell addresses like the B5 and C5 in the last example. Instead, while you are entering the formula you would enter the equal sign and then click the mouse to tell Excel which cell contents to use in the formula. Specifically, the formula =B5/C5 in cell D5 would be entered as:

1. Make cell D5 the active cell by clicking on it, or moving the active cell indicator to D5 using the keyboard arrow keys.
2. Enter the equal sign [=].
3. Use the mouse and click on cell B5. Excel will show B5 in the formula and, at this point, the formula will read =B5, as shown in Figure 1.21. Notice that the mouse pointer looks like a large plus sign when you are in Point mode.
4. Type the division operator [/].
5. Use the mouse and click on cell C5. Excel will show C5 in the formula. At this point, the formula will read =B5/C5, as shown in Figure 1.22.
6. Press [Enter] to complete the formula.

Notes:

1. Pressing an operator such as the [/] key tells Excel that you are done pointing at B5, so you can then use the mouse or arrow keys to point at cell C5. Technically, typing the operator jumps Excel out of Point mode back into Enter mode.
2. While you are entering or editing a formula, Excel shows the cells being used in the formula in color-coded boxes (see Figure 1.22). This visual indication of the cells that are being used in a calculation is very helpful when checking your worksheet for errors.

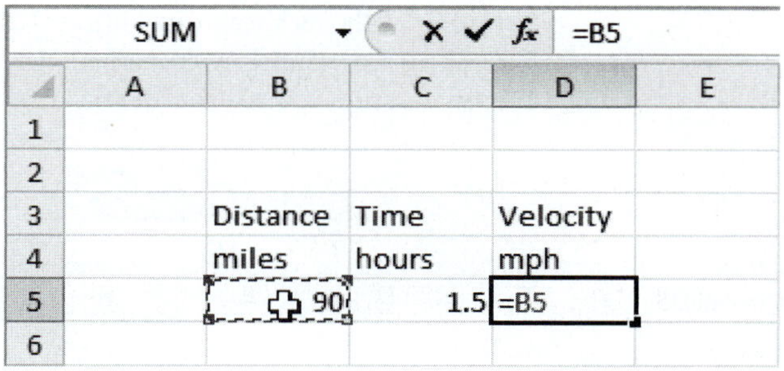

Figure 1.21

Using the mouse to point at cell B5 when needed in a formula.

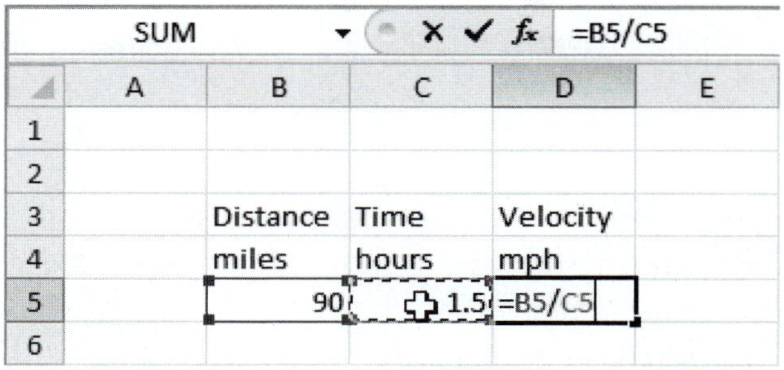

Figure 1.22

Using the mouse to point at cell C5 when needed in a formula.

1.4.4 Copying Cell Ranges to Perform Repetitive Calculations

It is common to perform the same calculation for each value in a column or row of data. Fortunately, Excel makes it easy to perform this type of repetitive calculation; you do not have to type the equation over and over again. Instead, simply copy the formula you typed (in cell D5 in Example 1.1) to all of the cells that need to calculate a velocity.

EXAMPLE 1.2

If you have a column of distances (cells B5 through B9) and another column of times (cells C5 through C9) as shown in Figure 1.23, simply copy the formula in cell D5 to cells D6 through D9.

In Excel, a group of adjacent cells, such as D6 through D9, is called a *cell range* and written as D6:D9.

The procedure for copying the contents of cell D5 to D6:D9 is as follows:

1. Select the cell (or cells, if more than one) containing the formula(s) to be copied. In this example, we simply need to select cell D5.
2. Copy the contents of cell D5 to the Windows clipboard in either of the following ways:
   - Using the copy button on the Ribbon: **Home/Clipboard/Copy**.
   - Using the keyboard shortcut [Ctrl-c]
     Excel uses a dashed border to show the cell that has been copied to the clipboard, as shown in Figure 1.24.
3. Select the beginning cell of the destination range, cell D6, by clicking on cell D6 or using the down arrow key.
4. Select the entire destination range by dragging the mouse to cell D9 (or, hold the [Shift] key and click on cell D9 or hold the [Shift] key and use the down arrow key to move to cell D9). Excel uses a heavy border to show the selected range of cells, as shown in Figure 1.25.
5. Paste the information stored on the Windows clipboard (the contents of cell D5, from step 1) into the destination cells (D6:D9) by:

Figure 1.23

Preparing to copy the velocity formula in cell D5 to cells D6:D9.

	D5			f_x	=B5/C5	
	A	B	C	D	E	
1						
2						
3		Distance	Time	Velocity		
4		miles	hours	mph		
5		90	1.5	60		
6		120	6			
7		75	3			
8		80	2			
9		135	9			
10						

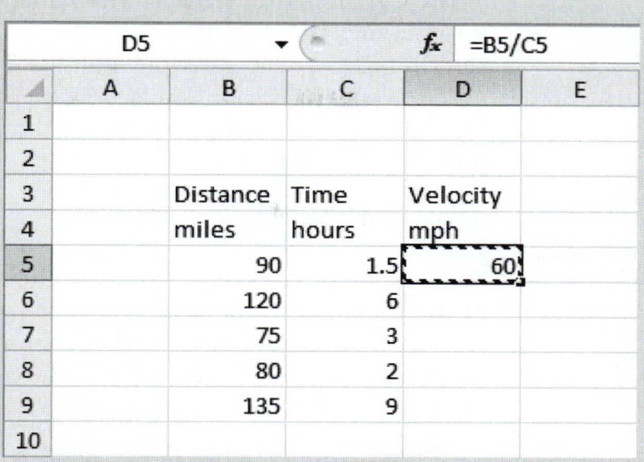

Figure 1.24
A cell that has been copied to the clipboard is shown with a dashed border.

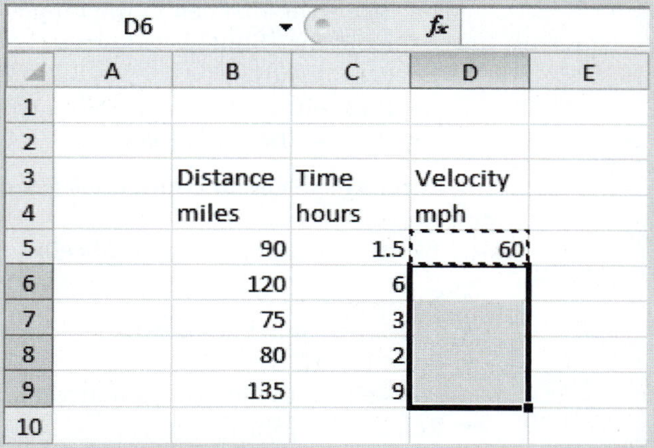

	D5	▾ ◦		f_x	=B5/C5	
⊿	A	B	C	D	E
1					
2					
3		Distance	Time	Velocity	
4		miles	hours	mph	
5		90	1.5	60	
6		120	6		
7		75	3		
8		80	2		
9		135	9		
10					

Figure 1.25
A heavy border is used to indicate a selected range of cells.

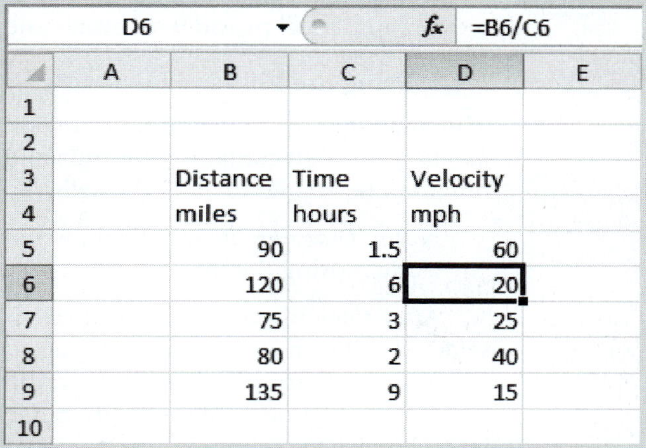

	D6	▾ ◦		f_x		
⊿	A	B	C	D	E
1					
2					
3		Distance	Time	Velocity	
4		miles	hours	mph	
5		90	1.5	60	
6		120	6		
7		75	3		
8		80	2		
9		135	9		
10					

Figure 1.26
Completing the velocity calculations.

	D6	▾ ◦		f_x	=B6/C6	
⊿	A	B	C	D	E
1					
2					
3		Distance	Time	Velocity	
4		miles	hours	mph	
5		90	1.5	60	
6		120	6	20	
7		75	3	25	
8		80	2	40	
9		135	9	15	
10					

- Using the paste button on the Ribbon: **Home/Clipboard/Paste**, or
- Using the keyboard shortcut [Ctrl-v]

The result of pasting the velocity formulas into cells D6:D9 is shown in Figure 1.26.

Notice in Figure 1.26 that the velocity formula in cell D6 is =B6/C6; that is, the formula uses the distances and times in row 6. When Excel copied the formula in cell D5 (=B5/C5) to cell D6, it automatically incremented the row numbers. This is called *relative addressing* and is a big part of why electronic spreadsheets are so easy to use. Relative addressing is discussed further in Section 1.4.6, but first a little more on copying and pasting.

Selecting Multiple Cells for Copying

When multiple cells are to be copied, select the source range of cells by clicking on the first cell in the range. Then use the mouse to drag the cursor to the other end of the range (or hold down the shift key while using the mouse or arrow keys to select the cell at the other end of the range). Do not click on the little box at the lower-right corner of a cell. This is the *fill handle*, which is useful for several things, but not for selecting a range of cells. Its use will be described in Section 1.4.5.

Note: There is a fast way to select a column of values by using the [End] key.

- Select the first cell in the range
- Hold the [Shift] key down
- Press [End]
- Press either the up or down arrow key
- Release the [Shift] key

Excel will select all contiguous filled cells in the column, stopping at the first empty cell.

To select a contiguous row of cells:

- Select an end cell
- Hold the [Shift] key down
- Press [End]
- Press either the left or right arrow key
- Release the [Shift] key

It is pretty straightforward to copy and paste using the procedures listed in this section, but Excel provides an even easier approach; it's called the *fill handle*.

1.4.5 Using the Fill Handle

The small square at the bottom-right corner of the active cell border is called the *fill handle*. If the cell contains a formula and you grab and drag the fill handle, the formula will be copied to all of the cells you select in the drag operation.

Note: *Dragging the mouse* simply implies holding the left-mouse button down while moving the mouse.

After selecting the source cell (D5 in our example), grab the fill handle of the selected cell with the mouse and drag, as shown in Figure 1.27.

The destination range will be outlined as you drag, as shown in Figure 1.27. Continue dragging until the entire range (cell range D5:D9) is outlined. The mouse icon changes to a plus symbol when the fill handle is in use.

When you have selected the desired destination range, release the mouse. Excel will copy the contents of the original cell to the entire destination range. The result is shown in Figure 1.28.

The little square icon next to the fill handle after the copy is a link to a pop-up menu that allows you to modify how the copy process is carried out. If you click on

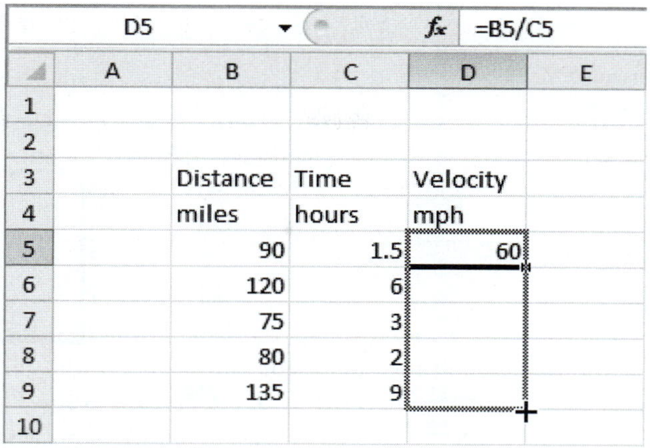

Figure 1.27
Using the fill handle to copy and paste in a single operation.

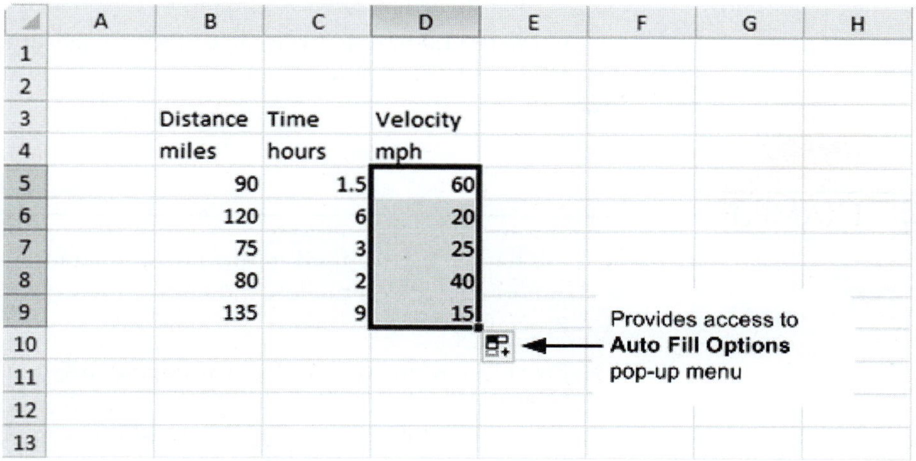

Figure 1.28
After using the fill handle, a pop-up menu of fill options is available.

the icon, a pop-up menu will appear with some options for how the copy using the fill handle should be completed (see Figure 1.29).

The **Copy Cells** option (default) copies both the contents and the formatting of the source cell. Other options allow you to copy only the contents or only the formatting.

The fill handle can be used for other purposes as well:

- To fill a range with a series of values incremented by one, place the first value (not formula) in a cell and then drag the fill handle with the left-mouse button to create the range of values.
- If you want a range of values with an increment other than one, enter the first two values of the series in adjacent cells, select both cells as the source, and drag the fill handle with the left-mouse button.
- If you want a nonlinear fill, enter the first two values of the series in adjacent cells, select both cells as the source, and drag the fill handle by using the right-mouse button. When you release the mouse, a menu will be displayed giving a variety of fill options.

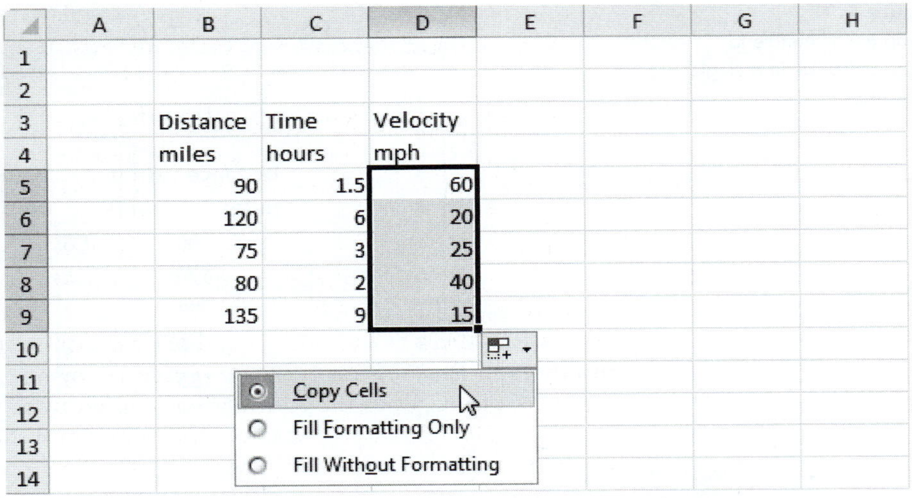

Figure 1.29
The Auto Fill Options pop-up menu after using the fill handle to copy and paste.

◢	A	B	C	D	E	F	G	H
1								
2								
3		Distance	Time	Velocity				
4		miles	hours	mph				
5		90	1.5	60				
6		120	6	20				
7		75	3	25				
8		80	2	40				
9		135	9	15				
10								
11			⊙ Copy Cells					
12			○ Fill Formatting Only					
13			○ Fill Without Formatting					
14								

Using the fill handle to create series of data values is very handy, but the operations listed above also work with times and dates, as Illustrated in the following example.

EXAMPLE 1.3

The fill handle can be used to quickly set up a class schedule showing hours from 8:00 am to 4:00 pm on the left and days of the week across the top.

Procedure:

1. Enter "8 am" in cell B3. (Excel will recognize that the entry is a time and display "8:00 am.")
2. Use the fill handle and drag from cell B3 to cell B11.
3. Enter "Monday" in cell C2.
4. Use the fill handle and drag from cell C2 to cell G2.

The result is shown in Figure 1.30.

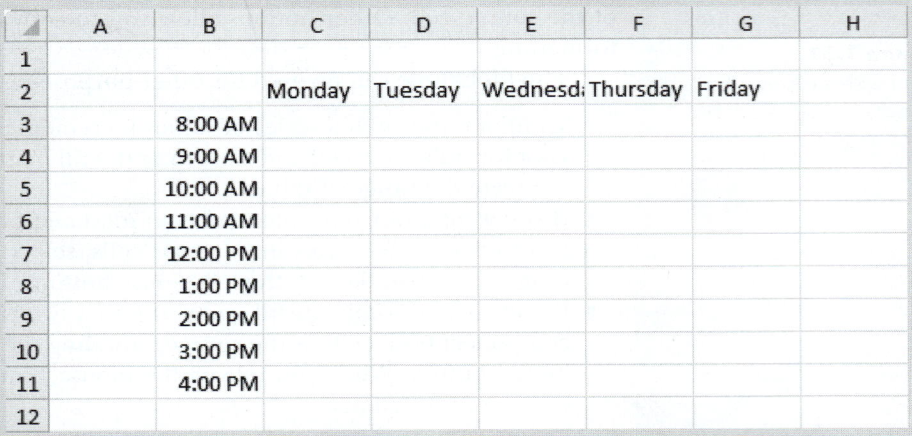

Figure 1.30
Using the fill handle to create a class schedule.

◢	A	B	C	D	E	F	G	H
1								
2			Monday	Tuesday	Wednesd;	Thursday	Friday	
3		8:00 AM						
4		9:00 AM						
5		10:00 AM						
6		11:00 AM						
7		12:00 PM						
8		1:00 PM						
9		2:00 PM						
10		3:00 PM						
11		4:00 PM						
12								

The next step would be to enter your classes in the appropriate cells, and some formatting is needed to display all of "Wednesday"—but that's the subject of Chapter 2.

1.4.6 Relative and Absolute Addressing

Returning to the velocity calculation example (Example 1.2), we used the fill handle to copy the formula entered in cell D5 to cells D6:D9. The result is shown in Figure 1.31.

At this point, the contents of cell D5 have been copied, with significant modifications, to cells D6:D9. As you can see in the formula bar in Figure 1.31, cell D6 contains the formula =B6/C6.

As the formula =B5/C5 in cell D5 was copied from row 5 to row 6, the row numbers in the formula were incremented by one, so the formula in cell D6 is =B6/C6. Similarly, as the formula was copied from row 5 to row 7, the row numbers in the formula were incremented by two and the formula =B7/C7 was stored in cell D7. So, the velocity calculated in cell D7 uses the distance and time values from row 7, as desired. This automatic incrementing of cell addresses during the copy process is called *relative cell addressing* and is an important feature of electronic spreadsheets.

> **Note:** If you had copied the formula in D5 across to cell E5, the column letters would have been incremented. If you had copied the formula in D5 diagonally to cell E9, both the row numbers and column letters would have been incremented.

- Copying down increments row numbers
- Copying across increments column letters
- Copying diagonally increments both row numbers and column letters

Sometimes you don't want relative addressing; you want the cell address to be copied unchanged. This is called *absolute cell addressing*.

You can make any address absolute in a formula by including dollar signs in the address, as $B$5. The nomenclature $B$5 tells Excel not to automatically increment either the B or the 5 while copying. Similarly, $B5 tells Excel it is OK to increment the 5, but not the B during a copy, and B$5 tells Excel it is OK to increment the B, but not the 5. During a copy, any row or column designation preceded by a $ is left alone. One common use of absolute addressing is building a constant into your calculations, as illustrated in Example 1.4.

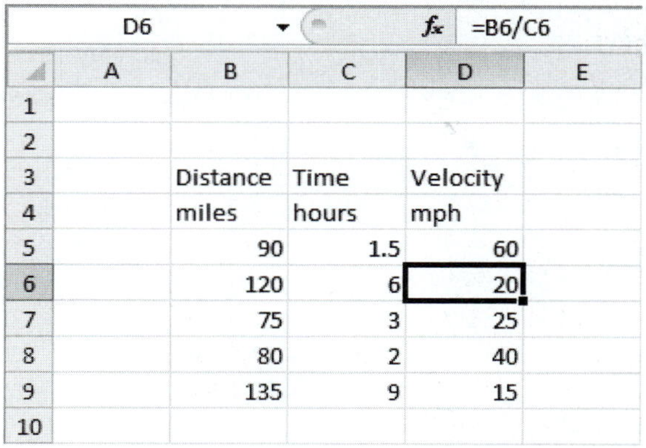

Figure 1.31
The result of copying the velocity formula in cell D5 to cells D6:D9.

D6	▼		f_x =B6/C6	
	A	B	C	D	E

	A	B	C	D	E
1					
2					
3		Distance	Time	Velocity	
4		miles	hours	mph	
5		90	1.5	60	
6		120	6	20	
7		75	3	25	
8		80	2	40	
9		135	9	15	
10					

EXAMPLE 1.4

Modify the worksheet developed for Example 1.2 to display the velocities in feet per second. The conversion factor between miles per hour (mph) and feet per second (fps) is 1.467 fps/mph.

First, we need to get the conversion factor onto the worksheet. To help keep things organized, I typically place constants and parameter values near the top of the worksheet, where they are easy to find. This has been done in Figure 1.32.

Figure 1.32
Adding the conversion factor to the worksheet.

D5				f_x	=B5/C5

	A	B	C	D	E
1	Factor:	1.467	fps/mph		
2					
3		Distance	Time	Velocity	
4		miles	hours	mph	
5		90	1.5	60	
6		120	6	20	
7		75	3	25	
8		80	2	40	
9		135	9	15	
10					

Notice that the label (cell A1), value (cell B1), and units (cell C1) are in three different cells. The value must be in a cell by itself so that Excel treats it as a value (a number) rather than as a label.

Next, we use the conversion factor in the formulas in column D. We change (edit or reenter) one formula to include the conversion factor (and change the units on velocity in cell D4).

Notice (Figure 1.33) that an absolute address has been used for the conversion factor in the formula in cell D5. The $B$1 will not be changed when this formula is copied.

Finally, use the fill handle to copy the formula to cells D6:D9. The result is shown in Figure 1.34. Notice that after copying the formula in cell D5 to cell D6, the new formula still references the conversion factor in cell B1. The dollar signs on $B$1 told Excel not to increment the cell address when the formula was copied.

Figure 1.33
Editing the velocity formula to include the conversion factor.

D5				f_x	=B5/C5*$B$1

	A	B	C	D	E
1	Factor:	1.467	fps/mph		
2					
3		Distance	Time	Velocity	
4		miles	hours	fps	
5		90	1.5	88.0	
6		120	6		
7		75	3		
8		80	2		
9		135	9		
10					

Figure 1.34
The resulting velocities, in feet per second.

	D6			f_x	=B6/C6*$B$1
	A	B	C	D	E
1	Factor:	1.467	fps/mph		
2					
3		Distance	Time	Velocity	
4		miles	hours	fps	
5		90	1.5	88.0	
6		120	6	29.3	
7		75	3	36.7	
8		80	2	58.7	
9		135	9	22.0	
10					

FLUID STATICS

The pressure at the bottom of a column of fluid is caused by the mass of the fluid, m, being acted on by the acceleration due to gravity, $g = 9.8$ m/s^2. The resulting force is called the weight of the fluid, F_W and can be calculated as

$$F_W = mg \qquad (1.1)$$

which is a specific version of Newton's law when the acceleration is due to the earth's gravity. The pressure at the bottom of the column is the force divided by the area of the bottom of the column, A:

$$P = \frac{F_W}{A} = \frac{mg}{A} \qquad (1.2)$$

The mass of the fluid can be calculated as the density, ρ, times the fluid volume, V.

$$P = \frac{F_W}{A} = \frac{mg}{A} = \frac{\rho V g}{A} \qquad (1.3)$$

The fluid volume is calculated as area of the column, A, times its height, h.

$$P = \frac{F_W}{A} = \frac{mg}{A} = \frac{\rho V g}{A} = \frac{\rho A h g}{A} = \rho h g \qquad (1.4)$$

Determine the pressure at the bottom of a column of mercury 760 mm high. The specific gravity of mercury is 13.6, meaning that mercury is 13.6 times as dense as water.

Do not consider any imposed pressure (e.g., air pressure) on the top of the column in this problem.

The solution is shown in Figure 1.35.

The equations in cells C9 through C12 are as follows:

```
C9:      =C5/1000
C10:     =C6*1000
C11:     =C10*C9*C4   or,   ρ·h·g
C12:     =C11/101300; the conversion factor between pascals
         and atmospheres is 101300 Pa/atm
```

APPLICATIONS

Figure 1.35
Determining the pressure at the bottom of a column of fluid.

	A	B	C	D	E
	C11			f_x	=C10*C9*C4
1	Pressure at the Bottom of a Column of Fluid				
2					
3	Information from Problem Statement				
4	Grav. Acceleration:		9.8	m/s²	
5	Column Height:		760	mm	
6	Specific Gravity:		13.6		
7					
8	Calculated Values				
9	Column Height (SI):		0.76	m	
10	Density:		13600	kg/m³	
11	Pressure:		101293	Pa, or N/m²	
12	Pressure:		1.00	atm	
13					

PRACTICE!

Create the worksheet shown in Figure 1.35 and use it to determine the pressure at the bottom of the columns of fluid specified below. [Do not consider any imposed pressure (e.g., air pressure) on the top of the column.]

a. A column of water 10 m high (SG$_{water}$ = 1.0) [Answer: 0.97 atm]
b. A column of seawater 11,000 m high. This is the depth of the Marianas Trench, the deepest spot known in the earth's oceans (SG$_{seawater}$ = 1.03) (ignore the variation in water density with pressure). [Answer: 1090 atm]

Letting Excel Add the Dollar Signs

You don't actually have to enter the dollar signs used to indicate absolute cell addresses by hand. If you enter cell address B1 and then press the [F4] key, Excel will automatically enter the dollar signs for you. Pressing [F4] once converts B1 to $B$1. Pressing the [F4] key multiple times changes the number and arrangement of the dollar signs, as follows:

- Press [F4] once $B$1
- Press [F4] a second time   B$1
- Press [F4] a third time $B1
- Press [F4] a fourth time   B1

In practice, you simply press [F4] until you get the dollar signs where you want them.

If you use the mouse to point to a cell in a formula, you must press [F4] right after you click on the cell. You can also use [F4] while editing a formula; just move the edit cursor to the cell address that needs the dollar signs and then press [F4].

Using Named Cells

Excel supports *named cells*, allowing you to assign descriptive names to individual cells or ranges of cells. Then you can use the names, rather than the cell addresses, in formulas. For example, the cell containing the conversion factor in the Example 1.4 (cell B1 in Figure 1.34) might be given the name, **ConvFactor**. This name could then be used in the velocity formulas instead of $B$1.

To give a single cell a name, click on the cell (cell B1 in this example) and enter the name in the Name box at the left side of the formula bar, as illustrated in Figure 1.36.

The name can then be used in place of the cell address in the velocity formulas, as in the formula in cell D5 in Figure 1.37.

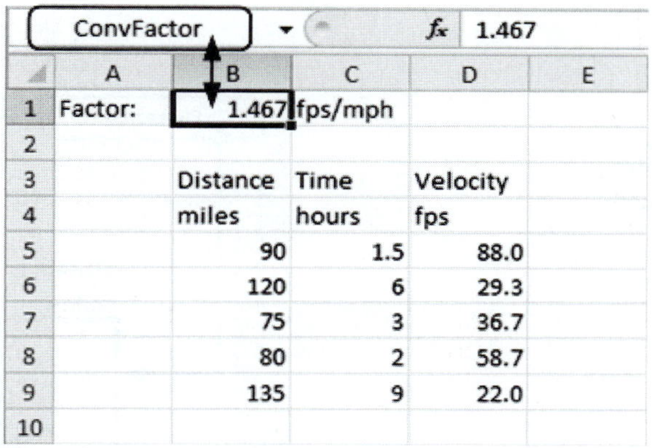

Figure 1.36

Cell B1 has been assigned the name, ConvFactor.

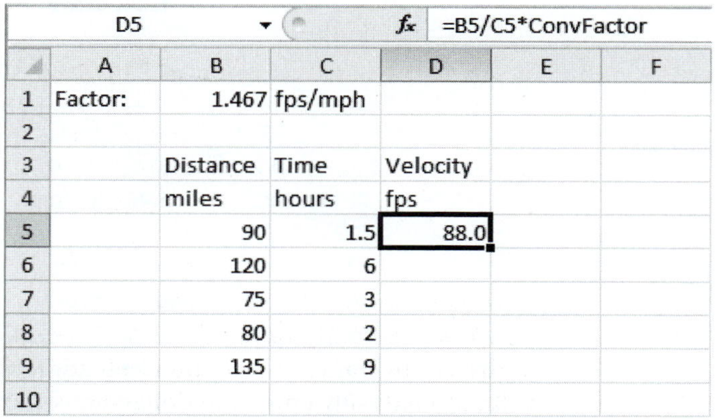

Figure 1.37

Using a named cell in a formula.

Using named cells in your formulas can make them easier to comprehend.

When the formula in cell D5 is copied to cells D6 through D9, the cell name is included in the formulas, as shown in cell D6 in Figure 1.38.

Notice that ConvFactor copied in the same way that $B$1 had copied previously. A cell name acts as an absolute cell address and is not modified when a formula is copied.

Naming Cell Ranges

It is also possible to assign names to cell ranges, such as the columns of distance values. To assign a name to a cell range, first select the range and then enter the name in the name box as shown in Figure 1.39.

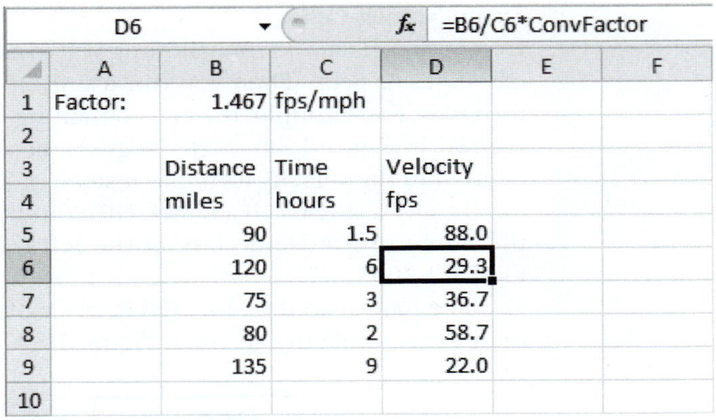

Figure 1.38
Named cells act as
absolute cell addresses
when copied.

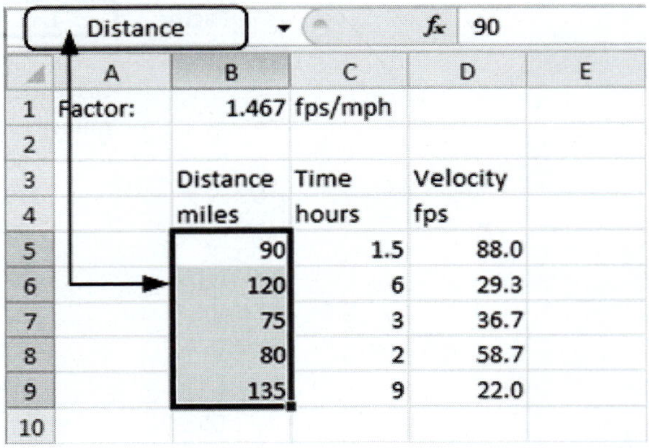

	D6		▾		f_x	=B6/C6*ConvFactor	
	A	B	C	D	E	F
1	Factor:	1.467	fps/mph			
2						
3		Distance	Time	Velocity		
4		miles	hours	fps		
5		90	1.5	88.0		
6		120	6	29.3		
7		75	3	36.7		
8		80	2	58.7		
9		135	9	22.0		
10						

Figure 1.39
Naming a cell range.
Here, cell range B5:B9 has
been named "Distance."

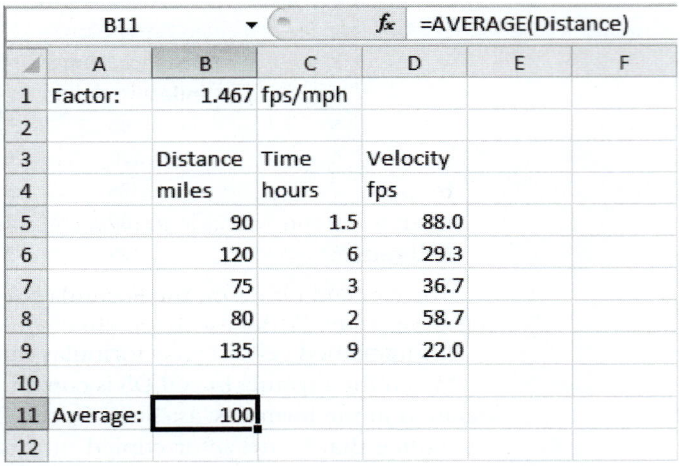

	Distance		▾		f_x	90
	A	B	C	D	E
1	Factor:	1.467	fps/mph		
2					
3		Distance	Time	Velocity	
4		miles	hours	fps	
5		90	1.5	88.0	
6		120	6	29.3	
7		75	3	36.7	
8		80	2	58.7	
9		135	9	22.0	
10					

Figure 1.40
Using a cell range name
in a function.

	B11		▾		f_x	=AVERAGE(Distance)	
	A	B	C	D	E	F
1	Factor:	1.467	fps/mph			
2						
3		Distance	Time	Velocity		
4		miles	hours	fps		
5		90	1.5	88.0		
6		120	6	29.3		
7		75	3	36.7		
8		80	2	58.7		
9		135	9	22.0		
10						
11	Average:	100				
12						

The name can then be used in place of the cell range, as in the
=AVERAGE(Distance) formula in cell B11 in Figure 1.40. (**AVERAGE** is Excel's
built-in function for calculating the arithmetic average of a set of values.)

If you have assigned a name and decide to remove it, use Ribbon commands
Formulas/Name Manager to see a list of the defined names in the worksheet.
[Excel 2003: Insert/Name/Define.]

Figure 1.41
The Name Manager
dialog.

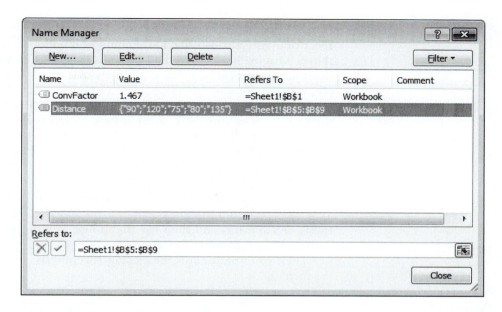

Select the name to remove, and then click the Delete button. Any formula that was using the deleted name will show the error message, **#NAME?**, since the name in the formula will no longer be recognized.

Notice in Figure 1.41 that both the named cell (ConvFactor) and the named cell range (Distance) show "Workbook" as their *scope*. This means that the name will be recognized throughout the workbook; that is, you can use these names in any worksheet in the workbook. By default, any name you create using the Name Box will have a workbook scope.

1.4.8 Editing the Active Cell

You can enter *edit mode* to change the contents of the active cell either by double-clicking the cell or by selecting the cell and then pressing [F2]. Be sure to press [Enter] when you are done editing the cell contents, or your next mouse click will change your formula.

> **Note:** Changing a formula that was previously copied to another range of cells does not cause the contents of the other cells to be similarly modified. If the change should be made to each cell in the range, edit one cell, then re-copy the cell range.

You can edit either on the Formula bar or right in the active cell if this option is activated. (Excel's default is to allow editing in cells.) The editing process is the same whether you edit directly in the cell or use the Formula bar. About the only time it makes a difference is when you want to edit a long formula in a cell near the right edge of the Excel window; then the extra space in the Formula bar is useful.

Excel's defaults are to show the Formula bar and allow editing directly in cells. However, if you are using a shared computer, someone could have turned these features off.

- To activate or deactivate editing in the active cell, use the **File** tab (Office button in Excel 2007) and select **Options/Advanced** panel and use the checkbox labeled **Allow editing directly in cells**.
- To show or hide the Formula bar, use Ribbon options **View/[Show/Hide]** and use the **Formula Bar** checkbox.

1.4.9 Using Built-In Functions

Electronic spreadsheets like Excel come with built-in functions to do lots of handy things. Originally, they handled business functions, but newer versions of spreadsheet programs also have many additional functions useful to engineers. For example, you can compute the arithmetic average and standard deviation of a column (or row) of values using Excel's **AVERAGE** and **STDEV** functions. These functions work on a range of values, so we'll have to tell the spreadsheet which values to include in the computation of the average and standard deviation.

In the last example (Figure 1.40), we calculated the average distance. Without using cell names, the formula could be written as =AVERAGE(B5:B9) as shown in cell B11 in Figure 1.42.

Again, you don't normally enter the entire formula from the keyboard. Typically, you would type the equal sign, the function name, and the opening parenthesis =AVERAGE(, and then use the mouse to indicate the range of values to be included in the formula (see Figure 1.43).

Then you would enter the closing parenthesis and press [Enter] to complete the formula.

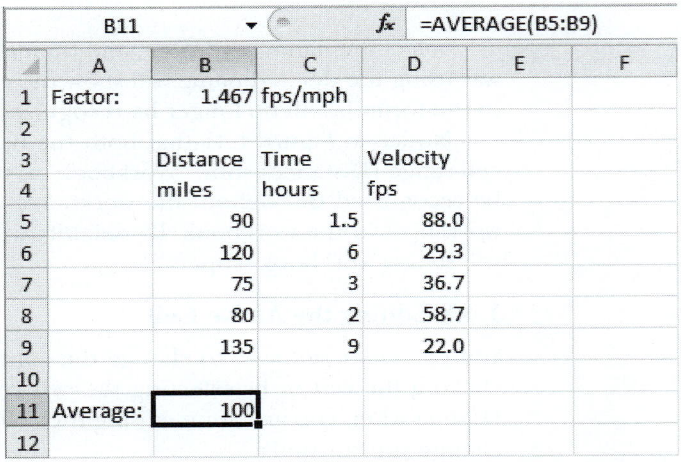

Figure 1.42
Using the **AVERAGE** function.

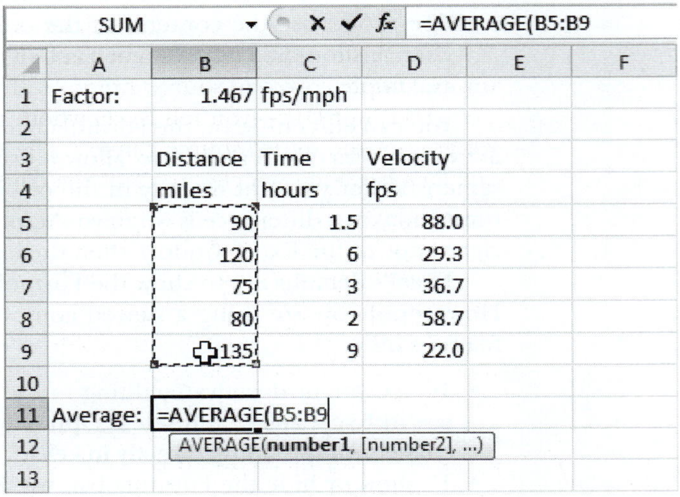

	A	B	C	D	E	F
		B11		f_x =AVERAGE(B5:B9)		
1	Factor:	1.467	fps/mph			
2						
3		Distance	Time	Velocity		
4		miles	hours	fps		
5		90	1.5	88.0		
6		120	6	29.3		
7		75	3	36.7		
8		80	2	58.7		
9		135	9	22.0		
10						
11	Average:	100				
12						

Figure 1.43
Using the mouse to select the cell range for the **AVERAGE** function.

	A	B	C	D	E	F
		SUM		✗ ✓ f_x =AVERAGE(B5:B9		
1	Factor:	1.467	fps/mph			
2						
3		Distance	Time	Velocity		
4		miles	hours	fps		
5		90	1.5	88.0		
6		120	6	29.3		
7		75	3	36.7		
8		80	2	58.7		
9		135	9	22.0		
10						
11	Average:	=AVERAGE(B5:B9				
12		AVERAGE(**number1**, [number2], ...)				
13						

Figure 1.44
Using the **STDEV** function.

	B12		▼	f_x	=STDEV(B5:B9)	
	A	B	C	D	E	F
1	Factor:	1.467	fps/mph			
2						
3		Distance	Time	Velocity		
4		miles	hours	fps		
5		90	1.5	88.0		
6		120	6	29.3		
7		75	3	36.7		
8		80	2	58.7		
9		135	9	22.0		
10						
11	Average:	100				
12	Std. Dev.:	26.2				
13						

Figure 1.45
Copying cells B11:B12
to columns C and D.

	C12		▼	f_x	=STDEV(C5:C9)	
	A	B	C	D	E	F
1	Factor:	1.467	fps/mph			
2						
3		Distance	Time	Velocity		
4		miles	hours	fps		
5		90	1.5	88.0		
6		120	6	29.3		
7		75	3	36.7		
8		80	2	58.7		
9		135	9	22.0		
10						
11	Average:	100	4.3	46.944		
12	Std. Dev.:	26.2	3.2	26.8		
13						

To compute the standard deviation of the distance values, use the same process with Excel's **STDEV** function, as shown in Figure 1.44.

To compute the averages and standard deviations of the Time and Velocity columns, simply copy the source range, B11:B12, to the destination range C11:D11. (Yes, D11, not D12—you tell Excel where to copy the top-left corner of the source range, not the entire source range.) The final worksheet is shown in Figure 1.45; cell C12 has been selected to show the contents.

1.4.10 Error Messages in Excel

Sometimes Excel cannot recognize the function you are trying to use (e.g., when the function name is misspelled) or cannot perform the requested math operation (e.g., because of a divide by zero). When this happens, Excel puts a brief error message in the cell to let you know that something went wrong. Some common error messages are listed in Table 1.2.

When these error messages are displayed, they indicate that Excel detected an error in the formula contained in the cell. The solution involves fixing the formula or the values in the other cells that the formula references.

Table 1.2 Excel Error Messages

Message	Meaning
#DIV/0	Attempted to divide by zero.
#N/A	Not available. There is an **NA** function in Excel that returns #N/A, meaning that the result is "not available." Some Excel functions return #N/A for certain errors, such as failures in using a lookup table. Attempts to do math with #N/A values also return #N/A.
#NAME?	Excel could not recognize the name of the function, cell, or cell range you tried to use.
#NUM!	Not a valid number. A function or math operation returned an invalid numeric value (e.g., a value too large to be displayed). This error message is also displayed if a function fails to find a solution. (For example, the **IRR** function uses an iterative solution to find the internal rate of return and may fail.)
#REF!	An invalid cell reference was encountered. For example, this error message is returned from the **VLOOKUP** function if the column index number (offset) points at a column outside of the table range.
#VALUE!	This error can occur when the wrong type of argument is passed to a function, or when you try to use math operators on something other than a value (such as a text string).

Sometimes Excel will detect an error when you press [Enter] to save a formula in a cell. When Excel detects this type of error, it pops up a message box indicating that there is an error in the formula. For example, the following formula for area of a circle has unbalanced parentheses:

$$=\text{PI}()*(B3/2\char`\^2 \quad \text{should be} \quad =\text{PI}()*(B3/2)\char`\^2$$

If you attempt to enter this formula in a cell, Excel will detect the missing parenthesis and display a message box indicating the problem and proposing a correction (see Figure 1.46).

If the proposed correction is correct, simply click the **Yes** button and let Excel correct the formula. However, if the proposed solution is not correct, click the **No** button and fix the formula by hand. In this example, the proposed correction would place the closing parenthesis in the wrong place, so the formula must be corrected by hand. The corrected equation is shown in Figure 1.47.

Figure 1.46
Excel will (often) propose corrections when errors are detected in your formulas.

Microsoft Excel found an error in the formula you entered. Do you want to accept the correction proposed below?

=PI()*(B3/2^2)

• To accept the correction, click Yes.
• To close this message and correct the formula yourself, click No.

[Yes]   [No]

Figure 1.47
The corrected formula for the area of a circle.

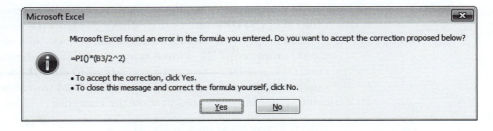

B4		f_x	=PI()*(B3/2)^2		
	A	B	C	D	E	F
1	Area of a Circle					
2						
3	Diameter:	7	cm			
4	Area:	38.5	cm^2			
5						

1.5 ORGANIZING YOUR WORKSHEETS

Computer programs, including Excel worksheets, typically contain the following standard elements:

- **Titles**: including a problem description and identifying the author(s)
- **Input Values**: values entered by the user when the worksheet is used
- **Formulas**: the equations that calculate results based on the input values
- **Results**: the computed values

Your worksheets will be easier to create and use if you group these elements together (when possible) and develop a standard placement of these items on your worksheets. These elements can be placed anywhere, but, for small problems, a common placement puts titles at the top followed by input values further down the page. Formulas and results are placed even further down the page. This layout is illustrated in Figure 1.48.

One feature of this approach is that the information flows down the page, making the worksheet easier to read. One drawback of this approach is that the results are separated from the input values, making it harder to prepare a report showing the input values and the calculated results together.

The layout shown in Figure 1.49 puts the input values next to the results, but can be harder to read, because the information flow is not simply from top to bottom.

The layout illustrated in Figure 1.50 is often used when the input values include columns of data values.

Figure 1.48
A common worksheet layout for small problems.

◢	A	B	C	D	E	F	G	H
1								
2		Title, Author, Date						
3								
4								
5		Input Values (usually from the problem statement)						
6								
7								
8		Formulas (calculated values)						
9								
10								
11		Results						
12								
13								

Figure 1.49
A modified layout that keeps the results near the input values.

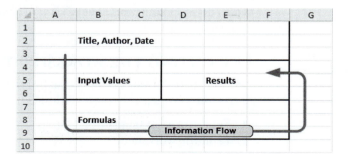

Figure 1.50
A worksheet layout suitable for working with columns of data values.

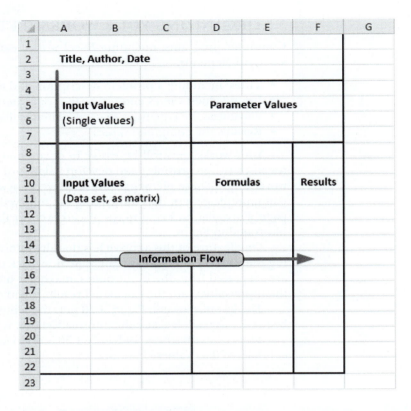

Figure 1.51
Large problems can be organized by using multiple worksheets.

When the problem requires a lot of space on the worksheet (several screens), it can be convenient to put the input values, formulas, and results on different worksheets within the same workbook, as illustrated in Figure 1.51.

There is no single layout that works for all problems, but taking time to organize your worksheets and being consistent with your layouts will make your life easier as you develop your worksheets. Organized worksheets are also easier for others to understand and use.

1.6 PRINTING THE WORKSHEET

Printing an Excel worksheet is usually a two- or three-step process. For small worksheets, the two-step process includes:

1. Set the area to be printed
2. Print the worksheet

For larger worksheets, or when you want to modify print options, include an additional step:

1. Set the area to be printed
2. Set print options using the Print dialog (Excel 2007: Print Preview screen)
3. Print the worksheet

1.6.1 Setting the Print Area

By default, Excel will print the rectangular region of the currently selected worksheet that includes all nonempty cells. If you want all of the cells that you have used to print, then you can skip the "set print area" step.

To specify exactly which cells to print, use the mouse to select the region of the spreadsheet that should be printed (see Figure 1.52). Then use Ribbon options **Page Layout/Page Setup/Print Area/Set Print Area** [Excel 2003: File/Print Area/ Set Print Area].

Figure 1.52
Setting the print area.

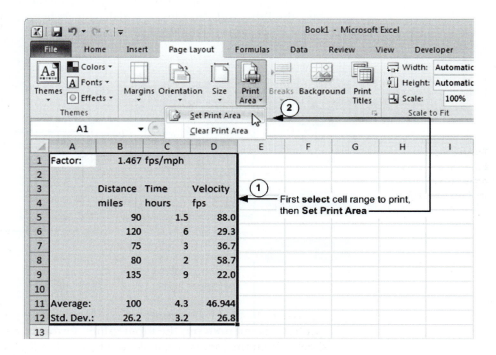

Once a print area has been set, Excel shows the region that will be printed with dashed lines, as shown in Figure 1.53. If multiple pages are to be printed, Excel also shows the page breaks with dashed lines.

1.6.2 Printing Using Current Options

Once the desired print area has been set, you can print the worksheet as follows:

- Excel 2010: **File tab/Print**
- Excel 2007: **Office/Print**
- Excel 2003: **File/Print**

Figure 1.53
Once a print area has been set, dashed lines show the print area.

◢	A	B	C	D	E	F	G	H
1	Factor:	1.467	fps/mph					
2								
3		Distance	Time	Velocity				
4		miles	hours	fps				
5		90	1.5	88.0				
6		120	6	29.3				
7		75	3	36.7				
8		80	2	58.7				
9		135	9	22.0				
10								
11	Average:	100	4.3	46.944				
12	Std. Dev.:	26.2	3.2	26.8				
13								

◄— Dashed lines indicate area to be printed

The appearance of the Print dialog has changed significantly in Excel 2010, since the Print and Print Preview dialogs have been combined. Images here (e.g., Figure 1.54) are from the Print dialog in Excel 2010.

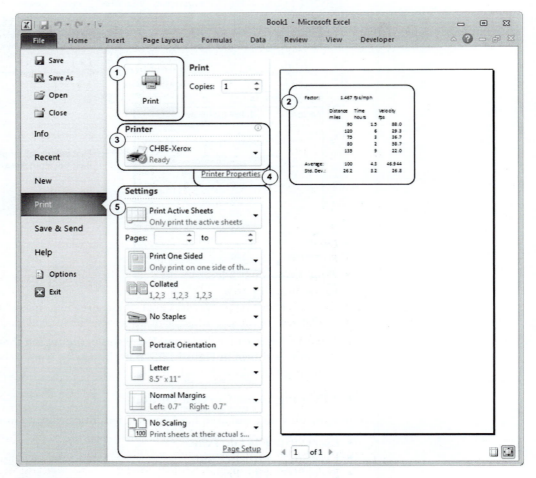

Figure 1.54
The Print dialog (the appearance of this dialog may vary depending on the features available on your printer).

The following items are indicated in Figure 1.54:

1. The **Print** button—click this button to send the data to your printer.
2. The Print Preview area—a preview of your printout is displayed at the right side of the Print dialog.
3. The Printer that will be used—use the drop-down list to select another printer.
4. **Printer Properties** link—click this link to access the printer's control dialog.
5. Printer **Settings**—common printer settings are available directly from the Print dialog in Excel 2010. In previous versions, these were accessed via the printer's control dialog or from the Ribbon's Page Layout tab. These options are still available in Excel 2010.

Note: You can customize the Quick Access Toolbar to include a **Quick Print** button. If you use the Quick Print button, the currently selected area of the current worksheet will print without displaying the Print dialog.

1.6.3 Changing Printing Options

There are several ways to change printing options with Excel; here we focus on using the Ribbon's **Page Layout** tab (Figure 1.55) to control how the worksheet will print.

Figure 1.55
The Ribbon's Page Layout tab.

Page Layout Tab/Page Setup Group

The **Page Setup** group allows you to set or modify the following print features:

- *Margins:*   Choose predefined or custom margins for the printed pages.
- *Orientation:*   Select portrait or landscape printing.
- *Size:*   Choose the desired paper size.
- *Print Area:*   First select the desired cell region, then use the Print Area button to set the print area.
- *Breaks:*   Insert page breaks; Excel will insert them automatically when needed.
- *Background:*   Choose a background image for your worksheet.
- *Print Titles:*   Select cells that contain titles that should be included on every printed page.

One of the few features that is not directly accessible from the Page Layout tab is the ability to add a header and/or footer to a printed worksheet. Headers and footers are text lines that appear in the margins of the printout that are used to provide information about the worksheet, like file name, page number, and print date.

To add a header or footer, click the **Expand** button at the bottom-right corner of the **Page Setup** group to open the Page Setup dialog shown in Figure 1.56.

You can use the drop-down lists to select from standard headers and footers, or you can click the **Custom Header …** or **Custom Footer …** buttons to create your own. Clicking the **Custom Header …** button opens the Header dialog shown in Figure 1.57.

Figure 1.56
The Page Setup dialog,
Header/Footer Tab.

Figure 1.57
The Header dialog.

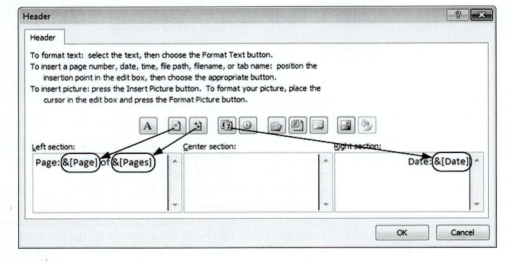

You can type into any of the sections (left, center, or right) to include text in the header. But the buttons just above the entry fields provide an easy way to include:

- Current page number (set while printing)
- Number of pages
- Current date (while printing)
- Current time (at start of printing)
- File path (drive and folder(s))
- File name
- Worksheet name

Using a header and/or footer is an easy way to add page numbers to your printed worksheets. In Figure 1.57, the header has been set to indicate the current page number (e.g., Page: 1 of 4) on the left and the date on which the worksheet was printed on the right.

Page Layout Tab/Scale to Fit Group

A handy printing feature is the ability to scale your worksheet to fit to a limited number of pages. There are many times when it is convenient to force a worksheet to print on a single sheet or paper. To do so, use the Ribbon's **Page Layout** tab and **Scale to Fit** group. Set the **Width** and **Height** options (see Figure 1.58) to "1 page" to force the printed worksheet to fit onto a single sheet of paper.

Figure 1.58
Set the Width and Height options to "1 page" to force the printed worksheet to fit onto a single sheet of paper.

Page Layout Tab/Sheet Options Group

When the reader needs to understand the formulas used in a worksheet, it is very helpful to print the worksheet showing the gridlines and (column and row) headings. To instruct Excel to include gridlines and headings on the printout, use the **Print** checkboxes in the **Sheet Options** group, as shown in Figure 1.59.

Figure 1.59
Use the **Print** checkboxes for **Gridlines** and **Headings** (cell labels) when you want them to appear on the printout.

Changing Print Options for Multiple Worksheets

When you use Ribbon options to set print options, you are setting the print options only for the currently selected worksheet. If you need to print multiple worksheets in a workbook and want them to have the same margins (and other print options), first select all of the worksheets to be printed before setting the print options. The print options that you select will then apply to all selected worksheets.

To select multiple worksheets, hold the [Ctrl] key down while you click on each worksheet's tab (near the bottom of the Excel window).

CAUTION: Be sure to de-select multiple worksheets before editing any cells because any cell changes will also be applied to all selected worksheets.

1.7 SAVING AND OPENING WORKBOOKS, EXITING EXCEL

1.7.1 Saving the Workbook

To save the workbook as a file, press the **Save** button (looks like a diskette) on the Quick Access toolbar or use:

- Excel 2010: **File tab/Save**
- Excel 2007: **Office/Save**
- Excel 2003: **File/Save**

If the workbook has not been saved before, you will be asked to enter a file name in the Save As dialog box.

Notes:

1. In Windows, it is generally preferable not to add an extension to the file name. Excel will allow you to add an extension, but the file name will not appear in any future selection box unless Excel's default extension is used. If you do not include a file name extension, Excel will add its default extension, .xlsx.

2. Prior to Excel 2007, Excel's default file extension was .xls. The new file extension with Excel 2007 indicates that the latest version of Excel is saving files in a new file format. Files saved with the new file format and with the new .xlsx extension cannot be read by earlier versions of Excel. If you need to maintain compatibility with a previous version of Excel, use **Save As** and select **Excel 97-2003 Workbook** as the type of file format to use. This causes Excel to use the old file format and the old .xls extension.

1.7.2 Opening a Previously Saved Workbook

To open an existing Excel workbook:

- Excel 2010: **File tab/Open**
- Excel 2007: **Office/Open**
- Excel 2003: **File/Open**

Then select the workbook file you wish to use from the Open File dialog box.

1.7.3 Setting AutoRecover Options

By default, Excel saves a backup, called an *AutoRecover file*, of any workbook you are editing. This is done every 10 minutes. When you successfully save your work and exit Excel, the AutoRecover file is deleted. But, if something goes wrong (e.g., power failure) and you don't exit Excel cleanly, the next time you start Excel you will have a chance to recover most of your lost work via the AutoRecover file.

To verify or modify the AutoRecover settings, use the Excel Options dialog available using the **Options** button on the **File** tab [Excel 2007, use **Office/Excel Options**]. The AutoRecover settings can be modified using the **Save** panel on the Excel Options dialog.

1.7.4 Exiting Excel

Use the Close button (Figure 1.60) or

- Excel 2010: **File tab/Exit**
- Excel 2007: **Office/Exit Excel**
- Excel 2003: **File/Exit**

Figure 1.60
Use the Close button to exit Excel.

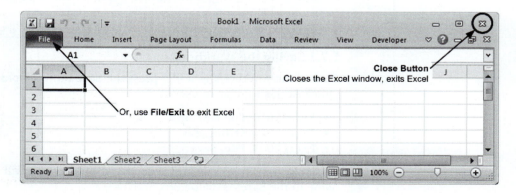

If there are open unsaved workbooks, Excel will ask if you want to save your changes before exiting.

FLUID STATICS

Manometers are tubes of fluid connected to two different locations (see Figure 1.61) that can be used to measure pressure differences between the two locations.

A general equation for manometers can be written as:

$$P_L + \rho_L g h_L = P_R + \rho_R g h_R + \rho_M g R$$

In this equation,

P_L is the pressure at the top of the left arm of the manometer,

P_R is the pressure at the top of the right arm of the manometer,

g is the acceleration due to gravity,

ρ_L is the density of the fluid in the upper portion of the left manometer arm,

ρ_R is the density of the fluid in the upper portion of the right manometer arm,

ρ_M is the density of the manometer fluid in the lower portion of the manometer,

h_L is the height of the column of fluid in the upper portion of the left manometer arm,

h_R is the height of the column of fluid in the upper portion of the right manometer arm,

R is the manometer reading, the height difference between the left and right columns of manometer fluid.

There are three commonly used types of manometers, and each has a particular purpose.

Sealed-End Manometer

A *sealed-end manometer* is used to determine an absolute pressure. The sealed-end manometer in Figure 1.62 can be used to determine the absolute pressure at the location marked P_L.

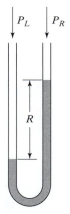

Figure 1.61
A manometer to measure the pressure difference between P_L and P_R.

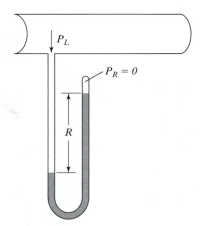

Figure 1.62
Sealed-end manometer.

To determine the absolute pressure, the space above the manometer fluid on the right side is evacuated, so the density of the vacuum is $\rho_R = 0$ and the pressure on the right is $P_R = 0$.

The general manometer equation can be simplified to

$$P_L = \rho_M g R - \rho_L g h_L$$

for a sealed-end manometer.

Problem: Water ($\rho_L = 1{,}000\,\text{kg/m}^3$) flows in the pipe connected to a sealed-end manometer. The height of the water column on the left side is $h_L = 35\,\text{cm}$. The manometer reading is 30 cm of mercury ($\rho_M = 13{,}600\,\text{kg/m}^3$). What is the absolute pressure, P_L?

The solution to this problem is shown in Figure 1.63.

Figure 1.63
The solution to a sealed-end manometer problem.

◢	A	B	C	D	E	F
1	Sealed-End Manometer					
2						
3		ρ_L	1000	kg/m³		
4		ρ_M	13600	kg/m³		
5		g	9.8	m/s²		
6		h_L	0.35	m		
7		R	0.3	m		
8						
9		P_L	36554	kg/m s² = N/m² = Pa		
10		P_L	0.3608	atm		
11						
12	Formulas Used					
13			C9:	=C4*C5*C7-C3*C5*C6		
14			C10:	=C9/101325		
15						

Open-End Manometer

A *gauge pressure* is a pressure measured relative to the current barometric pressure instead of perfect vacuum. An open-end manometer (Figure 1.64) is used to determine

Figure 1.64
Gauge pressure manometer.

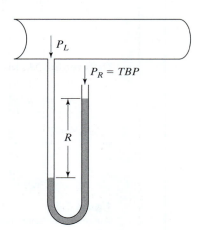

a gauge pressure; it is very similar to the sealed-end manometer except that the right side is left open to the atmosphere.

The open-end manometer shown in Figure 1.64 measures the pressure at the location marked P_L relative to the atmospheric pressure at P_R. (*TBP* stands for "today's barometric pressure.")

Assigning "today's barometric pressure" a value of zero ($P_R = 0$) causes the calculated pressure at P_L to be a gauge pressure. (The pressure due to the column of air on the right side is neglected.)

The general manometer equation again simplifies to

$$P_L = \rho_M g R - \rho_L g h_L$$

for an open-end manometer, with the understanding that P_L is now a gauge pressure, not absolute.

Problem: Air ($\rho_L \approx 0$) flows in the pipe connected to an open-ended manometer. The manometer reading is 12 cm of mercury ($\rho_M = 13{,}600 \, \text{kg/m}^3$). The barometric pressure when the manometer was read was 740 mm Hg. What is the gauge pressure, P_L?

The solution to this problem is shown in Figure 1.65.

Note that the barometric pressure (740 mm Hg) was not used in the problem. This value would be used to convert the calculated gauge pressure to an absolute pressure, as shown in Figure 1.66.

Figure 1.65

The solution to an open-end manometer problem.

◢	A	B	C	D	E	F
1	Open-End Manometer					
2						
3		ρ_M	13600	kg/m^3		
4		g	9.8	m/s^2		
5		R	0.12	m		
6						
7		P_L	15994	kg/m s^2 = N/m^2 = Pa		
8		P_L	0.1578	atm (gauge)		
9						
10	Formulas Used					
11			C7:	=C3*C4*C5		
12			C8:	=C9/101325		
13						

Differential Manometer

A differential manometer is used to determine the pressure difference between two locations, as illustrated in Figure 1.67.

For a differential manometer, the difference in column heights on the left and right sides is related to the manometer reading ($h_L - h_R = R$), and the general manometer equation simplifies to

$$P_L - P_R = R_g(\rho_M - \rho_F)$$

where ρ_F is the density of the fluid in the left and right arms of the manometer ($\rho_F = \rho_L = \rho_R$) (assumed equal).

Figure 1.66
Converting a gauge
pressure to absolute
pressure.

	A	B	C	D	E	F
1	**Open-End Manometer**					
2						
3		ρ_M	13600	kg/m³		
4		g	9.8	m/s²		
5		R	0.12	m		
6						
7		P_L	15994	kg/m s² = N/m² = Pa		
8		P_L	0.1578	atm (gauge)		
9						
10		P_{TBP}	740	mm Hg		
11		P_{TBP}	0.9737	atm		
12						
13		P_L	1.1315	atm (abs.)		
14						
15	**Formulas Used**					
16		C7:	=C3*C4*C5			
17		C8:	=C7/101325			
18		C11:	=C10/760			
19		C13:	=C8+C11			
20						

Problem: Water $(\rho_L = \rho_R = 1000\,\text{kg/m}^3)$ flows in a pipe connected to a differential manometer. The manometer reading is 30 cm of mercury $(\rho_M = 13{,}600\,\text{kg/m}^3)$. What is the pressure difference, $P_L - P_R$?

The solution to this problem is shown in Figure 1.68.

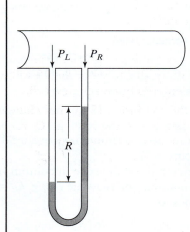

Figure 1.67
Differential manometer.

	A	B	C	D	E	F
1	**Differential Manometer**					
2						
3		ρ_F	1000	kg/m³		
4		ρ_M	13600	kg/m³		
5		g	9.8	m/s²		
6		R	0.3	m		
7						
8		P_L-P_L	37044	kg/m s² = N/m² = Pa		
9		P_L-P_L	0.3656	atm		
10						
11	**Formulas Used**					
12		C8:	=C6*C5*(C4-C3)			
13		C9:	=C8/101325			
14						

Figure 1.68
The solution to a differential
manometer problem.

KEY TERMS

Absolute cell addressing	Edit mode	Range of cells
Active cell	Electronic spreadsheet	Relative cell addressing
Automatic recalculation	Fill handle	Spreadsheet
AutoRecover file	Formula	Status bar
Built-in functions	Formula bar	Title bar
Cell address	Label	Value
Cell cursor	Name box	Workbook
Cells	Named cells	

SUMMARY

Cell Contents

Cells can contain labels (text), numbers, and equations (formulas).

Excel interprets what you type into a cell as follows:

- If the entry begins with an equal sign (or a plus sign followed by something other than a number), it is considered a formula.
- If the entry begins with a number, a currency symbol followed by a number, or a minus sign, Excel assumes that you are entering a value.
- If your entry cannot be interpreted as a formula or a value, Excel treats it as a label.

Cell References: Relative and Absolute Addressing

In formulas, cells are identified by using the cell's address, such as G18. Dollar signs are used to prevent the row index (18) and the column index (G) from incrementing if the formula is copied to other cells.

For example:

=$G18+2　Here, the column index G will not change if the formula is copied. If the formula is copied down the column, the 18 will automatically be incremented as the formula is entered into each new cell.

=G$18+2　In this case, the row index 18 will not change if the formula is copied. If the formula is copied across the row, the G will automatically be incremented as the formula is entered into each new cell.

=$G$18+2　In this case, neither the column index G nor the row index 18 will change if the formula is copied.

Selecting Cell Ranges

- To select a single cell, simply click on the cell, or use the arrow keys to locate the cell. The selected cell becomes the active cell.
- To select multiple cells in a rectangular region, click on a corner cell and move the mouse cursor to the opposite corner of the rectangular region. Or use the

arrow keys to select the first corner cell and hold the [Shift] key down while moving the cell cursor to the opposite corner of the rectangular region.

Copying Cell Ranges by Using the Clipboard

1. Select the source cell or range of cells.
2. Copy the contents of the source cell(s) to the clipboard by selecting **Edit/Copy**.
3. Select the destination cell or range of cells.
4. Paste the clipboard contents into the destination cell(s) by selecting **Edit/Paste**.

Copying Cell Ranges by Using the Fill Handle

1. Select the source cell or range of cells.
2. Click the fill handle with the mouse and drag it to copy the selected cell contents to the desired range.

Editing a Cell's Contents

Double-click the cell or select the cell and press [F2].

Printing a Spreadsheet

1. Select the area to be printed and then use Ribbon options: **Page Layout/Page Setup/Set Print Area** [Excel 2003: **File/Print Area/Set Print Area**].
2. Use the Print dialog, as
   - Excel 2010: **File tab/Print**
   - Excel 2007: **Office/Print**
   - Excel 2003: **File/Print**

Setting Print Options

Use Ribbon options: **Page Layout/Page Setup, Page Layout/Scale to Fit**, and **Page Layout/Sheet Options** to set print options.

Adding a Header or Footer to Printed Documents

1. Use the Expand button at the bottom-right corner of the **Page Layout/Page Setup** group to open the Page Setup dialog.
2. Choose the **Header/Footer** tab.
3. Click the **Custom Header ...** or **Custom Footer ...** button.
4. Create the desired header or footer.
5. Click **OK** to exit the dialog.

Saving a Workbook

Press the [Save] button (looks like a diskette) on the Quick Access toolbar, or use:

- Excel 2010: **File tab/Save**
- Excel 2007: **Office/Save**
- Excel 2003: **File/Save**

Opening a Previously Saved Workbook

Use the Open button as:

- Excel 2010: **File tab/Open**
- Excel 2007: **Office/Open**
- Excel 2003: **File/Open**

Exiting Excel

Click the Close button at the top-right corner of the Excel window, or:

- Excel 2010: **File tab/Exit**
- Excel 2007: **Office/Exit Excel**
- Excel 2003: **File/Exit**

PROBLEMS

1.1 Calculating an Average

In Section 1.4.7 Excel's **AVERAGE** function was introduced. Create an Excel worksheet similar to the one shown in Figure 1.69 to calculate the arithmetic average of the five values listed below.

$$3.6; 3.8; 3.5; 3.7; 3.6$$

First, calculate the average the long way, by summing the values and dividing by five.

$$\bar{x} = \frac{(x_1 + x_2 + x_3 + x_4 + x_5)}{5}$$

Then calculate the average using Excel's **AVERAGE** function by entering the following formula in a cell:

```
=AVERAGE(cell range)
```

Replace "cell range" by the actual addresses of the range of cells holding the five values (e.g., the cell range is B4:B8 in Figure 1.69).

Figure 1.69
Checking Excel's
AVERAGE
function.

	A	B	C	D	E	F	G
1	Checking Excel's AVERAGE() Function						
2							
3		Values					
4		3.6					
5		3.8					
6		3.5					
7		3.7					
8		3.6					
9							
10	AVG$_1$:		<< computed by summing and dividing				
11	AVG$_2$:		<< computed using AVERAGE(B4:B8)				
12							

1.2 Determining Velocities (in mph and kph)

Some friends at the University of Calgary are coming south for Spring Break. Help them avoid a speeding ticket by completing a velocity conversion worksheet like the one shown in Figure 1.70. A conversion factor you might need is 0.62 mile/km.

Figure 1.70
Vehicles speeds in two-unit systems.

	A	B	C	D	E	F	G
1	Vehicle Speed Conversion Chart						
2							
3		Conversion Factor:		0.62	miles per kilometer		
4							
5		Canadian to US			US to Canadian		
6		Speed (KPH)	Speed (MPH)		Speed (MPH)	Speed (KPH)	
7		10			10		
8		20			15		
9		30			20		
10		40			25		
11		50			30		
12		60			35		
13		70			40		
14		80			45		
15		90			50		
16		100			55		
17		110			60		
18		120			65		
19		130			70		
20		140			75		
21							

1.3 Temperature Increase due to Incandescent Lighting

When energy is added to a fluid, the temperature of the fluid increases. An equation describing this phenomenon is

$$Q = M \, C_p \, \Delta T$$

where Q is the amount of energy added (joules)

M is the mass of the fluid (kg)

C_p is the heat capacity of the fluid (joules/kg K)

ΔT is the change in temperature (K, or °C)

A garage (24 ft × 24 ft × 10 ft) is illuminated by six 60-W incandescent bulbs. It is estimated that 90% of the energy to an incandescent bulb is dissipated as heat. If the bulbs are on for 3 hours, how much would the temperature in the garage increase because of the light bulbs (assuming no energy losses). Complete an Excel worksheet like the one illustrated in Figure 1.71 to answer this question. Potentially useful information:

- Air density (approximate): 1.2 kg/m^3
- Air heat capacity (approximate): 1000 joules/kg K
- 3.28 ft = 1 m

Figure 1.71
Garage temperature
change calculation.

	A	B	C	D	E	F	G
1	Temperature Change in a Garage When Lights Left On						
2							
3	Specified Information						
4		Number of Bulbs:		6			
5		Bulb Power:		60	W		
6	Bulb Percent Power Loss as Heat:			90%			
7		Bulbs on Time:		3	hrs		
8		Garage Air Volume:		5760	ft^3		
9		Air Density:		1.2	kg/m^3		
10		Air Heat Capacity:		1000	joules/kg K		
11							
12	Calculated Information						
13		Total Bulb Power:			W		
14	Total Bulb Power Lost as Heat:				W		
15	Total Bulb Power Lost as Heat:				joules/second		
16	Total Bulb Energy Lost as Heat:				joules		
17		Garage Air Volume:			m^3		
18		Garage Air Mass:			kg		
19		Temperature Change:			K		
20							

1.4 Savings from Using CFL Bulbs

Compact fluorescent light (CFL) bulbs have been available for years; they are very efficient, but a little pricey. A CFL bulb that puts out as much light as a 60-W incandescent bulb might cost $10, compared to about $1 for the incandescent bulb. But CFL bulbs are expected to last (on average) 15,000 hours, compared to about 1000 hours for an incandescent bulb. So it is easy to see that you would need 15 incandescent bulbs (total cost $15) to last the 15,000 hours that you would get from one ($10) CFL bulb; you save $5 and a lot of climbing ladders to replace all those incandescent bulbs.

But there's more. A CFL bulb that puts out as much light as a 60-W incandescent bulb will use about 13 W or power. According to the US Energy Information Administration (www.eia.doe.gov), residential electricity costs average about $0.10 per kilowatt/hour (1 kW/h = 3600 kW/s = 3600 kJ). Over the 15,000 hours that one CFL bulb is expected to last, how much will you save on power if you replace an incandescent bulb in your home with a CFL bulb?

Develop an Excel worksheet something like the one shown in Figure 1.72 to find the answer.

Figure 1.72
Savings by replacing one
incandescent bulb with a
CFL bulb.

	A	B	C	D	E	F	G
1	Savings from One CFL Bulb						
2				CFL Bulb	Incand. Bulb		
3		Power Consumption:		13	60	W	
4		On Time:		15000	15000	hours	
5		Cost of Electricity:	$	0.10	$ 0.10	per kw-hr	
6		Total Energy Cost:				(dollars)	
7							
8		Savings:		(dollars)			
9							

1.5 Comparing Cell Phone Services

Anna is thinking of changing her cell phone service, and she is comparing three plans:

1. Plan 1 has a $20/month access fee, unlimited nights and weekends, 300 anytime minutes plus $0.29/minute for minutes over 300, text messages cost $0.10 each, and roaming costs $0.39/minute whenever she leaves the state.
2. Plan 2 has a $40/month access fee, unlimited nights and weekends, 750 anytime minutes plus $0.29/minute for minutes over 750, text messages cost $0.05 each, and roaming costs $0.39/minute whenever she leaves the state.
3. Plan 3 has a $60/month access fee, unlimited nights and weekends, 1500 anytime minutes plus $0.19/minute for minutes over 1500, text messages are included in the anytime minutes (1 minute per message), and roaming is free.

Her primary concern is during the summer when she is away from college, out of the cell phone company's state, and spending a lot of time communicating with her friends. Looking at her bills from last summer, and trying to predict what will likely happen this summer, she anticipates the following monthly cell phone usage during the summer months:

- 400 minutes nights and weekends
- 500 anytime minutes
- 370 text messages
- 150 roaming minutes

Develop an Excel worksheet something like the one shown in Figure 1.73 to determine which plan is the best for Anna.

Figure 1.73
Comparing the expected costs of cell phone plans.

	A	B	C	D	E	F	G	H	I
1	Comparing Cell Phone Services								
2			Anna's	Allowed					
3			Expectation	Free	Paid	Cost each	Total Cost		
4	PLAN 1								
5	Night and Weekend Minutes:		400	UNLIMITED	0	$ -	$ -		
6	Anytime Minutes:		500	300	200	$ 0.29	$ 58.00		
7	Text Messages:		370	0	370	$ 0.10	$ 37.00		
8	Roaming Minutes:		150	0	150	$ 0.39	$ 58.50		
9	Access Fee:						$ 20.00		
10						PLAN 1 TOTAL COST:	$ 173.50	per month	
11									
12	PLAN 2								
13	Night and Weekend Minutes:								
14	Anytime Minutes:								
15	Text Messages:								
16	Roaming Minutes:								
17	Access Fee:								
18						PLAN 2 TOTAL COST:		per month	
19									
20	PLAN 3								
21	Night and Weekend Minutes:								
22	Anytime Minutes:								
23	Text Messages:								
24	Roaming Minutes:								
25	Access Fee:								
26						PLAN 3 TOTAL COST:		per month	
27									
28	It looks like Plan 3 is the best one for Anna right now.								
29									

2 Using Excel's Ribbon

Objectives

By the end of this chapter, you will be able to

- Use the Ribbon to access Excel features
- Cut and paste within a worksheet
- Change the appearance of characters displayed in a cell by controlling
  - font style, size, and color
  - cell borders and background colors
  - text alignment
  - special numeric formats (currency, percentage, thousand separators)
  - displayed precision (number of decimal places)
- Use the format painter to copy an existing format to a new cell range

- Change the width of a column or the height of a row
- Hide columns or rows
- Use predefined styles to change the appearance of cells
- Use conditional formatting to highlight cells containing values that meet specified criteria
- Define a range of cells as a table
- Use formatting to make your worksheets easier to read
- Lock cells to prevent unwanted alterations to cell contents

Excel 2010 continues to use the redesigned menu system called the Microsoft Office *Ribbon* to provide easier access to features. A lot of the more accessible features are related to formatting your worksheet. In this chapter, we focus on the formatting options available in Excel 2010.

2.1 NAVIGATING THE RIBBON

In Office 2007, Microsoft consolidated the menus, toolbars, and many of the dialog boxes used in previous editions by adding the *Ribbon* near the top of the window (see Figure 2.1). The Ribbon provides easy access to commonly used and advanced features through a series of *tabs*, such as the Home tab, Insert tab, and so on. Tabs provide access to related *groups* of features. The Excel features we explore in this chapter are located on the Home tab of the ribbon.

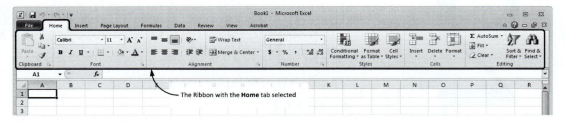

Figure 2.1
The Ribbon with the Home tab selected.

The Home tab of the Ribbon is divided into the following groups:

- *Clipboard*—provides quick access to the cut, copy, and paste functions as well as the format painter.
- *Font*—allows you to change the font type, size, and appearance as well as cell borders and colors.
- *Alignment*—allows you to change the way displayed text and values are aligned
- *Number*—provides quick access to the number formatting options
- *Styles*—allows you to apply a style to a range of cells, format a range of cells as a table, or apply conditional formatting to a range of cells.
- *Cells*—allows you to easily insert or delete a range of cells, hide or unhide columns or rows, and protect cell contents by locking the cells.
- *Editing*—provides access to some commonly used functions to sum, count, and average cell values, allows you to fill a range of cells with a series of values.

The groups on the Ribbon's Home tab provide a convenient way to organize the content of this chapter. We begin with the Clipboard group.

2.2 USING THE CLIPBOARD GROUP

The Clipboard group on the Ribbon's Home tab (Figure 2.2) provides access to:

- **Paste** button and **Paste Option** menu
- **Cut** button
- **Copy** button
- **Format Painter** button and **Format Painter toggle** switch

Figure 2.2
The Clipboard group on the Ribbon's Home tab.

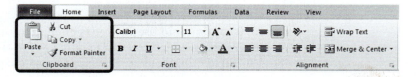

The *Windows clipboard* is a temporary storage location that allows you to move or duplicate information without retyping it. When you copy the contents of a cell or a range of cells, the original cells are left unchanged, but a copy of the contents is placed on the clipboard. You can then move the cell cursor to a different location and paste the copied information from the clipboard back into an Excel worksheet.

The clipboard is a Windows feature, not just an Excel feature. This means that the content stored on the clipboard becomes available to any Windows program, not just the program in which it was created. The Windows clipboard is commonly used to move information between programs as well as within one program. Moving information between programs is presented in Chapter 10; here, we show how to use the clipboard within an Excel workbook.

2.2.1 Using the Copy Button

When the **Copy** button is pressed, a copy of the currently selected information is placed on the Windows clipboard. The currently selected information is typically a cell, or a range of cells, but other information can be copied to the clipboard. You can copy a graph, for example.

The **Copy** button is accessed in Excel 2010 and Excel 2007 using Ribbon options **Home/Clipboard/Copy**. In older versions of Excel, use menu options Edit/Copy. You can also use a keyboard shortcut, [Ctrl-c], by holding down the Ctrl key while pressing the c key.

The copy process can be summarized as follows:

1. Select the information that you want copied to the clipboard.
2. Press the Copy button to copy the selected information.

Selecting Information to Be Copied

Selecting information is usually accomplished using the mouse. To select a single cell, simply click on it.

To select a cell range, click in a corner cell, and then drag (hold the left mouse button down while moving the mouse) the mouse icon to the opposite corner of the cell range. This is illustrated in Figure 2.3.

A selected range of cells is indicated (see Figure 2.3) by a heavy border; most of the selected cells have a colored background. The cell that was first clicked is shown without the colored background; that cell is still considered the active cell even though a range of cells has been selected. A cell range is indicated with a colon between the first and last cell addresses. For example, the selected cell range in Figure 2.3 would be described as B3:C7.

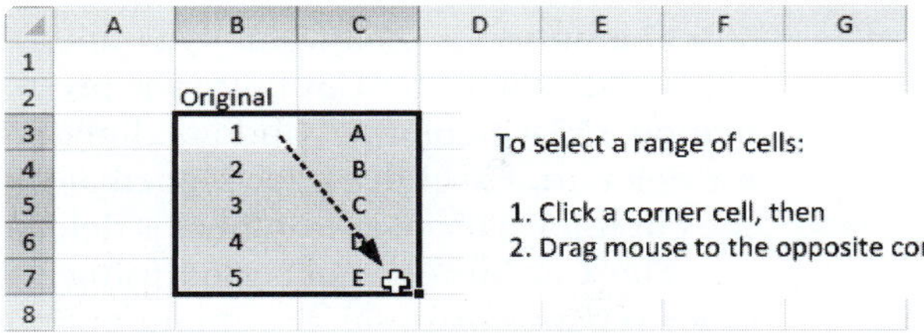

Figure 2.3

Selecting a cell range.

Checking the Contents of the Clipboard

In the recent versions of Excel, it is easy to see what is on the clipboard, just as the **Expand** control at the bottom-right corner of the Clipboard group (Figure 2.4).

Clicking the Clipboard group's **Expand** control opens the *Clipboard pane* so that you can see the contents of the clipboard. You may be surprised what you find on the clipboard because the Clipboard pane will display information copied from other programs as well, not just Excel.

Notice in Figure 2.4 that once a cell range (or cell) has been copied to the clipboard, the selection is indicated in the worksheet with a dashed border. That dashed border is a reminder that the information is on the clipboard and ready to be pasted back into the worksheet in another location.

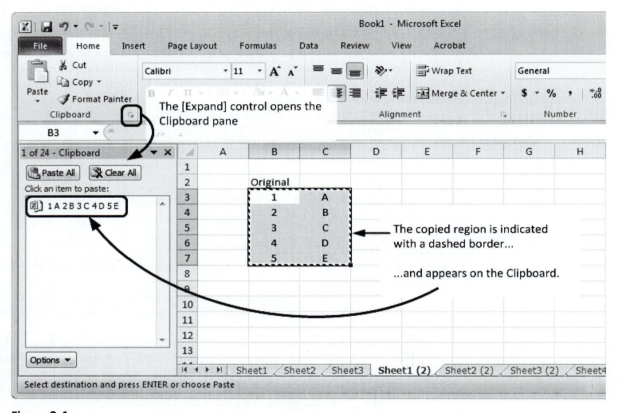

Figure 2.4
Using the Clipboard group's **Expand** control to open the Clipboard pane to view the Clipboard contents.

2.2.2 Using the Cut Button

The **Cut** button is used when you want to move (not duplicate) information to the clipboard. When the **Cut** button is pressed, a copy of the currently selected information is placed on the Windows clipboard, and the selected information is marked for deletion. Excel does not immediately delete the selection, but when you paste the cut information, the original is deleted.

The **Cut** button is accessed using Ribbon options **Home/Clipboard/Cut**. In older versions of Excel, use menu options Edit/Cut. You can also use a keyboard shortcut, [Ctrl-x].

The cut process can be summarized as follows:

1. Select the information that you want copied to the clipboard and delete (cut).
2. Press the **Cut** button to copy the selected information. The information will not be deleted until the paste operation is completed.

2.2.3 Using the Paste Button

The **Paste** button is used to copy information on the Windows clipboard to a selected location on an Excel worksheet.

The **Paste** button is accessed using Ribbon options **Home/Clipboard/Paste**. In older versions of Excel, use menu options Edit/Paste. You can also use a keyboard shortcut, [Ctrl-v].

The paste process (assuming information is available on the clipboard) can be summarized as follows:

1. Select the location (in a worksheet) where you want the clipboard information to be inserted.
2. Press the **Paste** button to copy the selected information from the clipboard to the worksheet.

In Figure 2.4, cell range B3:C7 has been copied to the clipboard. We need to tell Excel where to put the copy when we paste the information from the clipboard back into the worksheet. In Figure 2.5, we have selected cell E3; this tells Excel that the top-left corner of the cell range on the clipboard should go in cell E3 during the paste operation.

Note: You need to indicate just the top-left corner of the paste destination cell range, not the entire range.

Click the **Paste** button on the Ribbon's **Home** tab, or use the keyboard shortcut [Ctrl-v] to paste from the clipboard into the worksheet. The result is shown in Figure 2.6.

Figure 2.5

Indicating where the top-left corner of the copied cell range will be located.

Figure 2.6

The result of the paste operation.

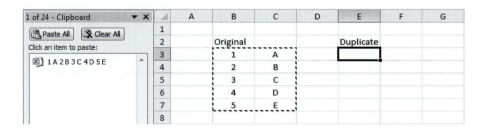

Notice that the contents of the original cell range (B3:C7) have been duplicated in cell range E3:F7; this is indicated with a label "1" in Figure 2.6. But the original, copied cell range is still indicated with a dashed border (label "2"); this is a reminder that the copied information is still available on the clipboard so that you can paste the same information into the worksheet multiple times, if needed.

Label "3" in Figure 2.6 is pointing out an icon that appears after the information is pasted into the cells. This icon gives you access to the Paste Options pop-up menu, as shown in Figure 2.7.

Figure 2.7
The Paste Options pop-up menu.

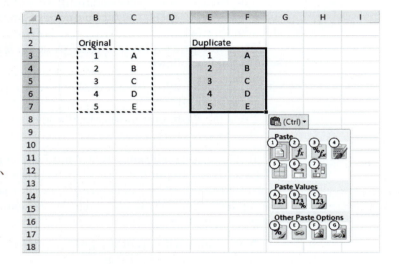

The Paste Options pop-up menu allows you to control the formatting of the pasted information. This is particularly useful when copying formulas because the pop-up menu allows you to indicate whether the copied formulas or just the numerical results (the values) should be pasted into the destination cells.

Note: The Paste Options pop-up menu is context sensitive; that is, the menu will provide different options depending on the type of information (e.g., number, formula, label) that is pasted.

There are 14 options for how to paste information into Excel cells! The option marked "1" is the default method, and that works most of the time. Option "A" is also useful when you want only values (not formulas) pasted into cells. The other options are less commonly needed. The options are listed in Table 2.1.

2.2.4 Using the Paste Options Menu

Another way to control how the information on the clipboard gets pasted into the worksheet is to use the *Paste Options menu*, which is accessed using the button directly below the **Paste** button (Figure 2.8). That is, use Ribbon options **Home/Clipboard/Paste (menu)** Excel 2003: Edit/Paste Special.

The Paste Options menu, shown in Figure 2.9, allows you to instruct Excel how to carry out the paste operation. The various options are listed in Table 2.1.

The default option, labeled **Paste**, simply pastes the currently selected contents of the clipboard using Excel's defaults. The defaults are to paste formulas (not values) and apply the formatting assigned to the source cell. Selecting the **Paste** option

Table 2.1 Paste options

1	Paste using Excel defaults
2	Paste formulas
3	Paste formulas and number formatting
4	Paste; keeping source formatting
5	Paste with no borders
6	Paste and keep source column widths
7	Transpose values when pasting
A	Paste values only (no formulas or formatting)
B	Paste values and number formatting
C	Paste values with source formatting
D	Paste formatting only (no values or formulas)
E	Paste link
F	Paste picture
G	Paste linked picture

Note: Numbers and letters refer to annotations in Figure 2.7.

Figure 2.8
The lower **Paste** button opens the Paste Options menu.

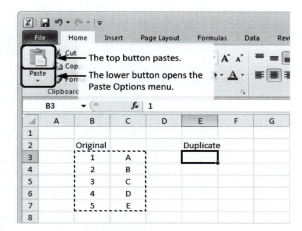

Figure 2.9
The Paste Options menu.

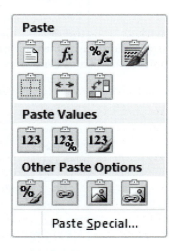

(top-left icon) from the Paste Options menu is equivalent to using the (upper) **Paste** button on the Ribbon.

The other options allow you to control how the information is pasted into a worksheet:

- *Paste:*   Pastes using Excel's defaults (paste formulas and use the source cell's formatting).
- *Formulas:*   If any of the copied cell contents included a formula, the formula (not the evaluated result) will be pasted into the destination cell(s).

  Note: If the copied formula contained relative cell addresses (without $), those cell addresses will be adjusted when the formula is pasted into the worksheet. Absolute addresses (with $) are not adjusted during the paste operation. This will be illustrated in Example 2.1.

- *Paste Values:*   If any of the copied contents included a formula, the evaluated result of the formula (not the equation) will be pasted into the destination cell(s).
- *No Borders:*   The copied cell contents (including formulas) and all formatting of the source cell(s) will be copied, except for borders.
- *Transpose:*   If a cell range was copied to the clipboard, using the **Transpose** option will cause the rows and columns in the cell range to be switched during the paste operation. This is discussed further in Section 4.4.4.
- *Paste Link:*   Instead of pasting the contents of the copied cells, the **Paste Link** operation places a formula in the destination cell(s) that points (links) back to the source cell. For example, if cell A3, containing the formula =12+2, is copied to the clipboard, and then the clipboard contents are pasted using **Paste Link** to cell C4, the **Paste Link** operation will place the formula =A3 into cell C4. Both cells A3 and C4 would display 14.
- *Paste Special...:*   The **Paste Special...** option opens the Paste Special dialog to allow even more control over the way the clipboard contents are pasted into the worksheet.
- *Paste as Picture:*   The **Paste as Picture** option allows you to paste the clipboard contents as an image. If a formula was copied to the clipboard, an image of the calculated result would be pasted with the **Paste as Picture** option.

Figure 2.10
The Paste Special dialog.

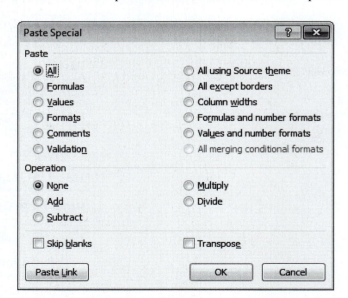

EXAMPLE 2.1

COPYING AND PASTING FORMULAS

In the worksheet shown in Figure 2.11, the formulas in cells C5 and D5 determine the surface area and volume of a sphere, using the radius in cell B5. Notice that the formula in cell C5 (shown in Figure 2.11) is based on the radius in cell B5,

```
=4*PI()*B5^2
```

We can tell that this formula uses relative addressing for cell B5, since there are no dollar signs on the B or the 5. The formula also uses the Excel function **PI** which tells Excel to use its built-in value of π (3.1415...).

Figure 2.11
Copying and pasting formulas.

C5 | fx =4*PI()*B5^2

	A	B	C	D	E	F
1	Area and Volume of Spheres					
2						
3		Radius	Area	Volume		
4		(cm)	(cm^2)	(cm^3)		
5		1	12.57	4.19		
6		2				
7		5				
8		12				
9						

We want to copy and paste the formulas to rows 6 through 8 to calculate the area and volume of the other three spheres. The dashed border around cells C5:D5 in Figure 2.11 indicates that the cells have already been copied to the clipboard. We just need to paste the formulas into the desired rows. We will do this using three different methods just to show some possible Excel techniques.

Paste Method 1: Paste the formulas into cells C6 and D6.

A two-step process is used to select the destination and then paste.

1. Select cell C6 to indicate the paste destination (Figure 2.12).
2. Use Ribbon options **Home/Clipboard/Paste** or keyboard shortcut [Ctrl-v] to paste the clipboard contents into the destination cells.

Figure 2.12
Selecting the destination cell (method 1).

	A	B	C	D	E	F
1	Area and Volume of Spheres					
2						
3		Radius	Area	Volume		
4		(cm)	(cm^2)	(cm^3)		
5		1	12.57	4.19		
6		2				
7		5				
8		12				
9						

(continued)

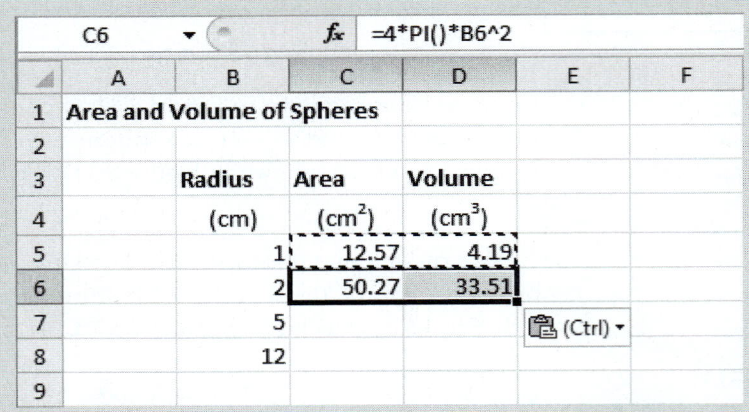

Figure 2.13
The result of the paste operation.

The result is shown in Figure 2.13.

Notice that the formula in cell C6 (shown in Figure 2.13) is

$$=4*PI()*B6^2$$

The "B5" in the original formula in cell C5 became "B6" when the formula was pasted below the original. The relative cell address used in the formula was adjusted in the paste process so that the new formula in cell C6 uses the radius just to the left (in the same relative position). The volume formula in cell D5 also uses a relative cell address (B5) for the radius, so the pasted volume formula also includes an adjusted cell address (B6):

D5 (original):   `=(4/3)*PI()*B5^3`
D6 (duplicate):   `=(4/3)*PI()*B6^3`

The dashed border around the source cells C5:D5 are a reminder that they are still available on the clipboard, so we could immediately repeat the two-step paste process to determine the area and volume of the 5 cm sphere—but there's a better way.

Paste Method 2: Paste the formulas into cells C6 through D8 (all three destination rows)

A two-step process is used to select the destination, and then paste, but we will paste the formulas into all three destination rows in the second step.

1. Select cells C6:C8 to indicate the paste destination (Figure 2.14).

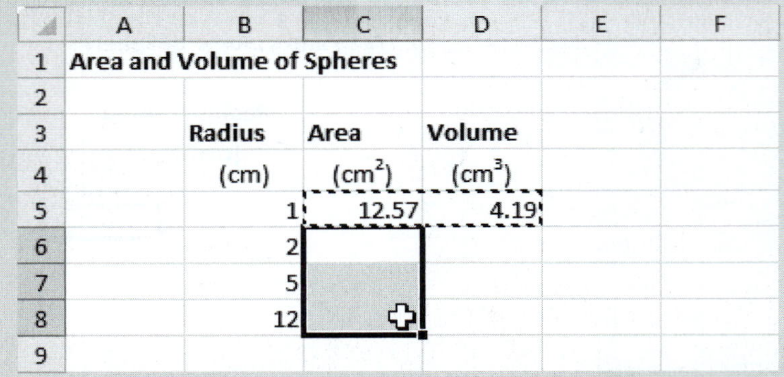

Figure 2.14
Select all three destination rows by selecting cells C6:C8.

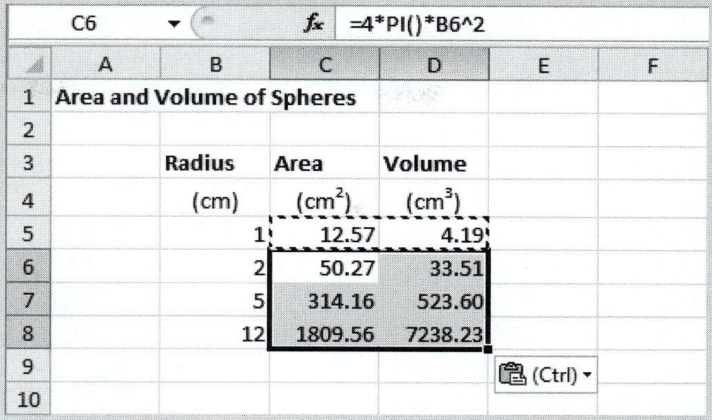

Figure 2.15
The result of pasting into three destination rows.

| C6 | | fx | =4*PI()*B6^2 | | |

	A	B	C	D	E	F
1	**Area and Volume of Spheres**					
2						
3		**Radius**	**Area**	**Volume**		
4		(cm)	(cm^2)	(cm^3)		
5		1	12.57	4.19		
6		2	50.27	33.51		
7		5	314.16	523.60		
8		12	1809.56	7238.23		
9					(Ctrl) ▾	
10						

Notice that only cells in column C have been selected to indicate the left side of the destination rows. Since the source (cells C5:D5) is only one row high, by selecting three destination rows you are telling Excel to paste the clipboard contents into each destination row.

2. Paste the clipboard contents into the destination cells using Ribbon options **Home/Clipboard/Paste**, or keyboard shortcut [Ctrl-v].

The result is shown in Figure 2.15.

Again, the paste operation has adjusted the relative cell address so that each formula uses the radius value on the correct row:

Original	C5:	`=4*PI()*B5^2`	D5:	`=(4/3)*PI()*B5^3`
Duplicates	C6:	`=4*PI()*B6^2`	D6:	`=(4/3)*PI()*B6^3`
	C7:	`=4*PI()*B7^2`	D7:	`=(4/3)*PI()*B7^3`
	C8:	`=4*PI()*B8^2`	D5:	`=(4/3)*PI()*B8^3`

In the distant past, this method was the usual way that formulas were copied in Excel, but the Fill Handle provides an even easier way to copy and paste formulas.

Paste Method 3: Duplicating formulas using the Fill Handle

The *Fill Handle* is the little black square in the bottom-right corner of the heavy border that shows a selected cell or cell range. In Figure 2.16, the original area and volume formulas have been selected (but not copied to the clipboard).

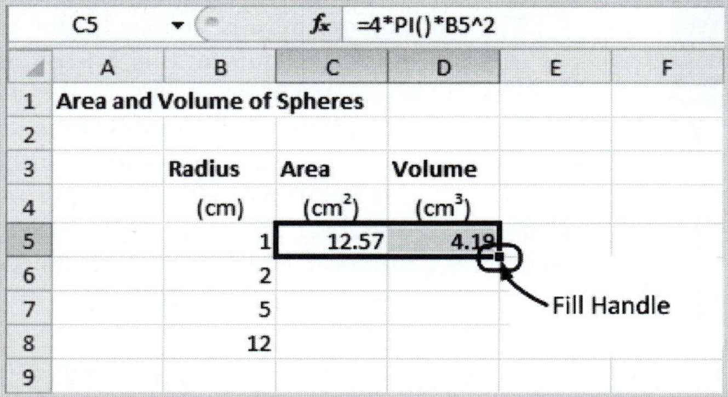

Figure 2.16
Selecting the cells to be duplicated.

| C5 | | fx | =4*PI()*B5^2 | | |

	A	B	C	D	E	F
1	**Area and Volume of Spheres**					
2						
3		**Radius**	**Area**	**Volume**		
4		(cm)	(cm^2)	(cm^3)		
5		1	12.57	4.19		
6		2				
7		5			Fill Handle	
8		12				
9						

(*continued*)

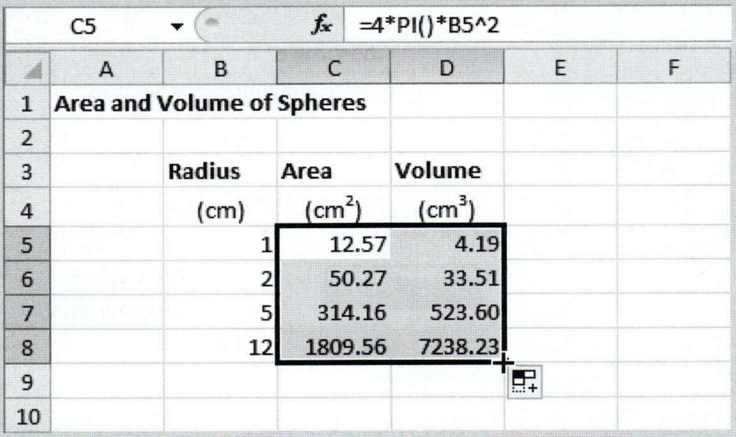

Figure 2.17
The result of duplicating the formulas using the Fill Handle.

To copy the formulas down the worksheet, just grab the Fill Handle with the mouse and drag it down three more rows. The formulas will be copied to the destination cells without ever using the clipboard.

There is yet another option when using the Fill Handle! Since the radius values were already entered into a column adjacent to the destination cells, you can select the original two formulas (C5:D5) and then double-click on the Fill Handle. Excel will automatically generate formulas for each radius value. The ability to double-click the Fill Handle to have Excel automatically complete a table can be very handy when you have many rows to fill with formulas.

LOAN AMORTIZATION TABLE

When you borrow money to buy a vehicle or a home, the bank may provide you with a *loan amortization table* showing how much you still owe each month. In this application, we'll show you how to create your own amortization table.

Consider a $25,000 loan to buy a new pickup. The loan has an annual percentage rate (APR) of 6%, and you will be making payments for 5 years. Create an amortization table showing how much is left to be paid after each payment.

The basic theory here is that each time you make a payment, you first have to pay the interest on the outstanding borrowed amount (called the *principal*); whatever doesn't go toward interest reduces the principal before the next payment.

The amount of interest depends on the length of the period between payments (typically 1 month) and the periodic (monthly) interest rate. In this example, the periodic interest rate is 6%/12 = 0.5% per month. The loan requires 5 × 12 = 60 payments of $483.32. The required payment amount can be determined using Excel's **PMT** function, which is presented in Section 8.3.2.

We begin by giving the worksheet a title and entering the basic loan data (Figure 2.18).

In step 2, we create headings for the amortization table and indicate that the principal before payment #1 is the full amount borrowed. Note that the formula in cell B14 is simply a link to the amount borrowed that was indicated in cell D3. That is, cell B14 contains the formula =D3 as shown in Figure 2.19.

D5		f_x	=D4/12		
A	**B**	**C**	**D**	**E**	**F**

	A	B	C	D	E	F
1	Loan Amortization Table					
2						
3		Amount Borrowed:		25000	dollars	
4		APR:		6.0%		
5		Periodic Interest Rate:		0.5%		
6		Term:		5	years	
7		Payments/Year:		12		
8		Payments:		60		
9		Payment Amount:		483.32	dollars	
10						

Figure 2.18
Loan amortization table, step 1.

B14		f_x	=D3		
A	**B**	**C**	**D**	**E**	**F**

	A	B	C	D	E	F
1	Loan Amortization Table					
2						
3		Amount Borrowed:		25000	dollars	
4		APR:		6.0%		
5		Periodic Interest Rate:		0.5%		
6		Term:		5	years	
7		Payments/Year:		12		
8		Payments:		60		
9		Payment Amount:		483.32	dollars	
10						
11		Principal			Principal	
12		Before	Interest	Paid on	After	
13	Payment	Payment	Payment	Principal	Payment	
14	1	25000				
15						

Figure 2.19
Step 2, creating table headings and identifying the initial principal.

For step 3, we calculate the interest on that principal by multiplying the "principal before payment" by the "periodic interest rate", or

$$C14: \quad =B14*\$D\$5$$

Note: The address of the periodic interest rate was made absolute in the formula above by using dollar signs as $D$5. By using absolute addresses whenever any of the input values in rows 3 through 9 is used in the table, the cell addresses for those input values will not change when the formula is copied down the table.

The calculation of interest payments based on periodic interest rates and remaining principal is termed *simple interest*. There are other ways to calculate interest payments, but simple interest is typically used for automobile and home loans.

	C14		▼	*fx*	=B14*$D$5	
◢	A	B	C	D	E	F
1	Loan Amortization Table					
2						
3		Amount Borrowed:		25000	dollars	
4		APR:		6.0%		
5		Periodic Interest Rate:		0.5%		
6		Term:		5	years	
7		Payments/Year:		12		
8		Payments:		60		
9		Payment Amount:		483.32	dollars	
10						
11		Principal			Principal	
12		Before	Interest	Paid on	After	
13	Payment	Payment	Payment	Principal	Payment	
14	1	25000	125			
15						

Figure 2.20
Step 3, calculating the interest in the first month.

Next (step 4), we subtract the interest payment from the total payment amount to determine how much was paid on principal in the first month.

D14: =$D$9-C14   (again, an absolute address was used for the payment amount)

	D14		▼	*fx*	=$D$9-C14	
◢	A	B	C	D	E	F
1	Loan Amortization Table					
2						
3		Amount Borrowed:		25000	dollars	
4		APR:		6.0%		
5		Periodic Interest Rate:		0.5%		
6		Term:		5	years	
7		Payments/Year:		12		
8		Payments:		60		
9		Payment Amount:		483.32	dollars	
10						
11		Principal			Principal	
12		Before	Interest	Paid on	After	
13	Payment	Payment	Payment	Principal	Payment	
14	1	25000	125	358.32		
15						

Figure 2.21
Step 4, determining how much was paid on principal with the first payment.

E14		▾	f_x	=B14-D14		
	A	B	C	D	E	F
1	Loan Amortization Table					
2						
3		Amount Borrowed:		25000	dollars	
4		APR:		6.0%		
5		Periodic Interest Rate:		0.5%		
6		Term:		5	years	
7		Payments/Year:		12		
8		Payments:		60		
9		Payment Amount:		483.32	dollars	
10						
11		Principal			Principal	
12		Before	Interest	Paid on	After	
13	Payment	Payment	Payment	Principal	Payment	
14	1	25000	125	358.32	24641.68	
15						

Figure 2.22
Step 5, determining the principal after the first payment.

In step 5, the principal after the payment is determined as $25,000 − $358.32 = $24,641.68, or

E14: =B14-D14

For step 6, we start the calculations for the second payment by increasing the payment number by one and using the payment #1 "principal after payment" as the "before payment" principal for payment #2.

A15: =A14+1
B15: =E14

Step 7. The last three calculations for payment #2 can be completed simply by copying the formulas in cells C14:E14 down to row 15. The results are shown in Figure 2.24.

Notice that the interest payment has decreased slightly for payment #2 because it was calculated using a slightly smaller principal.

Step 8. Copy the formula in cell A15 down another 58 rows to handle all 60 payments. The result is shown in Figure 2.25 with many rows (19 through 71) hidden. (How to hide rows is the topic of Section 2.7.6.)

Step 9. To complete the table, select the formulas in cells B15:E15 (as shown in Figure 2.25) and double-click on the Fill Handle. Excel will copy the formulas in row 15 (payment #2) down to all 58 remaining rows. The first five payments and last two payments are shown in Figure 2.26.

This example was included at this point in this chapter to illustrate how handy the Fill Handle is for completing tables (step 9). The appearance and readability of the amortization table could certainly be improved with some formatting, such as bolding the title and column headings, including dollar signs on currency values, and always presenting currency values to two decimal places (cents). These formatting topics are covered in the rest of this chapter.

A15	▼		f_x	=A14+1		
	A	**B**	**C**	**D**	**E**	**F**
1	Loan Amortization Table					
2						
3			Amount Borrowed:	25000	dollars	
4			APR:	6.0%		
5			Periodic Interest Rate:	0.5%		
6			Term:	5	years	
7			Payments/Year:	12		
8			Payments:	60		
9			Payment Amount:	483.32	dollars	
10						
11		Principal			Principal	
12		Before	Interest	Paid on	After	
13	Payment	Payment	Payment	Principal	Payment	
14	1	25000	125	358.32	24641.68	
15	2	24641.68				
16						

Figure 2.23
Step 6, starting the calculations for the second payment.

C15	▼		f_x	=B15*$D$5		
	A	**B**	**C**	**D**	**E**	**F**
1	Loan Amortization Table					
2						
3			Amount Borrowed:	25000	dollars	
4			APR:	6.0%		
5			Periodic Interest Rate:	0.5%		
6			Term:	5	years	
7			Payments/Year:	12		
8			Payments:	60		
9			Payment Amount:	483.32	dollars	
10						
11		Principal			Principal	
12		Before	Interest	Paid on	After	
13	Payment	Payment	Payment	Principal	Payment	
14	1	25000	125	358.32	24641.68	
15	2	24641.68	123.21	360.11	24281.57	
16						

Figure 2.24
Step 7, copy the formulas in cells C14:E14 down to row 15 to complete the calculations for payment #2.

	A	B	C	D	E	F
1	Loan Amortization Table					
2						
3			Amount Borrowed:	25000	dollars	
4			APR:	6.0%		
5			Periodic Interest Rate:	0.5%		
6			Term:	5	years	
7			Payments/Year:	12		
8			Payments:	60		
9			Payment Amount:	483.32	dollars	
10						
11		Principal			Principal	
12		Before	Interest	Paid on	After	
13	Payment	Payment	Payment	Principal	Payment	
14	1	25000	125	358.32	24641.68	
15	2	24641.68	123.21	360.11	24281.57	
16	3					
17	4					
18	5			Hidden Rows		
72	59					
73	60					
74						

Figure 2.25
Step 8, establishing the number of payments.

	A	B	C	D	E	F
1	Loan Amortization Table					
2						
3			Amount Borrowed:	25000	dollars	
4			APR:	6.0%		
5			Periodic Interest Rate:	0.5%		
6			Term:	5	years	
7			Payments/Year:	12		
8			Payments:	60		
9			Payment Amount:	483.32	dollars	
10						
11		Principal			Principal	
12		Before	Interest	Paid on	After	
13	Payment	Payment	Payment	Principal	Payment	
14	1	25000	125	358.32	24641.68	
15	2	24641.68	123.21	360.11	24281.57	
16	3	24281.57	121.41	361.91	23919.66	
17	4	23919.66	119.60	363.72	23555.93	
18	5	23555.93	117.78	365.54	23190.39	
72	59	959.44	4.80	478.52	480.92	
73	60	480.92	2.40	480.92	0.00	
74						

Figure 2.26
The completed amortization table (only 10 of 60 payments shown).

2.2.5 Using the Format Painter

The Microsoft Office programs include a handy feature called the *Format Painter*. In Excel 2010, it is part of the Clipboard Group on the Ribbon's **Home** tab, as shown in Figure 2.27.

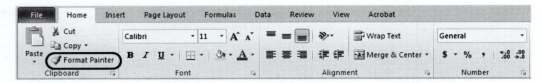

Figure 2.27
The Format Painter button.

The **Format Painter** is used to copy the format used in one cell to another cell or cell range, and nothing else (no values or formulas). This allows you to apply any number of formatting attributes to one cell and then apply all of those attributes to other cells at one time. For example, if you have a portion of your worksheet looking just the way you want it to look and have specified the following attributes:

• Numeric format
• Font type
• Font size
• Font color
• Fill color
• Border style
• Border thickness
• Border color

You do not have to set all of those attributes again in another portion of the worksheet; just use the Format Painter to copy the format from the previous portion of the worksheet and apply it to the new portion.

The **Format Painter** can be used in two ways:

• Single format application
• Multiple format application

Single Format Application

If you need to format one cell or one cell range to look just like an existing cell, then:

1. Click the cell with the desired formatting, to select it.
2. Click the **Format Painter** button.
3. Click the cell (or select the cell range) that is to be formatted.

The formatting of the first cell will be applied to the new cell or cell range.

Multiple Format Application

If you need to format multiple (noncontiguous) cells or multiple cell ranges to look just like an existing cell, then:

1. Click the cell with the desired formatting, to select it.
2. Double-click the **Format Painter** button to activate continuous format painting.
3. Click each cell (or select each cell range) that is to be formatted.
4. Click the **Format Painter** button to deactivate continuous format painting.

The formatting of the first cell will be applied to the each new cell or cell range.

2.3 USING THE FONT GROUP

The *Font group* on the Ribbon's Home tab (Figure 2.28) provides access to:

- Font typeface and point size selectors
- Font increase and decrease size buttons
- Bold, underline, and italics attribute toggle buttons
- Border application button and drop-down selection list
- Cell background color application and selection button
- Font color application and selection button
- Font group **Expand** button to open the Format Cells dialog

Figure 2.28
The font group on the Ribbon's **Home** tab.

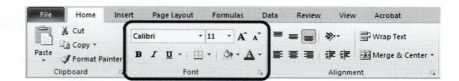

Many of the formatting options available through the Font group are very familiar to word processor users and will only be presented briefly here.

2.3.1 Font Typeface and Point Size

The default font in the last two versions of Microsoft Office is called Calibri (see Figure 2.28). The Calibri font is a simple font that is easy to read. In older versions of Excel, the Arial font was the default typeface. You can change the font type or size, either for a selected cell or range of cells, or for the entire workbook.

Changing the Default Typeface or Size Used for New Workbooks
Use the Excel Options dialog to change the default typeface and size. Access the Excel Options dialog as,

- Excel 2010: **File tab/Options**
- Excel 2007: **Office/Excel Options**
- Excel 2003: File/Options

Once the Excel Options dialog is open, select the **Popular** panel, and then change the font listed as **Use this font:** (to create new workbooks). This is illustrated in Figure 2.29.

The default typeface may appear as "Body Font" on the Excel Options dialog. Body font is simply the typeface used for the body of a document—the basic (default) font on your computer system.

Changing the Default Typeface or Size Used for a Currently Open Workbook
By default, text in a workbook is displayed using the *Normal style*. A *style* is a collection of attributes such as typeface and color, font size, attributes like bold and italics, cell background color, and many others. Any change you make to the Normal Style will automatically show up in the appearance of every cell that uses the Normal style, which is all cells when you first open the workbook. Changing the Normal style is covered in the next section of this chapter.

Figure 2.29
Changing the default
typeface and/or font size
for new workbooks.

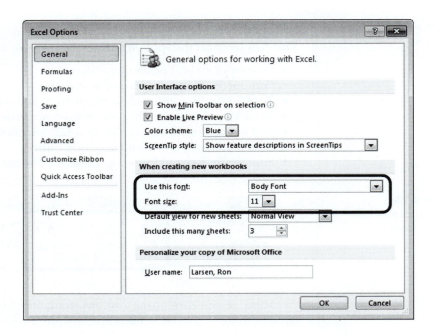

Changing the Typeface or Font Size for a Selected Cell or Range of Cells

The **Font** (style) and **Font Size** drop-down selectors (Figure 2.30) are used to change the typeface or font size of a selected cell or group of cells.

If we use the worksheet developed in Example 2.1 again, the readability could be improved by increasing the size of font used for the title. To do so:

1. Select the cell containing the title (cell A1 in Figure 2.31).
2. Choose a larger font size from the Font Size drop-down selection list.

In Figure 2.31, the title font size has been changed to 20 points; notice that the row height automatically adjusts when the font size is increased.

Figure 2.30
The Font and Font Size
drop-down selection lists,
with Font Size list
displayed.

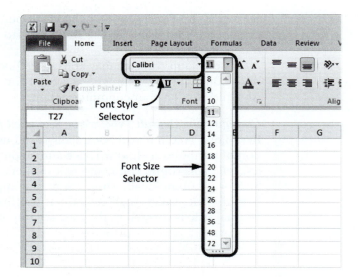

Figure 2.31
Increasing the size of the worksheet title.

	A	B	C	D	E	F
1	Area and Volume of Spheres					
2						
3		Radius	Area	Volume		
4		(cm)	(cm^2)	(cm^3)		
5		1	12.57	4.19		
6		2	50.27	33.51		
7		5	314.16	523.60		
8		12	1809.56	7238.23		
9						

2.3.2 Font Increase and Decrease Size Buttons

There are many instances when you might not care what point size is used for your fonts, you just want them to "look right" on the screen or printout. For these situations, the **Font Increase** and **Font Decrease** buttons (indicated in Figure 2.32) are very handy, simply keep clicking the buttons until the font is the size you want.

To increase the text size for the column headings in our example:

1. Select the column headings in cells B3:D3.
2. Click the **Font Increase** button a few times.

The result is shown in Figure 2.32.

Figure 2.32
The **Font Increase** and **Font Decrease** buttons.

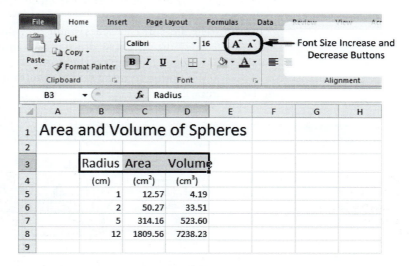

2.3.3 Bold, Underline, and Italics Attribute Toggle Buttons

The **Bold**, **Underline**, and **Italics** toggle buttons, indicated in Figure 2.33, allow you to apply or remove these font attributes to the text displayed in a selected cell or range of cells.

Figure 2.33
The Bold, Underline, and Italics toggle buttons.

Figure 2.34
After adding the bold attribute to the title in cell A1.

◢	A	B	C	D	E	F
1	**Area and Volume of Spheres**					
2						
3		Radius	Area	Volume		
4		(cm)	(cm^2)	(cm^3)		
5		1	12.57	4.19		
6		2	50.27	33.51		
7		5	314.16	523.60		
8		12	1809.56	7238.23		
9						

To activate the bold font attribute for the title in our example worksheet:

1. Select cell A1.
2. Click the **Bold** toggle button.

The result is shown in Figure 2.34.

The **Bold**, **Underline**, and **Italics** buttons are called *toggle* buttons because they toggle or switch back and forth between two states. Clicking the **Bold** button once, for example, activates the bold attribute for the text in the selected cell, and clicking it a second time deactivates the bold attribute. The **Underline**, and **Italics** toggle buttons work in the same manner to activate and deactivate their respective font attributes.

In Figure 2.35, you can see a small down-pointing arrow to the right of the **Underline** toggle button. Those down-arrows indicate that there is an option menu available. In this case, clicking on the down-arrow to the right of the **Underline** button gives you the option of using a single underline (the default) or a double underline (see Figure 2.35).

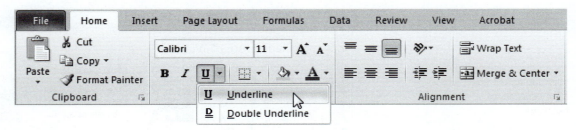

Figure 2.35
Underline options.

2.3.4 Cell Border

The **Border** button also has a companion menu, as shown in Figure 2.36.

When you click the **Border** button, the currently selected border style and location is applied to the selected cell or cell range. If the currently selected border style and location is not what you need, use the drop-down menu to choose the desired type of border.

Figure 2.36
The border selection list.

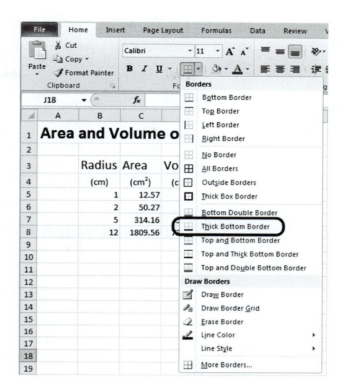

To add a thick border below the units in cells B4:D4 (to separate the column headings from the numeric values):

1. Select cells B4:D4 to select all of the cells displaying units.
2. Use the drop-down border selection list and choose **Thick Bottom Border** from the list of options.

The result is shown in Figure 2.37. Notice that the **Border** button now shows the **Thick Bottom Border** icon; the most recently used border style is always used when the **Border** button is used without the drop-down menu.

Figure 2.37
The example worksheet after adding a border below the column headings.

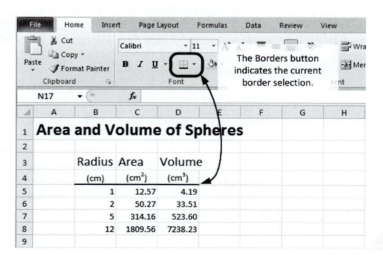

2.3.5 Cell Fill (Background) Color

The background color in a cell is called the *fill* color. The button and color selector used to set the fill color for a cell or range of cells is shown in Figure 2.38.

Figure 2.38
The button and color selector used to set the fill color for a cell or range of cells.

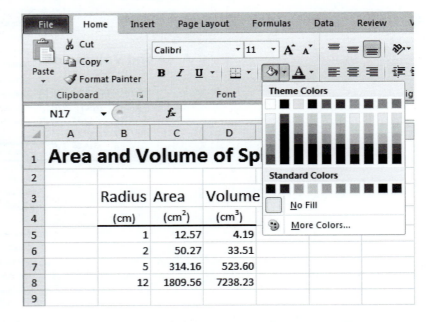

There are three ways to choose colors on the drop-down menu.

1. Near the top are the theme colors; every workbook has a theme that includes a color scheme. If you change the theme applied to the workbook, the choice of theme colors on the menu will change as well. Use Ribbon options **Page Layout/Themes (group)/Themes (button)** to change the applied theme.
2. Below the theme colors are some standard, common computer colors.
3. The **More Colors...** button opens the Colors dialog to allow you to select from a wide range of colors.

An important button on the colors drop-down menu is the **No Fill** button which is used to remove the background color from cells if you decide you don't want a colored background. (An alternative is to reapply the **Normal** style to a cell to remove the cell's background color. Use Ribbon options **Home/Styles/Cell Styles/Normal** to apply the **Normal** style.)

2.3.6 Font (Text) Color

The drop-down menu for choosing the font color (Figure 2.39) is very similar to the menu for choosing the cell fill color and provides the same three ways to choose a color: Theme Colors, Standard Colors, and opening the Colors dialog using the **More Colors...** button. A difference is that the font color is typically set to **Automatic** which generally means black. You can override automatic color selection by choosing a color. If you decide to remove color, you normally set the font color back to **Automatic**.

2.3.7 Format Cells Dialog

The small arrow button, called the **Expand** button at the bottom-right corner of the Font group (see Figure 2.40), can be used to open the Format Cells Dialog to the Font panel, as shown in Figure 2.41.

Figure 2.39
The button and color selector used to set the font (text) color for a cell or range of cells.

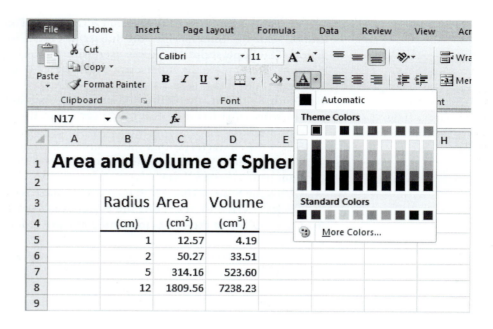

Figure 2.40
The Font group Expand button.

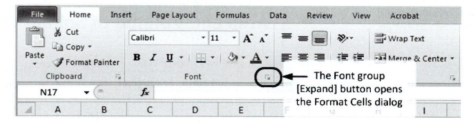

Figure 2.41
The Format Cells Dialog, **Font** panel.

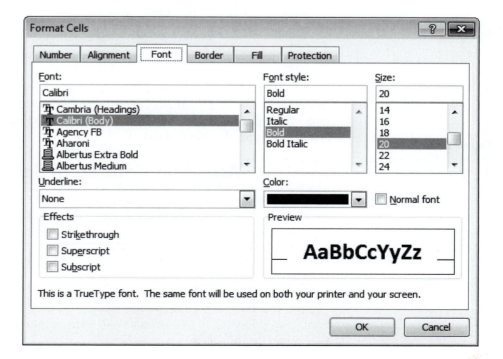

Most of the features of the Format Cells Dialog's **Font** panel are available on the Ribbon, but the **Superscript** and **Subscript** effects are only available from this dialog. The **Superscript** effect was used with the units in our example (shown in Figure 2.39) to create the superscript 2 and 3. Here's how:

1. Double-click on the cell containing the text that will have the superscript (or subscript) to enter edit mode (or select the cell and press [F2] to enter edit mode).
2. Select the character(s) that will be superscripts, as illustrated in Figure 2.42.
3. Click the Font group **Expand** button to open the Format Cells dialog to the **Font** panel (Figure 2.41).
4. Check the **Superscript** box to tell Excel to superscript the selected character(s).
5. Click the **OK** button to close the Format Cells dialog.
6. Click outside of the cell being edited to leave edit mode.

The character(s) with the superscript effect are raised and reduced in size, as shown in Figure 2.43.

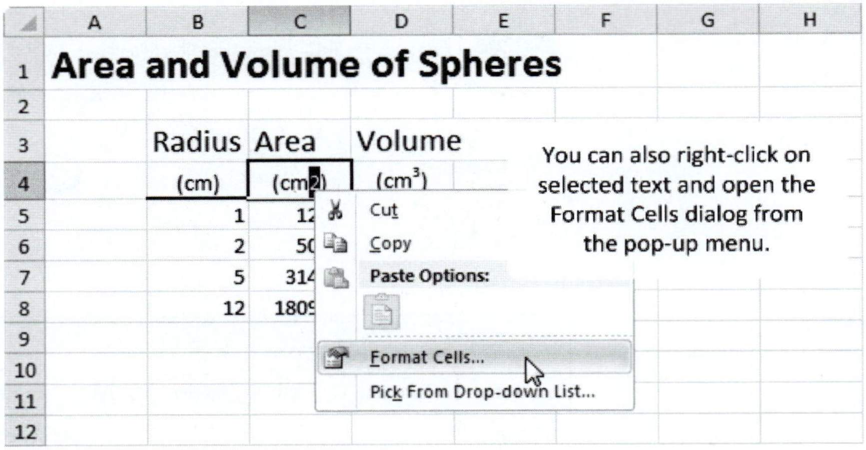

Figure 2.42
Select the character(s) that are to be superscripts.

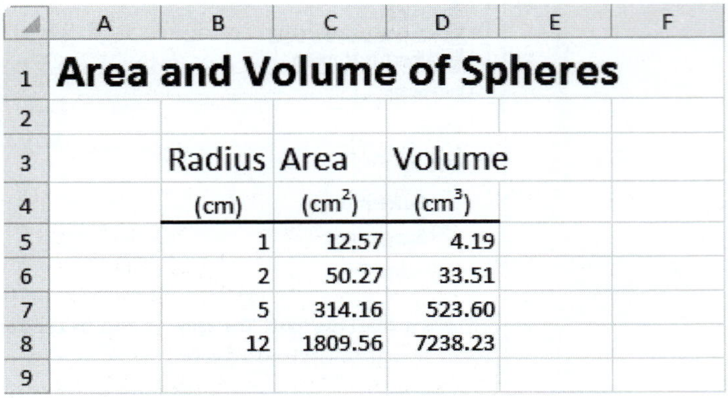

Figure 2.43
The result of superscripting the "2" in cm^2 (cell C4).

2.4 USING THE ALIGNMENT GROUP

The Alignment group on the Ribbon's **Home** tab (Figure 2.44) provides access to:

- Horizontal alignment buttons
- Vertical alignment buttons
- **Wrap Text** toggle button
- **Merge & Center** toggle button

Figure 2.44

The Alignment group on the Ribbon's **Home** tab.

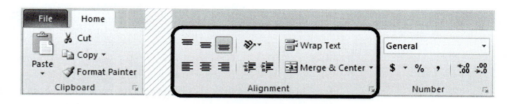

2.4.1 Horizontal Alignment Buttons

The *horizontal alignment* buttons in the Alignment group are indicated in Figure 2.45. These buttons allow you to left-, center-, or right-justify the displayed contents of a cell or range of cells. By default, Excel left-justifies text (labels) and right-justifies displayed values, but you can use the horizontal alignment buttons to override the default when desired.

Figure 2.45

The horizontal alignment buttons in the Alignment group.

when center, always wrap text

Note: Alignment is part of the Normal style; modify the Normal style if you want to change the default alignment.

As an example of using these horizontal alignment buttons, we will remove the centering of the column headings in the worksheet used in Example 2.1. The process is:

1. Select the cells containing the headings to be centered (cells B3:D4).
2. Click on the **Left Align** button (the left horizontal alignment button).

The result is shown in Figure 2.46. The difference in the heading labels between Figures 2.43 and 2.46 is not pronounced since the labels in row 3 nearly fill the cells anyway, but it is apparent that the units in row 4 are no longer centered beneath the labels.

2.4.2 Wrapping Text in Cells

When a label is too long to fit within a cell, Excel goes ahead and shows the entire label, as long as it doesn't interfere with the contents of another cell. An example of this is the worksheet title in Figure 2.46. The label, "Area and Volume of Spheres," is much longer than the space assigned to cell A1, but there is nothing else on row 1, so Excel shows the entire title.

When long labels do interfere, only the portion that fits in the cell is displayed. As an example of this, consider a modified version of the loan amortization table

Figure 2.46
The worksheet after left
aligning the column
headings in cells B3:D4.

developed earlier in this chapter. The modified version, shown in Figure 2.47, has
the column headings in individual cells—and they don't fit. This makes the column
headings hard to read. We can fix this by wrapping the text in the cells containing
these column headings (cells A11:E11).

The difference between this version and the table presented in Figure 2.26 is
the column headings. The column headings in cells B11 through D11 are:

A11: Payment

B11: Principal Before Payment

C11: Interest Payment

D11: Paid on Principal

E11: Principal After Payment

In the earlier example (Figure 2.26), we got around this problem by using
three cells for each heading. This time we will use *text wrapping*, which means the
text to be displayed on multiple text rows within the cells. To allow text wrapping
in the column headings,

1. Select the column headings that will be set to allow text wrapping (cells
   A11:E11).
2. Click the **Wrap Text** button in the Alignment group.

This process is indicated in Figure 2.48, with the result illustrated in Figure 2.49.

Figure 2.47
The loan amortization
table, with long column
headings in row 11.

Figure 2.48
The process used to allow text wrapping in selected cells.

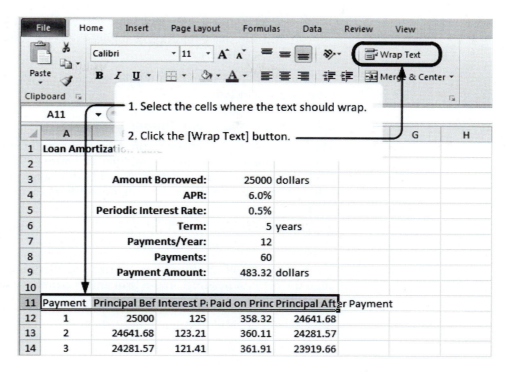

Text wrapping is used to allow long labels to be displayed in a cell by allowing multiple text lines in the cell. In Figure 2.49, you can see two features of Excel's method of text wrapping:

1. The height of row 11 was automatically increased to accommodate the headings. When text wrapping is requested, the row height will be increased to show all of the wrapped text.
2. By default, the vertical alignment of wrapped text is at the bottom of the cell. You can see this with the short headings, like the "Payment" in cell A11. When

Figure 2.49
The column headings after allowing text wrapping in cells A11:E11.

	A	B	C	D	E	F
1	Loan Amortization Table					
2						
3		Amount Borrowed:		25000	dollars	
4		APR:		6.0%		
5		Periodic Interest Rate:		0.5%		
6		Term:		5	years	
7		Payments/Year:		12		
8		Payments:		60		
9		Payment Amount:		483.32	dollars	
10						
11	Payment	Principal Before Payment	Interest Payment	Paid on Principal	Principal After Payment	
12	1	25000	125	358.32	24641.68	
13	2	24641.68	123.21	360.11	24281.57	
14	3	24281.57	121.41	361.91	23919.66	

three text lines are not needed, the text is placed at the bottom of the cell. You can override this default by changing the cell's vertical alignment. This is the topic of Section 2.4.3.

The **Wrap Text** switch is a toggle switch; it can also be used to deactivate text wrapping in selected cells.

2.4.3 Vertical Alignment Buttons

The *vertical alignment* buttons in the Alignment group are indicated in Figure 2.50. These buttons allow you to top-, middle-, or bottom-align the displayed contents of a cell or range of cells. By default, Excel bottom-aligns the contents displayed in a cell, but you can use the vertical alignment buttons to override the default when desired.

Note: Alignment is part of the Normal style; modify the Normal style if you want to change the default alignment.

As an example of using these horizontal alignment buttons, we will middle-align the column headings in the loan amortization worksheet. The process is:

1. Select the cells containing the headings to be aligned (cells A11:E11).
2. Click on the **Middle Align** button (the middle vertical alignment button).

The result is shown in Figure 2.51.

Figure 2.50
The vertical alignment buttons in the Alignment group.

Figure 2.51
After middle-aligning the headings in cells A11:E11.

	A	B	C	D	E	F
1	Loan Amortization Table					
2						
3		Amount Borrowed:		25000	dollars	
4		APR:		6.0%		
5		Periodic Interest Rate:		0.5%		
6		Term:		5	years	
7		Payments/Year:		12		
8		Payments:		60		
9		Payment Amount:		483.32	dollars	
10						
11	Payment	Principal Before Payment	Interest Payment	Paid on Principal	Principal After Payment	
12	1	25000	125	358.32	24641.68	
13	2	24641.68	123.21	360.11	24281.57	
14	3	24281.57	121.41	361.91	23919.66	

2.4.4 Centering Labels in Merged Cells

Excel allows multiple cells to be merged. Once the cells are merged, the *merged cells* are treated as a single cell for calculations. When you merge cells, no more than one cell should have any content. If you try to merge two or more cells that each have content, Excel will display a warning that some content will be lost during the merge.

One common use of a merged cell is to create a common heading for several columns in a table. For example, a data set might contain temperature readings from five digital thermometers (T1 through T5), each of which is sampled each second. The data set might look like Figure 2.52.

Since the data in columns C through H are all temperatures, a single "Temperatures (K)" heading over all six columns makes sense. The process for adding the common heading is:

1. Enter the text "Temperatures (K)" in (any) one of the cells C3 through H3.
2. Select cell range C3:H3.
3. Click the **Merge & Center** button in the Alignment group.

The process is illustrated in Figure 2.53, with the result shown in Figure 2.54.

In Figure 2.54, a border was added around the new heading to better illustrate the merged cells.

The **Merge & Center** button acts as a toggle button; if you select a previously merged cell and then click the **Merge & Center** button, the cells will be unmerged.

The **Merge & Center** button also has an associated drop-down menu. When the drop-down menu is opened, the following menu options are available:

- *Merge & Center:* Same as the **Merge & Center** button; selected cells are merged and the cell contents are displayed in the center (left to right) and using the currently selected vertical alignment (typically bottom-aligned).

◢	A	B	C	D	E	F	G	H	I
1	Temperature Data								
2									
3									
4	Sample #	Time (sec.)	T1	T2	T3	T4	T5	T avg	
5	1	0	243.2	246.8	241.2	242.2	244.1		
6	2	1	243.3	246.9	242.0	242.6	244.5		
7	3	2	243.4	247.0	242.8	243.0	244.9		
8	4	3	243.5	247.2	243.6	243.4	245.3		
9	5	4	243.6	247.3	244.4	243.8	245.7		
10	6	5	243.7	247.4	245.2	244.2	246.1		
11	7	6	243.8	247.5	246.0	244.6	246.5		
12	8	7	243.9	247.6	246.9	245.0	246.9		
13	9	8	244.0	247.8	247.7	245.4	247.3		
14	10	9	244.1	247.9	248.5	245.8	247.7		
15	11	10	244.2	248.0	249.3	246.2	248.1		
16									

Figure 2.52
Temperature data, before merging and cells.

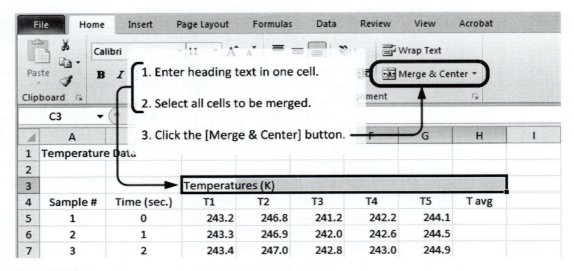

Figure 2.53
The process used to create a centered heading.

	A	B	C	D	E	F	G	H	I
1	Temperature Data								
2									
3			Temperatures (K)						
4	Sample #	Time (sec.)	T1	T2	T3	T4	T5	T avg	
5	1	0	243.2	246.8	241.2	242.2	244.1		
6	2	1	243.3	246.9	242.0	242.6	244.5		
7	3	2	243.4	247.0	242.8	243.0	244.9		
8	4	3	243.5	247.2	243.6	243.4	245.3		
9	5	4	243.6	247.3	244.4	243.8	245.7		
10	6	5	243.7	247.4	245.2	244.2	246.1		
11	7	6	243.8	247.5	246.0	244.6	246.5		
12	8	7	243.9	247.6	246.9	245.0	246.9		
13	9	8	244.0	247.8	247.7	245.4	247.3		
14	10	9	244.1	247.9	248.5	245.8	247.7		
15	11	10	244.2	248.0	249.3	246.2	248.1		
16									

Figure 2.54
The temperature data, with a centered heading over the temperature columns.

- *Merge Across:* Merges across selected columns, but does not merge rows. Cell content display is aligned according to the horizontal and vertical alignments currently selected for the merged cell.
- *Merge Cells:* Merges selected cells. Cell content display is aligned according to the horizontal and vertical alignments currently selected for the merged cell.
- *Unmerge Cells:* Separates previously merged cells. The cell content will be placed in the top-left cell of the unmerged cell range.

2.5 FORMATTING NUMBERS

The *Number group* on the Ribbon's **Home** tab (Figure 2.55) provides access numeric formats, including:

- Increase and Decrease Decimal buttons
- Currency and Accounting button and drop-down menu
- Percentage button
- Thousand Separator (Comma) button
- Named formats
  - General
  - Number
  - Currency
  - Accounting
  - Short Date
  - Long Date
  - Time
  - Percentage
  - Fraction
  - Scientific
  - Text
- Number group **Expand** button

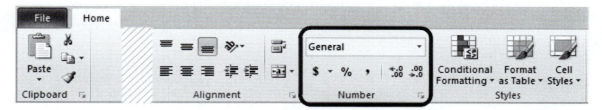

Figure 2.55
The Number group on the Ribbon's **Home** tab.

By default, the contents of the cells in a worksheet are displayed using the *General format*. The General format is a very flexible format that tries to display numbers in a readable form. Numbers around zero are presented in their entirety, but scientific notation is used when values are very large or very small.

Note: The General format is the default because it is specified as part of the Normal style; modify the Normal style if you want to change the default format used in cells.

Excel provides a number of predefined, named formats, but many of the best features of the named formats are now available directly on the Ribbon's **Home** tab and Number group. In older versions of Excel, these named formats are accessed using menu options **Format/Cells**. The number formatting options available directly from the Ribbon will be presented first.

2.5.1 Changing the Number of Displayed Decimal Places

One of the most common formatting needs for engineers is the ability to control the number of decimal places displayed on a calculated result. By default, Excel does not show trailing zeros, but if a calculated value is inexact, Excel will show as many decimal places as will fit into the cell. Leaving these extra digits makes it harder to read your worksheet and can make people think your results are far more accurate than they really are.

EXAMPLE 2.2

US CROSS-COUNTRY DRIVING DISTANCES

As an example of how a lot of decimal places can be displayed on inaccurate numbers, consider the driving distances between some US cities shown in Figure 2.56.

The values listed in miles are approximations from values listed at various Internet sites, and they are not very accurate. For example, the reported distance between New York and Los Angeles ranges from 2400 to 3000 miles and probably depends a lot on the route you take (and if there is a detour to Orlando, en route).

The values listed in kilometers were calculated from the values in miles by using the conversion factor 0.6214 miles per kilometer. One of the calculations is shown in the Formula bar in Figure 2.56. Excel displayed the calculated results with three decimal places, and someone might see those values and think those are highly precise values; but they were calculated using highly imprecise and inaccurate mileage values. We need to get rid of those extra decimal places to eliminate some of the confusion. Here's the process:

1. Select the cells containing values to be reformatted with fewer decimal places (F3:F5).
2. Click the **Decrease Decimal** button three times.

The process is illustrated in Figure 2.57, and the result is shown in Figure 2.58.

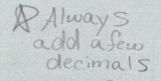

Figure 2.56
Distances between
US cities.

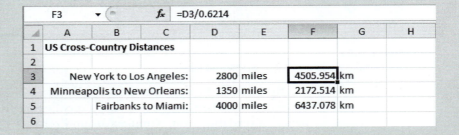

Figure 2.57
The process used to change
the number of displayed
decimal points.

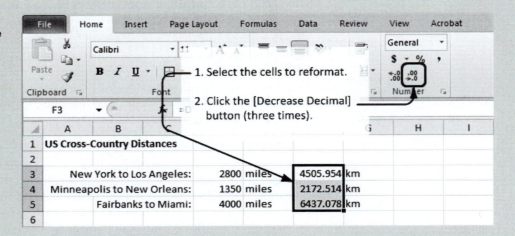

	A	B	C	D	E	F	G	H
1	US Cross-Country Distances							
2								
3	New York to Los Angeles:			2800	miles	4506	km	
4	Minneapolis to New Orleans:			1350	miles	2173	km	
5	Fairbanks to Miami:			4000	miles	6437	km	
6								

Figure 2.58
The result of changing the number of displayed decimal points in cells F3:F5.

★ Remember
( ) ★

While the kilometer values still suggest they are accurate to one kilometer, that's the best we can do using the **Decrease Decimal** button in the Number group on the Ribbon, and the new values are at least an improvement over the extreme number of decimal places initially presented.

There is a way to reduce the number of displayed digits even further, but it requires the use of an Excel function. We will wait until Chapter 3 to present most functions, but we'll present the **ROUND** function here to fix this table. If you have never used a function before, you may want to skip the next two paragraphs until after you've read Chapter 3.

Excel's **ROUND** function takes two arguments: the number to be rounded and the number of decimal places desired. It then returns the rounded number for display in the cell. For example, =ROUND(4505.954, 2) would return the distance from New York to Los Angeles rounded to 2 decimal places, or 4505.95 km. That's too many decimal places for this example; we want to round into the digits on the left side of the decimal point. To do that, we request a negative number of decimal points in the function call. The formula =ROUND(4505.954, -2) will return 4500 km. But we don't want to type the distances into the **ROUND** function, so instead we build the **ROUND** function into the calculation of the distances in kilometers, as illustrated in Figure 2.59.

The results shown in column F now more accurately reflect the low level of precision in these distances.

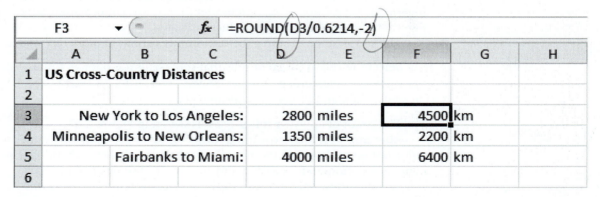

Figure 2.59
The result after rounding two digits left of the decimal point.

2.5.2 Adding Currency Symbols

Working with monetary values is a common practice in engineering, and Excel makes it easy to include currency symbols in calculations. The Number group on the **Home** tab of the Ribbon provides an **Accounting Number Format** button and a drop-down menu for additional currency and accounting formatting options.

The **Accounting Number Format** button is indicated in Figure 2.60.

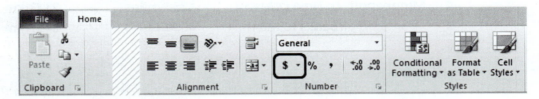

Figure 2.60
The **Accounting Number Format** button.

Excel provides two named formats for monetary values: *Accounting format* and *Currency format*. The difference is in the way the currency symbols are displayed. This is illustrated in Figure 2.61.

Figure 2.61
Contrasting the currency symbol location with the Accounting and Currency formats.

	A	B	C	D	E	F
1						
2	Accounting Format			Currency Format		
3		$ 12.32			$12.32	
4		$ 15.48			$15.48	
5		$ 125.21			$125.21	
6		$1,157.32			$1,157.32	
7						

The Accounting format aligns all of the currency symbols, which makes them easier to see and the values easier to read. Because of this, the Accounting format is more commonly used.

As an example of applying the Accounting format, let's return to the loan amortization table which could benefit from some improved formatting, especially for the monetary values. As a reminder, the mostly unformatted table is shown in Figure 2.62, with the monetary values indicated.

The simplest way to apply the Accounting format to monetary values is to:

1. Select the cell or range of cells that represents a monetary value.
2. Click the **Accounting Number Format** button to (at least in the US) add the dollar sign and display two decimal places (cents).

The result of applying this process is shown in Figure 2.63.

After applying the Accounting format to all of the monetary values in the loan amortization table, all of the monetary values are shown with a dollar sign, with a comma as a thousand separator, and shown with two decimal places.

Showing these values to two decimal places is appropriate in this case because loan values would be tracked to the penny. However, cents are not used in many high-value engineering situations. When you use the **Accounting Number Format**

Figure 2.62
Loan amortization table (truncated at 5 of 60 payments) with dollar values indicated.

	A	B	C	D	E	F
1	Loan Amortization Table					
2						
3		Amount Borrowed:		25000	dollars	
4		APR:		6.0%		
5		Periodic Interest Rate:		0.5%		
6		Term:		5	years	
7		Payments/Year:		12		
8		Payments:		60		
9		Payment Amount:		483.32	dollars	
10						
11	Payment	Principal Before Payment	Interest Payment	Paid on Principal	Principal After Payment	
12	1	25000	125	358.32	24641.68	
13	2	24641.68	123.21	360.11	24281.57	
14	3	24281.57	121.41	361.91	23919.66	
15	4	23919.66	119.60	363.72	23555.93	
16	5	23555.93	117.78	365.54	23190.39	

	A	B	C	D	E	F
1	Loan Amortization Table					
2						
3		Amount Borrowed:		$ 25,000.00		
4		APR:		6.0%		
5		Periodic Interest Rate:		0.5%		
6		Term:		5	years	
7		Payments/Year:		12		
8		Payments:		60		
9		Payment Amount:		$ 483.32		
10						
11	Payment	Principal Before Payment	Interest Payment	Paid on Principal	Principal After Payment	
12	1	$ 25,000.00	$ 125.00	$ 358.32	$ 24,641.68	
13	2	$ 24,641.68	$ 123.21	$ 360.11	$ 24,281.57	
14	3	$ 24,281.57	$ 121.41	$ 361.91	$ 23,919.66	
15	4	$ 23,919.66	$ 119.60	$ 363.72	$ 23,555.93	
16	5	$ 23,555.93	$ 117.78	$ 365.54	$ 23,190.39	

Figure 2.63
Loan amortization table with dollar values reformatted.

button, the values will always be shown with two decimal places, but you can then use the **Decrease Decimal** button to eliminate the extra decimal places.

2.5.3 Working with Percentages

The *Percentage format* has already been used in the loan amortization table, in cells D4 and D5. The value entered into cell D4 was 6%; Excel recognized the percent symbol and automatically applied the Percentage format to the cell and displayed the value with the percent symbol.

The APR value in cell D4 was used to calculate the periodic interest rate in cell D5, as

$$\text{D5:} \quad \texttt{=D4/12}$$

Since this is also a percentage, the **Percent** button was used to cause Excel to display the calculated result as a percentage, with the percent symbol. The **Percent** button is indicated in Figure 2.64.

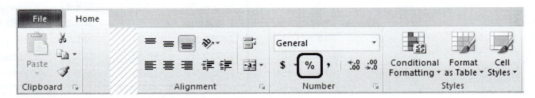

Figure 2.64
The **Percent** format button.

When you use the percent symbol, remember that it means "per *cent*" as in *cent*ury, or 100. Using the percent symbol is equivalent to dividing the value in front of the percent symbol by 100. So the 6% that appears in cell D4 of the loan amortization table in Figure 2.63 is numerically equivalent to $6/100 = 0.06$. If you remove the **Percentage** formatting on cell D4, and return to **General** formatting, the value will be shown as 0.06 and the rest of the calculations will be unaffected. To Excel, 6% and 0.06 are equivalent.

2.5.4 Using Commas as Thousand Separators

The **Accounting** format automatically included commas as thousand separators, as seen in Figure 2.63. Placing a comma every third digit can make large numbers easier to read. If you want to use commas as thousand separators on nonmonetary values, you can use the **Comma** style button indicated in Figure 2.65.

When you use the **Comma** style button, you are actually applying the **Accounting** format without a currency symbol. Most of the time, this works fine, but you could have problems if your worksheet includes either very large or very small values because the **Accounting** format will not switch over to scientific notation for

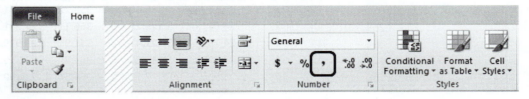

Figure 2.65
The **Comma** style button.

extreme values. Stick with **General** format or **Scientific** format for very large or very small values.

>**Note:** In some parts of the world, people use commas as decimal points and periods as thousand separators. By default, Excel uses your computer system's definitions for decimal points and thousand separators. You can override the default using the Excel Options dialog, **Advanced** panel, **Editing Options** section.

2.5.5 Using Named Formats

We have already presented several named formats:

* **General** format (Section 2.5)
* **Accounting** and **Currency** formats (Section 2.5.2)
* **Percentage** format (Section 2.5.3)

These and other named formats are available via a drop-down list of formats in the Number group of the Ribbon's **Home** tab. The location of the drop-down list of named formats is indicated in Figure 2.66.

Figure 2.66
The drop-down list of named formats.

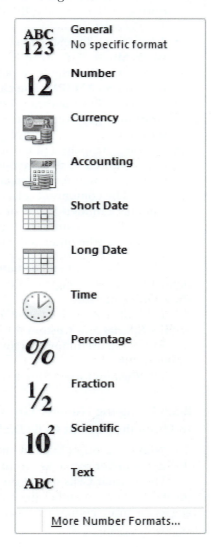

ABC 123	**General** No specific format
12	**Number**
	Currency
	Accounting
	Short Date
	Long Date
	Time
%	**Percentage**
½	**Fraction**
10^2	**Scientific**
ABC	**Text**

More Number Formats...

The named formats that have not been presented here include:

- *Number:* Similar to the General format, but with a specified number of decimal places displayed. When you use the **Increase Decimal** or **Decrease Decimal** buttons, you automatically switch the cell's formatting to Number format.
- *Scientific:* Numbers are presented as a mantissa and an exponent. For example, the values 20,000 could be written as a mantissa of 2 with an exponent of 4, or 2×10^4. In Excel, like most computer programs, the exponent is indicated by an E, so 20,000 written in Excel's scientific notation becomes "2.00 E +04."
- *Short Date:* Excel provides good support for working with dates in the last century. By representing dates and times as serial date/time codes, it is easy to do math with dates and times (see Section 3.14.3). The Short Date format causes a date to be displayed as **Month/Day/Year** (US default), **Day/Month/Year**, or **Year/Month/Day**, depending on where you live. The order of the day, month, and year is determined by the Region settings in Windows; it is not an Excel option. It can be modified using the Windows Control Panel. Example: 12/25/2010.
- *Long Date:* The Short Date format causes a date to be displayed as **Day of Week, Month Date, Year**. Example: Saturday, December 25, 2010.
- *Time:* Displays the time portion of a date/time code (see Section 3.14.3) as **HH:MM:SS AM/PM**. You can change the type of display (to 13:00:00 instead of 1:00:00 PM for example) by clicking the Number group **Expand** button (bottom-right corner of Number group) while a Time formatted cell is selected.
- *Fraction:* Displays a numeric value as a fraction. You can change the number of digits allowed in the fraction by clicking the Number group **Expand** button while a Fraction formatted cell is selected.
- *Text:* The contents of cells formatted as Text are treated as text, even if the cell contains a numeric value or a formula. Formulas in cells formatted as Text are not evaluated. Numbers in cells formatted as Text can be used in formulas in other cells (the other cell must not be formatted as Text).

The **More Number Formats** option at the bottom of the drop-down list of named formats (see Figure 2.66) opens the Format Cells dialog's **Number** panel, as shown in Figure 2.67.

The only categories (named numeric formats) that have not been presented are the **Special** and **Custom** formats. The **Special** format is locale dependent, but provides some useful formats for regionally specific values (such as formats for Zip Codes in the US).

The **Custom** format allows you to define your own special formats.

2.5.6 Using the Number Group's Expand Button to Open the Format Cells Dialog

The Number group's **Expand** button (see Figure 2.68) also opens the Format Cells dialog's **Number** panel, as shown in Figure 2.67.

The Format Cells dialog provides more formatting options than are available on the Ribbon's Number group (**Home** tab), but the Ribbon is a handier way to access the more commonly used formatting features.

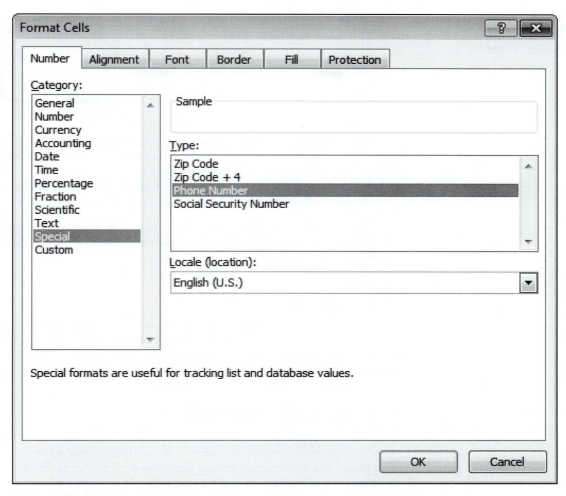

Figure 2.67
The Format Cells dialog, **Number** panel.

Figure 2.68
The Number group's **Expand** button.

2.6 USING THE STYLES GROUP

The *Styles group* on the Ribbon's Home tab (Figure 2.69) can be used to:

- Apply predefined styles to cells or cell ranges
- Create and modify styles
- Create a table within a worksheet
- Use conditional formatting to highlight parts of a data set

Figure 2.69
The Styles group on the Ribbon's **Home** tab.

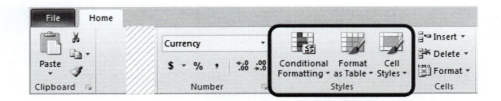

A *style* is a collection of attributes that can be applied to cells. They are usually used to create a "look" for certain portions of a worksheet. By default, all cells in a worksheet are displayed with the Normal style.

More specifically, a style defines the following attributes. The information in parentheses indicates how the Normal style is defined, by default.

- Numbers format (general format)
- Alignment (cell content is aligned with the bottom of the cell)
- Font (11 point Calibri)
- Border (none)
- Fill (none)
- Protection (locked)

It may surprise you that by default, the cells are *locked*. The fact that the cells are locked will not be noticed until the worksheet is *protected*, which is not done by default. If you choose to protect your worksheet, then the locked cells begin to function and prevent unauthorized editing. Protecting worksheets is presented in Section 2.7.

2.6.1 Using Predefined Cell Styles

While the **Normal** style, as the default, is the most commonly used, there are a number of other predefined styles available. If you have a wide-screen monitor, you may see some of the styles presented on the Ribbon, as shown in Figure 2.70.

If there is no room on the Ribbon to show the styles, then Excel shows a **Cell Styles** button, as shown in Figure 2.69. Click the **Cell Styles** button to open a selection box showing numerous styles (Figure 2.71).

By default, Excel 2010 and 2007 will show you what each style looks like on the currently selected cells as you move the mouse over the style selectors; this is called *Live Preview* and is activated by default. Live Preview is fairly resource intensive, and on slower computers you might want to turn it off. You can change the Live Preview default using the Excel Options dialog as follows:

- Excel 2010: **File tab/Options/General panel/Enable Live Preview**
- Excel 2007: **Office/Excel Options/Popular panel/Enable Live Preview**

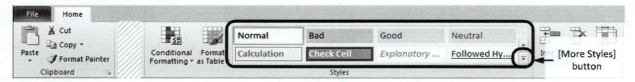

Figure 2.70
Styles available on the Ribbon (wide-screen monitors only).

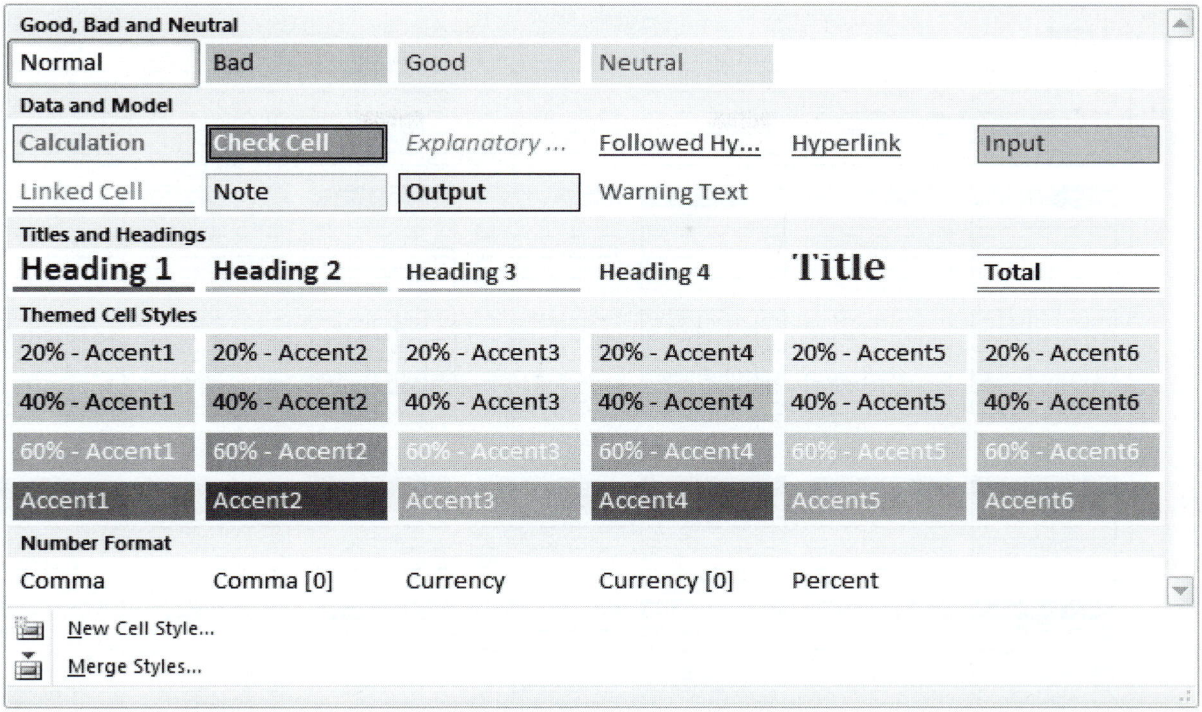

Figure 2.71
Predefined styles.

The process to apply a predefined style to one or more cells is:

1. Select the cells that are to be formatted using the predefined style.
2. Open the style selector using Ribbon options **Home/Styles/Cell Styles**.
3. Click on the style you wish to apply.

If you right-click on any of the predefined styles, you can modify them. You can also use the **New Cell Style...** button at the bottom of the style selector (Figure 2.71) to create your own style.

2.6.2 Defining a Table in a Worksheet

Clicking the **Format as Table** button opens a selector for predefined table styles, as shown in Figure 2.72. By default, Live Preview will show you what your table (selected cells) will look like as you move your mouse over the table style options.

While the **Format as Table** button in the Styles group does allow you to apply a table style to a selected range of cells, it does quite a bit more than just changing the appearance of the selected cells; it defines the selected cells as an *Excel table*.

An Excel table is a set of rows and columns containing related information (a data set) that Excel treats as separate from the rest of the worksheet. The key feature that is gained by defining a table is that you can sort and filter the data within the table without changing the rest of the worksheet. This can be very handy at times.

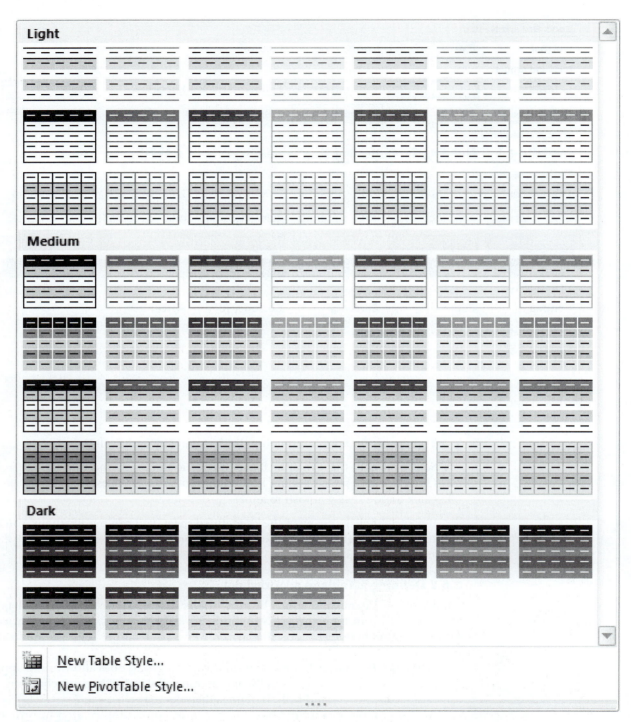

Figure 2.72
Predefined table styles.

A defined Excel table:

- Is kept largely separate from the rest of the worksheet.
- Has sorting and filtering enabled.
- Can have a *total row* automatically inserted that can do a lot more than just display the column totals. While it may be called a total row, you are not limited to just column totals. For any column, the total row can automatically display:
  - Average value
  - The number (count) of items in the column
  - The maximum value
  - The minimum value
  - The sum (total) of the values in the column
  - The sample standard deviation of the values in the column
  - The sample variance of the values in the column

Defining an Excel Table

As an example of working with an Excel table, consider the temperature data used in Section 2.4.4, shown again in Figure 2.73.

In Figure 2.73, cells A4:H15 have been selected; this cell range will become the Excel table. The Check column will be left outside of the table.

An Excel table can be defined in two ways:

Method 1: Using Ribbon options **Home tab/Styles/Format as Table**

1. Select the cells that will become the Excel table, including one row of headings (optional).
2. Click the **Format as Table** button in the **Home** tab's Styles group to open the Table styles selector.
3. Click on one of the predefined table styles.

	A	B	C	D	E	F	G	H	I	J	K
1	Temperature Data										
2											
3					Temperatures (K)						
4	Sample #	Time (sec.)	T1	T2	T3	T4	T5	T avg		Check	
5	1	0	243.2	246.8	241.2	242.2	244.1			1	
6	2	1	243.3	246.9	242.0	242.6	244.5			2	
7	3	2	243.4	247.0	242.8	243.0	244.9			3	
8	4	3	243.5	247.2	243.6	243.4	245.3			4	
9	5	4	243.6	247.3	244.4	243.8	245.7			5	
10	6	5	243.7	247.4	245.2	244.2	246.1			6	
11	7	6	243.8	247.5	246.0	244.6	246.5			7	
12	8	7	243.9	247.6	246.9	245.0	246.9			8	
13	9	8	244.0	247.8	247.7	245.4	247.3			9	
14	10	9	244.1	247.9	248.5	245.8	247.7			10	
15	11	10	244.2	248.0	249.3	246.2	248.1			11	
16											

Figure 2.73
Temperature data with cells A4:H15 selected.

Figure 2.74
Format as Table dialog.

Method 2: Using Ribbon options **Insert tab/Table**
1. Select the cells that will become the Excel table, including one row of headings (optional).
2. Click the **Table** button in the **Insert** tab's Tables group.

The primary difference between the two methods is that Method 1 gives you a choice of table styles; Method 2 does not.

Once you have completed the steps for one of the methods, Excel will display the Format as Table dialog (Figure 2.74) to verify the cell range that will be defined as a table, and to ask if your table has a row of headings, such as row 4 in our temperature data.

Click **OK** to finish defining the table. The formatted table is shown in Figure 2.75. Notice that the formatting does not apply outside of the cell range used to define the table (A4:H15).

	A	B	C	D	E	F	G	H	I	J	K
1	Temperature Data										
2											
3						Temperatures (K)					
4	Sample	Time (sec.	T1	T2	T3	T4	T5	T avg		Check	
5	1	0	243.2	246.8	241.2	242.2	244.1			1	
6	2	1	243.3	246.9	242.0	242.6	244.5			2	
7	3	2	243.4	247.0	242.8	243.0	244.9			3	
8	4	3	243.5	247.2	243.6	243.4	245.3			4	
9	5	4	243.6	247.3	244.4	243.8	245.7			5	
10	6	5	243.7	247.4	245.2	244.2	246.1			6	
11	7	6	243.8	247.5	246.0	244.6	246.5			7	
12	8	7	243.9	247.6	246.9	245.0	246.9			8	
13	9	8	244.0	247.8	247.7	245.4	247.3			9	
14	10	9	244.1	247.9	248.5	245.8	247.7			10	
15	11	10	244.2	248.0	249.3	246.2	248.1			11	
16											

Figure 2.75
The defined table.

Sorting an Excel Table

Also notice that each of the headings now has a drop-down menu button (indicated for heading T5 in Figure 2.75) on the right side of the heading cell. Clicking any of these buttons opens a menu of filtering and sorting options. The menu for heading T5 is shown in Figure 2.76.

If we select **Sort Largest to Smallest** on the T5 menu, the values in the T5 column (worksheet column G) will be arranged in descending order—but the table rows are

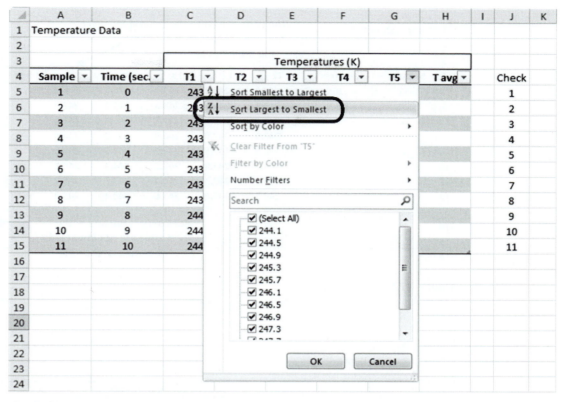

Figure 2.76
The filtering and sorting menu for heading T5.

kept together, so the rest of the table will be rearranged as well. The result of sorting on the T5 column is shown in Figure 2.77.

Notice:

1. The entire table has been rearranged, not just the T5 column. The rows in a table are kept together during a sort.
2. The button next to the T5 heading has changed; it now shows a small downward-pointing arrow as a reminder that the table has been sorted in descending order on the values in column T5.
3. The Check column (worksheet column J) has not been changed. The changes made to the table did not impact anything outside the table, and the Check column was outside of the cell range used to define the table (A4:H15).

Using the Total Row in an Excel Table

To insert a *total row* at the bottom of the Excel table, right-click anywhere on the table and select **Table/Totals Row** from the pop-up menu. An empty total row will be added to the table, as shown in Figure 2.78.

To use the total row, simply click in any cell in the total row, and a menu button will appear. Click the menu button and a list of options will be displayed, as shown in Figure 2.79.

In Figure 2.79, the total row for column T5 is being used to display the maximum value in that column (248.1 K).

	A	B	C	D	E	F	G	H	I	J	K
1	Temperature Data										
2											
3						Temperatures (K)					
4	Sample ▼	Time (sec. ▼	T1 ▼	T2 ▼	T3 ▼	T4 ▼	T5 ↓	T avg ▼		Check	
5	11	10	244.2	248.0	249.3	246.2	248.1			1	
6	10	9	244.1	247.9	248.5	245.8	247.7			2	
7	9	8	244.0	247.8	247.7	245.4	247.3			3	
8	8	7	243.9	247.6	246.9	245.0	246.9			4	
9	7	6	243.8	247.5	246.0	244.6	246.5			5	
10	6	5	243.7	247.4	245.2	244.2	246.1			6	
11	5	4	243.6	247.3	244.4	243.8	245.7			7	
12	4	3	243.5	247.2	243.6	243.4	245.3			8	
13	3	2	243.4	247.0	242.8	243.0	244.9			9	
14	2	1	243.3	24						10	
15	1	0	243.2	24						11	
16											

The control next to the column heading changes to show how the table has been sorted.

Values in cells outside of the defined table are not impacted when the table is resorted.

Figure 2.77
The result of filtering (descending order) on the T5 column.

	A	B	C	D	E	F	G	H	I	J	K
1	Temperature Data										
2											
3					Temperatures (K)						
4	Sample ▼	Time (sec. ▼	T1 ▼	T2 ▼	T3 ▼	T4 ▼	T5 ↓	T avg ▼		Check	
5	11	10	244.2	248.0	249.3	246.2	248.1			1	
6	10	9	244.1	247.9	248.5	245.8	247.7			2	
7	9	8	244.0	247.8	247.7	245.4	247.3			3	
8	8	7	243.9	247.6	246.9	245.0	246.9			4	
9	7	6	243.8	247.5	246.0	244.6	246.5			5	
10	6	5	243.7	247.4	245.2	244.2	246.1			6	
11	5	4	243.6	247.3	244.4	243.8	245.7			7	
12	4	3	243.5	247.2	243.6	243.4	245.3			8	
13	3	2	243.4	247.0	242.8	243.0	244.9			9	
14	2	1	243.3	246.9	242.0	242.6	244.5			10	
15	1	0	243.2	246.8	241.2	242.2	244.1			11	
16	Total								0		
17											

Figure 2.78
The Excel table with added total row.

Deactivating a Defined Excel Table

If you are done working with the data in a table and want to turn the table back into a simple cell range, you can. Here's how:

1. Right-click anywhere on the table. A pop-up menu will appear.
2. Select **Table/Convert to Range** from the pop-up menu. Excel will display a prompt window to make sure you really want to eliminate the table (but leaving the data).
3. Say **Yes** to the prompt "Do you want to convert the table to a normal range?"

	A	B	C	D	E	F	G	H	I	J	K
1	Temperature Data										
2											
3					Temperatures (K)						
4	Sample ▼	Time (sec. ▼	T1 ▼	T2 ▼	T3 ▼	T4 ▼	T5 ↓	T avg ▼		Check	
5	11	10	244.2	248.0	249.3	246.2	248.1			1	
6	10	9	244.1	247.9	248.5	245.8	247.7			2	
7	9	8	244.0	247.8	247.7	245.4	247.3			3	
8	8	7	243.9	247.6	246.9	245.0	246.9			4	
9	7	6	243.8	247.5	246.0	244.6	246.5			5	
10	6	5	243.7	247.4	245.2	244.2	246.1			6	
11	5	4	243.6	247.3	244.4	243.8	245.7			7	
12	4	3	243.5	247.2	243.6	243.4	245.3			8	
13	3	2	243.4	247.0	242.8	243.0	244.9			9	
14	2	1	243.3	246.9	242.0	242.6	244.5			10	
15	1	0	243.2	246.8	241.2	242.2	244.1			11	
16	Total							0 ▼			
17								None			
18								Average			
19								Count			
								Count Numbers			
20								Max			
21								Min			
								Sum			
22								StdDev			
								Var			
23								More Functions.			

Figure 2.79
The drop-down menu showing the options available for the total row.

When you eliminate the table definition, the formatting remains, and the cells contain the same values they held when the table was converted back to a range. That is, any sorting you did to the table remains. If the table was filtered, any data hidden by the filtering process is restored when the table definition is eliminated.

2.6.3 Using Conditional Formatting

With *conditional formatting*, particular format attributes are applied only if a certain condition is met. For example, you might want unsafe values to show up in bright red. Or, if you are the engineer in charge of quality control, you might want off-spec values to be highlighted so that they are easy to spot.

As an example of conditional formatting, consider again the temperature data from Section 2.4.4, shown again in Figure 2.80.

The temperature values have been selected. Next we will apply conditional formatting to make temperature values greater than 248 K stand out, since 248 K is considered the highest "safe" temperature for this freezer system. If the temperature gets too high, the contents of the freezer could go bad.

To apply conditional formatting:

1. Select the cells to which the conditional formatting should be applied (cells C5:G15, shown in Figure 2.80).
2. Use Ribbon options **Home/Styles/Conditional Formatting**, then menu options **Highlight Cell Rules/Greater Than...** to begin defining the condition when the

◢	A	B	C	D	E	F	G	H
1	Temperature Data							
2								
3			Temperatures (K)					
4	Sample #	Time (sec.)	T1	T2	T3	T4	T5	
5	1	0	243.2	246.8	241.2	242.2	244.1	
6	2	1	243.3	246.9	242.0	242.6	244.5	
7	3	2	243.4	247.0	242.8	243.0	244.9	
8	4	3	243.5	247.2	243.6	243.4	245.3	
9	5	4	243.6	247.3	244.4	243.8	245.7	
10	6	5	243.7	247.4	245.2	244.2	246.1	
11	7	6	243.8	247.5	246.0	244.6	246.5	
12	8	7	243.9	247.6	246.9	245.0	246.9	
13	9	8	244.0	247.8	247.7	245.4	247.3	
14	10	9	244.1	247.9	248.5	245.8	247.7	
15	11	10	244.2	248.0	249.3	246.2	248.1	
16								

Figure 2.80
Temperature data with temperature values (cells C5:G15) selected.

special format will be applied. These menu selections are shown in Figure 2.81. The Greater Than dialog will appear, as shown in Figure 2.82.

3. Excel tries to be helpful and fills in the fields in the Greater Than dialog. The value of 245.3 is the average of the selected temperature values. That's not the value we want. We want to know if there are any temperatures greater than 248 K, so we enter 248 in the Greater Than dialog, as shown in Figure 2.83.

4. A **Custom Format...** was specified (black fill with white text) so that it would show up well in this text. The resulting worksheet is shown in Figure 2.84.

The result indicates that there are four temperatures in the data set that are above the safety threshold. It's time to check the freezer.

Excel 2007 and 2010 come with a variety of predefined conditional formats to help show trends in data. Because of the Live Preview feature, the best way to learn about what the predefined conditional formats can do for you is to try them on some data.

Clearing Conditional Formatting

If you want to remove the conditional formatting, use one of the following Ribbon options:

- To remove conditional formatting from a selected range of cells:

  Home/Styles/Conditional Formatting/Clear Rules/Clear Rules from Selected Cells

- To remove conditional formatting from a current worksheet:

  Home/Styles/Conditional Formatting/Clear Rules/Clear Rules from Entire Sheet

Figure 2.81
Menu selections used to apply a "greater than" conditional formatting.

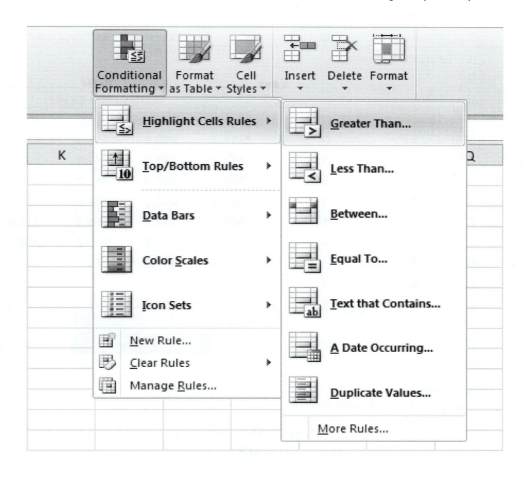

Figure 2.82
The Greater Than dialog with Excel's default values.

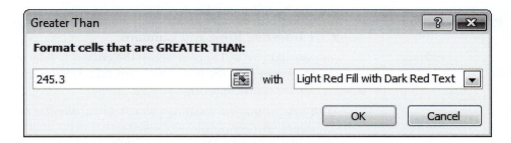

Figure 2.83
The completed Greater Than dialog.

⬛	A	B	C	D	E	F	G	H
1	Temperature Data							
2								
3			Temperatures (K)					
4	Sample #	Time (sec.)	T1	T2	T3	T4	T5	
5	1	0	243.2	246.8	241.2	242.2	244.1	
6	2	1	243.3	246.9	242.0	242.6	244.5	
7	3	2	243.4	247.0	242.8	243.0	244.9	
8	4	3	243.5	247.2	243.6	243.4	245.3	
9	5	4	243.6	247.3	244.4	243.8	245.7	
10	6	5	243.7	247.4	245.2	244.2	246.1	
11	7	6	243.8	247.5	246.0	244.6	246.5	
12	8	7	243.9	247.6	246.9	245.0	246.9	
13	9	8	244.0	247.8	247.7	245.4	247.3	
14	10	9	244.1	247.9	**248.5**	245.8	247.7	
15	11	10	244.2	**248.0**	**249.3**	246.2	**248.1**	
16								

Figure 2.84
The worksheet with conditional formatting to highlight temperatures above 248 K.

2.7 INSERTING, DELETING, AND FORMATTING ROWS AND COLUMNS

The *Cells group* on the Ribbon's Home tab provides access to buttons allowing you to insert, resize, and delete rows and columns. Many of these tasks have shortcuts that will also be presented here.

2.7.1 Inserting Rows and Columns

An inserted row will appear above the currently selected row, and an inserted column will appear to the left of the currently selected column.

To insert one new row:

1. Click on the row heading just below where the new row should be placed; this selects the entire row.
2. Use one of the following methods to insert a row:
   o Use Ribbon options **Home/Cells/Insert/Insert Sheet Rows**
   o Right-click on the selected row heading and choose **Insert** from the pop-up menu

To insert multiple new rows:

1. Drag the mouse across the row headings to select the number of rows to insert. The new rows will be inserted just above the selected rows.
2. Use one of the following methods to insert the new rows:
   o Use Ribbon options **Home/Cells/Insert/Insert Sheet Rows**
   o Right-click on the selected row heading and choose **Insert** from the pop-up menu

To insert one new column:

1. Click on the column heading just to the right of where the new column should be placed; this selects the entire column.
2. Use one of the following methods to insert a column:
   o Use Ribbon options **Home/Cells/Insert/Insert Sheet Columns**
   o Right-click on the selected column heading and choose **Insert** from the pop-up menu

To insert multiple new columns:

1. Drag the mouse across the column headings to select the number of columns to insert. The new columns will be inserted just to the left of the selected columns.
2. Use one of the following methods to insert the new columns:
   o Use Ribbon options **Home/Cells/Insert/Insert Sheet Columns**
   o Right-click on the selected row heading and choose **Insert** from the pop-up menu

2.7.2 Deleting Rows and Columns

To delete one or more rows from a worksheet:

1. Select the rows to be deleted by clicking (and dragging for multiple rows) on the row headings.
2. Use one of the following methods to delete the rows:
   o Use Ribbon options **Home/Cells/Delete/Delete Sheet Rows**
   o Right-click on a selected row heading and choose **Delete** from the pop-up menu

To delete one or more columns from a worksheet:

1. Select the columns to be deleted by clicking (and dragging for multiple columns) on the column headings.
2. Use one of the following methods to delete the columns:
   o Use Ribbon options **Home/Cells/Delete/Delete Sheet Columns**
   o Right-click on a selected column heading and choose **Delete** from the pop-up menu

2.7.3 Inserting a New Worksheet into the Current Workbook

You can use the following Ribbon options to insert a new worksheet into the current workbook:

Home/Cells/Insert/Insert Sheet

Alternatively, you can click the **Insert Worksheet** button on the worksheet tabs at the bottom of the Excel window (indicated in Figure 2.85).

2.7.4 Deleting the Currently Selected Worksheet

You can use the following Ribbon options to delete the current worksheet:

Home/Cells/Delete/Delete Sheet

Alternatively, you can right-click the worksheet's tab at the bottom of the Excel window and select **Delete** from the pop-up menu. This latter method can be used for any worksheet in the current workbook.

Figure 2.85
The **Insert Worksheet** button.

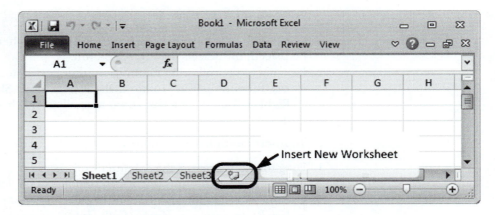

2.7.5 Adjusting Row Height and Column Width

Changing the displayed height of rows and width of columns is a common need and there are several ways to do each.

Changing row height with the mouse

The easiest way to adjust the height of a row is to grab the lower edge of the row heading and drag it to the size you want. When the mouse is positioned over the lower edge of the row heading, the mouse icon changes to a horizontal line with a vertical double-headed arrow through it, as shown in Figure 2.86. The row height is also displayed as the height is adjusted.

To change the height of several rows simultaneously:

1. Select the rows to be adjusted by dragging the mouse on the row headings.
2. Position the mouse over the lower edge of the bottom selected row (the icon will change as shown in Figure 2.86) and drag the edge to the desired height. The heights of all selected rows will be changed.

Changing row height with the Ribbon

To change the height of one or more rows using Ribbon options:

1. Select the rows to be adjusted.
2. Use Ribbon options **Home/Cells/Format/Row Height…**

Figure 2.86
The mouse icon while adjusting row height.

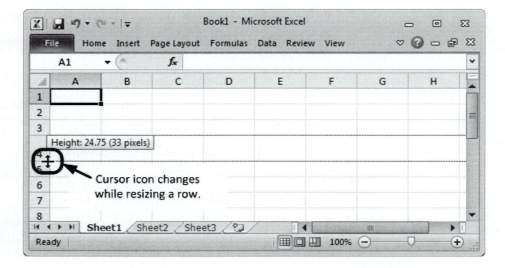

Figure 2.87
The Row Height dialog.

3. Enter the desired row height in the Row Height dialog, shown in Figure 2.87. A value of 15 is the default row height.

Changing column width with the mouse

To adjust the width of a column, grab the right edge of the column heading and drag it to the size you want. When the mouse is positioned over the right edge of the column heading, the mouse icon changes to a vertical line with a horizontal double-headed arrow through it, as shown in Figure 2.88. The column width is also displayed as the width is adjusted.

To change the width of several columns simultaneously:

1. Select the columns to be adjusted by dragging the mouse on the column headings.
2. Position the mouse over the right edge of the right-most selected row (the icon will change as shown in Figure 2.88) and drag the edge to the desired width. The widths of all selected columns will be changed.

Changing column width with the Ribbon

To change the height of one or more rows using Ribbon options:

1. Select the columns to be adjusted.
2. Use Ribbon options **Home/Cells/Format/Column Width...**
3. Enter the desired column width in the Column Width dialog, shown in Figure 2.89. A value of 8.43 is the default column width, but that can be changed with Ribbon options **Home/Cells/Format/Default Width...**

Figure 2.88
The mouse icon while adjusting column width.

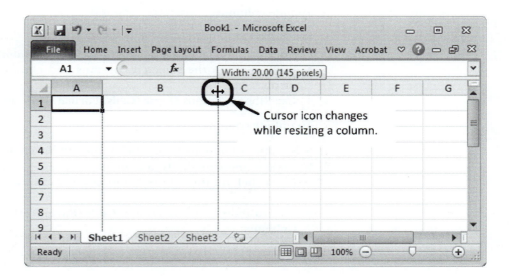

Figure 2.89
The Column Width dialog.

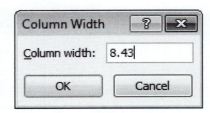

2.7.6 Hiding and Unhiding Rows and Columns

It is sometimes convenient to hide some of the rows and/or columns in a worksheet. This might be done when:

- The worksheet is very large and hiding unused rows or columns makes it easier to navigate to the areas that are being developed.
- You do not want parts of the worksheet to be seen by the end user. These might be company secrets, or just intermediate calculations that the end user is not interested in.

Hidden rows and *hidden columns* are still actively being kept up to date (recalculated) as the rest of the worksheet is created; they are simply not displayed. Hidden rows do not display when a worksheet is printed.

To hide rows:

1. Select the rows to be hidden.
2. Use one of these methods to hide the selected rows:
   - Right-click on the selected rows and choose **Hide** from the pop-up menu.
   - Use Ribbon options **Home/Cells/Format/Hide and Unhide/Hide Rows**

To unhide rows:

1. Select one row above and below the hidden rows.
2. Use one of these methods to unhide the selected rows:
   - Right-click on the selected rows and choose **Unhide** from the pop-up menu.
   - Use Ribbon options **Home/Cells/Format/Hide and Unhide/Unhide Rows**

To hide columns:

1. Select the columns to be hidden.
2. Use one of these methods to hide the selected columns:
   - Right-click on the selected columns and choose **Hide** from the pop-up menu.
   - Use Ribbon options **Home/Cells/Format/Hide and Unhide/Hide Columns**

To unhide columns:

1. Select one column above and below the hidden columns.
2. Use one of these methods to unhide the selected columns:
   - Right-click on the selected columns and choose **Unhide** from the pop-up menu.
   - Use Ribbon options **Home/Cells/Format/Hide and Unhide/Unhide Columns**

2.7.7 Renaming Worksheets

Many problems can be solved using a single worksheet, but when several worksheets are used it can be very helpful to give the worksheets meaningful names. To *rename* a worksheet, follow one of these methods:

Renaming a worksheet using the worksheet's tab

1. Double-click on a worksheet's tab. The name field on the tab will enter edit mode so that you can change the name.
2. Edit the name on the tab.
3. Click somewhere on the worksheet (away from the tab) to leave edit mode.

Renaming a worksheet using the Ribbon

1. Make sure the worksheet you wish to rename is the currently selected worksheet.
2. Use Ribbon options **Home/Cells/Format/Rename Sheet**. The name field on the worksheet's tab will enter edit mode.
3. Edit the name on the tab.
4. Click somewhere on the worksheet (away from the tab) to leave edit mode.

2.7.8 Protecting Worksheets

When a completed worksheet is made available to another person, it is often helpful to limit the other person's access to the calculations on the worksheet. This is done by *protecting* the worksheet.

As an example, consider the loan amortization table developed earlier (see Figure 2.90). Someone might want to use the table for their own car loan, and that would be OK, but we would want to protect the calculations in the table and only allow them to have access to the input values at the top of the worksheet.

In Figure 2.90, the input values that someone needs to enter are indicated. All of the other values in the table are calculated from those four numbers. We want to lock

Figure 2.90

The loan amortization table with unprotected cells indicated.

	A	B	C	D	E	F
1	Loan Amortization Table					
2						
3			Amount Borrowed:	$ 25,000.00		
4			APR:	6.0%		
5			Periodic Interest Rate:	0.5%		
6			Term:	5	years	
7			Payments/Year:	12		
8			Payments:	60		
9			Payment Amount:	$ 483.32		
10						
11		Payment	Principal Before Payment	Interest Payment	Paid on Principal	Principal After Payment	
12		1	$ 25,000.00	$ 125.00	$ 358.32	$ 24,641.68
13		2	$ 24,641.68	$ 123.21	$ 360.11	$ 24,281.57
14		3	$ 24,281.57	$ 121.41	$ 361.91	$ 23,919.66
15		4	$ 23,919.66	$ 119.60	$ 363.72	$ 23,555.93
16		5	$ 23,555.93	$ 117.78	$ 365.54	$ 23,190.39

down the worksheet, except for those four cells. This keeps inexperienced Excel users from accidentally messing up the table by entering values on top of the formulas.

In Section 2.6, it was pointed out that cells formatted with the Normal style are, by default, locked. You can't tell they are locked until the worksheet is protected, but they are. Before we protect the worksheet we need to unlock the four cells that we want people to be able to use (cells C3:C4, C6:C7).

Unlocking Cells

The procedure for unlocking one or more cells is:

1. Select the cell(s) to be unlocked. In Figure 2.92, cells C3:C4, C6:C7 were selected.
2. Use Ribbon options **Home/Cells/Format/Lock Cell**. This toggles the Lock Cell button to unlock the cells. You can tell whether or not a cell is locked by looking at the icon in front of the **Lock Cell** button (shown in Figure 2.91). If

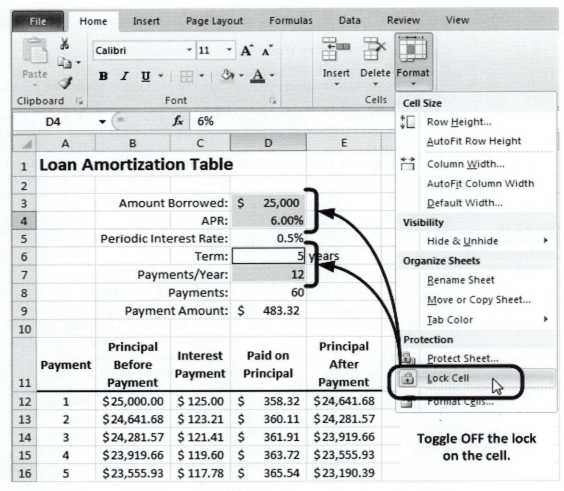

Figure 2.91
Unlocking cells.

the icon is selected (enclosed with a box), the cell is locked. The image shown in Figure 2.91 was taken just before the **Lock Cell** button was clicked to unlock cells C3:C4, C6:C7.

At this point, cells C3:C4, C6:C7 have been unlocked, but all cells are still accessible since the worksheet has not been protected.

Protecting the Worksheet

To protect the worksheet, use Ribbon options **Home/Cells/Format/Protect Sheet...** Excel 2003: Tools/Protection/Protect Sheet... The Protect Sheet dialog will appear, as shown in Figure 2.92.

Figure 2.92
The Protect Sheet dialog.

On this dialog you need to supply a password so that you can unprotect the sheet again, if needed. You should also review the items that users will be allowed to do in the worksheet. By default, users can select both locked and unlocked cells, and that's it. (They can also enter values into unlocked cells.)

When you enter a password and click **OK**, Excel will ask you to confirm the password by reentering the same password in the Confirm Password dialog, shown in Figure 2.93.

When you click **OK** on the Confirm Password dialog (and both passwords match), your worksheet has been protected, and only the four unlocked cells (C3:C4, C6:C7) can be edited.

As a service to people who will be using locked down worksheets, it helps if you provide some indication of which cells they are allowed to use. In Figure 2.94, the final loan amortization table is shown with borders around unlocked cells to help the user see that those are the cells they need to use.

Figure 2.93
The Confirm Password
dialog.

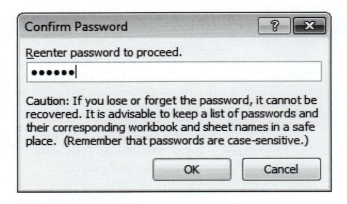

If a friend wants to see what the values look like for their loan, they just update
the input values (and they can't accidentally break the formulas). For example,
the loan amortization table for a $20,000 loan for 3 years at 7.25% is shown in
Figure 2.95.

	A	B	C	D	E	F
1	**Loan Amortization Table**					
2						
3			**Amount Borrowed:**	$ 25,000.00		
4			**APR:**	6.0%		
5			**Periodic Interest Rate:**	0.5%		
6			**Term:**	5	years	
7			**Payments/Year:**	12		
8			**Payments:**	60		
9			**Payment Amount:**	$ 483.32		
10						
11	Payment	Principal Before Payment	Interest Payment	Paid on Principal	Principal After Payment	
12	1	$25,000.00	$ 125.00	$ 358.32	$24,641.68	
13	2	$24,641.68	$ 123.21	$ 360.11	$24,281.57	
14	3	$24,281.57	$ 121.41	$ 361.91	$23,919.66	
15	4	$23,919.66	$ 119.60	$ 363.72	$23,555.93	
16	5	$23,555.93	$ 117.78	$ 365.54	$23,190.39	

Figure 2.94
The locked loan amortization table (showing 5 of 60 payments).

	A	B	C	D	E	F
1	Loan Amortization Table					
2						
3		Amount Borrowed:		$ 20,000.00		
4		APR:		7.3%		
5	Periodic Interest Rate:			0.6%		
6		Term:		3	years	
7		Payments/Year:		12		
8		Payments:		36		
9		Payment Amount:		$ 619.83		
10						
11	Payment	Principal Before Payment	Interest Payment	Paid on Principal	Principal After Payment	
12	1	$20,000.00	$ 120.83	$ 499.00	$19,501.00	
13	2	$19,501.00	$ 117.82	$ 502.01	$18,998.99	
14	3	$18,998.99	$ 114.79	$ 505.05	$18,493.95	
15	4	$18,493.95	$ 111.73	$ 508.10	$17,985.85	
16	5	$17,985.85	$ 108.66	$ 511.17	$17,474.68	

Figure 2.95

The loan amortization table for a $20,000 loan for 3 years at 7.25% (showing 5 of 60 payments).

KEYWORDS

Accounting format
Cell fill
Clipboard Group
Clipboard pane
Conditional formatting
Copy
Currency
Cut
Decrease decimal
    (button)
Excel Options
Excel table
Fill Handle
Font Attributes (bold,
    italics, underline)
Font Group
Format Painter
Fraction format

General format
Groups
Hidden rows and
    columns
Horizontal alignment
Increase decimal
    (button)
Live Preview
Locked cells
Long Date
Merged cells
Named formats
Normal style
Number group
Paste
Paste Options menu
Paste Values
Paste/Transpose

Percentage format
Protected worksheet
Rename worksheet
Ribbon
Scientific format
Short Date format
Style
Styles group
Subscript
Superscript
Tabs
Text format
Text wrapping
Time format
Total row (Excel table)
Vertical alignment
Windows clipboard

SUMMARY

Cut, Copy, and Paste

- Cut

  Copies cell contents to the clipboard and (ultimately) removes them from the worksheet.

  Ribbon: **Home/Clipboard/Cut**

  Keyboard: [Ctrl-x]

- Copy

  Copies cell contents to the clipboard and leaves them from the worksheet.

  Ribbon: **Home/Clipboard/Copy**

  Keyboard: [Ctrl-c]

- Paste

  Copies cell contents from the clipboard to the currently selected worksheet location.

  Ribbon: **Home/Clipboard/Paste**

  Keyboard: [Ctrl-v]

- Paste Options Menu

  The Paste Options menu, located just below the Paste button, provides access to additional ways to paste information from the clipboard, such as:
  - **Paste Values**—pastes the result of formulas, not the formulas themselves.
  - **Transpose**—interchanges rows and columns during the paste.
  - **Paste Special...**—opens the Paste Special dialog to allow greater control during the paste operation.

Format Painter

Copies the formatting applied to a cell, not the cell contents.

  Ribbon: **Home/Clipboard/Format Painter**

- Click the **Format Painter** button once to copy formatting from one cell to another cell or cell range.
- Double-click the **Format Painter** button once to copy formatting from one cell to multiple cells or cell ranges. Then click the **Format Painter** button again to deactivate format painting.

Fill Handle

The Fill Handle is the small black square at the bottom-right corner of a selected cell or group of cells. It can be used to:

- Copy selected cells down or across a worksheet.
- Fill cells with a series of values.
  - For series values incremented by one, enter the first two values in adjacent cells, select both, and drag the Fill Handle to create the series.
  - For series values incremented by values other than one, enter the first two values in adjacent cells, select both, and drag the Fill Handle to create the series.
  - For nonlinear series, drag the Fill Handle with the right mouse button; a menu will open with fill options.

Font Group

Use the Font Group on the Ribbon's Home tab to set or modify:

- Font type
- Font size
- Font color
- Font attributes (bold, italics, underline)
- Cell border
- Fill color (cell background color)

> Ribbon: **Home/Font**

Excel Default Options

Use the Excel Options dialog to set Excel's default values, such as:

- Default font type and size
- Number of worksheets in a new workbook
- Enabling iterative calculations
- Setting workbook AutoRecovery options
- Changing the date system (1900 or 1904)
- Activate or deactivate Add-Ins

> Access the Excel Options dial as follows:

- Excel 2010: **File tab/Options**
- Excel 2007: **Office/Excel Options**
- Excel 2003: **File/Options**

Superscripts and Subscripts

Use the Format Cells dialog to include subscripts or superscripts in text in cells.

1. Select the cell containing the text to be modified.
2. Activate Edit mode by double-clicking the cell, or pressing [F2].
3. Select the character(s) to be sub- or superscripted.
4. Click the Font group **Expand** button to open the Format Cells dialog.
5. Check the **Subscript** or **Superscript** box.
6. Click **OK** to close the Format Cells dialog.
7. Click outside of the cell to leave Edit mode.

> Ribbon: **Home/Font/Expand**

Horizontal and Vertical Alignment

Use the Horizontal and Vertical Alignment buttons on the Alignment group of the Ribbon's **Home** tab to adjust how information is displayed in cells.

Horizontal Alignment

- left-justify (default for text)
- center-justify
- right-justify (default for values)

Vertical Alignment

- top-justify
- middle-justify
- bottom-justify (default)

> Ribbon: **Home/Alignment**

Wrapping Text in Cells

When labels are too long to fit in a cell, word wrapping can be used to show the labels on multiple lines.

Ribbon: **Home/Alignment/Wrap Text**

Merged Cells

Cells on a worksheet can be merged; the *merged cells* are treated as a single cell for calculations.

The **Merge & Center** button is located in the Alignment group on the Ribbon's Home tab.

There is also a Merge Options menu available that allows the following options:

- Merge & Center
- Merge Across
- Merge Cells
- Unmerge Cells

Ribbon: **Home/Alignment/Merge & Center**, or **Home/Alignment/Merge (menu)**

Formatting Numbers

The Number group on the Ribbon's Home tab provides numeric formats, including:

- General
- Number
- Currency (available as a button)
- Accounting
- Short Date
- Long Date
- Time
- Percentage (available as a button)
- Fraction
- Scientific

Ribbon: **Home/Number**, then use the named format drop-down list

Number of Displayed Decimal Places

The Number group on the Ribbon's Home tab provides buttons to increase or decrease the number of displayed decimal places.

Ribbon: **Home/Number/Increase Decimal**, or **Home/Number/Decrease Decimal**

Working with Monetary Units

Excel provides two named formats for monetary values: *Accounting format* and *Currency format*. The difference is in the way the currency symbols are displayed. Accounting format aligns all of the currency symbols in a column, currency format does not. There is a shortcut button in the Number group for the currency format.

Ribbon: **Home/Number**, then use the named format drop-down list

Working with Percentages

Excel provides the *Percentage format* for working with percentages. When the percentage format is applied, the displayed value is increased by a factor of 100, and the percentage symbol is displayed.

Ribbon: **Home/Number/Percent Style**

Date and Time Formats

If Excel recognizes a cell entry as a date or time, it automatically converts it to a date/time code and applies a date or time format to the cell, as appropriate.

If you change the contents of a cell from a date or time to a number or text, you may need to manually set the cell format back to General format.

Ribbon: **Home/Number**, then use the named format drop-down list

Styles

All cells are initially formatted using the Normal style.

Excel provides a lot of built-in styles that can be applied to change the appearance of cells. To apply a built-in style:

1. Select the cells to be formatted with the style.
2. Select the style from the Styles group on the Ribbon's Home tab.

By default, Excel uses Live Preview to allow you to see how the styles will look before they are applied. As you move the mouse over the style options, Live Preview will show what the style will look like in the selected cells.

Ribbon: **Home/Styles** then choose a style from the available selection.

Excel Tables

An Excel table is a set of rows and columns containing related information (a data set) that Excel treats as separate from the rest of the worksheet. Sorting and filtering is easy within tables.

Defining an Excel Table

1. Select the cells that will become the Excel table.
2. Click the **Format as Table** button in the Home tab's Styles group.
3. Click on one of the predefined table styles.

Sorting and Filtering: Use the drop-down menus available with each column heading to sort or filter based on the values in that column.

Inserting a Total Row in an Excel Table

1. Right-click anywhere on the table.
2. Select **Table/Totals Row** from the pop-up menu.

Using the Total Row in an Excel Table

1. Click in the Total Row below the desired column.
2. Select the desired quantity (average, sum, etc.) from the drop-down list.

Inserting a Row

1. Click on the row heading just below where the new row should be placed.
2. Use one of the following methods to insert a row:
   - Use Ribbon options **Home/Cells/Insert/Insert Sheet Rows**
   - Right-click on the selected row heading and choose **Insert** from the pop-up menu

Inserting a Column

1. Click on the column heading just to the right of where the new column should be placed.

2. Use one of the following methods to insert a column:
- Use Ribbon options **Home/Cells/Insert/Insert Sheet Columns**
- Right-click on the selected column heading and choose **Insert** from the pop-up menu

Deleting Rows or Columns

1. Select the rows or columns to be deleted.

2. Right-click on a selected row or column heading and choose **Delete** from the pop-up menu.

Inserting a New Worksheet

You can use Ribbon options **Home/Cells/Insert/Insert Sheet**

Alternatively, you can click the Insert Worksheet button just to the right of the worksheet tabs at the bottom of the Excel window.

Adjusting Row Height and Column Width

With the mouse, grab the lower edge of the row heading, or right edge of a column heading, and drag it to the size you want.

Alternatively you can use Ribbon options **Home/Cells/Format/Row Height…** or **Home/Cells/Format/Column Width…**

Hiding Rows and Columns

1. Select the rows or columns to be hidden.

2. Right-click on the selected rows or columns and choose **Hide** from the pop-up menu.

Unhiding Rows

1. Select one row above and below the hidden rows.

2. Use one of these methods to unhide the selected rows:
- Right-click on the selected rows and choose **Unhide** from the pop-up menu.
- Use Ribbon options **Home/Cells/Format/Hide and Unhide/Unhide Rows**

Unhiding Columns

1. Select one column above and below the hidden columns.

2. Use one of these methods to unhide the selected columns:
- Right-click on the selected columns and choose **Unhide** from the pop-up menu.
- Use Ribbon options **Home/Cells/Format/Hide and Unhide/Unhide Columns**

Renaming Worksheets

1. Double-click on a worksheet's tab. The name field on the tab will enter edit mode so that you can change the name.

2. Edit the name on the tab.

3. Click somewhere on the worksheet (away from the tab) to leave edit mode.

Protecting Worksheets

Cells are, by default, locked; but the locking is not activated until the worksheet is protected. Before you protect the worksheet, you should unlock any cells that you want to be accessible afterward.

To unlock specific cells:

1. Select the cell(s) to be unlocked.
2. Use Ribbon options **Home/Cells/Format/Lock Cell** to toggle the **Lock Cell** button for the selected cells.

To protect a worksheet:
Use Ribbon options **Home/Cells/Format/Protect Sheet…**

11 3.3224

PROBLEMS

2.1 Paying Back Student Loans I

College students graduating from US universities often have accumulated $20,000 in loans. In recent years, the interest rate on those loans has been about 6% APR, and a common repayment plan is to pay the money back over 10 years. Such a loan would have a monthly payment of $222.04.

Create an amortization table similar to the one shown in Figure 2.96.
Be sure to include the following formatting features in your worksheet:

- Large, bold title
- Borders around the cells that require data entry (cells D3, D4, D6, and D7) in Figure 2.96

122.04
122.65

Figure 2.96
Student loan amortization table.

	A	B	C	D	E	F
1	**Loan Amortization Table**					
2						
3		Amount Borrowed:		$ 20,000.00		
4		APR:		6.0%		
5		Periodic Interest Rate:		0.5%		
6		Term:		10	years	
7		Payments/Year:		12		
8		Payments:		120		
9		Payment Amount:		$ 222.04		
10						
11	**Payment**	**Principal Before Payment**	**Interest Payment**	**Paid on Principal**	**Principal After Payment**	
12	1	$20,000.00	$ 100.00	$ 122.04	$19,877.96	
13	2	$19,877.96	$ 99.39	$ 122.65	$19,755.31	
14	3	$19,755.31	$ 98.78	$ 123.26	$19,632.04	
15	4	$19,632.04	$ 98.16	$ 123.88	$19,508.16	
16	5	$19,508.16	$ 97.54	$ 124.50	$19,383.66	

- Accounting format on all dollar amounts
- Use Percentage format on the APR and periodic interest rate
- For column headings
  - Text wrapping
  - Bold font
  - Centered headings
  - Heavy bottom border

Use your amortization table to determine:

a) Total amount paid on the loan.

b) Amount paid on interest.

2.2 Paying Back Student Loans II

The minimum monthly payment on the loan described in Problem 2.1 was $222.04, but there is (usually) no penalty for overpayment. Recalculate the loan amortization table assuming a monthly payment of $250.

a) How many months would it take to pay off the loan with the higher payment?

b) What is the total amount paid on the loan?

c) What percent of the total paid went toward interest?

2.3 Distances Between European Capitals

Perform an Internet search on "Travel Distances Between European Cities" and use the results to complete the grid shown in Figure 2.97.

Be sure to include the following formatting features in your worksheet:

- Large, bold title
- Border around the distance grid
- Center all headings and distance values
- Adjust column widths to fit all headings

2.4 Exponential Growth I

There is a legend that the inventor of chess asked for a small payment in return for the marvelous game he had developed: one grain of rice for the first square on the

	A	B	C	D	E	F	G	H	I	J
1	**Distances Between European Capitals (KM)**									
2										
3			Athens	Berlin	Bucharest	Copenhagen	London	Madrid	Rome	
4		Athens	0							
5		Berlin		0						
6		Bucharest			0				1140	
7		Copenhagen				0				
8		London					0			
9		Madrid						0		
10		Rome			1140				0	
11										

Figure 2.97

Distances between European capitals.

chess board, two for the second square, four for the third square, and so on. There are 64 squares on a chess board.

    a) How many grains of rice were placed on the 16th square?
    b) How many grains of rice were placed on the 64th square?
    c) How many grains of rice were placed on the chess board?

This problem is intended to give you some practice working with very large numbers in Excel. You may want to try using the scientific format. Creating a series of values from 1 to 64 for the "square number" column is a good place to use the Fill Handle.

2.5 Exponential Growth II

Bacteria grow by dividing, so one cell produces two, two cells produce four, and so on. This is another example of exponential growth. It is not uncommon for bacteria to double every hour; assuming, of course, that there is enough food about to sustain such growth. Use a worksheet something like the one shown in Figure 2.98 to find the answers to the following questions.

    a) How many hours does it take for one bacterium to turn into more than 10^{10} bacteria?

Figure 2.98
Calculating bacterial population.

	A	B	C	D
1	Bacterial Population			
2				
3		Hours	Population	
4		0	1	
5		1	2	
6		2	4	
7		...	...	
8				

3

Graphing with Excel

Objectives

After reading this chapter, you will know

- How to organize data in a worksheet to make graphing easy
- How to create an XY scatter graph
- How to modify an existing graph
  - ○ Adding additional data series to a graph

- ○ Modifying plot formatting
- ○ How to add trendlines to graphs
- ○ How to add error bars to graphs
- Two ways to print a graph
- How to use Web data in an Excel graph
- How to import text files into Excel for graphing

3.1 INTRODUCTION

An Excel worksheet is a convenient place to generate graphs. The process is quick and easy, and Excel gives you a lot of control over the appearance of the graph. Once a graph is created, you can then begin analyzing your data by adding a trendline to a graphed data set with just a few mouse clicks. The majority of Excel's trendlines are regression lines, and the equation of the best-fit curve through your data set is immediately available. Excel graphs are great tools for visualizing and analyzing data.

3.1.1 Nomenclature

- The good folks at Microsoft use the term *chart* rather than *graph*. Graph is a more commonly used term in engineering and is used here. Chart and graph can be treated as synonymous in this text.
- The points displayed on a graph are called *markers*.

- Microsoft Excel uses the term *trendline* (one word), but Microsoft Word's spell-checker claims it should be *trend line* (two words). While common usage agrees with Microsoft Word, I have used Excel's "trendline" in this text for consistency with the menu options presented by the Excel program.

3.2 GETTING READY TO GRAPH

You must have some data in the worksheet before creating a graph, and those data can come from a variety of sources. Creating a graph from values that were calculated within the worksheet is probably the most common, but data can also be imported from data files, copied from other programs (such as getting data from the Internet by using a browser), or perhaps even read directly from an experiment that uses automated data acquisition. Once the data, from whatever source, are in your worksheet, there are some things you can do to make graphing quicker and easier. That is the subject of this section on getting ready to create a graph.

Excel attempts to analyze your data to automatically create a basic graph. You can assist the process by laying out your data in a standard form. A typical set of data for plotting might look something like Figure 3.1.

Excel is fairly flexible, but the typical data layout for an XY scatter graph, such as the temperature and time data in Figure 3.1, includes the following elements:

- The data to be plotted on the *x* axis (Time) are stored in a single column.
  - No blank rows exist between the first *x* value and the last *x* value, or between the column heading and the first *x* value.
- The data to be plotted on the *y* axis (Temperature) are stored in a single column to the right of the column of *x* values.
  - No blank rows exist between the first *y* value and the last *y* value.
  - The series name ("Temp. (°C)") may be included in the cell directly above the first *y* value. If the top cell in the *y* values column contains text rather than

Figure 3.1
A data set for graphing.

	A	B	C
1	**Temperature vs. Time Data**		
2			
3	Time (sec.)	Temp. (°C)	
4	0	54.23	
5	1	45.75	
6	2	28.41	
7	3	28.30	
8	4	26.45	
9	5	17.36	
10	6	17.64	
11	7	9.51	
12	8	5.76	
13	9	8.55	
14	10	6.58	
15	11	4.62	
16	12	2.73	
17	13	2.91	
18	14	0.32	
19	15	1.68	
20			

Wondering how the degree symbol got into cell B3?

- From the numeric keypad (not the numbers at the top of the keyboard), press [Alt-0176]; that is, hold down the [Alt] key while pressing 0176 on the numeric keypad, or
- Use Insert/Symbol, and select the degree symbol from the list of available symbols.

Figure 3.2
The temperature and time data in rows.

	A	B	C	D	E	F	G	H
1	Temperature vs. Time Data							
2								
3	Time (sec.)	0	1	2	3	4	5	6
4	Temp. (°C)	54.23	45.75	28.41	28.30	26.45	17.36	17.64
5								

a number, Excel will use that text as the name of the series and will include that name in the graph's legend and title (if these are displayed).

Keeping data in columns is the most common practice, but rows will also work. The temperature and time data set shown above would look like Figure 3.2, if it were stored in rows. (Only the first seven values of each row have been shown.)

By putting your data into a standard form, you make it possible for the logic programmed into Excel to recognize the structure in your data and assist in creating the graph; this makes creating graphs in Excel easier.

3.3 CREATING AN XY SCATTER GRAPH

The majority of graphs used by engineers are *XY scatter graphs*, so preparing this type of graph is covered here. Other types of graphs available in Excel are described in Section 3.6.

Once you have a block of data in the worksheet, it can be plotted by following these steps:

1. Select the Data Range (including the series names, if desired).
2. Use Ribbon options **Insert/Charts/Scatter** to select the graph type and create the graph.
   [In Excel 2003: Start the Chart Wizard and select the chart type.]

At this point, Excel 2010 will display a basic graph on the worksheet, but the graph is missing some key features, such as axis labels. Complete the graph by using formatting options on the Ribbon, with these steps:

3. Choose a Quick Layout (**Chart Tools/Design/Chart Layouts/Quick Layout**)
4. Edit the axis labels and graph title

Each of these steps will be explained in more detail in the following paragraphs.

Step 1. Select the data range.   An XY scatter graph is made up of a set of data points that each has an *x* value and a *y* value, which determine the location of the point on the graph. To create an XY scatter plot in Excel, first select the two columns (or rows) of data to be graphed, as shown in Figure 3.3.

To select the data to be plotted, simply click on the cell containing the first *x* value and hold the left mouse button down as you move the mouse pointer to the cell containing the last *y* value. In the temperature vs. time data shown here, cells A3 through B19 have been selected for graphing.

Including the column headings (cells A3:B3) with the selected data (A3:B19) is optional, but Excel will use the heading in the graph's legend if included.

Step 2. Select the type of graph to be inserted on the worksheet.   Use Ribbon options **Insert/Charts/Scatter** to open the Scatter drop-down graph type selector (shown in Figure 3.3 and Figure 3.4).

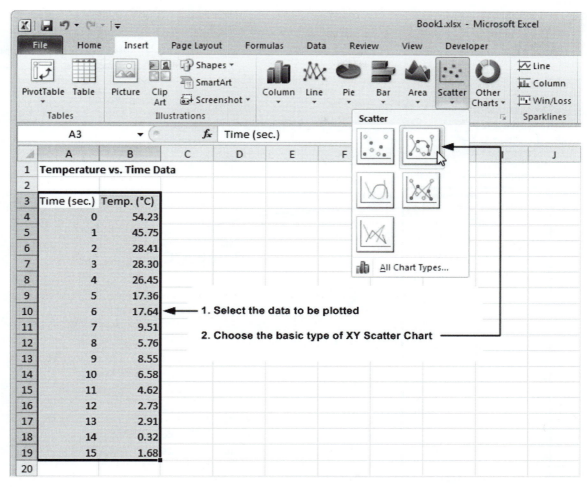

Figure 3.3
Select the data to be graphed, then choose the type of graph.

Excel provides five options for XY scatter graphs:

1. Markers (data points) only, no curves connecting the points in the *data series*.
2. Markers connected by *smoothed curves*. The curve will bend as needed to go through every data point.
3. No markers, just smoothed curves.
4. Markers connected by straight *line segments*.
5. No markers, just straight line segments.

When you choose one of the graph types (in Figure 3.3, markers connected with smoothed curves were selected), a basic graph of your selected data will be inserted on the worksheet, as shown in Figure 3.5.

Several important features have been indicated in Figure 3.5:

1. The basic graph (with markers connected with smoothed curves, as requested) has been inserted into the worksheet.
2. The data used to create the graph are indicated with colored borders. You can move these borders with the mouse to change the data range that appears on the graph.

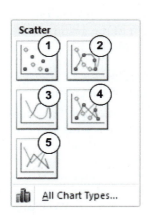

Figure 3.4
Scatter graph options.

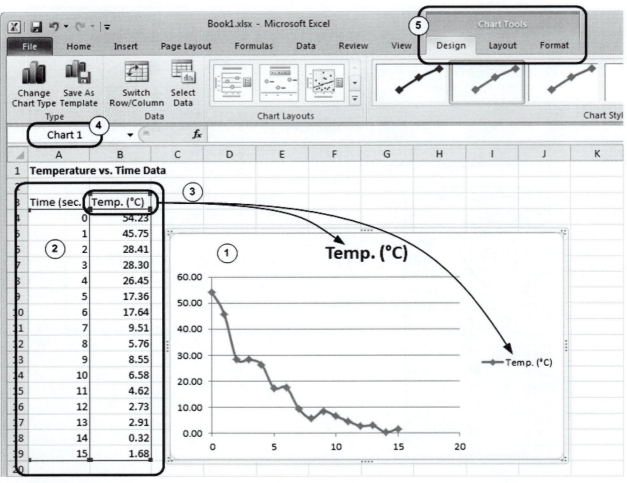

Figure 3.5
The basic graph inserted on the worksheet.

3. The column heading for the *y* values has been used as the *graph title* (above the graph) and appears on the *graph legend* (to the right of the graph).
4. The chart is an Excel *object* (this term is defined below) and has been given a name (Chart 1).
5. Three new tabs have appeared on the Ribbon. The **Chart Tools** tabs (**Design**, **Layout**, and **Format**) are used to modify the appearance of the graph.

An object, in a programming context, is an item that exists (has been created and stored in computer memory), has an identity (this graph is called "Chart 1"), and has properties that can be assigned values (e.g., diamond-shaped markers connected with smoothed curves, a location on the worksheet). The chart object has an associated data set (from cells A3:B19). Because the graph is a self-contained object, it can also be moved around, even between programs using copy and paste. We'll present moving graphs between programs (e.g., from Excel to Word) later in the chapter.

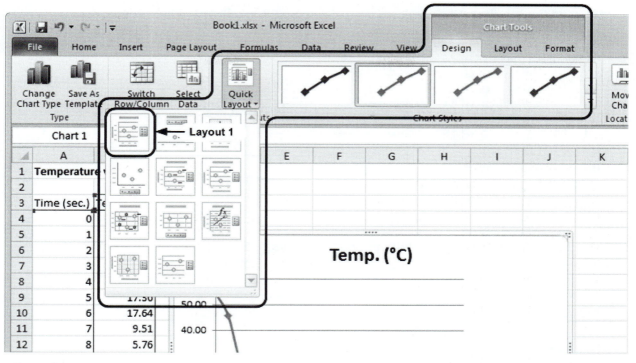

Figure 3.6
Select a graph layout.

Step 3. Choose a Quick Layout.   The basic graph shown in Figure 3.5 is missing axis labels, but this is easy to fix using the Chart Tools on the Ribbon. Use Ribbon options **Chart Tools/Design tab/Chart Layouts group/Quick Layout** to open the Quick Layouts selector panel, shown in Figure 3.6.

The Quick Layout selector panel packs a lot of information into a tiny space. For now we will simply use Layout 1 (selected in Figure 3.6), but the features of the available layouts are summarized later in this chapter. Layout 1 is adequate for most engineering graphs, and features can be activated or deactivated at any time.

When Layout 1 is selected, Excel applies the additional features (axis titles) to the graph, as shown in Figure 3.7.

Excel has added axis titles with the text "Axis Title"; something a little more descriptive is needed. We need to edit the axis titles.

Step 4. Edit the axis titles and graph title.   To edit the *axis titles*:

1. Make sure the graph is selected (a wide border is shown on a selected graph). Click on the graph to select it, if needed.
2. Click on the title to select it. A border around the title indicates that the title has been selected.
3. Select the text in the title (drag the mouse across it or triple-click on the text to select the entire text string).
4. Enter the desired text for the axis title.
5. Click outside the axis title, but inside the graph when you have finished editing the axis title.

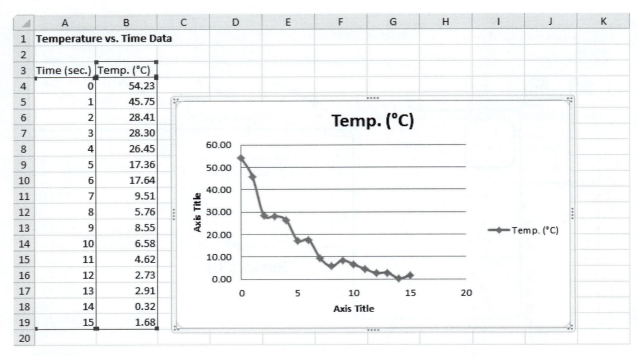

Figure 3.7
The XY scatter graph after Layout 1 has been applied.

The same process is used to edit the graph title. The result of editing the axis titles and graph title is shown in Figure 3.8.

The size and location of the graph are set using default values, but they can be changed with the mouse. A graph in edit mode is indicated by a thick border with groups of small dots (called *handles*) located around the border. Drag any of the handles to resize the graph and drag the borders between the handles to move the graph.

The colors used in the graph depend on the *theme* that has been applied to the workbook ("Office" theme by default). Themes are predefined collections of colors and font specifications. You can change the theme applied to your workbook using Ribbon options **Page Layout/ Themes (group)/Themes (button)** and then selecting a theme from the drop-down selector. Be aware that the theme impacts a lot more than just the selected graph; changing the theme will change the appearance (colors, font styles, font sizes) of everything in the workbook.

You can continue to edit the graph after it has been placed on the worksheet.

3.4 EDITING AN EXISTING GRAPH

Since Excel 2007, the Ribbon is the way to access and modify a graph's features. Whenever a graph has been selected, three **Chart Tools** tabs appear on the Ribbon. (If the graph is not selected, click anywhere on the graph to enter edit mode.) For now, each of the chart tabs will be presented briefly; then more information will be provided about how to use each tab in the next sections.

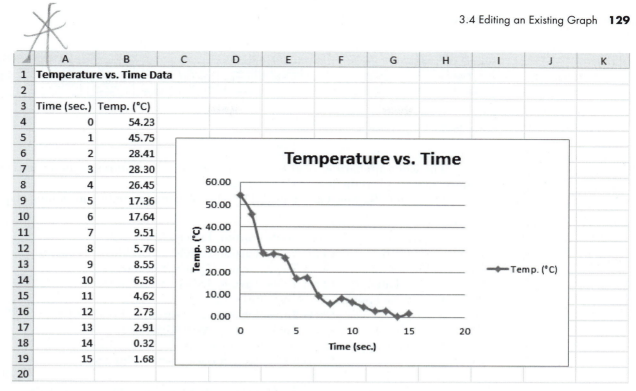

	A	B
1	Temperature vs. Time Data	
2		
3	Time (sec.)	Temp. (°C)
4	0	54.23
5	1	45.75
6	2	28.41
7	3	28.30
8	4	26.45
9	5	17.36
10	6	17.64
11	7	9.51
12	8	5.76
13	9	8.55
14	10	6.58
15	11	4.62
16	12	2.73
17	13	2.91
18	14	0.32
19	15	1.68
20		

Figure 3.8
The XY scatter graph after updating the axis titles.

Chart Tools/Design Tab (Figure 3.9)—this tab is primarily used when creating the graph. See Section 3.4.2 for more information on using this tab.

- Modify the way Excel is plotting your data
  - Change rows/columns   **Chart Tools/Design/Data/[Switch Row/Column]**
  - Select graph data   **Chart Tools/Design/Data/Select Data**
- Set or change the basic appearance of the graph
  - Choose a basic layout   **Chart Tools/Design/Chart Layouts/Quick Layouts**
  - Choose a chart style   **Chart Tools/Design/Chart Styles**

Under the **Design** tab, the **Chart Layouts** group provides quick access to 11 basic chart layouts. The layout selector that is displayed using the **Quick Layout** button (Ribbon options **Chart Tools/Design/Chart Layouts/Quick Layout**). The 11 layouts are displayed graphically as small chart icons, as illustrated in Figure 3.10.

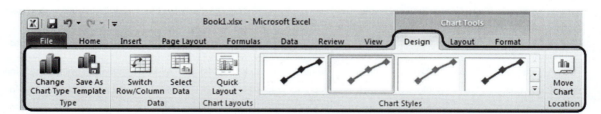

Figure 3.9
The **Chart Tools/Design** tab.

Figure 3.10
Available Quick Layouts for graphs (icons).

The icons are hard to read until you know the basic features of the various layouts. Table 3.1 lists the features of each layout.

Chart Tools/Layout Tab (Figure 3.11)—this tab is used to modify an existing graph.

- Add or modify labels
  - o Chart Title   **Chart Tools/Layout/Labels/Chart Title**
  - o Axis Titles   **Chart Tools/Layout/Labels/Axis Titles**
  - o Legend   **Chart Tools/Layout/Labels/Legend**
  - o Data Labels   **Chart Tools/Layout/Labels/Data Labels**
- Modify the appearance of axes and gridlines
  - o Axes   **Chart Tools/Layout/Axes (group)/Axes (button)**
  - o Gridlines   **Chart Tools/Layout/Axes/Gridlines**
- Modify the appearance of the plot area within the graph window
  - o Plot Area   **Chart Tools/Layout Background/Plot Area**
- Add trendlines and error bars
  - o Trendlines   **Chart Tools/Layout/Analysis/Trendline**
  - o Error bars   **Chart Tools/Layout/Analysis/Error Bars**
  -      **Chart Tools/Format Tab**
- Add or modify the graph's border and/or background
  - o Border/Background   **Chart Tools/Format/Shape Styles (selector)**

Table 3.1 Summary of Features of the Quick Layouts

Feature	Quick Layout										
	1	2	3	4	5	6	7	8	9	10	11
Graph Title	Top	Top	None	None	Top	Top	Top	Top	Top	None	None
X Axis Label	Yes	Yes	Yes	None	Yes	Yes	None	None	Yes	Yes	None
Y Axis Label	Yes	None	Yes	None	Yes	Yes	None	None	Yes	Yes	None
Legend	Right	Top	Right	Bottom	Right	Right	Right	Bottom	Right	Right	Right
X Axis Major Gridlines	No	No	Yes	No	No	No	Yes	No	No	Yes	No
X Axis Minor Gridlines	No	No	Yes	No	No	No	No	No	No	Yes	No
Y Axis Major Gridlines	Yes	No	Yes	No	Yes	Yes	Yes	Yes	Yes	Yes	Yes
Y Axis Minor Gridlines	No	No	Yes	No	No	No	No	No	Yes	Yes	No
Data Labels	None	All (x,y)	None	None	All (x,y)	All (x)	All (x,y)	None	None	None	None
Regression Line	No	No	Yes	No	No	No	No	No	Yes	No	No
Regression Equation	No	No	No	No	No	No	No	No	Yes	No	No
R^2 Value	No	No	No	No	No	No	No	No	Yes	No	No

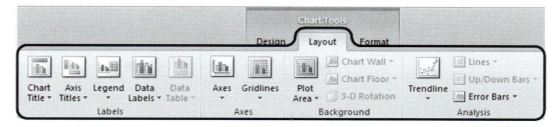

Figure 3.11
The **Chart Tools/Layout** tab (only a portion of this tab is shown here).

Note: If you are familiar with prior versions of Excel, you can still right-click on a feature and select "Format <feature>" from the pop-up menu or (in Excel 2010) double-click on the graph feature. This method opens the same dialog boxes that can be accessed from the Ribbon. But the Ribbon provides another quick way to access the dialogs:

- **Chart Tools/Layout/Current Selection/Format Selection**
- **Chart Tools/Format/Current Selection/Format Selection**

The **Current Selection** group is available from either the **Layout** tab or the **Format** tab (shown in Figure 3.12). If you click on the graph feature that you want to modify (or select the feature in the drop-down box in the **Current Selection** group) and then click the **Format Selection** button, the dialog box for that graph feature will open.

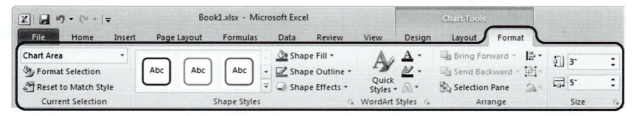

Figure 3.12
The **Chart Tools/Format** tab.

HINT

Once a Format dialog for a graph element is open, if you click on a different graph element Excel will automatically switch to the Format dialog for that graph element. For example, if you are changing the appearance of the x axis and want to also change the appearance of the y axis, don't close the Format x axis dialog. Instead, when you are done formatting the x axis, simply click on the y axis and Excel will display the Format y axis dialog.

3.4.1 Modifying the Appearance of the Plotted Curve

If you want to control the appearance of the plotted curve(s), Excel has been designed to have you choose a layout (**Chart Tools/Design/Chart Layouts/Quick Layouts**) and a style (**Chart Tools/Design/Chart Styles**) and that's it. Sometimes that's adequate; but often you need more control over the appearance of the plotted data. Here's how...

1. Click on the curve that you wish to modify to select it. (The one curve in Figure 3.13 has been selected.) The markers are highlighted to indicate that the curve has been selected.

   Note: Be careful to click only once on the curve; clicking once selects the entire curve, clicking a second time selects a single marker on the curve. (This allows you to change the appearance of a single marker if you wish to highlight one point on the curve.)

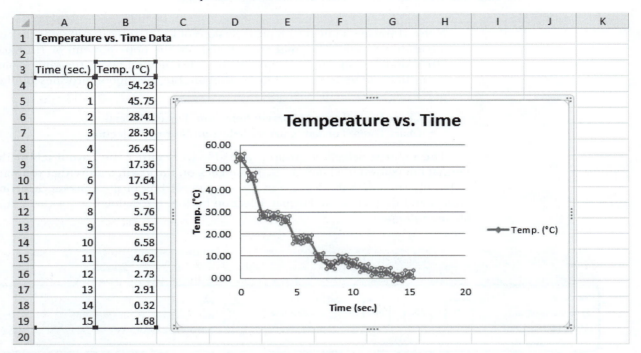

Figure 3.13
The graph with the plotted data curve selected.

2. Use one of the following (equivalent) Ribbon options:
   ○ **Chart Tools/Layout/Current Selection/Format Selection**
   ○ **Chart Tools/Format/Current Selection/Format Selection**
   Clicking the **Format Selection** button when the curve is selected will open the Format Data Series dialog shown in Figure 3.14.

   Note: In Excel 2010 (and Excel 2003, but not Excel 2007), you can simply double-click on the curve to open the Format Data Series dialog.

3. Choose the panel needed to make the desired change. The **Marker Options** panel is shown in Figure 3.14. This panel allows you to set the type and size of the marker—or choose **None** to deactivate markers altogether.
   The Format Data Series dialog panels listed below allow you to make the following common adjustments to the plotted data curves:
   ○ *Series Options:* Choose to plot the series on the primary (left) or secondary (right) *y* axis. This option is not available unless there are at least two curves on the graph.

Figure 3.14
The Format Data Series
dialog, **Marker Options**
panel.

o *Marker Options:* Set marker type and size.
o *Marker Fill:* Allows you to create filled and unfilled markers (e.g., ◆, ◇).
o *Line Color:* Set the line color and activate and deactivate the line between the markers.
o *Line Style:* Set the line width, solid or dashed style, activate or deactivate smoothing of the connecting lines.
o *Marker Line Color:* Set the color of the line used to draw the border of the marker. This is most useful for unfilled markers.
o *Marker Line Style:* Set the width of the line used to draw the border of the marker.

4. Once you have made the desired changes, **Close** the Format Data Series dialog. In Figure 3.15, the Format Data Series dialog was used to create large, unfilled markers without connecting lines.

3.4.2 Adding a Second Curve to the Plot

If there is another column of data to plot on the same graph, such as the predicted temperature values shown in column C in Figure 3.15, the easiest way to add a second curve to the plot is to:

1. Click on the graph to select it. Be sure not to click on the plotted data points; you need the graph selected, not the original data series. When the graph is selected, the data used to create the graph are surrounded by colored borders, as shown in Figure 3.16.
2. Drag the corner (handle) of the box containing the *y* values over one column so that both columns of *y* values are included in the box. This is illustrated in Figure 3.17.

Expanding the border around the *y* values is all that is required to add a second curve to the graph, but there's one catch: This method works only when the two columns of *y* values share the same set of *x* values. If the two sets of *y* values correspond to different sets of *x* values, a different approach must be used.

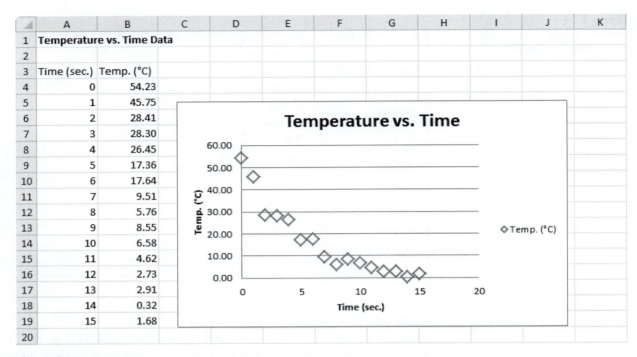

Figure 3.15
The modified graph.

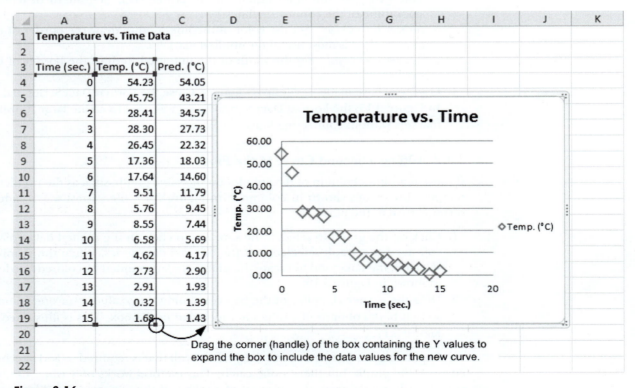

Drag the corner (handle) of the box containing the Y values to
expand the box to include the data values for the new curve.

Figure 3.16
The selected graph, ready to add an additional curve.

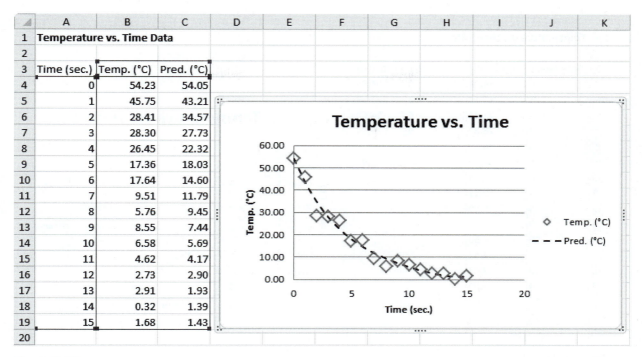

Figure 3.17
The second data series has been added to the graph.

The dashed line used for the predicted temperature values in Figure 3.17 is not Excel's default for the second curve on a graph. The appearance of the second data series was modified using the Format Data Series dialog as described in the previous section.

Adding a Second Data Series with Different x Values

It is not uncommon to have two related data sets that do not share the same *x* values. This happens, for example, when you repeat an experiment but don't record the data values at exactly the same times in each experiment. Figure 3.18 shows two temperature vs. time data sets, but the columns of times are quite different.

The process of graphing the two sets of data begins as before: the first data series is graphed by itself (as shown in Figure 3.18). Then, to add the second data series:

1. Use Ribbon options **Chart Tools/Design/Data/Select Data**. This opens the Select Data Source dialog, as shown in Figure 3.19.
2. The "Temp. (°C)" shown just below the **Add** button in Figure 3.19 is the name of the data series that is already plotted. We need to add the new data series to the graph. To do so, click the **Add** button to open the Edit Series dialog, as shown in Figure 3.20.
3. The Edit Series dialog needs three pieces of information in order to create the new data series: the series name (used in the legend, optional), the cell range containing the *x* values, and the cell range containing the *y* values.

    The small button at the right side of the Series name field (indicated in Figure 3.20) is supposed to look like an arrow pointing to a worksheet; clicking

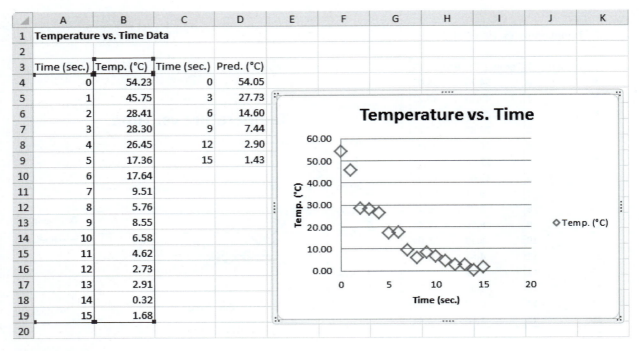

	A	B	C	D	E	F	G	H	I	J	K
1	**Temperature vs. Time Data**										
2											
3	Time (sec.)	Temp. (°C)	Time (sec.)	Pred. (°C)							
4	0	54.23	0	54.05							
5	1	45.75	3	27.73							
6	2	28.41	6	14.60							
7	3	28.30	9	7.44							
8	4	26.45	12	2.90							
9	5	17.36	15	1.43							
10	6	17.64									
11	7	9.51									
12	8	5.76									
13	9	8.55									
14	10	6.58									
15	11	4.62									
16	12	2.73									
17	13	2.91									
18	14	0.32									
19	15	1.68									
20											

Figure 3.18
Two temperature and time data sets.

Figure 3.19
The Select Data Source
dialog showing only one
data series.

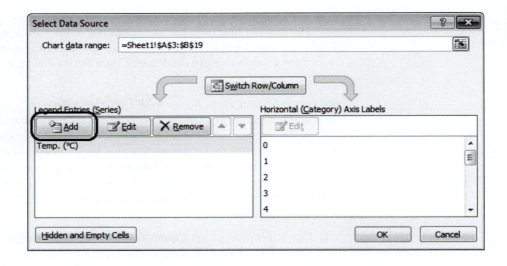

this button allows you to jump to the worksheet so that you can use the mouse
to select the cell containing the new series name (cell D3, see Figure 3.18).
When you do jump to the worksheet, yet another small Edit Series dialog opens,
as shown in Figure 3.21.

Wherever you click on the worksheet, that cell address appears in the Edit
Series dialog. In Figure 3.21, cell D3 was selected since that cell contains the
name of the new data series.

Figure 3.20
The Edit Series dialog.

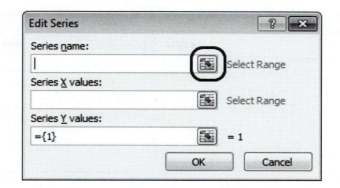

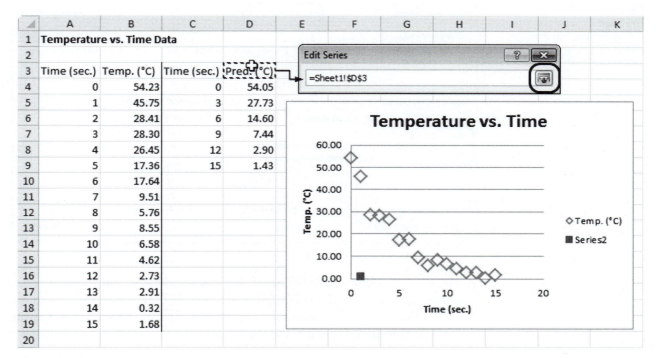

Figure 3.21
Pointing out the cell containing the new series name.

4. There is, again, a small button at the right side of the data entry field, indicated in Figure 3.21. This button will return back to the larger Edit Series dialog (shown in Figures 3.20 and 3.22). When you return to the larger Edit Series dialog, the information (cell address) gathered from the worksheet is automatically entered into the dialog.

5. The process of jumping to the worksheet is repeated two more times to point out (1) the cell range containing the new *x* values (cells C4:C9), and (2) the cell range containing the new *y* values (cells D4:D9). All of the necessary information has been collected in the Edit Series dialog shown in Figure 3.22.

6. Next, click **OK** to add the new data series to the Select Data Source dialog, as shown in Figure 3.23.

Figure 3.22
The Edit Series dialog (with
information needed to
create the new data series).

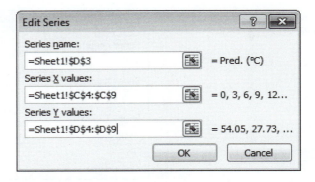

Figure 3.23
The Select Data Source
dialog showing two data
series.

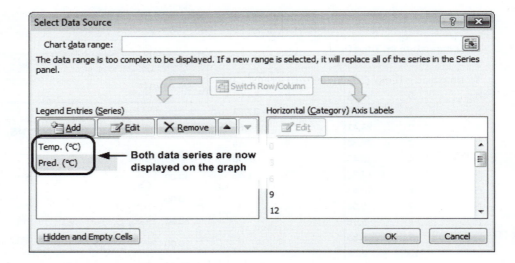

7. Finally, click **OK** to close the Select Data Source dialog and display the new curve on the graph. The result is shown in Figure 3.24. Again, the appearance of the new curve was modified using Ribbon options **Chart Tools/Layout/Current Selection/Format Selection** to access the Format Data Series dialog.

3.4.3 Changing the Appearance of the Plot using the Layout Tab

Once you have a graph with one or more curves (data series) displayed, you can use the **Layout** tab under **Chart Tools** to modify the appearance of the graph (Figure 3.25).

Changing Labels

The labels group (**Chart Tools/Layout/Labels**) is used primarily to activate or deactivate the various labels on a graph. If the labels are displayed, you do not need the Ribbon to edit the text displayed in the labels, simply:

1. Make sure the graph is selected.
2. Click on the label to select it.
3. Select the text in the label (drag the mouse across it or triple-click on the text to select the entire text string).
4. Enter the desired text for the label.
5. Click outside the label, but inside the graph when you have finished editing.

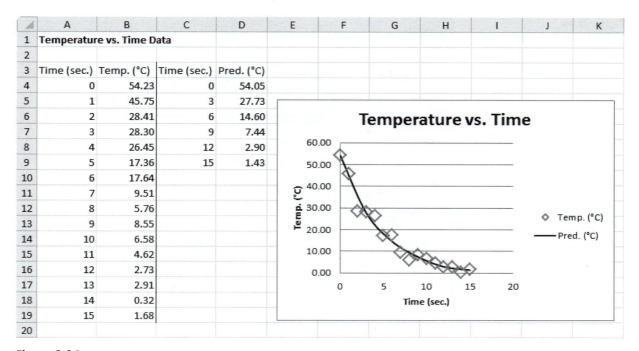

Figure 3.24
The graph with two data series plotted.

Figure 3.25
The **Chart Tools/ Layout** tab.

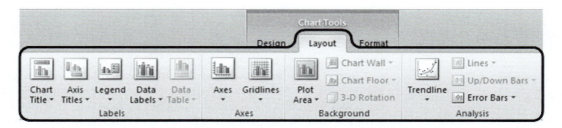

The exception to this is the legend. If the series names in the legend are coming from cells (such as cells B3 and D3 in Figure 3.24), then you cannot edit the legend directly. To change the legend, edit the cells that contain the series names and the legend will automatically be updated to reflect the changes.

Modify the Appearance of Axes and Gridlines

There are several ways to open the Format Axis dialog to make changes to the appearance of an axis or gridlines:

- Use Ribbon options **Chart Tools/Layout/Axes (group)/Axes (button)**, then select the desired axis to modify (primary or secondary, horizontal or vertical), then select **More Axis Options...**
- Click on the axis you want to modify to select it (click on the numbers, not the axis line), and then use Ribbon options **Chart Tools/Layout (or Format)/ Current Selection/Format Selection**.
- Right-click on the axis you want to modify (right-click on the numbers, not the axis line) and choose **Format Axis...** from the pop-up menu.
- In Excel 2010, double-click on the axis that you want to modify.

Figure 3.26
The Format Axis dialog.

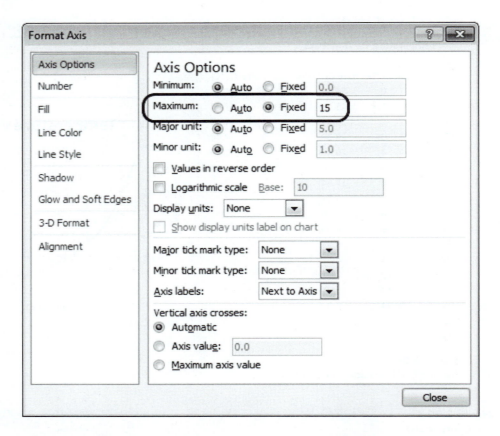

Whichever method you prefer, the Format Axis dialog shown in Figure 3.26 will open.

Again, there are a number of panels available on the Format Axis dialog. In Figure 3.26, the **Axis Options** panel has been used to change the **Maximum** value on the x axis from 20 (the autoscale value set by Excel) to 15, since 15 is the maximum x (time) value in the data sets. The result of this change is shown in Figure 3.27; the data fit the graph better with less wasted space.

To display vertical *gridlines*, use Ribbon options **Chart Tools/Layout/Axes/ Gridlines**, then choose **Primary Vertical Gridlines/Major Gridlines**. The graph with both horizontal and vertical gridlines is shown in Figure 3.28.

Figure 3.27
The result of rescaling the *x* axis.

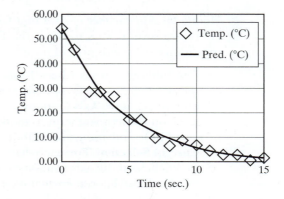

Figure 3.28
The graph after activating major vertical gridlines.

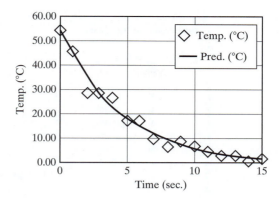

3.4.4 Adding a trendline to a graph

A *trendline* is a curve that goes through the data points (not through each point) to highlight the trend in the data. Most, but not all, of the trendlines available in Excel are best-fit regression lines (see Chapter 6 for more information on linear regression). Specifically, Excel provides the following trendline options:

- Linear (straight line)
- Exponential
- Logarithmic
- Polynomial
- Power
- Moving Average (this is the only nonregression trendline)

Some options may not be available for some data sets if the math involved is not valid for all values in the set. For example, ln(0) is not defined, so logarithmic fits are not possible for data sets containing *y* values of zero. Excel will display an error message if the trendline cannot be calculated.

Because the trendlines (except for the moving average) are best-fit regression lines, the *equations of the trendlines* and the R^2 *values* (indicating the "goodness of the fit") are available and can be displayed on the graph.

The quickest way to add a trendline to a graph is to right-click on the data series that the trendline should be fit to, and then choose **Add Trendline** from the series' pop-up menu (illustrated in Figure 3.29). (To simplify the graph, the predicted temperature values have been eliminated in Figure 3.29.)

Ribbon options could also be used to add a trendline. Use **Chart Tools/Layout/ Analysis/Trendline**, then select More Trendline Options… If more than one curve has been plotted, you will be asked to select the data series to fit with a trendline.

Whether you right-click the data set, or use the Ribbon, the Format Trendline dialog will open, as shown in Figure 3.30.

On the Format Trendline dialog, we have requested an exponential fit to the temperature data and have asked that the equation of the trendline and the R^2 value be printed on the graph. The result is shown in Figure 3.31.

In Figure 3.31, the equation of the trendline and R^2 value were relocated and increased in size to improve readability. The exponential curve seems to go through the data values fairly well, but the R^2 value of 0.876 isn't that great. We can try a different type of trendline, say a polynomial, to see if we can get a better fit. The result of applying a fourth-order polynomial trendline is shown in Figure 3.32.

Figure 3.29
Adding a trendline
using the data series'
pop-up menu.

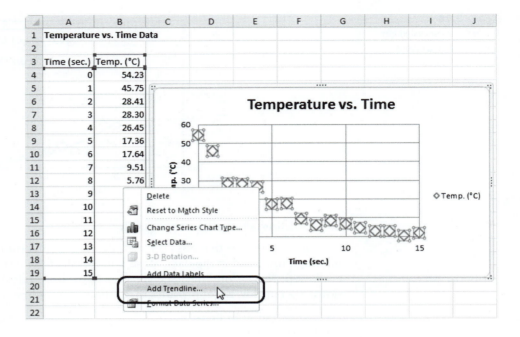

Figure 3.30
The Format Trendline
dialog.

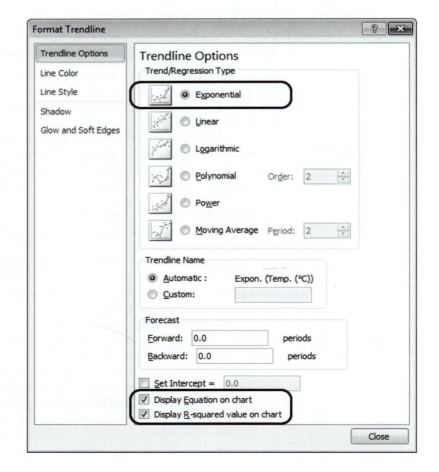

Figure 3.31

The graph with the exponential trendline, trendline equation, and R^2 value.

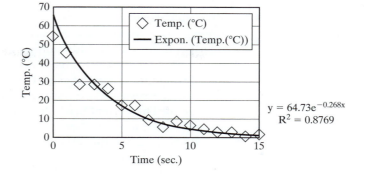

$y = 64.73e^{-0.268x}$
$R^2 = 0.8769$

Figure 3.32

The graph with the fourth-order polynomial trendline, trendline equation, and R^2 value.

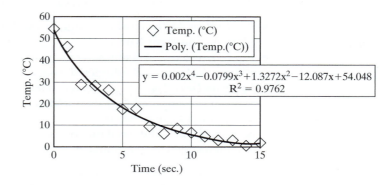

$y = 0.002x^4 - 0.0799x^3 + 1.3272x^2 - 12.087x + 54.048$
$R^2 = 0.9762$

The R^2 value is closer to 1 (1 is a perfect fit), so the fourth-order polynomial trendline is doing a better job of fitting the data than the exponential trendline did. But be careful not to infer too much from the trendline because this data is imperfect. A very high-order polynomial could probably wiggle through every data point and give an R^2 value of 1, but that won't help us interpret the data. It is OK to use trendlines to show trends in the data, but don't forget to go back to the theory behind the experiment to seek a fitting equation that makes sense. Trendlines often get used because they are easy, but the trendlines available in Excel may not be appropriate for analyzing every data set.

3.4.5 Adding Error Bars

NOTE

With Excel 2007, a bug appeared in Excel's handling of error bars, and it is still present in Excel 2010. When you use the Ribbon to open the error bar dialog, only the vertical error bars are available for editing on the Format Error Bars dialog, and Excel automatically adds horizontal error bars whether you want them or not! Fortunately there is a workaround for this:

- If you want to edit vertical error bars, use the expected Ribbon commands (**Chart Tools/Layout/Analysis/Error Bars**, then choose **More Error Bars Options...**). The Format Error Bars dialog will work correctly to allow you to generate vertical error bars. However, Excel will add horizontal error bars whether you want them or not, and even if you want horizontal error bars, they may not be calculated correctly for your situation. But we can make this work:

- ○ If you do not want horizontal error bars, click on any one of the horizontal error bars, then press the [Delete] key—problem solved.
- ○ If you want horizontal error bars, right-click on any one of the horizontal error bars and select **Format Error Bars...** from the pop-up menu. The Format Error Bars dialog will open and allow you to calculate the horizontal error bars using any of the available methods.
- ○ If you do not want vertical error bars, click on any one of the vertical error bars, then press the [Delete] key.

Excel can automatically add *x*- or *y-error bars*, calculated from the data in several ways:

- *Fixed Value:* A fixed value is added to and subtracted from each data value and plotted as an error bar.
- *Fixed Percentage:* A fixed percentage is multiplied by each data value, and the result is added to and subtracted from the data value and plotted as an error bar.
- *Standard Deviation:* The selected standard deviation of all of the values in the data set is computed and then added to and subtracted from each data value and plotted as an error bar.
- *Standard Error:* The standard error of all of the values in the data set is computed and then added to and subtracted from each data value and plotted as an error bar.

Additionally, you can calculate the values you want plotted as error bars by using the Custom error bar option.

To demonstrate how to add error bars to a graph, fixed-percentage (30%) error bars will be added to the temperature values (*y* values) shown in Figure 3.33.

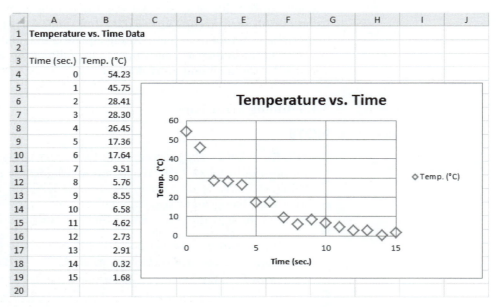

Figure 3.33
Temperature vs. time data.

Figure 3.34
The Format Error Bars
dialog.

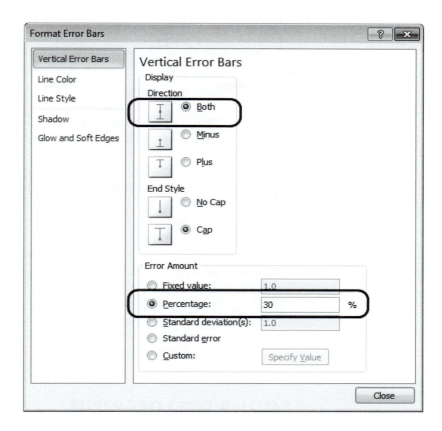

To bring up the Format Error Bars dialog, click on the graph to select it, then use Ribbon options **Chart Tools/Layout/Analysis/Error Bars**, then choose More Error Bar Options... from the menu. The Format Error Bars dialog will open as shown in Figure 3.34.

Select the **Both** option to display error bars both above and below the data marker and set the **Error Amount Percentage** to 30% as indicated in Figure 3.34. The graph with the vertical 30% error bars is shown in Figure 3.35.

Figure 3.35
The temperature vs. time
graph with 30% error bars.

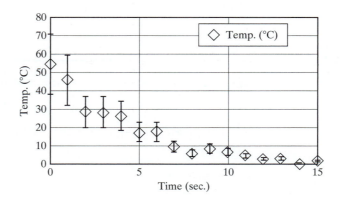

3.5 PRINTING THE GRAPH

There are two ways to get a graph onto paper:

1. Print the graph only.
2. Print the worksheet containing the graph.

Method 1. **Printing only the graph.** First, click the graph to select it. A border will be displayed around the edge of the graph to show that it has been selected. Then use:

- Excel 2010: **File tab/Print (panel)/Print (button)**
- Excel 2007: **Office/Print/Print (or Quick Print, or Print Preview)**
- Excel 2003: **File/Print**

The graph will, by default, be resized to fit the page. If you use Print Preview (Excel 2010 automatically includes a preview on the **Print** panel), you can change the margins to change the size of the printed graph.

Method 2. **Printing the worksheet containing the graph.** Include the graph when you set the print area with Ribbon options **Page Layout/Page Setup/Print Area/Set Print Area** [Excel 2003: File/Print Area/Set Print Area]. Then use:

- Excel 2010: **File tab/Print (panel)/Print (button)**
- Excel 2007: **Office/Print/Print (or Quick Print, or Print Preview)**
- Excel 2003: **File/Print**

The graph will be printed with the specified worksheet print area.

3.6 OTHER TYPES OF GRAPHS

Most of this chapter has focused on XY scatter graphs, the most common type of graph for engineering work; however, engineers use other types of graphs too. Other standard types include the following:

- Line graphs
- Column and bar graphs
- Pie charts
- Surface plots

CAUTION: A common error is to use a line graph when an XY scatter graph is needed. You will get away using a line graph if the *x* values in your data set are uniformly spaced. But if the *x* values are not uniformly spaced, your curve will appear distorted on the line graph. The following example is intended to illustrate how a line graph will misrepresent your data if your *x* values are not uniformly spaced.

EXAMPLE 3.2

The *x* values in the worksheet shown in Figure 3.33 are calculated as

$$x_{i+1} = 1.2 \cdot x_i \tag{3.1}$$

This creates nonuniformly spaced *x* values. The *y* values are calculated from the *x* values as

$$y_i = 3 \cdot x_i \tag{3.2}$$

so that the relationship between *x* and *y* is linear. On the XY scatter graph shown in Figure 3.36, the linear relationship is evident in the straight-line relationship between *x* and *y*.

However, when the same data are plotted on a line graph (shown in Figure 3.37), the relationship between *x* and *y* appears to be nonlinear.

The apparent nonlinear relationship between *x* and *y* in the line graph is an artifact of the line graph, because the actual *x* values were not used to plot the points on the graph. Excel's line graph simply causes the *y* values to be distributed evenly across the chart (effectively assuming uniform *x* spacing), and the *x* values (if included when the graph is created) are simply used as labels on the *x* axis. This can be misleading on a line graph, because the *x* values are displayed on the *x* axis, but they were not used to position the data points.

Figure 3.36
The data plotted on and XY scatter graph.

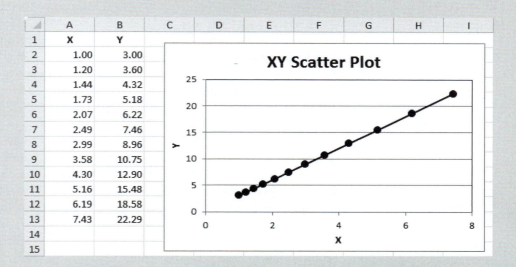

Figure 3.37
The same data plotted as a line graph.

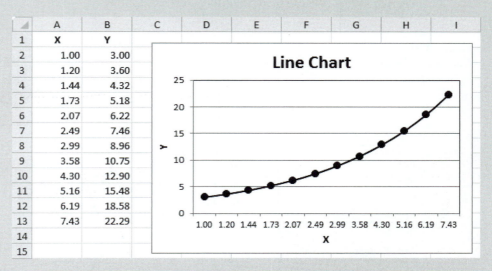

(continued)

Excel's line, column, bar, and pie charts all require only y values to create the graph. If you provide two columns (or rows) of data, the left column (or top row) will be used as labels for the graph. If you need to plot x and y values on a graph, you must use an XY scatter plot.

Surface Plots

A surface plot takes the values in a two-dimensional range of cells and displays the values graphically.

EXAMPLE 3.3

The surface plot in Figure 3.38 shows the value of $F(x,y) = \sin(x)\cos(y)$ for $-1 \le x \le 2$ and $-1 \le y \le 2$.

To create the plot, x values ranging from -1 to 2 were entered in column A, and y values ranging from -1 to 2 were entered in row 2, as shown in Figure 3.39. To create these series, the first two values were entered by hand, and then the Fill Handle was used to complete the rest of each series.

The first value for $F(x,y)$ is calculated in cell B3 using the formula =SIN($A3)*COS(B$2), as shown in Figure 3.40.

The dollar signs on the A in SIN($A3) and the 2 in COS(B$2) allow the formula to be copied to the other cells in the range (both x and y directions). When the formula is copied, the **SIN** functions will always reference x values in column A, and **COS** functions will always reference y values in row 3.

Next, the formula in cell B3 is copied to all of the cells in the range B3:Q18, to fill the two-dimensional array (Figure 3.41).

Then the array of values is selected before inserting a surface graph using Ribbon options **Insert/Charts/Other Charts** and selecting the **Wire-frame** icon in the **Surface** category (illustrated in Figure 3.42).

Figure 3.38
Surface plot of F(x,y) = sin(x) cos(y).

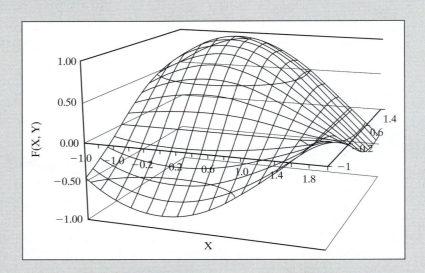

	A	B	C	D	E	F	G	H	I	J	K	L	M	N	O	P	Q	R
1		Y >>>																
2	X	-1.0	-0.8	-0.6	-0.4	-0.2	0.0	0.2	0.4	0.6	0.8	1.0	1.2	1.4	1.6	1.8	2.0	
3	-1.0																	
4	-0.8																	
5	-0.6																	
6	-0.4																	
7	-0.2																	
8	0.0																	
9	0.2																	
10	0.4																	
11	0.6																	
12	0.8																	
13	1.0																	
14	1.2																	
15	1.4																	
16	1.6																	
17	1.8																	
18	2.0																	
19																		

Figure 3.39
x and y values.

B3 f_x =SIN($A3)*COS(B$2)

	A	B	C	D	E	F	G	H	I	J	K	L	M	N	O	P	Q	R
1		Y >>>																
2	X	-1.0	-0.8	-0.6	-0.4	-0.2	0.0	0.2	0.4	0.6	0.8	1.0	1.2	1.4	1.6	1.8	2.0	
3	-1.0	-0.45																
4	-0.8																	
5	-0.6																	
6	-0.4																	
7	-0.2																	
8	0.0																	
9	0.2																	
10	0.4																	
11	0.6																	
12	0.8																	
13	1.0																	
14	1.2																	
15	1.4																	
16	1.6																	
17	1.8																	
18	2.0																	
19																		

Figure 3.40
The first calculated cell, cell B3.

(continued)

X	-1.0	-0.8	-0.6	-0.4	-0.2	0.0	0.2	0.4	0.6	0.8	1.0	1.2	1.4	1.6	1.8	2.0
-1.0	-0.45	-0.59	-0.69	-0.78	-0.82	-0.84	-0.82	-0.78	-0.69	-0.59	-0.45	-0.30	-0.14	0.02	0.19	0.35
-0.8	-0.39	-0.50	-0.59	-0.66	-0.70	-0.72	-0.70	-0.66	-0.59	-0.50	-0.39	-0.26	-0.12	0.02	0.16	0.30
-0.6	-0.31	-0.39	-0.47	-0.52	-0.55	-0.56	-0.55	-0.52	-0.47	-0.39	-0.31	-0.20	-0.10	0.02	0.13	0.23
-0.4	-0.21	-0.27	-0.32	-0.36	-0.38	-0.39	-0.38	-0.36	-0.32	-0.27	-0.21	-0.14	-0.07	0.01	0.09	0.16
-0.2	-0.11	-0.14	-0.16	-0.18	-0.19	-0.20	-0.19	-0.18	-0.16	-0.14	-0.11	-0.07	-0.03	0.01	0.05	0.08
0.0	0.00	0.00	0.00	0.00	0.00	0.00	0.00	0.00	0.00	0.00	0.00	0.00	0.00	0.00	0.00	0.00
0.2	0.11	0.14	0.16	0.18	0.19	0.20	0.19	0.18	0.16	0.14	0.11	0.07	0.03	-0.01	-0.05	-0.08
0.4	0.21	0.27	0.32	0.36	0.38	0.39	0.38	0.36	0.32	0.27	0.21	0.14	0.07	-0.01	-0.09	-0.16
0.6	0.31	0.39	0.47	0.52	0.55	0.56	0.55	0.52	0.47	0.39	0.31	0.20	0.10	-0.02	-0.13	-0.23
0.8	0.39	0.50	0.59	0.66	0.70	0.72	0.70	0.66	0.59	0.50	0.39	0.26	0.12	-0.02	-0.16	-0.30
1.0	0.45	0.59	0.69	0.78	0.82	0.84	0.82	0.78	0.69	0.59	0.45	0.30	0.14	-0.02	-0.19	-0.35
1.2	0.50	0.65	0.77	0.86	0.91	0.93	0.91	0.86	0.77	0.65	0.50	0.34	0.16	-0.03	-0.21	-0.39
1.4	0.53	0.69	0.81	0.91	0.97	0.99	0.97	0.91	0.81	0.69	0.53	0.36	0.17	-0.03	-0.22	-0.41
1.6	0.54	0.70	0.82	0.92	0.98	1.00	0.98	0.92	0.82	0.70	0.54	0.36	0.17	-0.03	-0.23	-0.42
1.8	0.53	0.68	0.80	0.90	0.95	0.97	0.95	0.90	0.80	0.68	0.53	0.35	0.17	-0.03	-0.22	-0.41
2.0	0.49	0.63	0.75	0.84	0.89	0.91	0.89	0.84	0.75	0.63	0.49	0.33	0.15	-0.03	-0.21	-0.38

Figure 3.41
Cell B3 has been copied to cells B3:Q18.

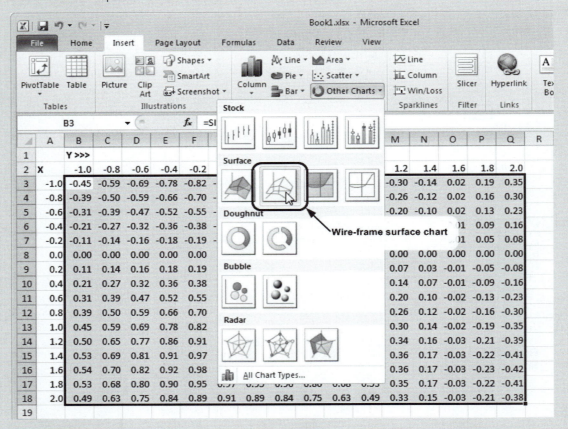

Figure 3.42
Inserting a surface graph into the worksheet.

The surface graph is inserted into a window on top of the data (see Figure 3.43), but it can be moved.

The basic graph has been inserted, but some better labels will help people understand what they are seeing. First, we add three axis labels and a title, as shown in Figure 3.44. This was accomplished from the Ribbon using **Chart Tools/Layout/Labels/Chart Title** and **Chart Tools/Layout/Labels/Axis Titles**.

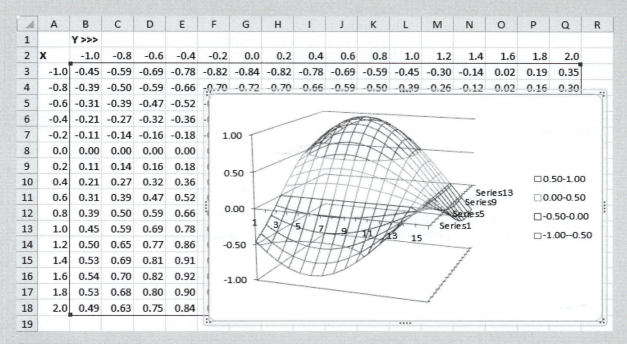

	A	B	C	D	E	F	G	H	I	J	K	L	M	N	O	P	Q	R
1		Y >>>																
2	X	-1.0	-0.8	-0.6	-0.4	-0.2	0.0	0.2	0.4	0.6	0.8	1.0	1.2	1.4	1.6	1.8	2.0	
3	-1.0	-0.45	-0.59	-0.69	-0.78	-0.82	-0.84	-0.82	-0.78	-0.69	-0.59	-0.45	-0.30	-0.14	0.02	0.19	0.35	
4	-0.8	-0.39	-0.50	-0.59	-0.66	-0.70	-0.72	-0.70	-0.66	-0.59	-0.50	-0.39	-0.26	-0.12	0.02	0.16	0.30	
5	-0.6	-0.31	-0.39	-0.47	-0.52													
6	-0.4	-0.21	-0.27	-0.32	-0.36													
7	-0.2	-0.11	-0.14	-0.16	-0.18													
8	0.0	0.00	0.00	0.00	0.00													
9	0.2	0.11	0.14	0.16	0.18													
10	0.4	0.21	0.27	0.32	0.36													
11	0.6	0.31	0.39	0.47	0.52													
12	0.8	0.39	0.50	0.59	0.66													
13	1.0	0.45	0.59	0.69	0.78													
14	1.2	0.50	0.65	0.77	0.86													
15	1.4	0.53	0.69	0.81	0.91													
16	1.6	0.54	0.70	0.82	0.92													
17	1.8	0.53	0.68	0.80	0.90													
18	2.0	0.49	0.63	0.75	0.84													
19																		

Figure 3.43
The surface graph inserted into the worksheet.

Figure 3.44
The surface graph with axis labels and title.

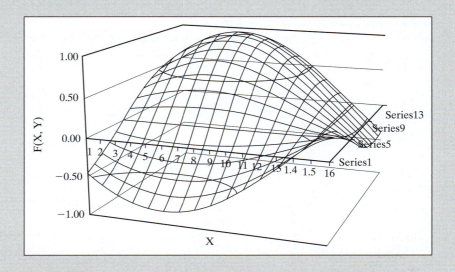

Next, we want to replace the labels on the x axis (1 ... 16) and the y axis (Series1 ... Series13) with something more meaningful. The x and y values both range from −1 to 2, so we want to get these values into the labels on the x and y axes.

Opening the Select Data Source dialog (**Chart Tools/Design/Data/Select Data**), we can quickly see where the labels on the x and y axes are coming from (Figure 3.45).

The x and y axis labels are coming from the series names (y axis) and horizontal axis labels (x axis) that Excel created when the surface graph was created. We need to use the **Edit** buttons located just above each set of labels to change the series names (one at a time) and horizontal axis labels (all at once) to more meaningful values. The result is shown in Figure 3.46.

And the updated graph is shown in Figure 3.47.

Figure 3.45
The Select Data Source dialog.

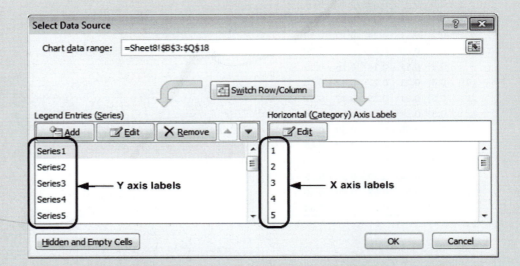

Figure 3.46
The Select Data Source dialog with updated x and y axis labels.

Figure 3.47
The surface plot with updated x and y axis labels.

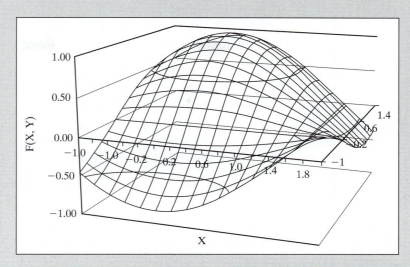

Note that the *x* and *y* values on the worksheet (column A and row 2) were never used to plot the points on the surface plot. The F(*x,y*) values in cells B3 through Q18 were plotted with uniform spacing in the *x* and *y* directions. This is a significant limitation of surface plotting in Excel.

APPLICATIONS

MATERIALS TESTING

Stress–Strain Curve I

Strength testing of materials often involves a *tensile test* in which a sample of the material is held between two mandrels while increasing force—actually, *stress* (i.e., force per unit area)—is applied. A stress vs. strain curve for a typical ductile material is shown in Figure 3.48.

During the test, the sample first stretches reversibly (A to B). Then irreversible stretching occurs (B to D). Finally, the sample breaks (point D).

Point C is called the material's *ultimate stress*, or *tensile strength*, and represents the greatest stress that the material can endure (with deformation) before coming

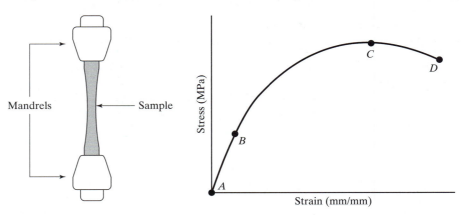

Figure 3.48
Tensile test.

apart. The *strain* is the amount of elongation of the sample (mm) divided by the original sample length (mm).

The reversible stretching portion of the curve (A to B) is linear, and the proportionality constant relating stress and strain in this region is called *Young's modulus*, or the *modulus of elasticity*.

Tensile test data on a soft, ductile sample are listed in Table 3.2 (available electronically at http://www.chbe.montana.edu/Excel).

To analyze the data we will want to:

1. Plot the tensile test data as a stress vs. strain graph.
2. Evaluate the tensile strength from the graph (or the data set).
3. Create a second graph containing only the elastic-stretch (linear) portion of the data.
4. Add a linear trendline to the new plot to determine the modulus of elasticity for this material.

First, the data are entered into the worksheet (Figure 3.49). Then, an XY scatter graph is prepared, as shown in Figure 3.50. The ultimate tensile stress can be read from the graph or the data set, as shown in Figure 3.50. Finally, another graph (Figure 3.51) is prepared, containing only the linear portion of the data (the first eight data points).

A linear trendline with the intercept forced through the origin has been added to the graph in Figure 3.48. We will use the slope from the equation for the trendline to compute the modulus of elasticity; the R^2 value, 1, provides reassurance that we have indeed plotted the linear portion of the test data.

From slope of the regression line, we see that the modulus of elasticity for this material is 1793 MPa, or 1.79 GPa.

Table 3.2 Tensile test data

Strain (mm/mm)	Stress (MPa)
0.000	0.00
0.003	5.38
0.006	10.76
0.009	16.14
0.012	21.52
0.014	25.11
0.017	30.49
0.020	33.34
0.035	44.79
0.052	53.29
0.079	57.08
0.124	59.79
0.167	60.10
0.212	59.58
0.264	57.50
0.300	55.42

	A	B	C	D
1	**Stress-Strain Curve I**			
2				
3		**Strain (mm/mm)**	**Stress (MPa)**	
4		0.000	0.00	
5		0.003	5.38	
6		0.006	10.76	
7		0.009	16.14	
8		0.012	21.52	
9		0.014	25.11	
10		0.017	30.49	
11		0.020	33.34	
12		0.035	44.79	
13		0.052	53.29	
14		0.079	57.08	
15		0.124	59.79	
16		0.167	60.10	
17		0.212	59.58	
18		0.264	57.50	
19		0.300	55.42	
20				

Figure 3.49
Strain and stress data.

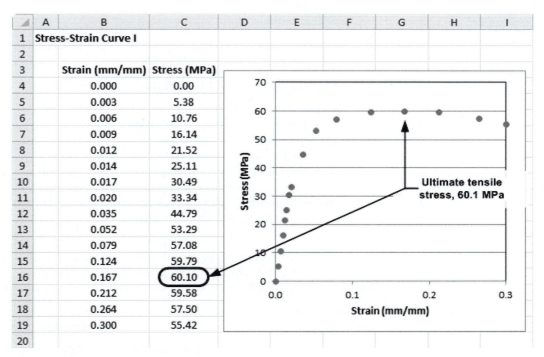

Figure 3.50
Stress–strain curve.

Figure 3.51
The linear portion of the stress–strain curve, with trendline.

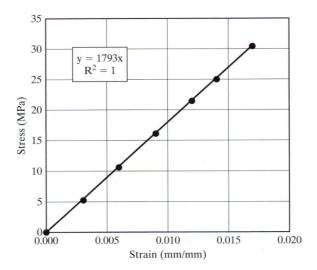

3.7 GRAPHING WEB DATA

The Internet's World Wide Web is becoming an increasingly common place to locate data, but those data may be in many different forms. Web pages may provide links to data files or embed the data as HTML tables. There is no single way to move Web data into Excel for graphing, but the following two methods frequently work:

1. Copy and paste; or
2. Save the data file, then import into Excel.

The first method is described in this section; the second method is introduced, then described more fully in Section 3.9.

3.7.1 Copying and Pasting Web Data

Data sets presented on the Web are often HTML tables. It is usually possible to copy the information from such tables and paste it into Excel, but you have to copy entire rows; you can't select portions. Then, when you want to paste the data into Excel, you might need to use Paste Special ... to instruct Excel to paste the values as text and ignore the HTML format information. The copy and paste process will be presented using the same temperature vs. time data used throughout this chapter. This data is available on the text's website in two forms: as an HTML table and as text file Ex_Data.prn. The text's website is located at

http://www.chbe.montana.edu/Excel

To copy the data from the HTML table, simply select all of the rows (and headings, if desired) and copy the data to the Windows clipboard. This is illustrated in Figure 3.52. Options for copying the selected data to the Windows clipboard include:

• Press [Ctrl-c]
• Right-click on the selected data and choose **Copy** from the pop-up menu

Then, you can paste the data into Excel. Frequently, you will find it necessary to use Ribbon options **Home/Clipboard/Paste/Paste Special...** to paste the data into the worksheet as **Text**. Pasting as **Text** tells Excel to ignore HTML formatting information and paste just the data into the cells. This approach was used to paste the data into the worksheet shown in Figure 3.53.

[Excel 2003: Edit/Paste Special…, then paste as "Text"]

Figure 3.52
Copying Web data.

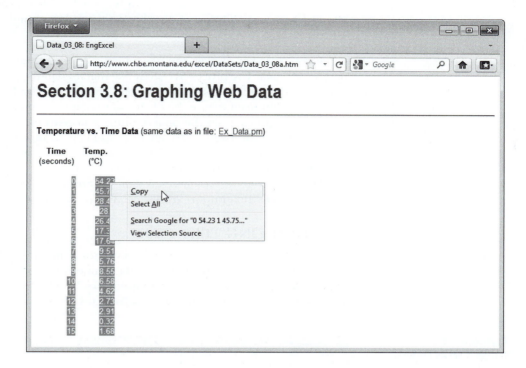

Figure 3.53
The Web data pasted into an Excel worksheet.

	A	B	C
1	0	54.23	
2	1	45.75	
3	2	28.41	
4	3	28.3	
5	4	26.45	
6	5	17.36	
7	6	17.64	
8	7	9.51	
9	8	5.76	
10	9	8.55	
11	10	6.58	
12	11	4.62	
13	12	2.73	
14	13	2.91	
15	14	0.32	
16	15	1.68	
17			

3.7.2 Importing Data Files from the Web

A Web page could provide a link to a data file, such as the link to file Ex_Data.prn from the text's website. What happens when you click on a link to a .prn file depends on how your browser is configured:

1. The browser might display the contents of the file on the screen.
2. The browser might present an option box asking whether you want to open the file or save it.

If your browser displays the file contents on the screen, you can try copying and pasting the data. The process is exactly like that used in the previous section. If copying and pasting the data fails (or if your browser will not display the data-file contents), then you might need to save the file to your computer and import the file into Excel.

If you right-click on the data-file link, a pop-up menu will offer a **Save Target As...** option (Microsoft Internet Explorer) or **Save Link As...** option (Mozilla Firefox), as shown in Figure 3.54. These options allow you to save a copy of the link's target (the data file) to your own computer.

Once the data file has been saved on your own computer, it can be imported into Excel by Excel's Text Import Wizard. That process is described in the next section.

3.8 IMPORTING TEXT FILES

Text files are a common way to move data from one program to another, and Excel is good at creating graphs that use data from other programs. Importing a text file is one way to get the data to be plotted into Excel. Excel provides a Text Import Wizard to make it easy to import data from text files.

Figure 3.54
Saving a linked file from the Web.

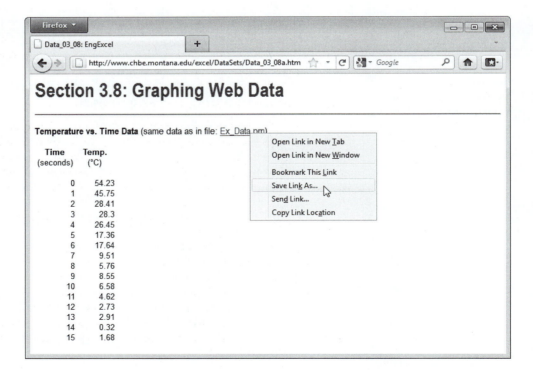

Two types of text files are used to store data:

- *delimited*
- *fixed width*

Delimited data has a special character, called a *delimiter*, between data values. Commas, spaces, and tabs are the most common delimiters, but any non-numeric character can be used. Quotes are frequently used as text-string delimiters.
The following is an example of comma-delimited data:

```
0, 54.23
1, 45.75
2, 28.41
```

Fixed-width files align the data values in columns and use character position to distinguish individual data values. The comma-delimited data shown previously would look quite different in a fixed-width data file. In the following example, the data have been written to the file with eight-character fields, using four decimal places (a couple of header lines have been included to show the layout of the data fields):

```
Field 1 Field 2
1234567812345678
0.0000 54.2300
1.0000 45.7500
3.0000 28.4100
```

Excel can read the data from either type of file, but Excel must know the format used in the data file before the data can be imported. The Text Import Wizard allows you to select the appropriate format as part of the import process. Fixed-width files were once very common, but delimited data seem to be more common at this time. Both types of data files are used regularly, and Excel's Text Import Wizard can handle either type of file.

3.8.1 Using the Text Import Wizard

The temperature–time data set used throughout this chapter consists of 16 temperature values measured at one-second intervals from 0 to 15 seconds. The values are available as a space-delimited text file called Ex_Data.prn. The file is available at the text's website, http://www.chbe.montana.edu/Excel. The following example assumes that the data file is available on a drive labeled M: in a folder called "Excel Data."
You begin importing a text file into Excel by attempting to open the file.

1. To open the Open dialog, use:
   - Excel 2010: **File tab/Open**
   - Excel 2007: **Office/Open**
   - Excel 2003: **File/Open**
2. Enter the name of the data file in the **File name** field, or, to select the data file from the files on the drive, change the displayed file type (near the bottom of the Open dialog) to **Text Files (*.prn; *.txt; *.csv)** and browse for the file. This is illustrated in Figure 3.55.

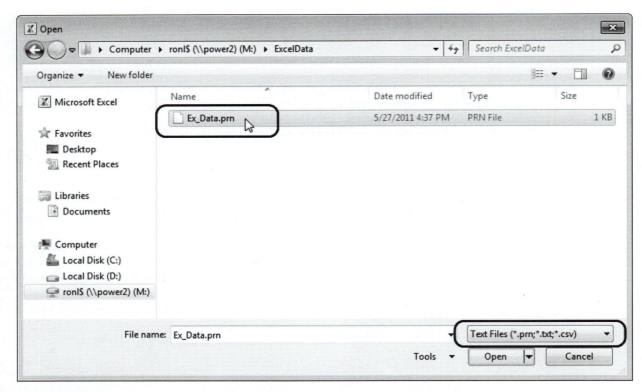

Figure 3.55
Change the file type to **Text Files (*.prn; *.txt; *.csv)** to display text data files.

3. Click the **Open** button to open the file.

When Excel attempts to open the file and finds that it is not saved as an Excel workbook, it starts the Text Import Wizard to guide you through the import process. The steps in the process are as follows:

Step 1. **Select the type of text file.** Excel will analyze the file contents and make a recommendation on the best way to import the data, but you should verify the data format. You can see how the data will import in the preview, as shown in Figure 3.56.

Excel determined that the data in Ex_Data.prn are delimited, which it is.

Notice that the Text Import Wizard allows you to begin importing data at any row by changing the value in the Start import at row: field. This is very useful if your data file contains heading or title information that you do not want to import into the worksheet.

Click **Next >** to move to Step 2 of the import process.

Step 2. **Select the type of delimiter(s).** If you selected **Delimited** data in the previous step, then you can choose the type(s) of delimiters. In file Ex_Data. prn, the values have leading spaces at the left of each line and spaces between two columns of values. When **Space** delimiter is checked (see Figure 3.57), Excel treats the spaces in the files as delimiters and adds lines in the **Data preview** panel to show how the values will be separated into columns.

Click **Next >** to go to the next step in the process.

Figure 3.56
The Text Import Wizard.

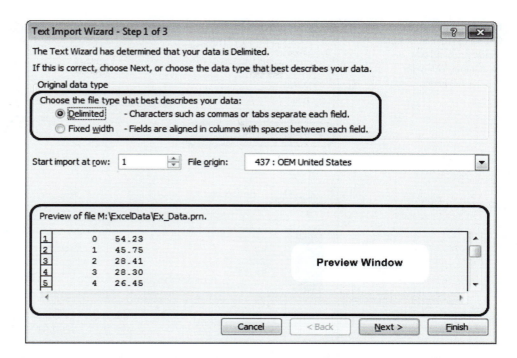

Figure 3.57
Choosing the delimiter(s).

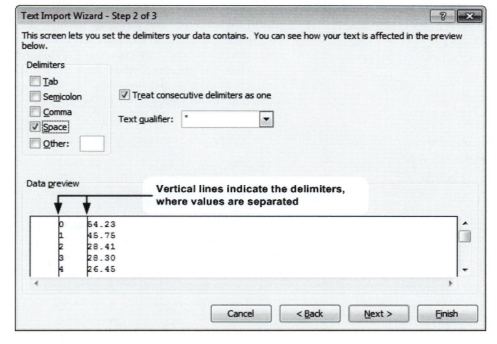

Step 3. Select the data to be imported and the data formats to be used. The third step allows you to tell Excel the number format you want to be used for each imported column, or you can choose not to import one or more columns. To select a column, click on the column heading. In Figure 3.58, the first column heading was selected, and then we chose **Do not import column**

Figure 3.58

Choosing whether or not to import each column, and the number format for imported columns.

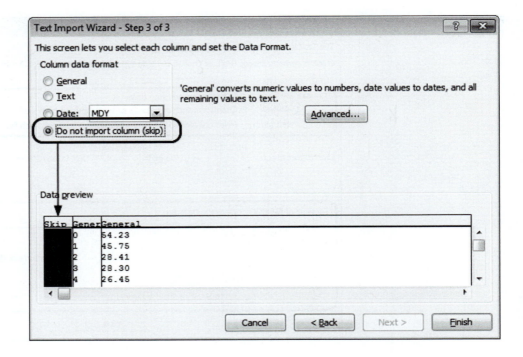

(**skip**) the column. This was done because the first column is empty. (Excel interpreted leading spaces on each line as an empty column.) The other two columns will be imported, using a **General** number format.

When the correct formats have been specified for each column to be imported, click **Finish** to complete the import process. The values are placed at the top-left corner of the worksheet, as shown in Figure 3.59.

Figure 3.59

The imported data.

	A	B	C
1	0	54.23	
2	1	45.75	
3	2	28.41	
4	3	28.3	
5	4	26.45	
6	5	17.36	
7	6	17.64	
8	7	9.51	
9	8	5.76	
10	9	8.55	
11	10	6.58	
12	11	4.62	
13	12	2.73	
14	13	2.91	
15	14	0.32	
16	15	1.68	
17			

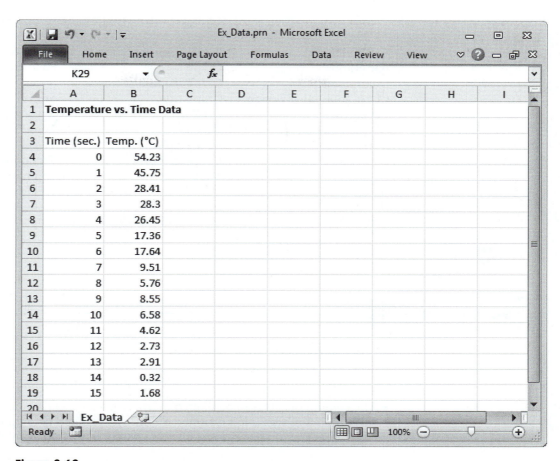

Figure 3.60
The imported data, moved down to allow a title and column headings to be inserted.

You can move the cells to a different location within the same worksheet or copy and paste them to another worksheet. In Figure 3.60, the values were moved down to make room for a title and column headings.

Note: Since file Ex_Data.prn was opened, Excel named the workbook shown in Figure 3.60 Ex_Data.prn. Excel will allow you to use the workbook just as you would a standard workbook (.xlsx) file, but any nonalphanumeric content (e.g., graphs) will be lost if the file is saved as a .prn file. Excel will show a warning if you attempt to save the file with a .prn extension. It is a good idea to immediately save the file with an Excel workbook file extension, .xlsx. To do this, use:

- Excel 2010: **File tab/Save As…**
- Excel 2007: **Office/Save As…**
- Excel 2003: **File/Save As…**

Then, select **Excel Workbook** as the file type.

KEY TERMS

Axis titles	Fixed width	R^2 value
Bar graphs	Graph	Smoothed curves
Chart	Graph legend	Strain
Chart Tools tabs (Design,	Graph title	Stress
Layout, and Format)	Gridlines	Surface plots
Column graphs	Handles	Tensile strength
Data series	Line graphs	Tensile test
Delimited	Line segments	Trendline
Delimiter	Markers	Ultimate stress
Equation of the	Modulus of elasticity	XY scatter graphs
trendline	Pie charts	Young's modulus
Error bars	Quick (Graph) Layout	

SUMMARY

Preparing to Plot an XY Scatter Graph

Excel will use the organization of your data to infer information necessary to create the XY scatter graph. To facilitate this, organize your data as follows:

- The column of values to be plotted on the *x* axis should be on the left.
- The column of values to be plotted on the *y* axis should be on the right.
- A single-cell heading at the top of each column may be used (optional).

   The cell heading on the *y* values (if used) will be used as the graph title and in the graph legend.

Creating an XY Scatter Plot from Existing Data

1. Select the data range (including the series names, if desired).
2. Use Ribbon options **Insert/Charts/Scatter** to select the graph type and create the graph.
3. Choose a Quick Layout (**Chart Tools/Design/Chart Layouts/Quick Layout**).
4. Edit the axis labels and graph title.

Editing an Existing Graph

Select the item you wish to modify, then use Ribbon options **Chart Tools/Layout/Format Selection** to open a dialog to edit the selected chart element. (With Excel 2010 you can also double-click a chart element to directly open the Format dialog for that element.) Items that can be modified include:

- Axes
- Gridlines
- Legend

- Markers and lines
- Plot area
- Titles

Add Another Curve or Modify Existing Data Series

Use Ribbon options **Chart Tools/Design/Data/Select Data**. This opens the Select Data Source dialog which is used to add a new data series or edit the existing data series.

Adding a Trendline to a Graph

Right-click on the data series that the trendline should be fit to and then choose **Add Trendline** from the series' pop-up menu. The following trendlines are available:

- Linear (straight line)
- Exponential
- Logarithmic
- Polynomial
- Power
- Moving average (nonregression)

Printing a Graph

There are two ways to print a graph:

1. Print the graph only—select the graph, then choose one of the printing options listed below.
2. Print the worksheet containing the graph—set the print area to include the graph, then use one of the printing options listed below.

Printing Options

- Excel 2010: **File tab/Print (panel)/Print (button)**—opens the Print dialog (including a preview panel) to allow you to set printer options before printing.
- Excel 2007:
  - **Office/Print/Print**—opens the Print dialog to allow you to set printer options before printing.
  - **Office/Print/Quick Print**—prints directly to the default printer using default printer settings.
  - **Office/Print/Print Preview**—shows what the printout will look like on the screen to allow you to change margins, scale to fit a page, etc., before printing.
- Excel 2003: **File/Print**—opens the Print dialog to allow you to set printer options before printing.

Available Graph Types

- XY scatter plots
- Line graphs
- Column and bar graphs
- Pie charts
- Surface plots

PROBLEMS

3.1 Stress–Strain Curve II

The following tabulated data represent stress–strain data from an experiment on an unknown sample of a white metal (modulus of elasticity values for various white metals are also listed):

    a. Graph the stress–strain data. If the data include values outside the elastic-stretch region, discard those values.

    b. Use a linear trendline to compute the modulus of elasticity for the sample.

    c. What type of metal was tested?

Stress	Strain		Material	Modulus of Elasticity
(mm/mm)	(MPa)			(GPa)
0.0000	0		Mg Alloy	45
0.0015	168		Al Alloy	70
0.0030	336		Ag	71
0.0045	504		Ti Alloy	110
0.0060	672		Pt	170
			SS	200

Note: The modulus of elasticity depends on the type of alloy or purity of a non-alloyed material. The values listed here are typical.

3.2 Tank Temperature During a Wash-Out

One evening, a few friends come over for a soak, and you discover that the water in the hot tub (Figure 3.61) is at 115°F (46°C)—too hot to use. As your friends turn on the cold water to cool down the tub, the engineer in you wants to know how long this is going to take, so you write an energy balance on a well-mixed tank (ignoring heat losses to the air). You end up with the following differential equation relating

Figure 3.61
Hot tub.

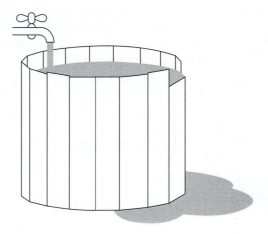

the temperature in the tank, T, to the temperature of the cold water flowing into the tank, T_{in}, the volume of the tank, V, and the volumetric flow rate of the cold water, $\dot{V}$:

$$\frac{dT}{dt} = \frac{\dot{V}}{V}(T_{in} - T).$$

(3.3)

Integrating, you get an equation for the temperature in the tank as a function of time:

$$T = T_{in} - (T_{in} - T_{init.})e^{\frac{-\dot{V}}{V}t}.$$

(3.4)

If the initial temperature $T_{init.}$ is 115°F, the cold water temperature is 35°F (1.7°C), and the volume and volumetric flow rate are 3000 liters and 30 liters per minute, respectively,

a. Calculate the expected water temperature at 5-minute intervals for the first 60 minutes after the flow of cold water is established.
b. Plot the water temperature in the hot tub as a function of time.
c. Calculate how long it should take for the water in the tub to cool to 100°F (37.8°C).
d. Explain whether a hot tub is really a well-mixed tank. If it is not, will your equation predict a time that is too short or too long? Explain your reasoning.

3.3 Fluid Statics: Manometer

Manometers used to be common pressure-measurement devices, but, outside of laboratories, electronic pressure transducers are now more common. Manometers are sometimes still used to calibrate the pressure transducers.

In the calibration system shown in Figure 3.62, the mercury manometer on the right and the pressure transducer on the left are both connected to a piston-driven pressure source filled with hydraulic oil ($\rho = 880 \, \text{kg/m}^3$). The bulbs connected to the transducer and the right side of the manometer are both evacuated ($\rho = 0$).

During the calibration, the piston is moved to generate a pressure on both the manometer and the transducer. The manometer reading R is recorded, along with the output of the pressure transducer A (assuming a 4- to 20-mA output current from the transducer).

Figure 3.62
Calibrating pressure transducers.

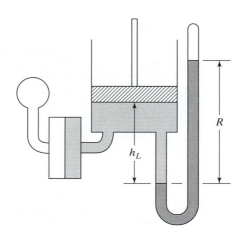

Consider the following calibration data:

a. Calculate pressures from the manometer readings.
b. Create a calibration table and graph showing the transducer output (mA) as a function of measured pressure.

	Calibration Data		
Piston Setting	hL (mm Oil)	Manometer Reading (mm Hg)	Transducer Output (mA)
1	300	0	4.0
2	450	150	5.6
3	600	300	7.2
4	750	450	8.8
5	900	600	10.4
6	1050	750	13.0
7	1200	900	13.6
8	1350	1050	15.2
9	1500	1200	16.8
10	1650	1350	18.4
11	1800	1500	20.0

3.4 Thermocouple Calibration Curve

A type J (iron/constantan) thermocouple was calibrated by using the system illustrated in Figure 3.63. The thermocouple and a thermometer were dipped into a beaker of water on a hot plate. The power level was set at a preset level (known only as 1, 2, 3, ... on the dial) and the thermocouple readings were monitored on a computer screen. When steady state had been reached, the thermometer was read, and 10 thermocouple readings were recorded. Then the power level was increased and the process repeated.

The accumulated calibration data (steady-state data only) are as follows (available electronically at http://www.coe.montana.edu/che/Excel):

Power Setting	Thermometer (°C)	Thermocouple Average (mV)	Thermocouple Standard Deviation (mV)
0	24.6	1.264	0.100
1	38.2	1.841	0.138
2	50.1	3.618	0.240
3	60.2	3.900	0.164
4	69.7	3.407	0.260
5	79.1	4.334	0.225
6	86.3	4.506	0.212
7	96.3	5.332	0.216
8	99.8	5.084	0.168

Figure 3.63
Calibrating thermocouples.

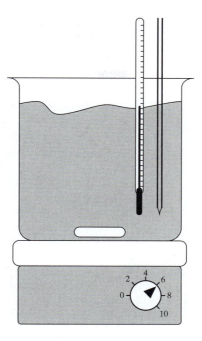

a. Plot the thermocouple calibration curve with temperature on the x axis and average thermocouple reading on the y axis.
b. Add a linear trendline to the graph and have Excel display the equation for the trendline and the value.
c. Use the standard deviation values to add error bars (± 1 standard deviation) to the graph.
d. The millivolt output of an iron/constantan thermocouple can be related to temperature by the correlation equation[1]

$$T = aV^b, \qquad (3.5)$$

where

T is temperature in °C,
V is the thermocouple output in millivolts,
a is 19.741 for iron/constantan, and
b is 0.9742 for iron/constantan.

Use this equation to calculate predicted thermocouple outputs at each temperature and add these to the graph as a second data series. Do the predicted values appear to agree with the average experimental values?

3.5 Resistance Temperature Detector

The linear temperature coefficient α of a resistance temperature detector (RTD) is a physical property of the metal used to make the RTD that indicates how the

---

[1]Thermocouple correlation equation from *Transport Phenomena Data Companion*, L.P.B.M. Janssen and M.M.C.G. Warmoeskerken, Arnold DUM, London, 1987, p. 20.

electrical resistance of the metal changes as the temperature increases. The equation relating temperature to resistance[1] is

$$R_T = R_0[1 + \alpha T],$$ (3.6)

or

$$R_T = R_0 + (R_0\alpha)^T \quad \text{(in linear regression form)},$$ (3.7)

where

R_T is the resistance at the unknown temperature, T,
R_0 is the resistance at 0°C (known, one of the RTD specifications), and
α is the linear temperature coefficient (known, one of the RTD specifications).

The common grade of platinum used for RTDs has an α value equal to 0.00385 ohm/ohm/°C (sometimes written simply as $0.00385°\text{C}^{-1}$), but older RTDs used a different grade of platinum and operated with $\alpha = 0.003902°\text{C}^{-1}$, and laboratory grade RTDs use very high-purity platinum with $\alpha = 0.003923°\text{C}^{-1}$. If the wrong α value is used to compute temperatures from RTD readings, the computed temperatures will be incorrect.

The following data show the temperature vs. resistance for an RTD:

a. Use Excel to graph the data and add a trendline to evaluate the linear temperature coefficient.
b. Is the RTD of laboratory grade?

Temperature	Resistance
°C	Ohms
0	100.0
10	103.9
20	107.8
30	111.7
40	115.6
50	119.5
60	123.4
70	127.3
80	131.2
90	135.1
100	139.0

3.6 Experimentally Determining a Value for π

The circumference and diameter of a circle are related by the equation

$$C = \pi D$$ (3.8)

A graph of diameter (x axis) and circumference (y axis) for a number of circles of various sizes should produce a plot with a slope of π.

- Find at least six circular objects of various sizes.
- Measure the diameter and circumference of each circle.

- Create an XY scatter plot with diameter on the *x* axis and circumference on the *y* axis.
- Add a linear trendline through your data.
  - Force the intercept through the origin.
  - Display the equation and R^2 value on the graph.
  a. What is the experimentally determined value of π?
  b. What is the percent error in your experimental result compared with the accepted value of 3.141593?

$$\% \text{ error} = \frac{\text{measured value} - \text{true value}}{\text{true value}} \times 100$$

3.7 Predicting Wind Speed

As part of an experiment looking into how energy losses depend on wind speed, some data were collected on air velocity at three distances from a portable fan. The data are shown in Table 3.3.

The data were collected with a tape measure and a hand-held anemometer, and there are mixed unit systems in the data in Table 3.3.

Create an Excel worksheet containing:

- a. The original data (from Table 3.3).
- b. The distance data converted inches to meters [2.54 cm/inch].
- c. A plot of the distance (*x* axis) and air velocity (*y* axis) values.
- d. An exponential trendline through the data, with the trendline equation and R^2 value shown on the graph.

Use the trendline equation to predict air velocity values at 0.3 m and 0.6 m.

Table 3.3 Air velocity at three distances from a portable fan

Distance (in.)	Air Velocity (m/s)
8	2.0
8	1.9
8	2.0
8	2.2
8	2.1
16	1.3
16	1.2
16	1.1
16	1.2
16	1.2
32	0.5
32	0.4
32	0.5
32	0.5
32	0.5

3.8 Using a Pie Chart

A pie chart is used to show how a complete entity is divided into parts. The data in Table 3.4 show how the total cost of bringing a new product to market can be attributed to various aspects of the development process.

Table 3.4 Costs associate with new product development

Category	As Budgeted	Actual
Research	$1,200,000	$1,050,000
Patenting	$87,000	$89,000
Development	$1,600,000	$2,400,000
Legal	$32,000	$104,000
Marketing	$134,000	$85,000
Packaging	$48,000	$36,000
TOTAL:	$3,101,000	$3,764,000

The budgeted and actual costs associated with bringing a new product to market are listed in the worksheet shown in Figure 3.64. The "As Budgeted" *pie chart* is shown and was created by:

1. Selecting cells B4:C9.
2. Using Ribbon Options **Insert/Charts/Pie** and choosing the first pie chart option.

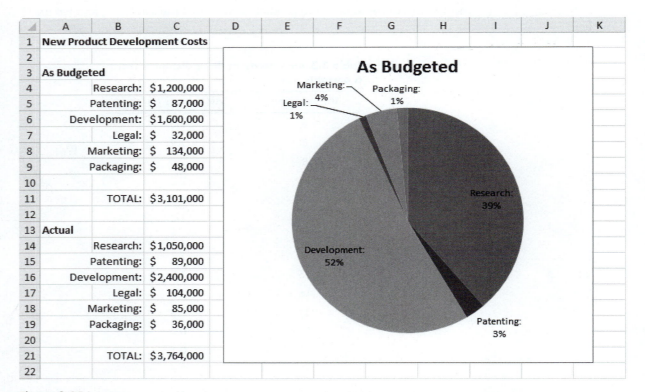

Figure 3.64
Budgeted and actual product development costs.

3. Using Ribbon Options **Chart Tools/Design/Chart Layouts** and selecting the first chart layout.
4. Editing the chart title.

Create an Excel worksheet similar to Figure 3.64 that shows the As Budgeted and Actual data values, and two pie charts, one for the "as budgeted" values (as shown) and another for the "actual" values.

3.9 Using a Column Chart

The data shown in Figure 3.64 can also be plotted with a *column chart* (commonly called a *bar chart*, but a bar chart is different in Excel), as shown in Figure 3.65. A pie chart is used to show how costs are distributed, whereas a column chart focuses attention on the differences between "as budgeted" and "actual" costs for each category. The column chart was created by:

1. Selecting cells B3:D9.
2. Using Ribbon Options **Insert/Charts/Column** and choosing the first column chart option.
3. Moving the legend to allow the plot area to be larger.

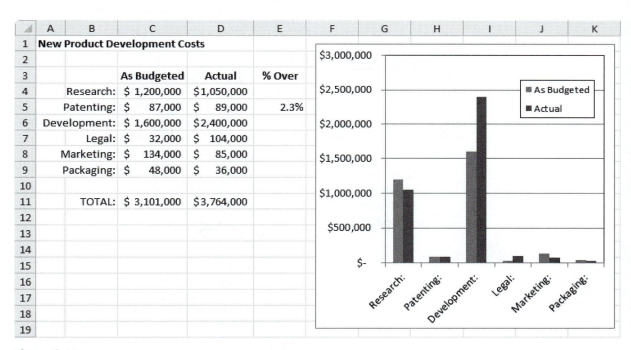

Figure 3.65
Budgeted and actual product development costs as column chart.

Recreate the worksheet shown in Figure 3.65 and complete the "% over" column to show much each category was over budget (or under budget, if the percentage is negative).

3.10 Safety of an Old Bridge

An old bridge over a river was constructed with steel beams and wood planking. The owner tried to test the safety of the bridge by measuring deflections in the main span under various loadings. Specifically, a truck with a 1000-gallon water tank

Table 3.5 Span deflections

Load (lbs)	Deflection (mm)
5400	3.2
5400	3.2
5400	3.1
7400	4.8
7400	4.9
7400	4.8
9400	7.6
9400	7.5
9400	7.7
11,400	10.8
11,400	10.9
11,400	11.1
13,400	Broke

was driven back and forth across the bridge and the deflection in the main span was measured as the truck went across. The load was increased by adding water to the tank. The experiment was stopped when the water truck fell into the river.

The collected data are listed in Table 3.5. Plot the data using an XY scatter graph and use a trendline to predict the deflection that the bridge was unable to withstand. Try all of the regression trendlines available in Excel.

a. Which trendlines do not fit this data?
b. Which type of trendline appears to be the best fit to this data?
c. What is your estimate of the deflection that the bridge could not withstand?
d. How much does the predicted deflection value change as you change the type of trendline used?

CHAPTER

4 Excel Functions

Objectives

After reading this chapter, you will

- Know how to use Excel's built-in functions
- Know the common functions for
  ○ Basic math operations
  ○ Computing sums
  ○ Trigonometric calculations
  ○ Advanced math operations

- Be aware of Excel's specialized functions that may be useful in some engineering disciplines
- Be familiar with the quick reference to the Excel functions included in the summary at the end of this chapter

4.1 INTRODUCTION TO EXCEL FUNCTIONS

In a programming language, the term *function* is used to mean a piece of the program dedicated to a particular calculation. A function accepts input from a list of *arguments* or *parameters*, performs calculations using the values passed in as arguments, and then returns a value or a set of values. Excel's functions work the same way and serve the same purposes. They receive input from an argument list, perform a calculation, and return a value or a set of values.

Excel provides a wide variety of functions that are predefined and immediately available. This chapter presents the functions most commonly used by engineers.

Functions are used whenever you want to:

- Perform the same calculations multiple times, using different input values.
- Reuse the calculation in another program without retyping it.
- Make a complex program easier to comprehend by assigning a section a particular task.

4.2 EXCEL'S BUILT-IN FUNCTIONS

Excel provides a wide assortment of built-in functions. Several commonly used classes of functions are as follows:

- Elementary math functions
- Trigonometric functions
- Advanced math functions
- Matrix math functions (not described in this chapter)
- Functions for financial calculations (not described in this chapter)
- Functions for statistical calculations (not described in this chapter)
- Date and time functions
- String functions
- Lookup and reference functions
- File-handling functions
- Functions for working with databases (not described in this chapter)

The functions that are not described in this chapter are presented elsewhere in the text. And, if Excel does not provide a necessary built-in function, you can always write your own functions, directly from Excel, by using Visual Basic for Applications (VBA).

4.2.1 Function Syntax

Built-in functions are identified by a *name* and usually require an *argument list*. They use the information supplied in the argument list to compute and return a value or set of values. One of the simplest functions in Excel, but still a very useful one, is **PI()**. When this function is called (without any arguments inside the parentheses), it returns the value of π.

In the Figure 4.1, cell B2 is assigned the value π by the formula =PI().

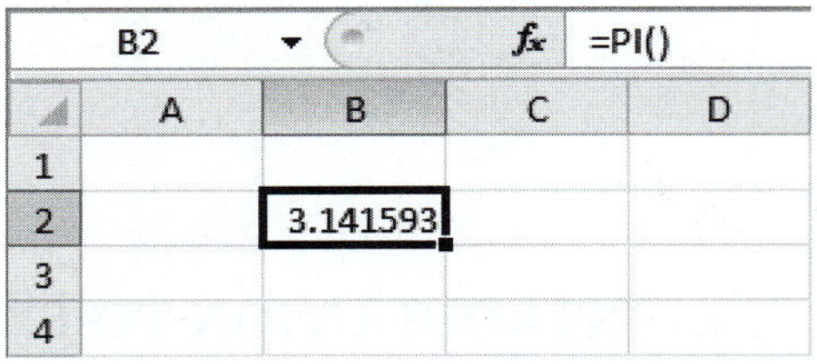

Figure 4.1
Using the **PI** function.

Functions can also be built into formulas. In Figure 4.2, the function **PI** is used in a formula to calculate the area of a circle.

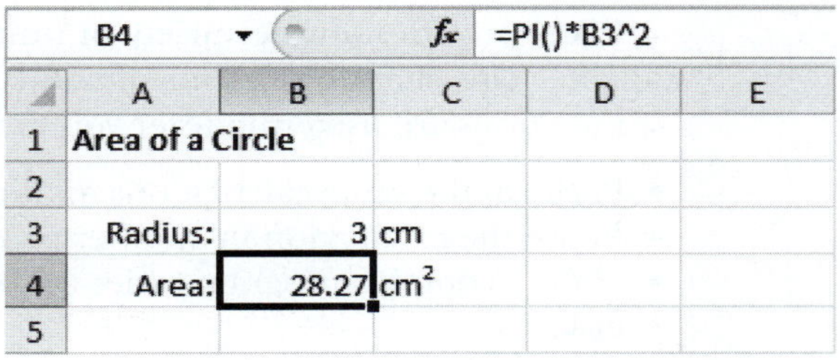

Figure 4.2
Using a function within a formula.

Most functions take one or more values as arguments. An example of a function that takes a single value as its argument is the factorial function, **FACT(x)**. The factorial of 4 is 24 ($4 \times 3 \times 2 \times 1 = 24$). This can be computed by using the **FACT** function, as seen in Figure 4.3.

Figure 4.3
Using the **FACT** function.

	A	B	C	D
		B4		f_x =FACT(B3)
1	Calculating Factorial of X			
2				
3	X:	4		
4	Factorial of X:	24		
5				

Some functions take a range of values as arguments, such as the **SUM(range)** function which calculates the sum of all of the values in the cell range specified as the function's argument. In Figure 4.4, the **SUM** function is used to compute the sum of three values.

Figure 4.4
Using the **SUM** function
to compute the sum of the
values in a range of cells.

	A	B	C	D	E
		B7		f_x =SUM(B3:B5)	
1	Sum of a Column of Values				
2					
3		10			
4		12			
5		14			
6					
7	Total:	36			
8					

An easy way to enter a function that takes a range of cells as an argument is to type in the function name and the opening parenthesis and then use the mouse to highlight the range of cells to be used as an argument. For the preceding example, this would be done as:

1. Type =sum(in cell B6.
2. Then use the mouse to select cells B3:B5.

The worksheet will look like Figure 4.5.

In Figure 4.5, the dashed line around cells B3:B5 indicates a range of cells currently selected, and the selected range is automatically included in the formula in cell B7. When you type the opening parenthesis and then move the mouse, Excel jumps into Point mode and expects you to use the mouse to point out the cells that will be used as arguments in the function.

The *ScreenTip* below the formula lets you know what arguments the function requires. Once the complete range has been selected, type the closing parenthesis and press [Enter], or simply press [Enter], and Excel will automatically add the final parenthesis.

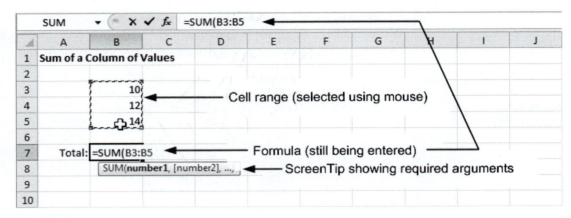

Figure 4.5
Entering the **SUM** function.

The cells to be included in the summation do not have to be next to each other on the worksheet; in Figure 4.6 two cell ranges appear in the **SUM** function.

Figure 4.6
Using noncontiguous cell ranges as arguments in a function.

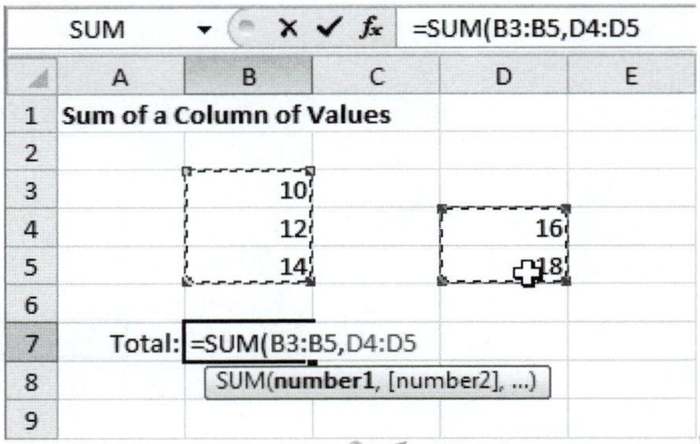

To select noncontiguous cells, you first select the first portion of the cell range (B3:B5) with the mouse. Then, either type a comma or hold down the control key while you select the second range (D4:D5) with the mouse. This completed formula is shown in Figure 4.7.

Figure 4.7
The completed summation of noncontiguous cells.

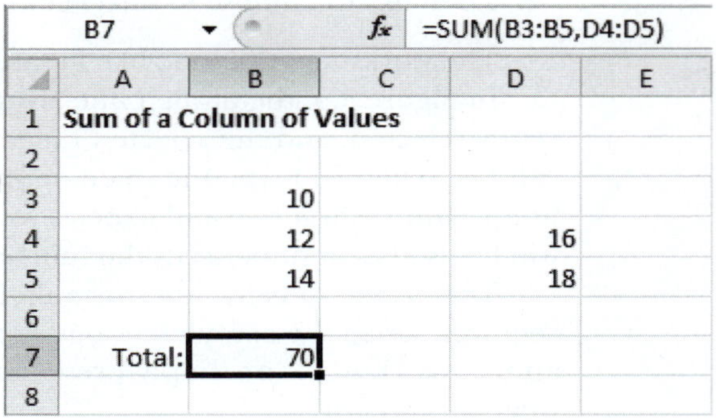

PROFESSIONAL SUCCESS

When do you use a calculator to solve a problem, and when should you use a computer? The following three questions will help you make this decision:

1. Is the calculation long and involved?
2. Will you need to perform the same calculation numerous times?
3. Do you need to document the results for the future, either to give them to someone else or for your own reference?

A "yes" answer to any of these questions suggests that you consider using a computer. Moreover, a "yes" to the second question suggests that you might want to write a reusable function to solve the problem.

4.3 USING THE CONVERT FUNCTION TO CONVERT UNITS

Excel provides an interesting function called **CONVERT** to convert units. The **CONVERT** function receives (as an argument) a value in certain units and then calculates and returns an equivalent value in different units. The **CONVERT** function has the syntax

$$CONVERT(\texttt{value, from\_units, to\_units})$$

For example, =CONVERT(1, "in", "cm") returns 2.54; this is illustrated in Figure 4.8.

from in. → cm

Figure 4.8
Converting inches to centimeters.

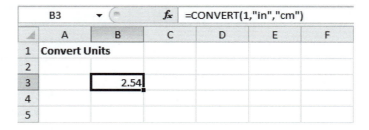

The **CONVERT** function can be useful for converting a table from one set of units to another. As an example, the fractional part sizes (in inches) are converted to centimeters in Figure 4.9.

Figure 4.9
Converting units on a table of part sizes.

C3		ƒₓ	=CONVERT(B3,"in","cm")			
	A	B	C	D	E	F
1	Parts Sizes					
2		(inches)	(cm)			
3		3 1/8	7.94			
4		4 1/10	10.41			
5		1/4	0.64			
6		6 3/8	16.19			
7						

Note: Cells B3:B6 in Figure 4.9 are formatted using Fraction format, one of the less commonly used numeric formats.

Excel looks up the unit abbreviations in a list, so you can only use the abbreviations in Excel's list. These are available in Excel's help system (press [F1] and search on **CONVERT**). A list of commonly used units is provided in Table 4.1.

Table 4.1 Common unit abbreviations for the CONVERT function

Unit Type	Unit	Abbreviation
Mass	Gram	g
	Pound	lbm
Length	Meter	m
	Mile	mi
	Inch	in
	Foot	ft
	Yard	yd
Time	Year	yr
	Day	day
	Hour	hr
	Minute	mn
	Second	sec
Pressure	Pascal	Pa
	Atmosphere	atm
	mm of Mercury	mmHg
Force	Newton	N
	Dyne	dyn
	Pound force	lbf
Energy or Work	Joule	J
	Erg	e
	Foot-pound	flb
	BTU	BTU
Power	Horsepower	HP
	Watt	W
Temperature	Degree Celsius	C
	Degree Fahrenheit	F
	Degree Kelvin	K
Volume	Quart	qt
	Gallon	gal
	Liter	l

When using the **CONVERT** function, it is important to remember the following:

- The unit names are case sensitive.
- SI unit prefixes (e.g., "k" for kilo) are available for metric units. In the following example, the "m" prefix for milli can be used with the meter name "m" to report the part sizes in millimeters, as shown in Figure 4.10.

Figure 4.10
Reporting part sizes
in millimeters.

	A	B	C	D	E	F
	C3	▼	f_x =CONVERT(B3,"in","mm")			
1	Parts Sizes					
2		(inches)	(mm)			
3		3 1/8	79.38			
4		4 1/10	104.14			
5		1/4	6.35			
6		6 3/8	161.93			
7						

- A list of prefix names is included in Excel's help system for the **CONVERT** function.
- Combined units such as gal/mn (gallons per minute) are not supported.

4.4 SIMPLE MATH FUNCTIONS

Most common math operations beyond multiplication and division are implemented as functions in Excel. For example, to take the square root of four, you would use =SQRT(4). SQRT() is Excel's square root function.

4.4.1 Common Math Functions

Table 4.2 presents some common mathematical functions in Excel.

Table 4.2 Excel's common mathematical functions

Operation	Function Name
Square Root	**SQRT(x)**
Absolute Value	**ABS(x)**
Factorial	**FACT(x)**
Summation	**SUM(range)**
Greatest Common Divisor	**GCD(x1, x2, ...)**
Least Common Multiple	**LCM(x1, x2, ...)**

Many additional mathematical operations are available as built-in functions. To see a list of the functions built into Excel, first click on an empty cell and then click on the **Insert Function** button on the Formula bar. This is illustrated in Figures 4.11 and 4.12.

Figure 4.11
Use the **Insert Function**
button to obtain a list
of available functions.

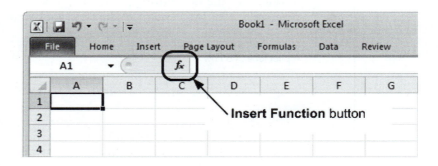

Figure 4.12
The Insert Function dialog.

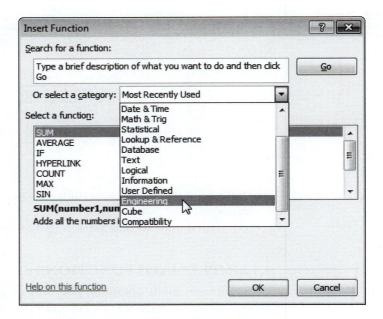

Clicking the **Insert Function** button opens the Insert function dialog, shown in Figure 4.12. The Insert Function dialog provides access to all of Excel's built-in functions.

Note: Excel's help files include descriptions of every built-in function, including information about the required and optional arguments.

APPLICATION

FLUID MECHANICS

Fluid Velocity from Pitot Tube Data

A *pitot tube* is a device that allows engineers to obtain a *local velocity* (velocity value in the immediate vicinity of the pitot tube). A pitot tube measures a pressure drop, but that pressure drop can be related to local velocity with a little math. The theory behind the operation of the pitot tube comes from Bernoulli's equation (without the usual potential energy terms) and relates the change in kinetic energy to the change in fluid pressure.

$$\frac{P_a}{\rho} + \frac{u_a^2}{2} = \frac{P_b}{\rho} + \frac{u_b^2}{2}. \tag{4.1}$$

In Figure 4.13, part of the flow hits point *a*, which is a dead end tube; there is nowhere for the flow hitting point *a* to go, so it has to come to a stop; this is called *stagnation*. Nearby, at point *b*, the flow goes right past the point *b* and does not

Figure 4.13
A pitot tube in a flow stream.

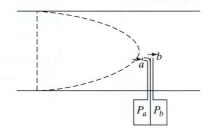

slow down at all; it is flowing at the *local free stream velocity*. The pressure measured at point *b* is called the *free stream pressure*. At point *a*, when the flowing fluid came to a stop, the kinetic energy in the fluid had to turn into something else because the total energy carried by the fluid has to be conserved. The kinetic energy of the moving fluid at point *a* is transformed to pressure energy as the velocity goes to zero. (You've felt this if you have ever stood in a river and felt the water pushing at you.) The pitot tube measures a higher pressure at point *a* than it does at point *b*, and the pressure difference can be used to determine the local free stream velocity at point *b*.

When the velocity at point *a* has been set to zero, and Bernoulli's equation is rearranged to solve for the local velocity at point *b*, we get

$$u_b = \sqrt{\frac{2}{\rho}(p_a - p_b)}. \tag{4.2}$$

If we used a pitot tube with a real flowing fluid with a specific gravity 0.81, and the pressure transducer indicated the pressure difference $p_a - p_b = 0.25$ atm, what was the local velocity at point *b*? A worksheet designed to solve this problem is shown in Figure 4.14. The formulas used in column C in the following have been displayed in column F.

The fluid is moving past the pitot tube at 7.9 m/s.

Figure 4.14
Solving for local velocity at *b*.

C8			f_x	=SQRT((2/C6)*C4)			
	A	B	C	D	E	F	G
1	**Pitot Tube**						
2						**Formulas in Column C**	
3	Pressure Difference:		0.25	atm			
4	Pressure Difference SI:		25331	Pa		=CONVERT(C3,"atm","Pa")	
5	Specific Gravity:		0.81				
6	Density:		810	kg/m³		=C5*1000	
7							
8	Velocity at b:		7.91	m/s		=SQRT((2/C6)*C4)	
9							

PRACTICE!

If the velocity at point *b* is doubled, what pressure difference would be measured across the pitot tube? [Answer: 1 atm]

4.5 COMPUTING SUMS

Calculating the sum of a row or column of values is a common worksheet operation. Excel provides the **SUM** function for this purpose, but also tries to simplify the process even further by putting the [AutoSum] button on the Ribbon's Home tab. Both approaches to calculating sums will be presented.

4.5.1 The SUM Function

The **SUM(range)** function receives a range of cell values as its argument, computes the sum of all of the numbers in the range, and returns the sum. The use of the **SUM** function is illustrated for several situations in Figures 4.15 through 4.18.

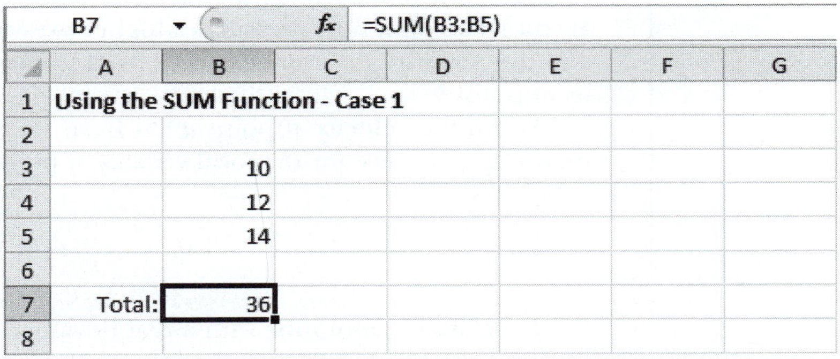

Figure 4.15
Basic use of the **SUM** function.

	B7	▼	f_x	=SUM(B3:B5)		
	A	B	C	D	E	F	G
1	Using the SUM Function - Case 1						
2							
3		10					
4		12					
5		14					
6							
7	Total:	36					
8							

If the 12 becomes "twelve," that is, if one of the cells being summed contains text rather than a numeric value, the non-numeric value is ignored. This is illustrated in Figure 4.16.

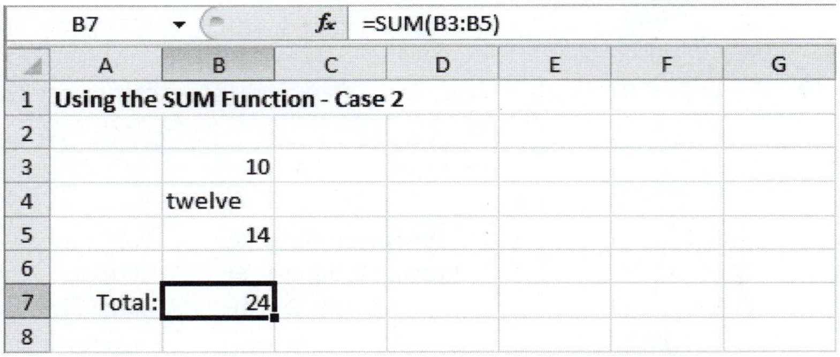

Figure 4.16
The **SUM** function ignores empty cells and cells containing text.

	B7	▼	f_x	=SUM(B3:B5)		
	A	B	C	D	E	F	G
1	Using the SUM Function - Case 2						
2							
3		10					
4	twelve						
5		14					
6							
7	Total:	24					
8							

The values to be summed don't have to be in a column, they can be in a row, a rectangular cell range, or a set of noncontiguous cells. These situations are illustrated in Figure 4.17.

4.5.2 The AutoSum Button

The **AutoSum** button on the Ribbon's Home tab is designed to seek out and find a column or row of values to sum. If you select the cell at the bottom of a column of numbers or to the right of a row of numbers and then press the **AutoSum** button, Excel will automatically include the entire column or row in the **SUM** function.

Note: There is a keyboard shortcut for the **AutoSum** button: [Alt =]; hold the [Alt] key down and press the [=] key.

For example, to sum the values in an expense report, first enter the expenses and select the cell immediately below the list of values. Then click the **AutoSum** button. This is illustrated in Figure 4.18.

Figure 4.17
The **SUM** function can handle any arrangement of cells.

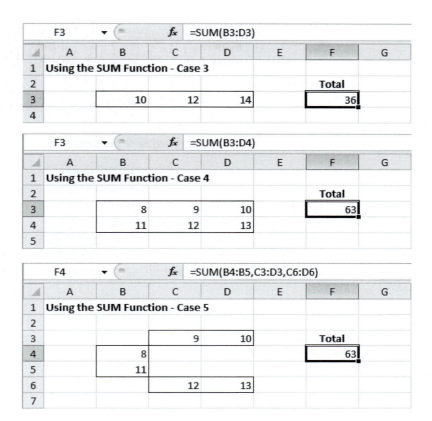

Note: The **AutoSum** button is on the far right side of the Ribbon's Home tab, so a section of the worksheet has been omitted from Figure 4.18 to show both the values to be added and the **AutoSum** button in the same figure.

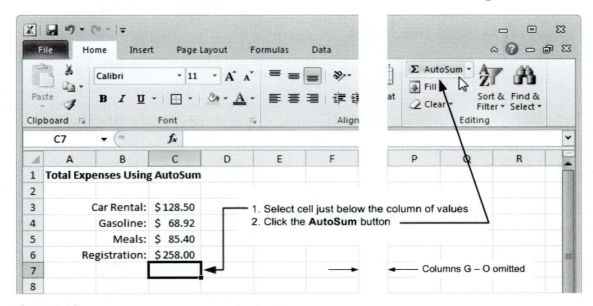

Figure 4.18
Using the **AutoSum** button.

When you click the **AutoSum** button, Excel shows you the nearly completed summation formula (shown in Figure 4.19) so that you can be sure that the correct cells were included.

Figure 4.19
Excel shows the nearly completed summation formula when the **AutoSum** button is pressed.

SUM	▼	✕ ✔ *fx*	=SUM(C3:C6)			
	A	B	C	D	E	F
1	Total Expenses Using AutoSum					
2						
3		Car Rental:	$ 128.50			
4		Gasoline:	$ 68.92			
5		Meals:	$ 85.40			
6		Registration:	$ 258.00			
7			=SUM(C3:C6)			
8			SUM(**number1**, [number2], ...)			
9						

If the summation is correct, press [Enter] to complete the calculation. The result is shown in Figure 4.20.

Figure 4.20
The completed summation, with an added label in cell B7 and formatting (bottom border) on cell C6 to improve clarity.

	A	B	C	D	E	F
1	Total Expenses Using AutoSum					
2						
3		Car Rental:	$ 128.50			
4		Gasoline:	$ 68.92			
5		Meals:	$ 85.40			
6		Registration:	$ 258.00			
7		TOTAL:	$ 540.82			
8						

More than just summations . . .

Like many of the Ribbon's features, the **AutoSum** button acts like both a button and a drop-down menu. If you click the small down arrow at the right side of the **AutoSum** button, the menu shown in Figure 4.21 will be displayed.

Figure 4.21
The **AutoSum** button's drop-down menu.

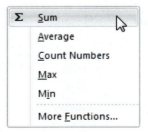

This menu provides quick access to some useful functions that work with cell ranges.

4.5.3 Logarithm and Exponentiation Functions

Excel's *logarithm* and *exponentiation functions* are listed in Table 4.3.

Excel also provides functions for working with logarithms and exponentiation of *complex numbers* (Table 4.4).

Table 4.3 Logarithm and exponentiation functions

Function Name	Operation
EXP(x)	Returns e raised to the power x
LN(x)	Returns the natural log of x
LOG10(x)	Returns the base-10 log of x
LOG(x, base)	Returns the logarithm of x to the specified base

Table 4.4 Functions for working with complex numbers

Function Name	Operation
IMEXP(x)	Returns the exponential of complex number x
IMLN(x)	Returns the natural log of complex number x
IMLOG10(x)	Returns the base-10 log of complex number x
IMLOG2(x)	Returns the base-2 logarithm of complex number x

PRACTICE!

Try out Excel's functions. First try these easy examples:

a. =SQRT(4)
b. =ABS(-7)
c. =FACT(3)
d. =LOG 10(100)

Then try these more difficult examples and check the results with a calculator:

a. =FACT(20)
b. =LN(2)
c. =EXP(-0.4)

4.6 TRIGONOMETRIC FUNCTIONS

Excel provides all of the common *trigonometric functions*, such as **SIN(x)**, **COS(x)**, and **SINH(x)**. The x in these functions is an angle, measured in radians. If your angles are in degrees, you don't have to convert them to radians by hand; Excel provides a **RADIANS** function to convert angles in degrees to radians. This is illustrated in Figure 4.22.

Similarly, the **DEGREES** function takes an angle in radians and returns the same angle in degrees. Also, the **PI** function is available whenever π is required in a calculation. Both of these functions have been used in the example in Figure 4.23.

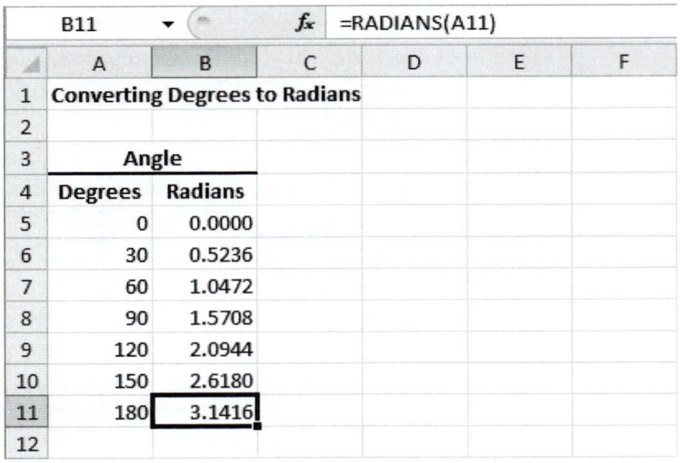

Figure 4.22
Using the **RADIANS**
function.

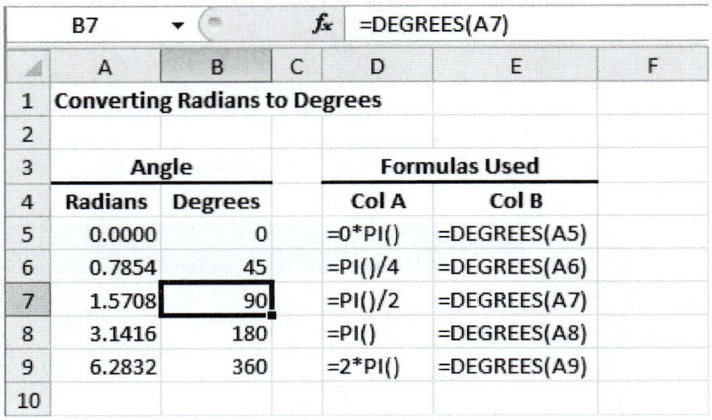

	B11	▼	fx	=RADIANS(A11)		
	A	B	C	D	E	F
1	Converting Degrees to Radians					
2						
3		Angle				
4	Degrees	Radians				
5	0	0.0000				
6	30	0.5236				
7	60	1.0472				
8	90	1.5708				
9	120	2.0944				
10	150	2.6180				
11	180	3.1416				
12						

Figure 4.23
Using the **DEGREES**
function.

	B7	▼	fx	=DEGREES(A7)		
	A	B	C	D	E	F
1	Converting Radians to Degrees					
2						
3		Angle			Formulas Used	
4	Radians	Degrees		Col A	Col B	
5	0.0000	0		=0*PI()	=DEGREES(A5)	
6	0.7854	45		=PI()/4	=DEGREES(A6)	
7	1.5708	90		=PI()/2	=DEGREES(A7)	
8	3.1416	180		=PI()	=DEGREES(A8)	
9	6.2832	360		=2*PI()	=DEGREES(A9)	
10						

4.6.1 Standard Trigonometric Functions

Excel's trigonometric functions are listed in Table 4.5.

Table 4.5 Trigonometric functions

Function Name	Operation
SIN(x)	Returns the sine of x
COS(x)	Returns the cosine of x
TAN(x)	Returns the tangent of x

To test these functions, try sin(30°), which should equal 0.5. The 30° may be converted to radians as a preliminary step, as illustrated in Figure 4.24.

But the conversion to radians and the calculation of the sine can be combined in a single formula, as shown in Figure 4.25.

Figure 4.24
Testing Excel's **SIN** function.

	C5		▾	●		f_x	=SIN(B5)		
	A	B	C	D	E	F			
1	Testing the Sine Function, SIN( ) - Case 1								
2									
3		X	SIN(X)						
4	Degrees:	30							
5	Radians:	0.5236	0.5						
6									

Figure 4.25
Combining the **SIN** and **RADIANS** functions in a formula.

	C4		▾	●		f_x	=SIN(RADIANS(B4))		
	A	B	C	D	E	F			
1	Testing the Sine Function, SIN( ) - Case 2								
2									
3		X	SIN(X)						
4	Degrees:	30	0.5						
5									

KINEMATICS

Projectile Motion I

A projectile is launched at an angle of 35° from the horizontal with velocity equal to 30 m/s. Neglecting air resistance and assuming a horizontal surface, determine how far away from the launch site the projectile will land.

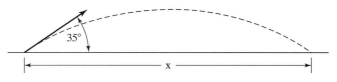

Figure 4.26
Projectile motion.

To answer this problem, we will need:

1. Excel's trigonometry functions to handle the 35° angle
2. Equations relating distance to velocity and acceleration

When the velocity is constant, as in the horizontal motion of our particle (since we're neglecting air resistance), the distance traveled is simply the initial horizontal velocity times the time of flight.

$$x(t) = v_{ox}\, t \tag{4.3}$$

What keeps the projectile from flying forever is gravity. Since the gravitational acceleration is constant, the vertical distance traveled becomes

$$y(t) = v_{oy}\, t + \frac{1}{2}gt^2 \tag{4.4}$$

Because the projectile ends up back on the ground, the final value of y is zero (a horizontal surface was specified), so equation (4.4) can be used to determine the time of flight, t.

The initial velocity is stated (30 m/s at the angle 35° from the horizontal). We can compute the initial velocity components in the horizontal and vertical directions with Excel's trigonometry functions as shown in Figure 4.27.

◢	A	B	C	D	E	F
1	Projectile Motion I					
2						
3		Angle:	35	degrees		
4		V_0:	30	m/s		
5						
6		V_{0y}:	17.2	m/s		
7		V_{0x}:	24.6	m/s		
8						

Figure 4.27
Determining the initial horizontal and vertical velocity components.

In Figure 4.27, cells C6 and C7 contain the following formulas:

C6: =C4*SIN(RADIANS(C3))
C7: =C4*COS(RADIANS(C3))

The **RADIANS** function has been used to convert 35° to radians for compatibility with Excel's trigonometric functions.

We now use equation (4.4) to solve for the time of flight in Figure 4.28.

C12			f_x	=2*C6/C11		
◢	A	B	C	D	E	F
1	Projectile Motion I					
2						
3		Angle:	35	degrees		
4		V_0:	30	m/s		
5						
6		V_{0y}:	17.2	m/s		
7		V_{0x}:	24.6	m/s		
8						
9	Time of Flight					
10		y:	0	m		
11		g:	9.8	m/s^2		
12		t:	3.51	s		
13						

Figure 4.28
Solving for time of flight.

The time of flight can then be used in equation (4.3) to find the horizontal distance traveled. This is illustrated in Figure 4.29.

| C16 | | f_x | =C7*C12 |

	A	B	C	D	E	F
7		V_{0x}:	24.6	m/s		
8						
9	**Time of Flight**					
10		y:	0	m		
11		g:	9.8	m/s^2		
12		t:	3.51	s		
13						
14	**Horizontal Travel Distance**					
15						
16		x:	86.3	m		
17						

Figure 4.29
Determine the horizontal distance travelled.

PRACTICE!

If the launch angle is changed to 55°:

1. What is the time of flight?
2. How far away will the projectile land?
3. What is the maximum height the projectile will reach? (Without air resistance, the maximum height is attained at half the time of flight.)

[Answers: 5 seconds, 86 m, 61 m]

4.6.3 Inverse Trigonometric Functions

Excel's *inverse trigonometric functions* are listed in Table 4.6. These functions return an angle in radians (the **DEGREES** function is available if you would rather see the result in degrees).

Table 4.6 Inverse trigonometric functions

Function Name	Operation
ASIN(x)	Returns the angle (between $-\pi/2$ and $\pi/2$) that has a sine value equal to x
ACOS(x)	Returns the angle (between 0 and π) that has a cosine value equal to x
ATAN(x)	Returns the angle (between $-\pi/2$ and $\pi/2$) that has a tangent value equal to x

STATICS

Resolving Forces

If one person pulls to the right on a rope connected to a hook imbedded in a floor, using the force 400 N at 20° from the horizontal, while another person pulls to the left on the same hook, but using the force 200 N at 45° from the horizontal, what is the net force on the hook? The situation is illustrated in Figure 4.30.

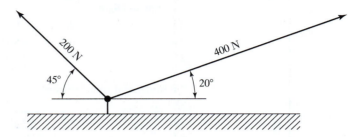

Figure 4.30
Forces on a fixed point.

Because both people are pulling up, their vertical contributions combine; but one is pulling left and the other right, so they are (in part) counteracting each other's efforts. To quantify this distribution of forces, we can calculate the horizontal and vertical components of the force being applied by each person. These components are illustrated in Figure 4.31. Excel's trigonometric functions are useful for calculating components of forces.

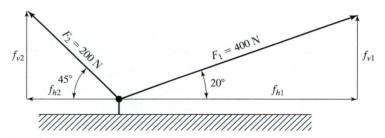

Figure 4.31
Horizontal and vertical components of the applied forces.

The 400-N force from person 1 resolves into a *vertical component*, f_{v1}, and a *horizontal component*, f_{h1}. The magnitudes of these force components can be calculated as follows:

$$f_{v1} = 400 \sin(20°) = 136.8 \text{ N},$$
$$f_{h1} = 400 \cos(20°) = 375.9 \text{ N}. \tag{4.5}$$

Figure 4.32 illustrates how these component forces are calculated using Excel.

Similarly, the 200-N force from the person on the left can be resolved into component forces as

$$f_{v2} = 200 \sin(45°) = 141.4 \text{ N},$$
$$f_{h2} = 200 \cos(45°) = 141.4 \text{ N}. \tag{4.6}$$

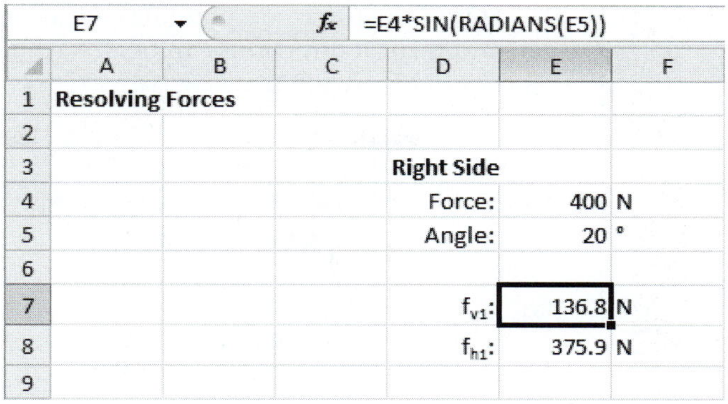

Figure 4.32
Calculating force components (right side) in Excel.

These calculations have been added to the worksheet in Figure 4.33.

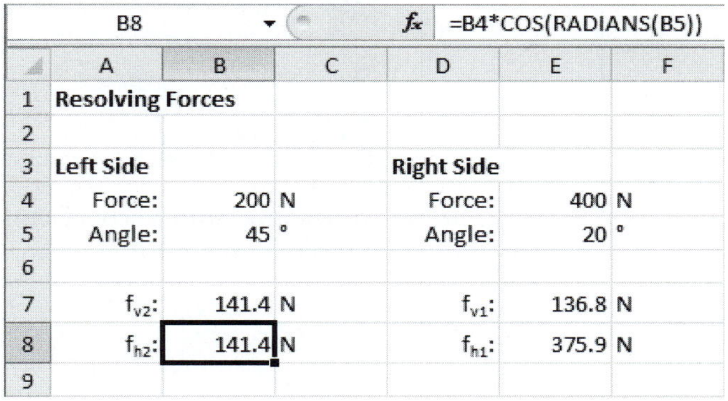

Figure 4.33
Calculating force components (left side) in Excel.

Actually, force component f_{h2} would usually be written as $f_{h2} = -141.421$ N, since it is pointed in the negative x direction. If all angles had been measured from the same position (typically counterclockwise from horizontal), the angle on the 200-N force would have been 135°, and the signs would have taken care of themselves, as shown in the updated worksheet in Figure 4.34.

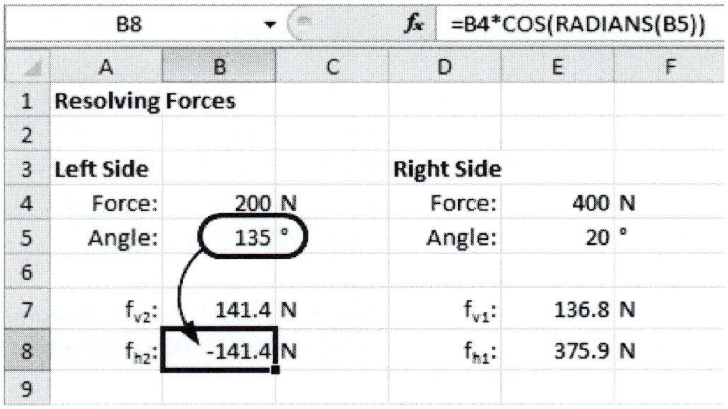

Figure 4.34
Calculating force components using angles measured from the same position.

Once the force components have been computed, the net force in the horizontal and vertical directions can be determined. This has been done in the worksheet shown in Figure 4.35.

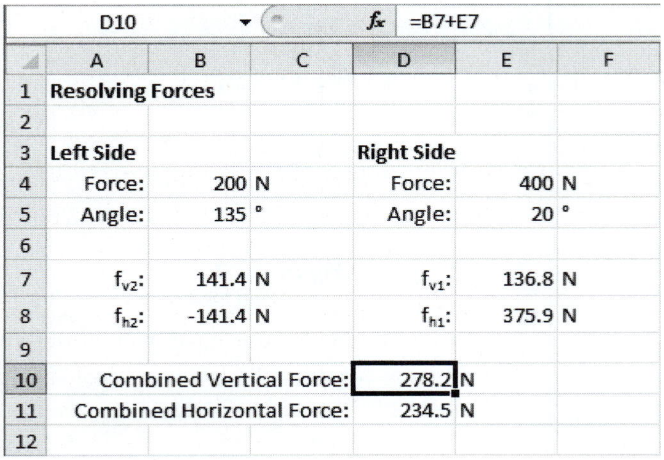

	D10			f_x	=B7+E7	
	A	B	C	D	E	F
1	**Resolving Forces**					
2						
3	**Left Side**			**Right Side**		
4	Force:	200	N	Force:	400	N
5	Angle:	135	°	Angle:	20	°
6						
7	f_{v2}:	141.4	N	f_{v1}:	136.8	N
8	f_{h2}:	-141.4	N	f_{h1}:	375.9	N
9						
10	Combined Vertical Force:			278.2	N	
11	Combined Horizontal Force:			234.5	N	
12						

Figure 4.35
Calculating net forces in the horizontal and vertical directions.

The net horizontal and vertical components can be recombined to find a combined net force on the hook, F_{net}, at angle θ:

$$F_{net} = \sqrt{f_h^2 + f_v^2} = 363.84 \text{ N},$$

$$\theta = a\tan\left(\frac{f_v}{f_h}\right) = 49.88°. \tag{4.7}$$

These calculations are shown in Figure 4.36.

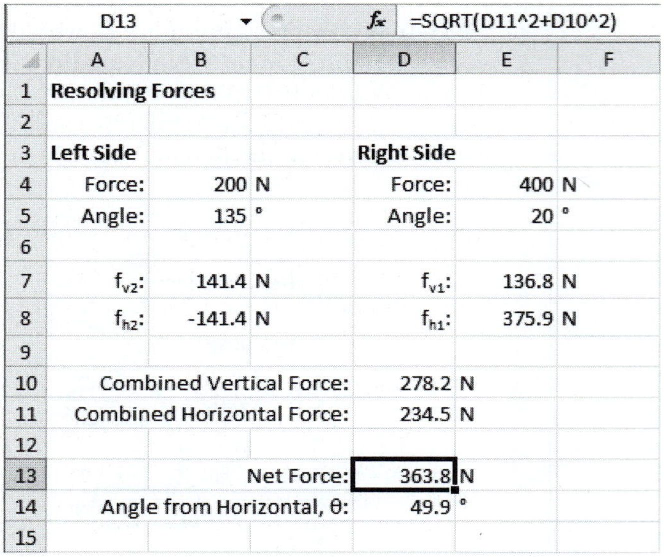

	D13			f_x	=SQRT(D11^2+D10^2)	
	A	B	C	D	E	F
1	**Resolving Forces**					
2						
3	**Left Side**			**Right Side**		
4	Force:	200	N	Force:	400	N
5	Angle:	135	°	Angle:	20	°
6						
7	f_{v2}:	141.4	N	f_{v1}:	136.8	N
8	f_{h2}:	-141.4	N	f_{h1}:	375.9	N
9						
10	Combined Vertical Force:			278.2	N	
11	Combined Horizontal Force:			234.5	N	
12						
13	Net Force:			363.8	N	
14	Angle from Horizontal, θ:			49.9	°	
15						

Figure 4.36
Calculating the net force and resultant angle.

The formulas in cells D13 and D14 are:

```
D13: =SQRT(D10^2+D11^2)
D14: =DEGREES(ATAN(D11/D10))
```

At the beginning of this application example the question was asked: What is the net force on the hook? The answer is 363.8 N at an angle of 49.9° from horizontal.

4.6.4 Hyperbolic Trigonometric Functions

Excel provides the common *hyperbolic trigonometric* and *inverse hyperbolic trigonometric functions*; they are listed in Table 4.7.

Table 4.7 Hyperbolic trigonometric functions

Function Name	Operation
SINH (x)	Returns the hyperbolic sine of x
COSH (x)	Returns the hyperbolic cosine of x
TANH (x)	Returns the hyperbolic tangent of x
ASINH (x)	Returns the inverse hyperbolic sine of x
ACOSH (x)	Returns the inverse hyperbolic cosine of x
ATANH (x)	Returns the inverse hyperbolic tangent of x

PRACTICE!

Use Excel's trigonometric functions to evaluate each of the following:

a. $\sin(\pi/4)$
b. $\sin(90°)$ (Don't forget to convert to radians.)
c. $\cos(180°)$
d. $\text{asin}(0)$
e. $\text{acos}(0)$

4.7 ADVANCED MATH FUNCTIONS

Some of the built-in functions in Excel are pretty specialized. The advanced math functions described here will be very useful to engineers in certain disciplines and of little use to many others.

4.7.1 Logical Functions

Excel provides the *logical functions* listed in Table 4.8.

A simple test of these functions is shown in Figure 4.37, where two unequal values are tested with an **IF** function to see if x_1 is less than x_2.

In this example, Excel tests to see whether 3 is less than 4. Because the test is true, the value **TRUE** is returned.

The following example (Figure 4.38) shows how a worksheet might be used to monitor the status of a tank being filled.

Table 4.8 Logical functions

Function Name	Operation
IF(*test, Tvalue, Fvalue*)	Performs the operation specified by the test argument and then returns *Tvalue* if the test is true, *Fvalue* if the test is false
TRUE()	Returns the logical value TRUE
FALSE()	Returns the logical value FALSE
NOT(*test*)	Reverses the logic returned by the test operation. If test returns TRUE, then NOT (*test*) returns FALSE
AND(*x1, x2, ...*)	Returns TRUE if all arguments are true, FALSE if any argument is false
OR(*x1, x2, ...*)	Returns TRUE if any argument is true, FALSE if all arguments are false

Figure 4.37
Comparing two values with an **IF** function.

	A	B	C	D	E	F
1						
2		x_1:	3			
3		x_2:	4			
4						
5		test:	TRUE			
6						

C5 f_x =IF(C2<C3,TRUE(),FALSE())

B9 f_x =IF(B5>B4,TRUE(),FALSE())

	A	B	C	D	E	F
1	**Tank Monitor**					
2						
3	Tank Operating Capacity:	1200	liters			
4	Tank Maximum Capacity:	1350	liters			
5	Actual Volume in Tank:	800	liters			
6	Is the tank filling?	TRUE				
7					**Formulas Used**	
8	Is the tank full to operating capacity?	FALSE			=IF(B5>=B3,TRUE(),FALSE())	
9	Is the tank overflowing?	FALSE			=IF(B5>B4,TRUE(),FALSE())	
10						
11	Operator Action:	No action required			=IF(B9,"SHUT THE VALVE!","No action required")	
12						

Figure 4.38
A worksheet to monitor a tank-filling operation.

In this example,

- The **IF** function in cell B8 checks to see whether the volume in the tank (B5) has reached or exceeded the operating volume (B3).
- The **IF** function in cell B9 checks to see whether the volume in the tank (B5) has reached or exceeded the tank capacity (B4).

If the volume has exceeded the tank capacity (i.e., if the tank is overflowing), the **IF** function in cell B11 tells the operator to shut the valve. To test this, let's make the actual volume in the tank equal to the tank capacity (1350 liters). The result is shown in Figure 4.39.

	A	B	C	D	E	F
1	**Tank Monitor**					
2						
3	Tank Operating Capacity:	1200	liters			
4	Tank Maximum Capacity:	1350	liters			
5	Actual Volume in Tank:	1350	liters			
6	Is the tank filling?	TRUE				
7					**Formulas Used**	
8	Is the tank full to operating capacity?	TRUE			=IF(B5>=B3,TRUE(),FALSE())	
9	Is the tank overflowing?	TRUE			=IF(B5>=B4,TRUE(),FALSE())	
10						
11	Operator Action:	SHUT THE VALVE!			=IF(B9,"SHUT THE VALVE!","No action required")	
12						

Figure 4.39
What happens when the tank starts overflowing …

PRACTICE!

How would you modify the preceding worksheet to give the operators instructions in the following situations?

1. If the volume in the tank reaches or exceeds the operating volume, tell the operators to shut the valve. [ANSWER: =IF(B8, "Shut the valve.", "none required")]
2. If the volume in the tank is less than 200 liters and the tank is not filling, tell the operators to open the valve. [ANSWER: =IF(AND(B5<200,NOT(B6)),"Open the valve", "none required")]

4.8 ERROR FUNCTION

Excel provides two functions for working with *error functions*: **ERF** and **ERFC** (**Table 4.9**). **ERF**(x) returns the error function integrated between 0 and x, defined as

$$\text{ERF}(x) = \frac{2}{\sqrt{\pi}} \int_0^x e^{-t^2} \, dt. \tag{4.8}$$

Or you can specify two integration limits, as ERF($x1$, $x2$):

$$\text{ERF}(x_1, x_2) = \frac{2}{\sqrt{\pi}} \int_{x_1}^{x_2} e^{-t^2} \, dt. \tag{4.9}$$

The complementary error function, **ERFC**(x), is also available:

$$\text{ERFC}(x) = 1 - \text{ERF}(x) \tag{4.10}$$

Table 4.9 Error function functions

Function Name	Operation
ERF(x)	Returns the error function integrated between 0 and x
ERF(x1,x2)	Returns the error function integrated between x_1 and x_2.
ERFC(x)	Returns the complementary error function integrated between x and ∞

4.9 BESSEL FUNCTIONS

Bessel functions are commonly needed when integrating differential equations in cylindrical coordinates. These functions are complex and commonly available in tabular form. Excel's functions for using Bessel and modified Bessel functions are listed in Table 4.10.

Table 4.10 Bessel functions

Function Name	Operation
BESSELJ(x,n)	Returns the Bessel function, J_n, of x (N is the order of the Bessel function)
BESSELY(x,n)	Returns the Bessel function, Y_n, of x
BESSELI(x,n)	Returns the modified Bessel function, I_n, of x
BESSELK(x,n)	Returns the modified Bessel function, K_n, of x

The worksheet shown in Figure 4.40 uses the **BESSELJ(x,n)** function to create graphs of the $J_0(x)$ and $J_1(x)$ Bessel functions.

In Figure 4.40, the **BESSELJ** function has been used as follows:

```
Cell B4:  =BESSELJ(A4,0)
Cell C4:  =BESSELJ(A4,1)
```

4.10 WORKING WITH COMPLEX NUMBERS

Excel does provide functions for handling *complex numbers,* although working with complex numbers by using built-in functions is cumbersome at best. Still, if you occasionally need to handle complex numbers, the functions listed in Table 4.11 are available.

4.11 WORKING WITH BINARY, OCTAL, AND HEXADECIMAL VALUES

Excel provides the functions listed in Table 4.12 for converting *binary, octal, decimal,* and *hexadecimal* values.

In the next example (Figure 4.41), the decimal value 43 has been converted to binary, octal, and hexadecimal. In each case, eight digits were requested, and Excel added leading zeros to the result. The formulas used in column C are listed in column E.

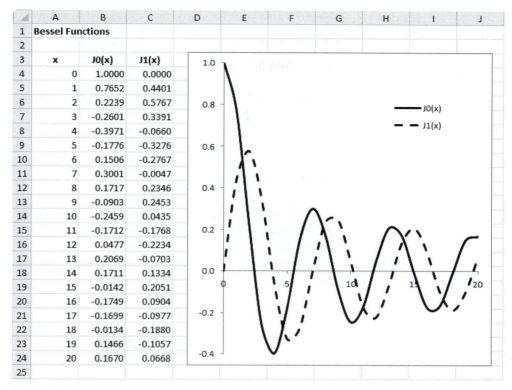

	A	B	C
1	**Bessel Functions**		
2			
3	**x**	**J0(x)**	**J1(x)**
4	0	1.0000	0.0000
5	1	0.7652	0.4401
6	2	0.2239	0.5767
7	3	-0.2601	0.3391
8	4	-0.3971	-0.0660
9	5	-0.1776	-0.3276
10	6	0.1506	-0.2767
11	7	0.3001	-0.0047
12	8	0.1717	0.2346
13	9	-0.0903	0.2453
14	10	-0.2459	0.0435
15	11	-0.1712	-0.1768
16	12	0.0477	-0.2234
17	13	0.2069	-0.0703
18	14	0.1711	0.1334
19	15	-0.0142	0.2051
20	16	-0.1749	0.0904
21	17	-0.1699	-0.0977
22	18	-0.0134	-0.1880
23	19	0.1466	-0.1057
24	20	0.1670	0.0668
25			

Figure 4.40
$J_0(x)$ and $J_1(x)$ Bessel functions.

Table 4.11 Functions for complex numbers

Function Name	Operation
COMPLEX(real, img, suffix)	Combines real and imaginary coefficients into a complex number. The suffix is an optional text argument allowing you to use a "j" to indicate the imaginary portion ("i" is used by default)
IMAGINARY(x)	Returns the imaginary coefficient of complex number x
IMREAL(x)	Returns the real coefficient of complex number x
IMABS(x)	Returns the absolute value (modulus) of complex number x
IMCOS(x)	Returns the cosine of complex number x
IMSIN(x)	Returns the sine of complex number x
IMLN(x)	Returns the natural logarithm of complex number x
IMLOG10(x)	Returns the base-10 logarithm of complex number x
IMLOG2(x)	Returns the base-2 logarithm of complex number x
IMEXP(x)	Returns the exponential of complex number x
IMPOWER(x,n)	Returns the value of complex number x raised to the integer power n
IMSQRT(x)	Returns the square root of complex number x
IMSUM(x1, x2, ...)	Adds complex numbers
IMSUB(x1, x2)	Subtracts two complex numbers
IMPROD(x1, x2, ...)	Determines the product of up to 29 complex numbers
IMDIV(x1, x2)	Divides two complex numbers

Table 4.12 Functions for converting binary, octal, decimal, and hexadecimal values

Function Name	Operation
BIN2OCT(*number, places*)	Converts a binary number to octal. Places can be used to pad leading digits with zeros
BIN2DEC(*number*)	Converts a binary number to decimal
BIN2HEX(*number, places*)	Converts a binary number to hexadecimal
DEC2BIN(*number, places*)	Converts a decimal number to binary
DEC2OCT(*number, places*)	Converts a decimal number to octal
DEC2HEX(*number, places*)	Converts a decimal number to hexadecimal
OCT2BIN(*number, places*)	Converts an octal number to binary
OCT2DEC(*number*)	Converts an octal number to decimal
OCT2HEX(*number, places*)	Converts an octal number to hexadecimal
HEX2BIN(*number, places*)	Converts a hexadecimal number to binary
HEX2OCT(*number, places*)	Converts a hexadecimal number to octal
HEX2DEC(*number*)	Converts a hexadecimal number to decimal

Figure 4.41
Converting numbers.

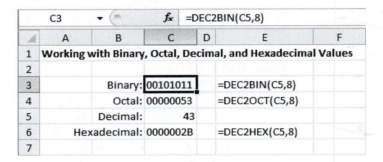

4.12 MISCELLANEOUS FUNCTIONS

Here are a few built-in functions that, although less commonly used, can be very handy on occasion.

4.12.1 Working with Random Numbers

The **RAND** and **RANDBETWEEN** functions generate *random numbers*. The **RAND** function returns a value greater than or equal to zero and less than 1. The **RANDBETWEEN**(*low, high*) function returns a value between the *low* and *high* values you specify.

Excel also provides a *random number tool* that will create columns of random numbers with unique properties, such as a column of normally distributed random numbers with a specific mean and standard deviation. This random number tool is part of Excel's Analysis ToolPak Add-In which is installed, but not activated when Excel is installed. You activate Add-Ins from the Excel Options dialog, as

- Excel 2010: **File tab/Options**
- Excel 2007: **Office/Excel Options**

Once the Excel Options dialog is open (illustrated in Figure 4.42), use the **Add-Ins** panel. The available Excel Add-Ins will vary greatly from one computer to the next, so your screen may look quite different.

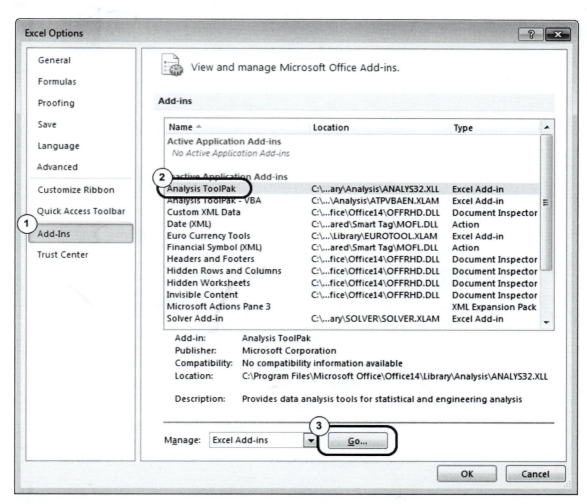

Figure 4.42
Activating the Analysis ToolPak Add-In.

The Add-Ins panel lists Active Application Add-Ins and Inactive Application Add-Ins.

- If the Analysis ToolPak is in the active list, you're good to go.
- If the Analysis ToolPak is not on any list, it wasn't installed; you will have to get out the CD's to install the Analysis ToolPak.
- If it is in the inactive list, you need to activate it. To activate the Analysis ToolPak:
  1. Click on the **Analysis ToolPak** list item (the name in the list).
  2. Click the **Go ...** button.
     These steps will cause the Add-Ins dialog to open, as shown in Figure 4.43.

Figure 4.43
The Add-Ins dialog.

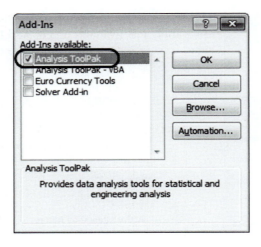

3. Check the box labeled **Analysis ToolPak.**
4. Click the **OK** button.

The Ribbon will then be updated, and a new button will appear on the **Data** tab in the Analysis group as the **Data Analysis** button. This is shown in Figure 4.44.

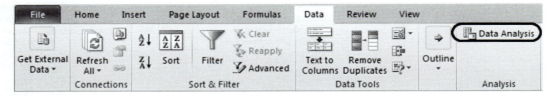

Figure 4.44
Ribbon options **Data/Analysis/Data Analysis** access the Analysis ToolPak.

Clicking on the **Data Analysis** button opens the Data Analysis dialog (Figure 4.45) and displays a list of available analysis tools. Many of these tools are very useful to engineers.

Figure 4.45
The Analysis Tools available through the Data Analysis dialog.

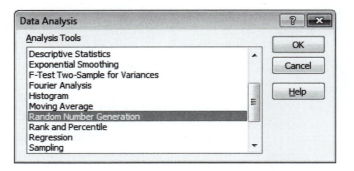

If you select **Random Number Generation** (as in Figure 4.45) and click the **OK** button, the Random Number Generation dialog will open (Figure 4.46). This dialog provides a quick way to create sets of random numbers with various distributions.

Figure 4.46
The Random Number
Generation dialog.

Example: Normally Distributed Random Values

If you want to test some of Excel's statistical functions, you might want to create a set of values that are normally distributed (the "bell curve") with a specific mean and standard deviation.

To generate a set of 15 random numbers, normally distributed with a mean of 5 and a standard deviation of 1, the Random Number Generation dialog should be filled out as shown in Figure 4.47.

Figure 4.47
Using the Random Number
Generation dialog.

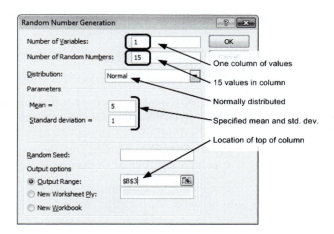

The result (after clicking **OK**) is a set of 15 values placed in the requested position on the worksheet. In Figure 4.48, the mean and standard deviation have been computed to verify that the random number generator did generate values with the requested characteristics.

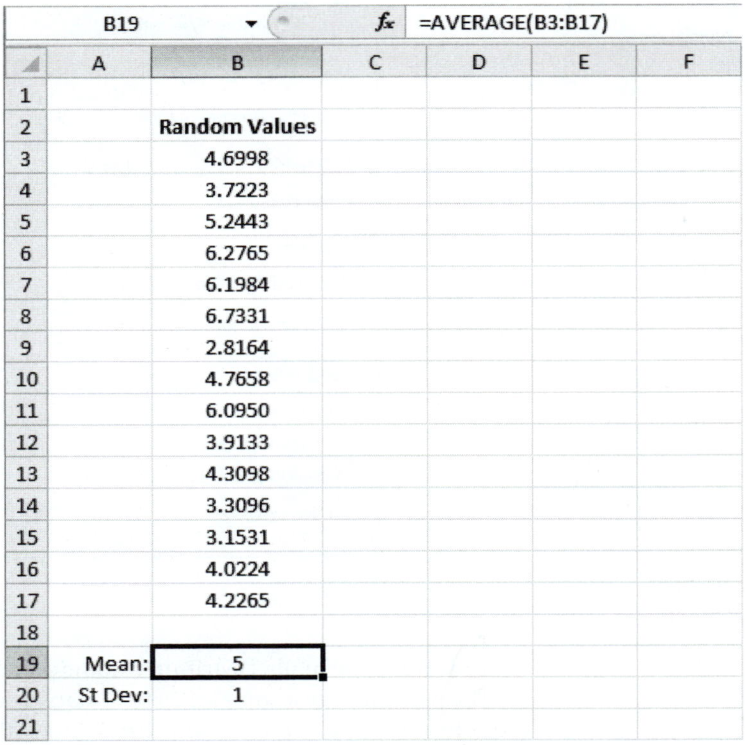

Figure 4.48
The calculated random numbers.

	A	B	C	D	E	F
		B19		f_x	=AVERAGE(B3:B17)	
1						
2		**Random Values**				
3		4.6998				
4		3.7223				
5		5.2443				
6		6.2765				
7		6.1984				
8		6.7331				
9		2.8164				
10		4.7658				
11		6.0950				
12		3.9133				
13		4.3098				
14		3.3096				
15		3.1531				
16		4.0224				
17		4.2265				
18						
19	Mean:	5				
20	St Dev:	1				
21						

Note: The actual mean and standard deviation of the 15 values shown here are 4.6 and 1.2, respectively. Using a small data set, like the 15 values created in this example, you will not get precisely the requested mean and standard deviation (except perhaps, to one significant digit) because these are random values. As the size of the data set gets larger, the data set mean and standard deviation will be closer to the specified values.

4.12.2 Rounding Functions

Excel provides several functions for rounding numbers.

- **ROUND**(*Number, Digits*)—rounds the *Number* to the specified number of Digits.
- **ROUNDUP**(*Number, Digits*)—rounds the *Number* up (away from zero) to the specified number of Digits.
- **ROUNDDOWN**(*Number, Digits*)—rounds the *Number* down (toward zero) to the specified number of Digits.

For example, to round π (3.14159 ...) to three digits, the following formula could be used:

$$=ROUND(PI(),3)$$

The result would be 3.142.

Excel also provides functions that round away from zero to the next *odd* or *even* value.

- **ODD**(*Number*)—rounds the *Number* up (away from zero) to the next odd integer value.

- **EVEN**(*Number*)—rounds the *Number* up (away from zero) to the next even integer value.

If we round π to the next even value using the **EVEN** function, we should get 4. This is illustrated in Figure 4.49.

Note: The **ODD** function rounds zero to 1; **EVEN**(*0*) returns a zero.

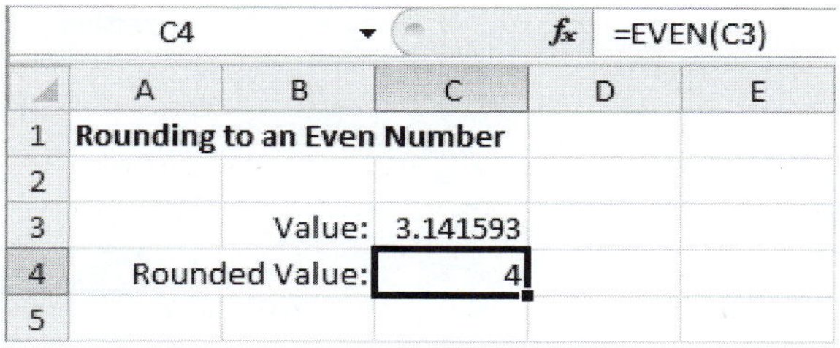

Figure 4.49
Rounding to an even integer using the **EVEN** function.

4.12.3 Date and Time Functions

Excel provides a number of *date* and *time functions* that read the calendar and clock on your computer to make date and time information available to your worksheets. The date and time is stored as a *date–time code*, such as 6890.45834. The number to the left of the decimal point represents the number of days after the defined start date. By default, the start date is January 1, 1900, for PCs, and January 1, 1904, for Macintosh computers. You can tell Excel to use a non-default date system by modifying the Excel options. For example, to use Macintosh-style date–time codes on a PC, first open the Excel Options dialog, as:

- Excel 2010: **File tab/Options**
- Excel 2007: **Office/Excel Options**

Then select the **Advanced** panel. Find the **When Calculating this Workbook** section and check **Use 1904 Date System**.

The date–time code 6890.45833 represents a unique moment in history; the eleventh hour ($11/24 = 0.45833$) of the eleventh day, of the eleventh month of 1918: the time of the signing of the armistice that ended World War I. November 11, 1918 is 6890 days after the PC start date of 1/1/1900.

You can find the date code for any date after 1900 (or 1904 if the 1904 date system is in use) by using the **DATE**(*year, month, day*) function. Using the **DATE** function with Armistice Day returns the date value, 6890, as shown in Figure 4.50.

Excel tries to make date codes more readable to the user by formatting cells containing date codes with a *Date format*. The worksheet in the preceding example actually displayed 11/11/18 in cell B8 until the formatting for that cell was changed from date format to Number format.

If you enter a date or time into a cell, Excel converts the value to a date–time code, but formats the cell to display the date or the time. Some examples are shown in Figure 4.51.

Date–time codes (also called *serial date values*) are used to allow Excel to perform calculations with dates and times. For example, Excel can calculate the number of days that were left until Christmas on August 12, 1916, by using the **DATE** function (the result is 135 days):

```
=DATE(1916,12,25)-DATE(1916,8,12)
```

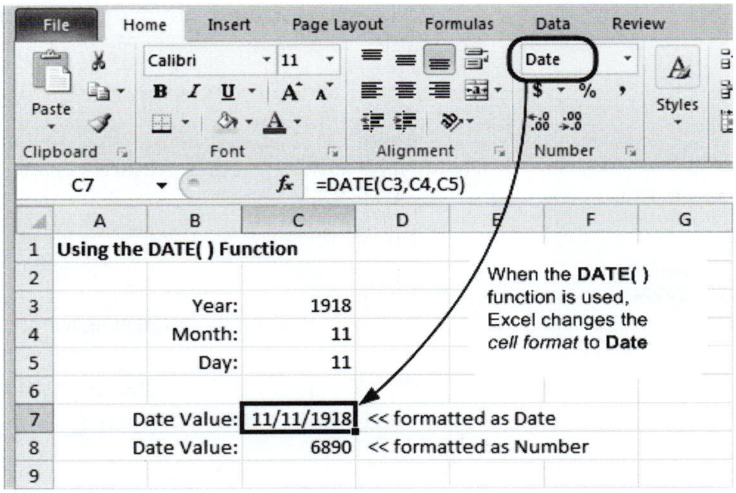

Figure 4.50
Using the **DATE** function to return a date value from a date.

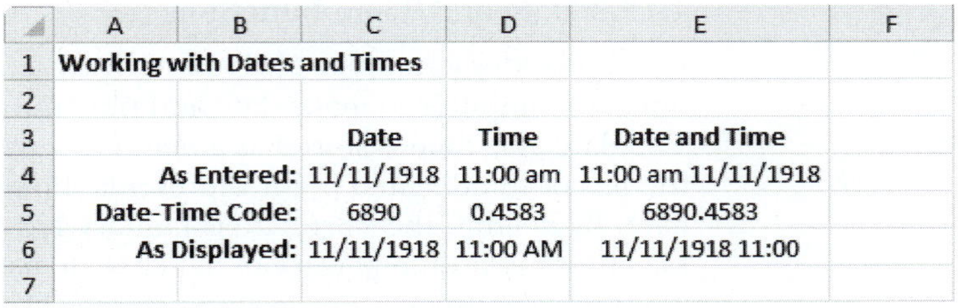

	A	B	C	D	E	F
1	**Working with Dates and Times**					
2						
3			**Date**	**Time**	**Date and Time**	
4		**As Entered:**	11/11/1918	11:00 am	11:00 am 11/11/1918	
5		**Date-Time Code:**	6890	0.4583	6890.4583	
6		**As Displayed:**	11/11/1918	11:00 AM	11/11/1918 11:00	
7						

Figure 4.51
Examples of how Excel converts and displays dates and times.

Excel provides a number of functions for working with dates and times. Some of these functions are listed in Table 4.13.

Table 4.13 Functions for working with dates and times

Function Name	Operation
TODAY()	Returns today's date code. Excel displays the date with a date format
DATE(*year, month, day*)	Returns the date code for the specified *year, month,* and *day.* Excel displays the date code with a date format (shows the date, not the date code). Change the cell format to number or general to see the date code
DATEVALUE(*date*)	Returns the date code for the *date,* which is entered as text such as "12/25/1916" or "December 25, 1916." The quotes are required. Excel displays the date code with a general format (i.e., shows the date code, not the date)
NOW()	Returns the current time code. Excel displays the time with a time format
TIME(*hour, minute, second*)	Returns the time code for the specified *hour, minute,* and *second.* Excel displays the time code with a time format (shows the time, not the time code). Change the cell format to number or general to see the time code
TIMEVALUE(*time*)	Returns the date code for the *time,* which is entered as text such as "11:00 am." The quotes are required. Excel displays the time code with a general format (i.e., shows the time code, not the time)

Table 4.14 Functions for extracting portions of date–time codes

Function Name	Operation
YEAR (*date code*)	Returns the year referred to by the date represented by the *date code*. For example, =YEAR(6890) returns 1918
MONTH (*date code*)	Returns the month referred to by the date represented by the *date code*. For example, =MONTH(6890) returns 11
DAY (*date code*)	Returns the day referred to by the date represented by the *date code*. For example, =DAY(6890) returns 11
HOUR (*time code*)	Returns the hour referred to by the time represented by the *time code* or date–time code. For example, =HOUR(6890.45833) returns 11. So does =HOUR(0.45833)
MINUTE (*time code*)	Returns the minute referred to by the time represented by the *time code* or date–time code
SECOND (*time code*)	Returns the second referred to by the time represented by the *time code* or date–time code

Excel also provides a series of functions for pulling particular portions from a date–time code (Table 4.14).

When using Excel's date–time codes, remember the following:

• Date–time codes work only for dates on or after the starting date (1/1/1900 or 1/1/1904).
• Using four-digit years avoids ambiguity. By default, 00 through 29 is interpreted to mean 2000 through 2029, while 30 through 99 is interpreted to mean 1930 through 1999.
• Excel tries to be helpful when working with date–time codes. You can enter a date as text when the function requires a date–time code, and Excel will convert the date to a date–time code before sending it to the function. This is illustrated in Figure 4.52.

Figure 4.52
Sending dates into functions that use date–time codes.

Note: Cell C3 in Figure 4.52 was formatted as Text before entering the date to keep Excel from immediately recognizing and converting the date.

• Whenever Excel interprets an entry in a cell as a date, it changes the format to Date format and changes the cell contents to a date code. This can be frustrating if the cell contents happen to look like a date, but are not intended to be interpreted as such. For example, if several part sizes are indicated in inches as 1/8, 1/10, and 1/16, they would appear in a worksheet as shown in Figure 4.53.

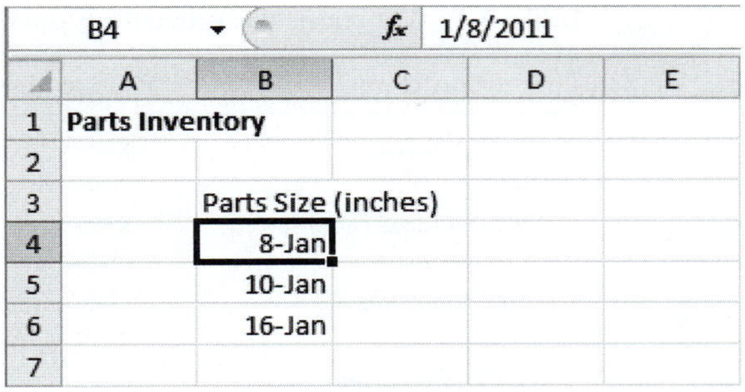

Figure 4.53
How Excel can incorrectly interpret values as dates.

Excel interpreted the 1/8 typed into cell B4 as a date (January 8th of the current year), converted the entry to a date code, and changed the cell formatting to display the date. Because Excel has changed the cell contents to a date code, simply fixing the cell format will not give you the fraction back. You need to tell Excel that it should interpret the entry as a fraction by setting the format for cells B4:B6 to Fraction before entering the values in the cell.

4.12.4 Text-Handling Functions

Excel's standard *text-manipulation functions* are listed in Table 4.15.

4.12.5 Lookup and Reference Functions

Because of the tabular nature of an Excel worksheet, lists of numbers are common and Excel provides ways to look up values in tables. Excel's *lookup functions* are summarized in Table 4.16.

Table 4.15 Text-handling functions

Function Name	Operation
CHAR (*number*)	Returns the character specified by the code *number*
CONCATENATE (*text1, text2*)	Joins several text items into one text item
	(The & operator can also be used for concatenation)
EXACT (*text1, text2*)	Checks to see if two text values are identical. Returns **TRUE** if the text strings are identical, otherwise **FALSE** (case sensitive)
FIND (*text_to_find, text_to_search, start_pos*)	Finds one text value within another (case sensitive)
LEFT (*text, n*)	Returns the leftmost *n* characters from a text value
LEN (*text*)	Returns the number of characters in text string, *text*
LOWER (*text*)	Converts *text* to lowercase
MID (*text, start_pos, n*)	Returns *n* characters from a text string starting at *start_pos*
REPLACE (*old_text, start_pos, n, new_text*)	Replaces *n* characters of *old_text* with *new_text* starting at *start_pos*
RIGHT (*text, n*)	Returns the rightmost *n* characters from a text value
SEARCH (*text_to_find, text_to_search, start_pos*)	Finds one text value within another (not case sensitive)
TRIM (*text*)	Removes spaces from *text*
UPPER (*text*)	Converts *text* to uppercase

Table 4.16 Lookup and reference functions

Function Name	Operation
VLOOKUP	Vertical lookup
(*value, table, N, matchType*)	Looks up a value in the left column of a *table,* jumps to column *N* of the table (the matched column is column 1), and returns the value in that location
	If *matchType* is set to **TRUE,** the lookup will fail unless the match is exact. If *matchType* is omitted or set to **FALSE,** Excel will use the next highest value as the match
HLOOKUP	Horizontal Lookup
(*value, table, N, matchType*)	Looks up a value in the top row of a *table,* jumps down to row *N* of the same table (same column), and returns the value in that location

The worksheet example in Figure 4.54 uses the **VLOOKUP** function to determine letter grades for students from a grade table.

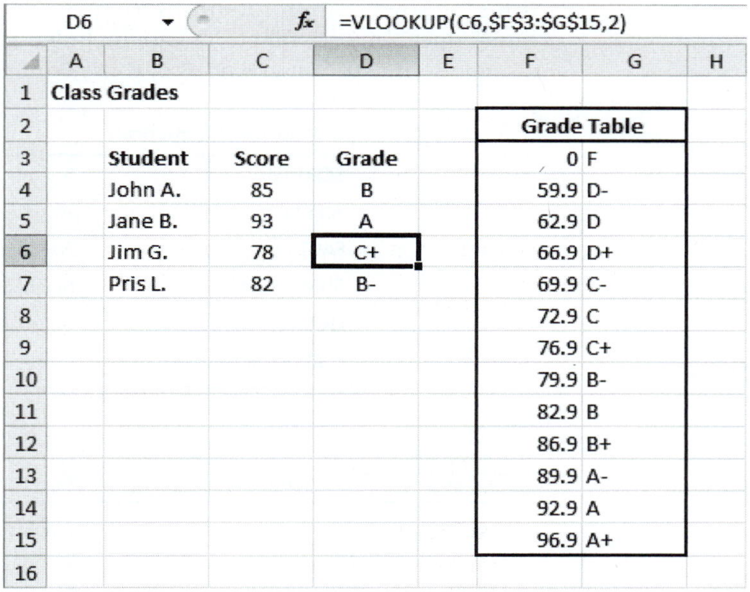

Figure 4.54
Looking up grades.

KEY TERMS

- Argument list
- Arguments
- AutoSum button
- Bessel functions
- Binary
- Complex numbers
- Date format
- Date functions
- Date–time code
- Decimal
- Error functions
- Even
- Exponentiation functions
- Function
- Function name
- Hexadecimal
- Horizontal component
- Hyperbolic trigonometric functions
- Inverse hyperbolic trigonometric functions
- Inverse trigonometric functions

SUMMARY

A *function* is a reusable equation that accepts arguments, uses the argument values in a calculation, and returns the result(s). In this section, the commonly used built-in functions in Excel are collected and displayed in tabular format. Also, descriptions of some other functions (matrix math functions, statistical functions, time-value-of-money functions) are included in the section in an attempt to provide a single source for descriptions of those commonly used functions.

Common math functions

SQRT (*x*)	Square root
ABS (*x*)	Absolute value
FACT (*x*)	Factorial
SUM (*range*)	Summation
GCD (*x1, x2, ...*)	Greatest common divisor
LCM (*x1, x2, ...*)	Least common multiple

Log and Exponentiation	Operation
EXP (*x*)	Returns e raised to the power *x*
LN (*x*)	Returns the natural log of *x*
LOG10 (*x*)	Returns the base-10 log of *x*
LOG (*x, base*)	Returns the logarithm of *x* to the specified base

Trigonometry Functions	Operation
SIN (*x*)	Returns the sine of *x*
COS (*x*)	Returns the cosine of *x*
TAN (*x*)	Returns the tangent of *x*
RADIANS (*x*)	Converts *x* from degrees to radians
DEGREES (*x*)	Converts *x* from radians to degrees

Inverse Trignometric Functions	Operation
ASIN (*x*)	Returns the angle (between $-\pi/2$ and $\pi/2$) that has a sine value equal to *x*
ACOS (*x*)	Returns the angle (between 0 and π) that has a cosine value equal to *x*
ATAN (*x*)	Returns the angle (between $-\pi/2$ and $\pi/2$) that has a tangent value equal to *x*

Hyperbolic Trignometric Functions	Operation
SINH (*x*)	Returns the hyperbolic sine of *x*
COSH (*x*)	Returns the hyperbolic cosine of *x*
TANH (*x*)	Returns the hyperbolic tangent of *x*
ASINH (*x*)	Returns the inverse hyperbolic sine of *x*
ACOSH (*x*)	Returns the inverse hyperbolic cosine of *x*
ATANH (*x*)	Returns the inverse hyperbolic tangent of *x*

Logical Functions	Operation
IF(*test, Tvalue, Fvalue*)	Performs the operation specified by the test argument and then returns *Tvalue* if the *test* is true, or *Fvalue* if the *test* is false
TRUE ()	Returns the logical value TRUE
FALSE ()	Returns the logical value FALSE
NOT (*test*)	Reverses the logic returned by the test operation. If *test* returns TRUE, then NOT (*test*) returns FALSE
AND (*x1, x2, ...*)	Returns TRUE if all arguments are true or FALSE if any argument is false
OR (*x1, x2, ...*)	Returns TRUE if any argument is true or FALSE if all arguments are false

Error Functions	Operation
ERF (*x*)	Returns the error function integrated between 0 and *x*
ERF (*x1, x2*)	Returns the error function integrated between x_1 and x_2.
ERFC (*x*)	Returns the complementary error function integrated between *x* and ∞.

Bessel Functions	Operation
BESSELJ (*x, n*)	Returns the Bessel function, J_n, of *x* (*n* is the order of the Bessel function)
BESSELY (*x, n*)	Returns the Bessel function, Y_n, of *x*
BESSELI (*x, n*)	Returns the modified Bessel function, I_n, of *x*
BESSELK (*x, n*)	Returns the modified Bessel function, K_n, of *x*

Complex Number Functions	Operation
COMPLEX(*real, img, suffix*)	Combines real and imaginary coefficients into a complex number. The *suffix* is an optional text argument in case you wish to use a "j" to indicate the imaginary portion ("i" is used by default)
IMAGINARY (*x*)	Returns the imaginary coefficient of complex number *x*
IMREAL (*x*)	Returns the real coefficient of complex number *x*
IMABS (*x*)	Returns the absolute value (modulus) of complex number *x*
IMCOS (*x*)	Returns the cosine of complex number *x*
IMSIN (*x*)	Returns the sine of complex number *x*
IMLN (*x*)	Returns the natural logarithm of complex number *x*
IMLOG10 (*x*)	Returns the base-10 logarithm of complex number *x*
IMLOG2 (*x*)	Returns the base-2 logarithm of complex number *x*
IMEXP (*x*)	Returns the exponential of complex number *x*
IMPOWER (*x, n*)	Returns the value of complex number *x* raised to the integer power *n*
IMSQRT (*x*)	Returns the square root of complex number *x*
IMSUM (*x1, x2, ...*)	Adds complex numbers
IMSUB (*x1, x2*)	Subtracts two complex numbers
IMPROD (*x1, x2, ...*)	Computes the product of up to 29 complex numbers
IMDIV (*x1, x2*)	Divides two complex numbers

Conversion Functions	Operation
BIN2OCT(*number, places*)	Converts a binary *number* to octal. *Places* can be used to pad leading digits with zeros
BIN2DEC (*number*)	Converts a binary *number* to decimal
BIN2HEX (*number, places*)	Converts a binary *number* to hexadecimal
DEC2BIN (*number, places*)	Converts a decimal *number* to binary
DEC2OCT (*number, places*)	Converts a decimal *number* to octal
DEC2HEX (*number, places*)	Converts a decimal *number* to hexadecimal
OCT2BIN (*number, places*)	Converts an octal *number* to binary
OCT2DEC (*number*)	Converts an octal *number* to decimal
OCT2HEX (*number, places*)	Converts an octal *number* to hexadecimal
HEX2BIN (*number, places*)	Converts a hexadecimal *number* to binary
HEX2OCT (*number, places*)	Converts a hexadecimal *number* to octal
HEX2DEC (*number*)	Converts a hexadecimal *number* to decimal

Date and Time Functions	Operation
TODAY()	Returns today's date code. Excel displays the date with a date format
DATE(*year, month, day*)	Returns the date code for the specified *year, month*, and *day*. Excel displays the date code with a date format (i.e., shows the date, not the date code). Change the cell format to number or general to see the date code

DATEVALUE (*date*)	Returns the date code for the *date,* which is entered as text such as "12/25/1916" or "December 25, 1916." The quotes are required. Excel displays the date code with a general format (i.e., shows the date code, not the date)
NOW ()	Returns the current time code. Excel displays the time with a time format
TIME(*hour, minute, second*)	Returns the time code for the specified *hour, minute,* and *second.* Excel displays the time code with a time format (i.e., shows the time, not the time code). Change the cell format to number or general to see the time code
TIMEVALUE (*time*)	Returns the date code for the *time,* which is entered as text such as "11:00 am." The quotes are required. Excel displays the time code with a general format (i.e., shows the time code, not the time)
YEAR (*date code*)	Returns the year referred to by the date represented by the *date code.* Example: =YEAR(6890) returns 1918
MONTH (*date code*)	Returns the month referred to by the date represented by the *date code.* Example: =MONTH(6890) returns 11
DAY (*date code*)	Returns the day referred to by the date represented by the *date code.* Example: =DAY(6890) returns 11
HOUR (*time code*)	Returns the hour referred to by the time represented by the *time code* or date–time code. Example: =HOUR (6890.45833) returns 11. So does =HOUR(0.45833)
MINUTE (*time code*)	Returns the minute referred to by the time represented by the *time code* or date–time code
SECOND (*time code*)	Returns the second referred to by the time represented by the *time code* or date–time code

Text Functions	Operation
CHAR (*number*)	Returns the character specified by the code *number*
CONCATENATE (*text1, text2*)	Joins several text items into one text item (The & operator can also be used for concatenation)
EXACT (*text1, text2*)	Checks to see whether two text values are identical. Returns TRUE if the text strings are identical, FALSE otherwise (case sensitive)
FIND (*text_to_find, text_to_search, start_pos*)	Finds one text value within another (case sensitive)
LEFT (*text, n*)	Returns the leftmost *n* characters from a text value
LEN (*text*)	Returns the number of characters in text string, *text*
LOWER (*text*)	Converts *text* to lowercase
MID (*text, start_pos, n*)	Returns *n* characters from a *text* string starting at *start_pos*
REPLACE (*old_text, start_pos, n, new_text*)	Replaces *n* characters of *old_text* with *new_text* starting at *start_pos*
RIGHT (*text, n*)	Returns the rightmost *n* characters from a *text* value
SEARCH (*text_to_find, text_to_search, start_pos*)	Finds one text value within another (not case sensitive)
TRIM (*text*)	Removes spaces from *text*
UPPER (*text*)	Converts *text* to uppercase

Function Name	Operation
VLOOKUP	Vertical lookup
(*value, table, N, matchType*)	Looks up a *value* in the left column of a *table*, jumps to column *N* of the table (to the right of the matched column), and returns the value in that location
	If *matchType* is set to TRUE, the lookup will fail unless the match is exact. If *matchType* is omitted or set to FALSE, Excel will use the next higher value as the match
HLOOKUP	Horizontal lookup
(*value, table, N, matchType*)	Looks up a *value* in the top row of a *table*, jumps down to row *N* of the same table (same column), and returns the value in that location

Statistical Functions	Description
AVERAGE (*range*)	Calculates the arithmetic average of the values in the specified *range* of cells
STDEV (*range*)	Calculates the sample standard deviation of the values in the specified *range* of cells
STDEVP (*range*)	Calculates the population standard deviation of the values in the specified *range* of cells
VAR (*range*)	Calculates the sample variance of the values in the specified *range* of cells
VARP (*range*)	Calculates the population variance of the values in the specified *range* of cells

Time-Value-of-Money Functions	Description
FV(*iP, Nper, Pmt, PV, Type*)	Calculates a future value, *FV*, given a periodic interest rate, i_p, the number of compounding periods, *N*, a periodic payment, *Pmt*, an optional present value, *PV*, and an optional code indicating whether payments are made at the beginning or end of each period, *Type*: *Type* = 1 means payments at the beginning of each period. *Type* = 0 or omitted means payments at the end of each period
PV(*iP, Nper, Pmt, FV, Type*)	Calculates a present value, *PV*, given a periodic interest rate, i_p, the number of compounding periods, *N*, a periodic payment, *Pmt*, an optional future value, *FV*, and an optional code indicating whether payments are made at the beginning or end of each period, *Type*
PMT(*iP, Nper, PV, FV, Type*)	Calculates the periodic payment, *Pmt*, equal to a given present value, *PV*, and (optionally) future value, *FV*, given a periodic interest rate, i_p, the number of compounding periods, *N*. Type is an optional code indicating whether payments are made at the beginning (*Type* =1) or end (*Type* =0) of each period

Internal Rate of Return and Depreciation Functions	Description
IRR(*Incomes, irrGuess*)	Calculates the internal rate of return for the series of incomes and expenses:
	Incomes is a range of cells containing the incomes (and expenses as negative incomes) (required)
	IrrGuess is a starting value, or guess value, for the iterative solution required to calculate the internal rate of return, optional
NPV (*iP, Inc1, Inc2, ...*)	Calculates the net present value, *NPV*, of a series of incomes (*Inc1, Inc2, ...*) given a periodic interest rate, i_P:
	Inc1, Inc2, etc., can be values, cell addresses, or cell ranges containing incomes (and expenses as negative incomes)
VDB (*Cinit, S, NSL, Perstart, Perend, FDB, NoSwitch*)	Calculates depreciation amounts using various methods (straight line, double-declining balance, MACRS)
	Cinit is the initial cost of the asset (required)
	S is the salvage value, required (set to zero for MACRS depreciation)
	NSL is the service life of the asset, required
	Perstart is the start of the period over which the depreciation amount will be calculated, required
	Perstop is the end of the period over which the depreciation amount will be calculated, required
	FDB is declining balance percentage, optional (Excel uses 200% if omitted)
	NoSwitch tells Excel whether to switch to straight-line depreciation when the straight-line depreciation factor is larger than the declining-balance depreciation factor, optional (set to FALSE to switch to straight line, TRUE to not switch; the default is FALSE)

Matrix Math Functions	Description
MMULT(*M1, M2*)	Multiplies matrices *M1* and *M2*. Use [Ctrl–Shift–Enter] to tell Excel to enter the matrix function into every cell of the result matrix
TRANSPOSE(*M*)	Transposes matrix *M*
MINVERSE(*M*)	Calculates the inverse of matrix *M* (if possible)
MDETERM(*M*)	Calculates the determinant of matrix *M*

PROBLEMS

4.1 Trigonometric Functions

Devise a test to demonstrate the validity of the following common trigonometric formulas. What values of A and B should be used to test these functions thoroughly?

a. $\sin(A + B) = \sin(A)\cos(B) + \cos(A)\sin(B)$

b. $\sin(2A) = 2\sin(A)\cos(A)$

c. $\sin^2(A) = \dfrac{1}{2} - \dfrac{1}{2}\cos(2A)$

Note: In Excel, $sin^2(A)$ should be entered as $=(\texttt{sin}(\texttt{A}))\texttt{\^{}}\texttt{2}$. This causes $\sin(A)$ to be evaluated first and that result to be squared.

4.2 Basic Fluid Flow

A commonly used rule of thumb is that the average velocity in a pipe should be about 1 m/s or less for "thin" fluids (viscosity about water). If a pipe needs to deliver 6,000 m^3 of water per day, what diameter is required to satisfy the 1-m/s rule?

4.3 Projectile Motion II

Sports programs' "shot of the day" segments sometimes show across-the-court baskets made just as (or after) the final buzzer sounds. If a basketball player, with three seconds remaining in the game, throws the ball at a 45° angle from 4 feet off the ground, standing 70 feet from the basket, which is 10 feet in the air, as illustrated in Figure 4.55:

  a. What initial velocity does the ball need to have in order to reach the basket?
  b. What is the time of flight? and
  c. How much time will be left in the game after the shot?

Ignore air resistance in this problem.

Figure 4.55
Throwing a basketball.

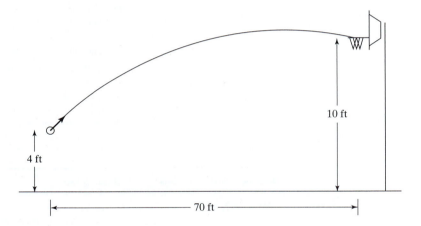

4.4 Pulley I

A 200-kg mass is hanging from a hook connected to a pulley, as shown in the accompanying figure. The cord around the pulley is connected to the overhead support at two points as illustrated in Figure 4.56.

Figure 4.56
Pulley supports.

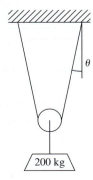

What is the tension in each cord connected to the support if the angle of the cord from vertical is

    a. 0°?
    b. 5°?
    c. 15°?

4.5 Force Components and Tension in Wires I

A 150-kg mass is suspended by wires from two hooks, as shown in the accompanying figure. The lengths of the wires have been adjusted so that the wires are each 50° from horizontal, as illustrated in Figure 4.57. Assume that the mass of the wires is negligible.

Figure 4.57
Forces and tensions in wires I.

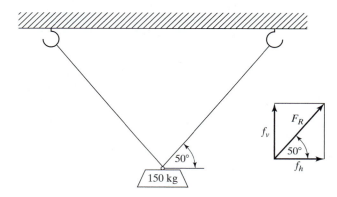

    a. Two hooks support the mass equally, so the vertical component of force exerted by either hook will be equal to the force resulting from 75 kg being acted on by gravity. Calculate this vertical component of force, f_v on the right hook. Express your result in Newtons.
    b. Compute the horizontal component of force, f_h, by using the result obtained in part (a) and trigonometry.
    c. Determine the force exerted on the mass in the direction of the wire F_R (equal to the tension in the wire).
    d. If you moved the hooks farther apart to reduce the angle from 50° to 30° would the tension in the wires increase or decrease? Why?

4.6 Force Components and Tension in Wires II

If two 150-kg masses are suspended on a wire as shown in Figure 4.58, such that the section between the loads (wire B) is horizontal, then wire B is under tension, but is doing no lifting. The entire weight of the 150-kg mass on the right is being held up by the vertical component of the force in wire C. In the same way, the mass on the left is being supported entirely by the vertical component of the force in wire A. What is the tension on wire B?

Figure 4.58
Forces and tensions in wires II.

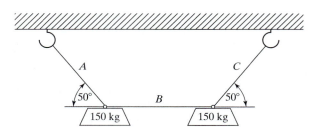

4.7 Force Components and Tension in Wires III

If the hooks shown in the previous problem are pulled farther apart, as illustrated in Figure 4.59, the tension in wire B will change, and the angle of wires A and C with respect to the horizontal will also change.

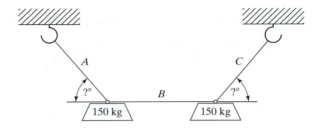

If the hooks are pulled apart until the tension in wire B is 2000 N, compute:

    a. The angle between the horizontal and wire C.
    b. The tension in wire C.

How does the angle in part (a) change if the tension in wire B is increased to 3000 N?

4.8 Finding the Volume of a Storage Bin I

A fairly common shape for a dry-solids storage bin is a cylindrical silo with a conical collecting section at the base where the product is removed (see Figure 4.60).

Figure 4.60
Storage silo.

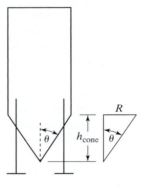

To calculate the volume of the contents, you use the formula for a cone, as long as the height of product, h, is less than the height of the conical section, h_{cone}:

$$V = \tfrac{1}{3}\pi r_h^2 h \qquad \text{if } h < h_{cone}, \qquad (4.11)$$

Here, r_h is the radius at height h and can be calculated from h by using trigonometry:

$$r_h = h_{cone} \tan(\theta). \qquad (4.12)$$

If the height of the stored product is greater than the height of the conical section, the equation for a cylinder must be added to the volume of the cone:

$$V = \tfrac{1}{3}\pi R^2 h_{cone} + \pi R^2 (h - h_{cone}) \qquad \text{if } h > h_{cone}. \qquad (4.13)$$

If the height of the conical section is 3 meters, the radius of the cylindrical section is 2 meters, and the total height of the storage bin is 10 meters, what is the maximum volume of material that can be stored?

4.9 Finding the Volume of a Storage Bin II

Consider the storage bin described in the previous problem.

    a. Calculate the angle θ as shown in the diagram.

    b. For a series of h values between 0 to 10 meters, calculate r_h values and bin volumes.

    c. Plot the volume vs. height data.

4.10 Pulley Problem II

A 200-kg mass is attached to a pulley, as shown in Figure 4.61.

Figure 4.61

Four-pulley system.

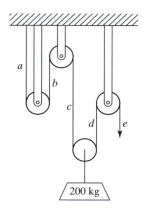

    a. What force must be exerted on cord e to keep the mass from moving?

    b. When the mass is stationary, what is the tension in cords a through e?

    c. Which, if any, of the solid supports connecting the pulleys to the overhead support is in compression?

4.11 Nonideal Gas Equation

The Soave–Redlich–Kwong (SRK) equation is a commonly used equation of state that relates the absolute temperature T, the absolute pressure P, and the molar volume $\hat{V}$ of a gas under conditions in which the behavior of the gas cannot be considered ideal (e.g., moderate temperature and high pressure). The SRK equation[1] is

$$P = \frac{RT}{(\hat{V} - b)} - \frac{\alpha a}{\hat{V}(\hat{V} + b)} \tag{4.14}$$

where R is the ideal gas constant and α, a, and b are parameters specific to the gas, calculated as follows:

$$a = 0.42747\frac{(RT_C)^2}{P_C}$$

$$b = 0.08664\frac{RT_C}{P_C}$$

$$m = 0.48508 + 1.55171\omega - 0.1561\omega^2 \tag{4.15}$$

$$T_r = \frac{T}{T_C}$$

$$\alpha = \left[1 + m\left(1 - \sqrt{T_r}\right)\right]^2$$

---

[1]*From Elementary Principles of Chemical Processes*, 3d ed., R. M. Felder and R. W. Rousseau, New York: Wiley, 2000.

Here,

T_C is the critical temperature of the gas,
P_C is the critical pressure of the gas, and
ω is the Pitzer acentric factor.

Each of these is readily available for many gases.
Calculate the pressure exerted by 20 gram-moles of ammonia at 300 K in a 5-liter container, using

    a. the ideal gas equation,
    b. the SRK equation.

The required data for ammonia are tabulated as follows:

$$
\begin{aligned}
T_C &= 405.5\text{K}, \\
P_C &= 111.3 \text{ atm}, \\
\omega &= 0.25
\end{aligned}
\qquad (4.16)
$$

4.12 Projectile Motion III

Major-league baseball outfield fences are typically about 350 feet from home plate. Home run balls need to be at least 12 feet off the ground to clear the fence. Calculate the minimum required initial velocity (in miles per hour or kilometers per hour) for a home run, for baseballs hit at the angles 10°, 20°, 30°, and 40° from the horizontal.

5 Matrix Operations in Excel

Objectives

After reading this chapter, you will know

- How matrices fit into Excel worksheets
- How to define and name arrays in Excel
- How to carry out standard matrix math operations in Excel:
  - Multiplying matrices by scalar values
  - Matrix addition
  - Matrix multiplication
  - Transposing matrices
  - Inverting a matrix
- How to find the determinant of a matrix
- How to solve systems of simultaneous linear equations

5.1 INTRODUCTION

A *matrix* is a collection of related values, and matrices show up frequently in engineering calculations. Sets of simultaneous linear equations are typically solved using matrix methods, and these occur in numerous areas, including linear programming problems, and in solving differential equations.

Matrix manipulations are a natural for Excel worksheets; the worksheet grid provides a natural home for the columns and rows of a matrix. All standard spreadsheet programs provide the standard matrix math operations. Excel goes a step further and allows many matrix operations to be performed by using *array functions* rather than menu commands. This makes the matrix operations "live"; they will automatically be recalculated if any of the data is changed.

5.2 MATRICES, VECTORS, AND ARRAYS

Excel uses the term *array* to refer to a collection of values organized in rows and columns that should be kept together. For example, when arrays are used in formulas any mathematical operation performed on any element of the array is performed on each element of the array. Also, Excel will not allow you to delete a portion of an array; the entire array must be deleted, or the array must be converted to values using Ribbon options **Home/Clipboard/Paste (menu)/Paste Values** [Excel 2003: Edit/Paste Special/Values]. Functions designed to operate on or return arrays are called *array functions.*

Standard mathematics nomenclature calls a collection of related values organized in rows and columns a *matrix.* A matrix with a single row or column is called a *vector.* In Excel, both would be considered arrays.

Defining and Naming Arrays

You define an array in Excel simply by filling a range of cells with the contents of the array. For example, the 3×2 matrix

$$A = \begin{bmatrix} 1 & 3 \\ 7 & 2 \\ 8 & 11 \end{bmatrix}$$

can be entered into a 3×2 range of cells as shown in Figure 5.1.

Figure 5.1
Matrix A in cells B2:C4.

Naming a range of cells allows you to use the name (e.g., A) in place of the cell range (e.g., B2:C4) in array formulas. Excel's matrix math functions do not require *named arrays,* but naming the cell ranges that contain the arrays is commonly done. Using named arrays not only expedites entering array formulas, it also makes your worksheets easier to read and understand.

To give a name to the range of cells that hold an array:

1. Select the cells containing the array
2. Enter the desired name in the Name box at the left side of the Formula Bar.

In Figure 5.2, the array has been named "A."

Alternatively, a selected range can be assigned a name by using Ribbon options **Formulas/Defined Names/Define Name** [Excel 2003: Insert/Name/Define].

Figure 5.2
Cell range B2:C4 has been named "A."

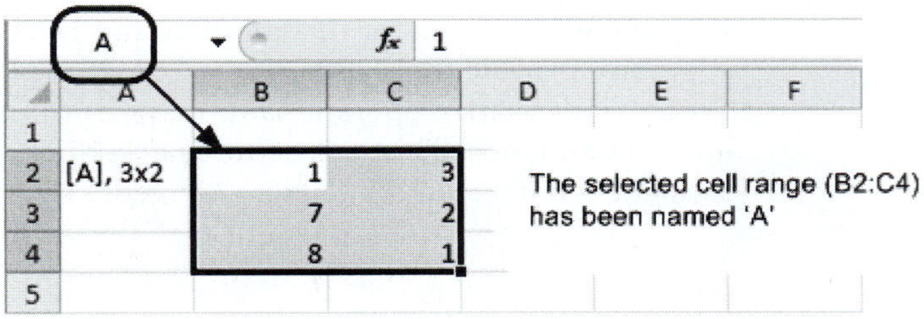

5.3 HOW EXCEL HANDLES MATRIX MATH

Excel provides mechanisms for performing each of the standard matrix operations, but they are accessed in differing ways:

- Addition and *scalar* multiplication are handled through either basic cell arithmetic or array math operations.
- Matrix transposition, multiplication, and inversion are handled by array functions.

The various worksheet programs handle matrix operations, such as transposition, multiplication, and inversion, in different ways. Some programs treat a matrix as a range of values, and matrix operations (from menus, not functions) produce a new range of values. Excel handles standard matrix operations using *array functions*. There are pros and cons to each approach. Excel's array functions require more effort on the user's part to perform matrix transposition, multiplication, and inversion, but, by using array functions, Excel can recalculate the resulting matrices if any of the input data change. When worksheet programs do not use functions for matrix transposition, multiplication, and inversion, they cannot automatically recalculate your matrix results when your worksheet changes.

There are a couple of common features of engineering work that suggest that the *automatic recalculation* provided by array functions is important:

1. Engineers frequently have to start preparing designs with estimates or "guesstimates" (educated guesses) of important parameters. When better values become available, the estimates are replaced with the values. Automatic recalculation allows you to immediately see the impact of the new value on the proposed design.
2. Engineers routinely develop designs that have adjustable parameters, at least in the early stages. Changing the parameter values to evaluate the impact on the design is common, and automatic recalculation is essential.

Because array functions support automatic recalculation, array functions will get most of the attention in this chapter.

5.4 BASIC MATRIX OPERATIONS

5.4.1 Adding Two Matrices

The two matrices to be added must be the same size. Begin by entering the two matrices to be added. Figure 5.3 shows two small matrices, [A] and [B], that will be used to demonstrate matrix addition.

Figure 5.3

Entering the matrices to be added.

▲	A	B	C	D	E
1	**Matrix Addition**				
2	Matrices must be the same size				
3					
4	[A], 3x2	1	3		
5		7	2		
6		8	1		
7					
8	[B], 3x2	4	8		
9		6	1		
10		0	5		
11					

The matrices are added element by element. For example, the top-left cell of the resultant matrix will hold the formula required to add 1 + 4 (cell B4 plus cell B8). The formula would be written as =B4+B8. When this formula is placed in cell B12, the result is displayed (Figure 5.4).

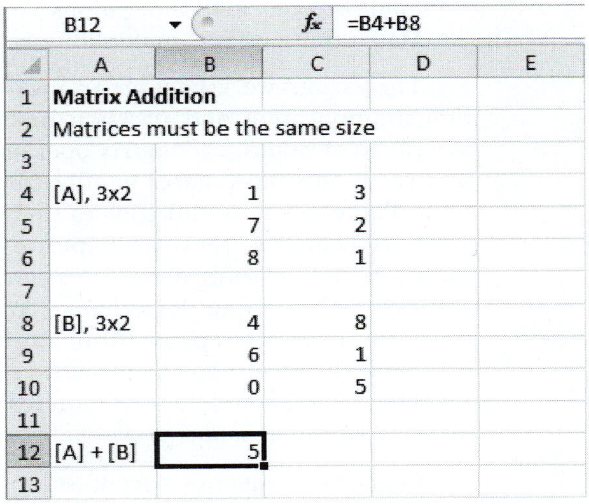

Figure 5.4
Beginning the addition process.

If the formula in cell B12 is now copied to cells B12:C14, Excel will add the [A] matrix and the [B] matrix element by element. The result is shown in Figure 5.5.

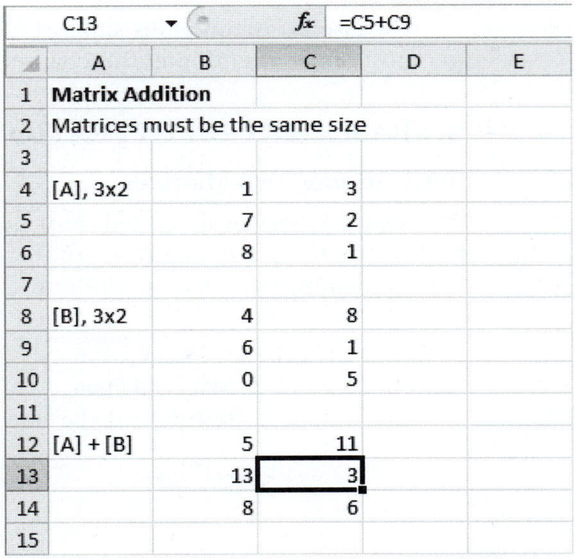

Figure 5.5
Adding matrices A and B.

Matrix Addition Using Array Math

If the cell ranges holding matrices [A] and [B] are named, then the array names can be used to perform the addition. In this example, the name "A" was applied to cell range B4:C6, and cells B8:C10 were named "B."

In Excel, any time *array math* is used, the size of the resulting array must be indicated before entering the array formula. The result of adding matrices [A] and [B] will be a 3 × 2 matrix, so a 3 × 2 region of cells (B16:C18) is selected, as shown in Figure 5.6.[1]

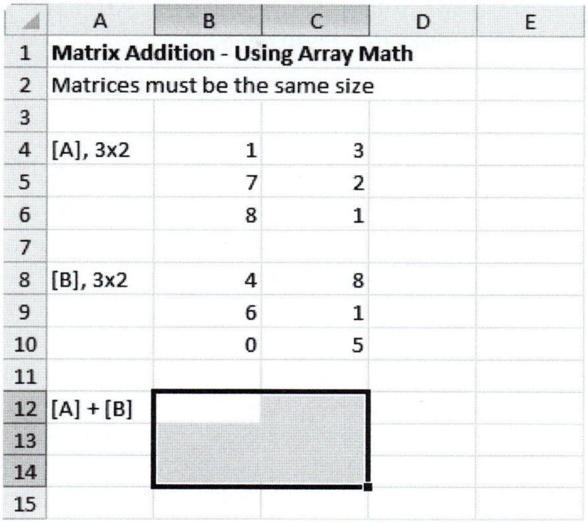

Figure 5.6

Indicate the size of the result before entering an array formula.

◢	A	B	C	D	E
1	Matrix Addition - Using Array Math				
2	Matrices must be the same size				
3					
4	[A], 3x2	1	3		
5		7	2		
6		8	1		
7					
8	[B], 3x2	4	8		
9		6	1		
10		0	5		
11					
12	[A] + [B]				
13					
14					
15					

Then the formula =A+B is entered in the top-left cell of the selected range, as shown in Figure 5.7.

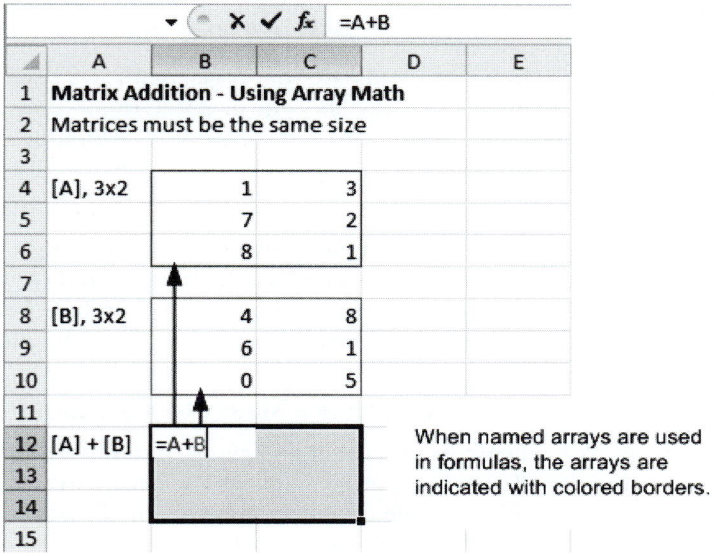

Figure 5.7

Enter the array formula in one cell of the selected array.

			× ✔ *fx*	=A+B	
◢	A	B	C	D	E
1	Matrix Addition - Using Array Math				
2	Matrices must be the same size				
3					
4	[A], 3x2	1	3		
5		7	2		
6		8	1		
7					
8	[B], 3x2	4	8		
9		6	1		
10		0	5		
11					
12	[A] + [B]	=A+B			
13					
14					
15					

When named arrays are used in formulas, the arrays are indicated with colored borders.

---

[1] Originally, in Figure 5.2, the A matrix was in cells B2:C4. When two rows were inserted for the headings in rows 1 and 2, Excel kept track of the new location of the A matrix, now in cells B4:C6.

Notice that, when the named arrays were entered in the formula, Excel put a box around the named arrays. This allows you to quickly see that the correct matrices are being added.

IMPORTANT! Excel requires a special character sequence when entering array formulas: [Ctrl-Shift-Enter], not just the [Enter] key. The [Ctrl-Shift-Enter] key combination tells Excel to fill the entire array (i.e., the selected region) with the formula, not just one cell. After the pressing of [Ctrl-Shift-Enter], the worksheet looks like Figure 5.8.

Figure 5.8

The result, using array math to add two matrices.

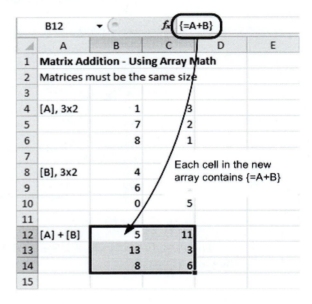

Each cell in the new array (B16:C18) contains the same formula, {=A+B}. The braces {} indicate that array math was used and that the result is an array. This means that the six cells in B16:C18 are considered a *collection*, and individual elements of the new array cannot be edited or deleted. However, if either matrix [A] or matrix [B] is changed, the new array will automatically be updated.

> **Note:** Named arrays are not required for array math in Excel; the same result could have been obtained by selecting the B16:C18 range, then typing in the array formula =B4:C6+B8:C10 and pressing [Ctrl-Shift-Enter]. The array formula {=B4:C6+B8:C10} would have been entered into every cell of the new array, and matrices [A] and [B] would have been added.

5.4.2 Multiplying a Matrix by a Scalar

Begin by entering the *scalar* (constant value) and the matrix that it is to multiply, as shown in Figure 5.9. It is not necessary to name the cell containing the scalar, but in this example cell B3 has been named S, for scalar.

Multiplying a matrix by a scalar simply requires you to multiply each element of the matrix by the scalar. Using cell math, you first multiple the scalar by one element of the matrix, usually the top-left element as shown in Figure 5.10.

By placing the formula =S*B5 in cell B9, you instruct Excel to multiply the contents of the cell named S (cell B3, the scalar) and the contents of cell B5 (the

Figure 5.9

Preparing to multiple matrix [A] by a scalar.

	A	B	C	D	E
1	**Scalar Multiplication**				
2					
3	Scalar, S:	10			
4					
5	[A], 3x2	1	3		
6		7	2		
7		8	1		
8					

Figure 5.10

Multiplying the scalar by the first element of the matrix.

B9			f_x	=S*B5	
	A	B	C	D	E
1	**Scalar Multiplication**				
2					
3	Scalar, S:	10			
4					
5	[A], 3x2	1	3		
6		7	2		
7		8	1		
8					
9	S [A]	10			
10					

top-left element of the matrix). Cell names in formulas act as absolute addresses, so the formula =S*B5 is equivalent to =$B$3*B5. The absolute address on the scalar is in preparation for copying the formula to other cells. When copied, the new formulas will continue to reference the scalar in cell B3.

Copying the formula in cell B9 to the range B9:C11 completes the process of multiplying the matrix [A] by the scalar, 10. The results are shown in Figure 5.11.

Figure 5.11

Copying the formula in cell B9 to cells B9:C11 to complete the multiplication.

C11			f_x	=S*C7	
	A	B	C	D	E
1	**Scalar Multiplication**				
2					
3	Scalar, S:	10			
4					
5	[A], 3x2	1	3		
6		7	2		
7		8	1		
8					
9	S [A]	10	30		
10		70	20		
11		80	10		
12					

Scalar Multiplication Using Array Math

The same result can be obtained using array math and a named array. If the cells holding the [A] matrix (B5:B7 in this example) are named "A," then the array name can be used to perform the scalar multiplication.

First, the size of the result matrix is indicated by selecting a 3×2 range of cells (B13:C15), and the array formula =S*A is entered in the top-left cell of the new array, as shown in Figure 5.12.

Figure 5.12

Starting the scalar multiplication using array math.

◢	A	B	C	D	E
1	Scalar Multiplication - Using Array Math				
2					
3	Scalar, S:	10			
4					
5	[A], 3x2	1	3		
6		7	2		
7		8	1		
8					
9	S [A]	=S*A			
10					
11					
12					

The formula is concluded by pressing [Ctrl-Shift-Enter] to tell Excel to fill the entire selected region with the array formula. Excel places the array formula {=S*A} in each cell in the result array, as illustrated in Figure 5.13.

Figure 5.13

The completed scalar multiplication using array math.

B9		f_x	{=S*A}		

◢	A	B	C	D	E
1	Scalar Multiplication - Using Array Math				
2					
3	Scalar, S:	10			
4					
5	[A], 3x2	1	3		
6		7	2		
7		8	1		
8					
9	S [A]	10	30		
10		70	20		
11		80	10		
12					

Note: Array math works with or without named arrays; you can also use cell range addresses. In this example, the array formula {=B3*B5:C7} could have been entered into the top-left cell of the result matrix as =B3*B5:C7, and then completed using [Ctrl-Shift-Enter].

5.4.3 Multiplying Two Matrices

In order to multiply two matrices, the number of columns in the first matrix must equal the number of rows in the second matrix.

Again, begin by entering the two matrices to be multiplied (Figure 5.14).

Figure 5.14

Preparing to multiply two matrices.

	A	B	C	D	E	F	G
1	**Matrix Multiplication - Using Array Math**						
2	Inside dimensions must match (2, 2 in this example)						
3	Product array size comes from outside dimensions (3x1 in this example)						
4							
5	[A], 3x2	1	3				
6		7	2				
7		8	1				
8							
9	[e], 2x1	4					
10		8					
11							

Begin the matrix multiplication process by indicating where the product matrix will go, as shown in Figure 5.15. You determine the dimensions of the product matrix from the outside dimensions of the matrices being multiplied. In this example, the product matrix will be 3 rows by 1 column.

Note: It is best to indicate the correct number of cells, but if you indicate too many cells, the multiplication will still work, but extra cells will display the #N/A error code. If you indicate too few cells, part of the product matrix will not be displayed.

Figure 5.15

Indicating the size of the product matrix.

	A	B	C	D	E	F	G
1	**Matrix Multiplication - Using Array Math**						
2	Inside dimensions must match (2, 2 in this example)						
3	Product array size comes from outside dimensions (3x1 in this example)						
4							
5	[A], 3x2	1	3				
6		7	2				
7		8	1				
8							
9	[e], 2x1	4					
10		8					
11							
12	[A][e], 3x1						
13							
14							
15							

Once you have indicated where the new matrix is to go, begin entering the matrix multiplication array function, **MMULT(first matrix, second matrix)**. After entering the equal sign, function name, and the opening parenthesis, enter the array names separated by a comma as illustrated in Figure 5.16.

Alternatively, you can use the mouse to indicate the cell ranges for each array (Figure 5.17), but the array names will not be used with this approach.

Figure 5.16
Indicating matrices to be multiplied by name.

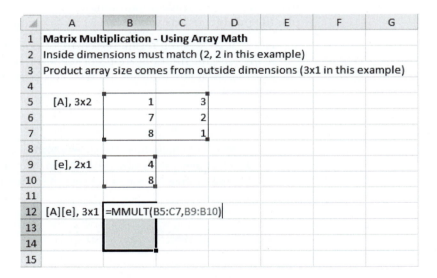

Figure 5.17
Indicating matrices to be multiplied by cell range.

Whether you use array names (Figure 5.16) or cell ranges (Figure 5.17), you complete the matrix multiplication formula in the same manner, by pressing [Ctrl-Shift-Enter], not just the [Enter] key. The **MMULT** function is an array function, and the [Ctrl-Shift-Enter] is used to enter the function into *all* of the cells from B12 to B14. The result (using array names) is shown in Figure 5.18.

Note: Pressing [Enter] instead of [Ctrl-Shift-Enter] would enter the formula only in cell B12, and only one element of the product matrix would be displayed.

Note: Excel places the formulas for multiplying the matrices in the result cells, not just the final values. Because of this, any changes in the first two matrices will automatically cause the product matrix to be updated.

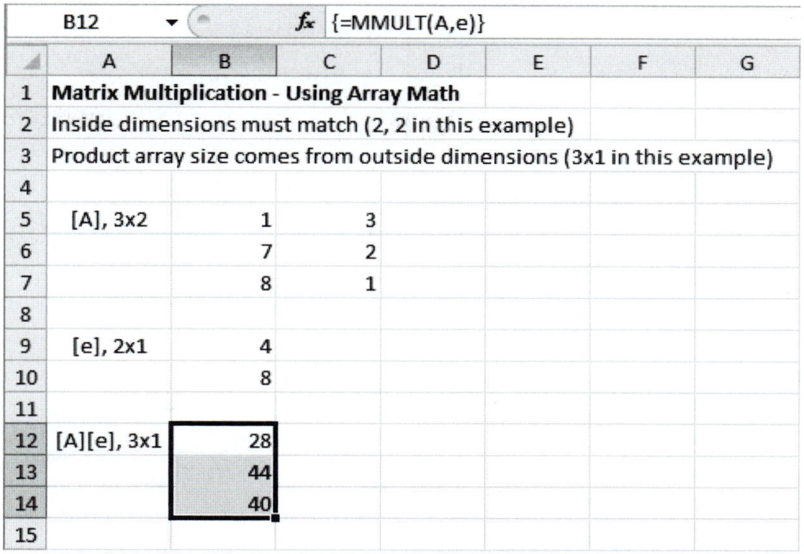

Figure 5.18

The result of multiplying two matrices.

	B12	▼		f_x	{=MMULT(A,e)}		
◢	A	B	C	D	E	F	G
1	**Matrix Multiplication - Using Array Math**						
2	Inside dimensions must match (2, 2 in this example)						
3	Product array size comes from outside dimensions (3x1 in this example)						
4							
5	[A], 3x2	1	3				
6		7	2				
7		8	1				
8							
9	[e], 2x1	4					
10		8					
11							
12	[A][e], 3x1	28					
13		44					
14		40					
15							

Process Summary: Matrix Multiplication

1. Enter the matrices to be multiplied; the inside dimensions must be the same.
2. Give each matrix a name by selecting the matrix cells and using the Name box at the left side of the Formula bar (optional).
3. Determine the size of the product matrix from the outside dimensions of the matrices being multiplied.
4. Select the cells that will hold the product matrix.
5. Enter the matrix multiplication formula =MMULT(first_matrix, second_matrix) in one cell of the destination cell range. Indicate the first and second matrices either by name or using cell addresses.
6. Complete the multiplication by pressing [Ctrl-Shift-Enter] to cause Excel to place the formula in each cell of the product matrix.

PRACTICE!

Can the following matrices be multiplied, and, if so, what size will the resulting matrices be?

$$[A]_{3 \times 2}[B]_{2 \times 2} \qquad [C]_{3 \times 3}[D]_{1 \times 3} \qquad [D]_{1 \times 3}[C]_{3 \times 3} \qquad (5.1)$$

[ANSWERS: yes, 3×2; no; yes, 1×3]

EXAMPLE 5.1

The matrices shown here are to be multiplied as [A][G].

1. Is this multiplication possible?
2. If possible…

(continued)

a. What is the size of the product matrix?

b. What is the result of the multiplication?

$$A = \begin{bmatrix} 1 & 3 \\ 7 & 2 \\ 8 & 11 \end{bmatrix} \qquad G = \begin{bmatrix} 1 & 2 & 3 & 4 \\ 5 & 6 & 7 & 8 \end{bmatrix}$$

The proposed multiplication is $[A]_{3\times2}$ $[G]_{2\times4}$. Since the inside dimensions, the twos, are equal, this multiplication is possible. The size of the product matrix, from the outside dimensions, will be 3×4. The result is shown in Figure 5.19.

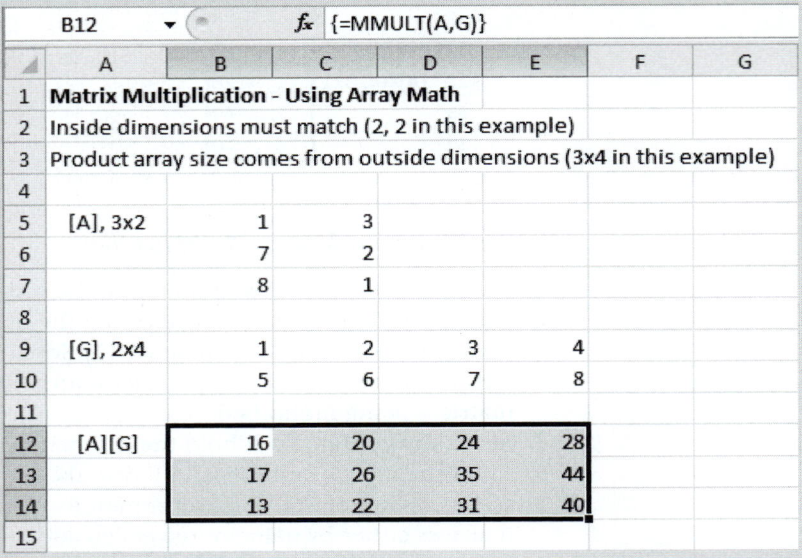

Figure 5.19

The result of multiplying the [A] and [G] matrices.

5.4.4 **Transposing a Matrix**

Any matrix can be transposed. To *transpose* a matrix, simply interchange the rows and columns. You can transpose a matrix in two ways: as values or by using array function **TRANSPOSE**. Using values is simpler, but the result will not be automatically recalculated if the input data change. Both methods are described here.

Using PASTE SPECIAL to Transpose as Values

The process in Excel can be summarized as follows:

1. Enter the original matrix.
2. Begin the transposition process by selecting the cells containing the matrix and copying the matrix to the Windows clipboard using Ribbon options **Home/ Clipboard/Copy** [Excel 2003: Edit/Copy].
3. Indicate the cell that will contain the top-left corner of the result matrix.
4. Open the Paste Special dialog using Ribbon options **Home/Paste (menu)/ Paste Special...** [Excel 2003: Edit/Paste Special...].
5. Select Values in the Paste section and check the Transpose checkbox near the bottom of the dialog.
6. Click the [OK] button to close the dialog box and create the transposed matrix (as values, not array functions).

Using a Matrix Function to Transpose a Matrix

Using the **TRANSPOSE** array function, the process is as follows:

1. Enter the original matrix (Figure 5.20).

Figure 5.20

Preparing to transpose matrix A.

◢	A	B	C	D	E
1	**Matrix Transpose - Using Array Math**				
2	Interchange rows and columns				
3	Any size matrix				
4					
5	[A], 3x2	1	3		
6		7	2		
7		8	1		
8					

2. Indicate where the result should be placed, showing the exact size of the transposed matrix (Figure 5.21).

Figure 5.21

Indicating where the transposed matrix will be placed.

◢	A	B	C	D	E
1	**Matrix Transpose - Using Array Math**				
2	Interchange rows and columns				
3	Any size matrix				
4					
5	[A], 3x2	1	3		
6		7	2		
7		8	1		
8					
9	[A-trans], 2x3				
10					
11					

3. Enter the formula =TRANSPOSE(matrix) in the top-left cell on the result matrix (as shown in Figure 5.22). The matrix to be transposed can be indicated by name, or cell range.

Figure 5.22

Entering the **TRANSPOSE** formula.

◢	A	B	C	D	E
1	**Matrix Transpose - Using Array Math**				
2	Interchange rows and columns				
3	Any size matrix				
4					
5	[A], 3x2	1	3		
6		7	2		
7		8	1		
8					
9	[A-trans], 2x3	=transpose(A)			
10					
11					

4. Complete the transpose operation by pressing [Ctrl-Shift-Enter] to transpose the matrix and place the formula in every cell in the result matrix (Figure 5.23).

Figure 5.23

Completing the transpose operation.

	A	B	C	D	E
	B9	▾	f_x {=TRANSPOSE(A)}		
1	**Matrix Transpose – Using Array Math**				
2	Interchange rows and columns				
3	Any size matrix				
4					
5	[A], 3x2	1	3		
6		7	2		
7		8	1		
8					
9	[A-trans], 2x3	1	7	8	
10		3	2	1	
11					

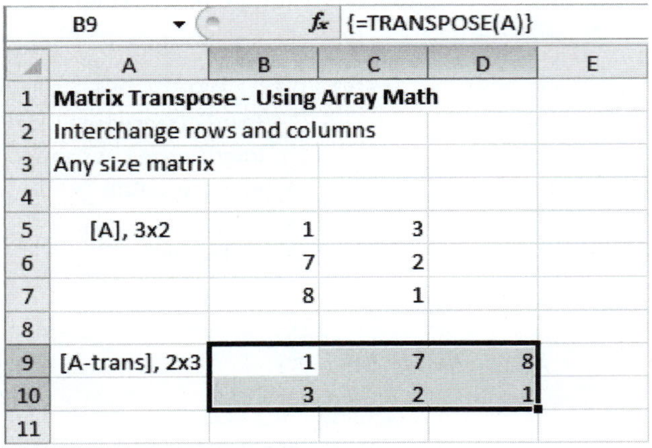

PRACTICE!

Transpose the following matrices.

$$[6 \quad 1 \quad 4]$$

$$\begin{bmatrix} 1 & 1.2 \\ 3 & 6.1 \\ 4 & 2.3 \end{bmatrix}$$

$$\begin{bmatrix} 1 & 0 & 0 \\ 0 & 1 & 0 \\ 0 & 0 & 1 \end{bmatrix} \tag{5.2}$$

ANSWERS: $\begin{bmatrix} 6 \\ 1 \\ 4 \end{bmatrix}, \begin{bmatrix} 1 & 3 & 4 \\ 1.2 & 6.1 & 2.3 \end{bmatrix}, \begin{bmatrix} 1 & 0 & 0 \\ 0 & 1 & 0 \\ 0 & 0 & 1 \end{bmatrix}.$

5.4.5 Inverting a Matrix

Only square matrices (number of rows equal to number of columns) can possibly be inverted, and not even all square matrices can actually be inverted. (They must be *nonsingular* to be inverted.)

The procedure to invert a nonsingular, square matrix in Excel, using the **MINVERSE** array function, is as follows:

1. Enter the matrix to be inverted, and name it if desired.
2. Indicate where the inverted matrix should be placed and the correct size (same size as original matrix).
3. Enter the **MINVERSE(matrix)** array function. The matrix to be inverted can be indicated by name, or cell range.
4. Press [Ctrl-Shift-Enter] to enter the array function in all the cells making up the result matrix.

As an example of this procedure, we will invert the [J] matrix, $J = \begin{bmatrix} 2 & 3 & 5 \\ 7 & 2 & 4 \\ 8 & 11 & 6 \end{bmatrix}$.

After the first three steps, the worksheet looks like Figure 5.24.

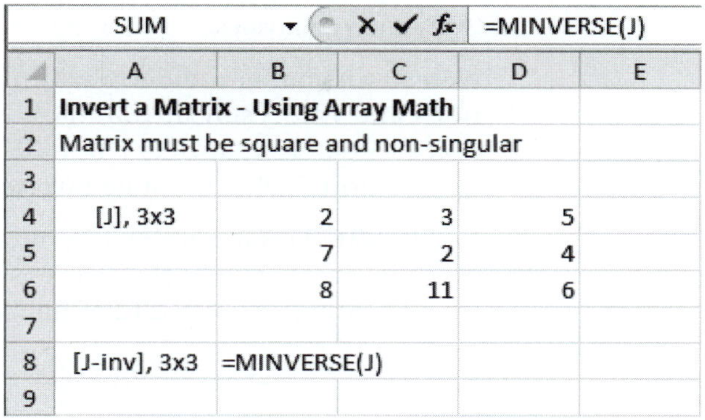

SUM | × ✓ fx | =MINVERSE(J)

	A	B	C	D	E
1	**Invert a Matrix - Using Array Math**				
2	Matrix must be square and non-singular				
3					
4	[J], 3x3	2	3	5	
5		7	2	4	
6		8	11	6	
7					
8	[J-inv], 3x3	=MINVERSE(J)			
9					

Figure 5.24
Inverting the [J] matrix, just before completing the array formula.

We complete the inversion process by pressing [Ctrl-Shift-Enter] to enter the array formula into every cell in the result matrix. The result is shown in Figure 5.25.

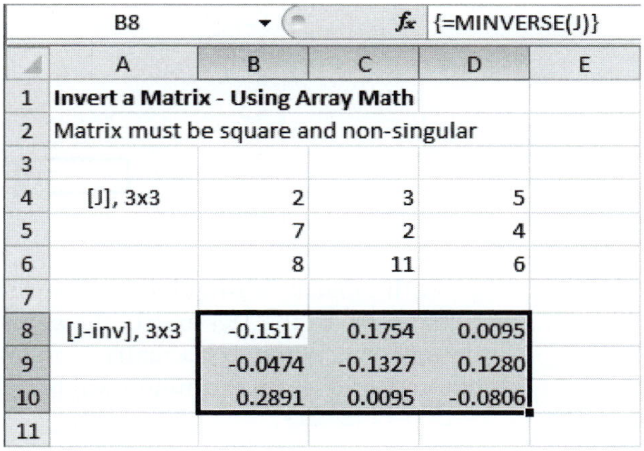

B8 | fx | {=MINVERSE(J)}

	A	B	C	D	E
1	**Invert a Matrix - Using Array Math**				
2	Matrix must be square and non-singular				
3					
4	[J], 3x3	2	3	5	
5		7	2	4	
6		8	11	6	
7					
8	[J-inv], 3x3	-0.1517	0.1754	0.0095	
9		-0.0474	-0.1327	0.1280	
10		0.2891	0.0095	-0.0806	
11					

Figure 5.25
The inverted matrix.

Again, it is not necessary to name the arrays prior to using array math. Figure 5.26 illustrates how to invert the [J] matrix using cell range address instead of naming the array.

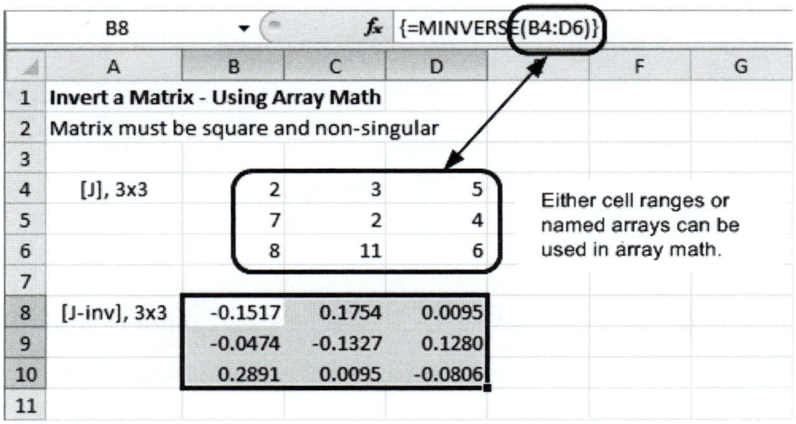

B8 | fx | {=MINVERSE(B4:D6)}

	A	B	C	D	E	F	G
1	**Invert a Matrix - Using Array Math**						
2	Matrix must be square and non-singular						
3							
4	[J], 3x3	2	3	5			
5		7	2	4			
6		8	11	6			
7							
8	[J-inv], 3x3	-0.1517	0.1754	0.0095			
9		-0.0474	-0.1327	0.1280			
10		0.2891	0.0095	-0.0806			
11							

Figure 5.26
The inverted matrix using the cell range address.

5.4.6 Matrix Determinant

The *determinant* of a matrix is a single value, calculated from a matrix, which is often used in solving systems of equations. One of the most straightforward ways to use the determinant is to see whether a matrix can be inverted. If the determinant is zero, the matrix is *singular* and cannot be inverted. You can calculate a determinant only for square matrices.

> **Note:** Calculating a determinant for a large matrix requires a lot of calculations and can result in round-off errors on digital computers. A very small, nonzero determinant(e.g., 1×10^{-14}) is probably a round-off error, and the matrix most likely cannot be inverted.

Excel's **MDETERM(matrix)** array function is used to compute determinant values. In the previous section, the [J] matrix was inverted, so it must have had a nonzero determinant. We can use the **MDETERM** function to verify that, as illustrated in Figure 5.27.

Figure 5.27
Calculating the determinant of [J].

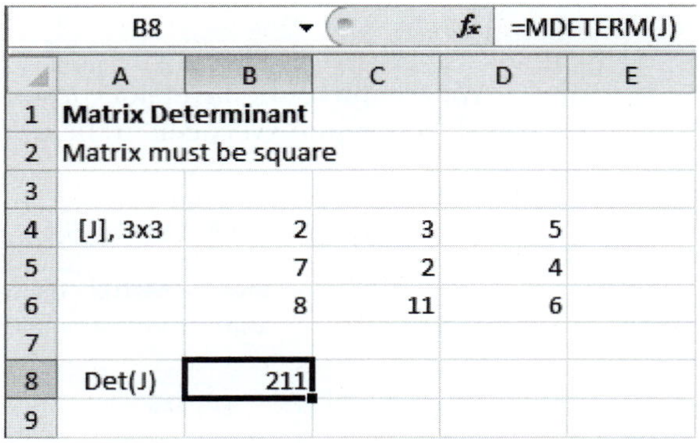

If the matrix is singular (and therefore cannot be inverted), the determinant will be zero. In the [K] matrix shown in Figure 5.28, the second row is the same as the first row. Whenever a matrix contains two identical rows the matrix cannot be inverted, and the determinant will be zero.

Figure 5.28
Calculating the determinant of singular matrix [K].

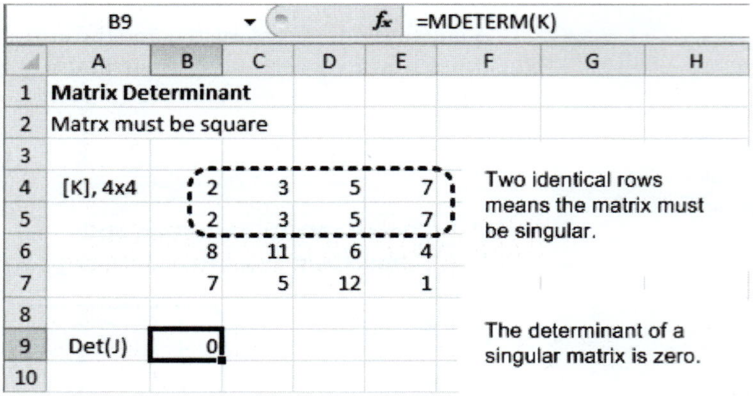

A matrix is singular (and cannot be inverted) if

* any row (or column) contains all zeros;

- any two rows (or columns) are identical;
- any row (or column) is equal to a *linear combination* of other rows (or columns).

When two or more rows are multiplied by constants and then added, the result is a linear combination of the rows. When a linear combination of two or more rows is equal to another row in the matrix, the matrix is singular.

5.5 SOLVING SYSTEMS OF LINEAR EQUATIONS

One of the most common uses of matrix operations is to solve *systems of linear algebraic equations*. The process of solving simultaneous equations by using matrices works as follows:

1. Write the equations in matrix form (*coefficient matrix* multiplying an *unknown vector,* equal to a *right-hand-side vector*).

$$[C] \, [x] = [r]$$

2. Invert the coefficient matrix.
3. Multiply both sides of the equation by the inverted coefficient matrix.

The result of step 3 is a solution matrix containing the answers to the problem.

In order for inverting the coefficient matrix to be possible, it must be nonsingular. In terms of solving simultaneous equations, this means that you will be able to invert the coefficient matrix only if there is a solution to the set of equations. If there is no solution, the coefficient matrix will be singular.

Consider the following three equations in three unknowns:

$$3x_1 + 2x_2 + 4x_3 = 5,$$
$$2x_1 + 5x_2 + 3x_3 = 17,$$
$$7x_1 + 2x_2 + 2x_3 = 11. \tag{5.3}$$

Step 1. Write the Equations in Matrix Form.   The unknowns are $x_1, x_2,$ and x_3, which can be written as the vector of unknowns, $[x]$:

$$[x] = \begin{bmatrix} x_1 \\ x_2 \\ x_3 \end{bmatrix}.$$

The coefficients multiplying the various x's can be collected in a coefficient matrix $[C]$:

$$[C] = \begin{bmatrix} 3 & 2 & 4 \\ 2 & 5 & 3 \\ 7 & 2 & 2 \end{bmatrix}.$$

PRACTICE!

Try multiplying $[C]$ times $[x]$ (symbolically) to see that you do indeed get back the left side of the preceding equation.

$$\text{ANSWER: } [C][x] = \begin{bmatrix} 3x_1 + 2x_2 + 4x_3 \\ 2x_1 + 5x_2 + 3x_3 \\ 7x_1 + 2x_2 + 2x_3 \end{bmatrix}.$$

The constants on the right side of the equations can be written as a right-hand-side vector $[r]$:

$$[r] = \begin{bmatrix} 5 \\ 17 \\ 11 \end{bmatrix}.$$

The three equations in three unknowns can now be written as

$$[C][x] = [r].$$

In a worksheet, the arrays can be entered as shown in Figure 5.29. In Excel, names C and r are reserved for "column" and "row" so the matrices have been named Coeff and rhs in the worksheet shown in Figure 5.29. So, for this example, the matrix equation is written as

$$[\text{Coeff}][x] = [\text{rhs}].$$

Figure 5.29

Entering the coefficient matrix and right-hand-side vector in Excel.

	A	B	C	D	E	F	G	H
1	Solving Simultaneous Linear Equations							
2								
3	[Coeff]	3	2	4		[rhs]	5	
4		2	5	3			17	
5		7	2	2			11	
6								

Step 2. Invert the Coefficient Matrix. Use the array function **MINVERSE** to invert the [Coeff] matrix as shown in Figure 5.30.

Figure 5.30

Inverting the coefficient matrix.

B7			f_x {=MINVERSE(Coeff)}					
	A	B	C	D	E	F	G	H
1	Solving Simultaneous Linear Equations							
2								
3	[Coeff]	3	2	4		[rhs]	5	
4		2	5	3			17	
5		7	2	2			11	
6								
7	[Coeff-inv]	-0.05	-0.05	0.179				
8		-0.22	0.282	0.013				
9		0.397	-0.1	-0.14				
10								

Step 3. Multiply Both Sides of the Equation by the Inverted [Coeff] Matrix, [Coeff$_{\text{inv}}$]

$$[\text{Coeff}_{\text{inv}}][\text{Coeff}][x] = [\text{Coeff}_{\text{inv}}][\text{rhs}]$$
$$[I][x] = [\text{Coeff}_{\text{inv}}][\text{rhs}]$$
$$[x] = [\text{Coeff}_{\text{inv}}][\text{rhs}]. \tag{5.4}$$

What's happening here is:

- Multiplying the inverted coefficient matrix [Coef matrix returns an identity matrix, called [I]. (Try
- Multiplying the [x] vector by an identity matr unchanged.
- So multiplying the inverted coefficient matrix [returns the [x] values, which is the solution.

The solution to the original three equations in three unknown. Figure 5.31.

Figure 5.31

Solving for the unknown vector, [x].

	G7	▾		f_x	{=MMULT(CoeffInv,rhs)}		
◢	A	B	C	D	E	F	G	H
1	**Solving Simultaneous Linear Equations**							
2								
3	[Coeff]	3	2	4		[rhs]	5	
4		2	5	3			17	
5		7	2	2			11	
6								
7	[Coeff-inv]	-0.05	-0.05	0.179		[x]	0.846	
8		-0.22	0.282	0.013			3.846	
9		0.397	-0.1	-0.14			-1.31	
10								

So the answers to the original three equations are

$$x_1 = 0.846$$
$$x_2 = 3.846$$
$$x_3 = -1.31.$$

MULTILOOP CIRCUITS I

Multiloop circuits are analyzed by using Kirchhoff's laws of voltage and current. Kirchhoff's voltage law says that

for a closed circuit (a loop), the algebraic sum of all changes in voltage must be zero.

Kirchhoff's current law says that

at any junction in a circuit, the input current(s) must equal the output current(s).

We will use these laws, with the circuit shown in Figure 5.32, to determine the three unknown currents, i_1 through i_3.

Figure 5.32

Resistances in parallel and series.

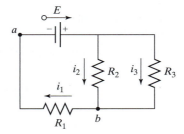

The voltage from the battery and the three resistances are as listed in Table 5.1.

Table 5.1 Specified values

E	12 volts
R_1	30 ohms
R_2	40 ohms
R_3	50 ohms

Applying the current law at point b gives one equation:

$$i_1 = i_2 + i_3 \tag{5.5}$$

Applying the voltage law to the left loop and the overall loop provides two more equations:

$$E - V_2 - V_1 = 0,$$
$$E - V_3 - V_1 = 0.$$

In terms of current and resistance, these are

$$E - i_2 R_2 - i_1 R_1 = 0,$$
$$E - i_3 R_3 - i_1 R_1 = 0.$$

We now have three equations for i_1, i_2, and i_3. Writing them in matrix form with constants on the right side of the equal sign yields

$$1 i_1 - 1 i_2 - 1 i_3 = 0,$$
$$R_1 i_1 + R_2 i_2 + 0 i_3 = E,$$
$$R_1 i_1 + 0 i_2 + R_3 i_3 = E.$$

The coefficients 1 and 0 have been included in the equations as a reminder to include them in the coefficient matrix:

$$C = \begin{bmatrix} 1 & -1 & -1 \\ R_1 & R_2 & 0 \\ R_1 & 0 & R_3 \end{bmatrix} \quad r = \begin{bmatrix} 0 \\ E \\ E \end{bmatrix}.$$

Substituting the known values, we get

$$C = \begin{bmatrix} 1 & -1 & -1 \\ 30 & 40 & 0 \\ 30 & 0 & 50 \end{bmatrix} \quad r = \begin{bmatrix} 0 \\ 12 \\ 12 \end{bmatrix}.$$

In Excel, the coefficient matrix and right-hand-side vector are entered as shown in Figure 5.33.

Figure 5.33
The coefficient matrix and right-hand-side vector.

	A	B	C	D	E	F	G	H
1	**Multiloop Circuits I**							
2								
3	[Coeff]	1	-1	-1		[rhs]	0	
4		30	40	0			12	
5		30	0	50			12	
6								

A quick check with Excel's **MDETERM** array function (Figure 5.34) shows that a solution is possible.

Figure 5.34
Checking for a nonzero
determinant.

	B7			f_x	=MDETERM(Coeff)		
	A	B	C	D	E	F	G	H
1	**Multiloop Circuits I**							
2								
3	[Coeff]	1	-1	-1		[rhs]	0	
4		30	40	0			12	
5		30	0	50			12	
6								
7	Determinant:	4700						
8								

The coefficient matrix is inverted by using the **MINVERSE** array function, as shown in Figure 5.35.

Figure 5.35
Inverting the
coefficient matrix.

	CoeffInv			f_x	{=MINVERSE(Coeff)}		
	A	B	C	D	E	F	G	H
1	**Multiloop Circuits I**							
2								
3	[Coeff]	1	-1	-1		[rhs]	0	
4		30	40	0			12	
5		30	0	50			12	
6								
7	Determinant:	4700						
8								
9	[Coeff-inv]	0.43	0.01	0.01				
10		-0.32	0.02	-0.01				
11		-0.26	-0.01	0.01				
12								

The currents are found by multiplying the inverted coefficient matrix and the right-hand-side vector, using Excel's **MMULT** array function. The result is shown in Figure 5.36.

The unknown currents have been found to be $i_1 = 0.23$, $i_2 = 0.13$, and $i_3 = 0.10$ amp, respectively.

Figure 5.36
The solution.

	G9			f_x	{=MMULT(CoeffInv,rhs)}		
	A	B	C	D	E	F	G	H
1	**Multiloop Circuits I**							
2								
3	[Coeff]	1	-1	-1		[rhs]	0	
4		30	40	0			12	
5		30	0	50			12	
6								
7	Determinant:	4700						
8								
9	[Coeff-inv]	0.43	0.01	0.01		[i]	0.23	
10		-0.32	0.02	-0.01			0.128	
11		-0.26	-0.01	0.01			0.102	
12								

WHEATSTONE BRIDGE I

A fairly commonly used circuit in instrumentation applications is the *Wheatstone Bridge*, shown in Figure 5.37.

Figure 5.37
Wheatstone Bridge.

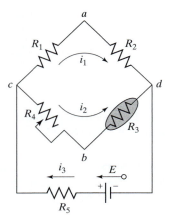

The bridge contains two known resistances, R_1 and R_2, and an adjustable resistance (i.e., a potentiometer), R_4. In use, the setting on the potentiometer is adjusted until points a and b are at the same voltage. Then the known resistances can be used to calculate an unknown resistance, shown as R_3 in Figure 5.36.

To develop the equations for finding currents i_1 through i_3, note that R_1 and the potentiometer, R_4, are connected at point c, and the circuit has been adjusted to have the same potential at points a and b. Therefore the voltage drops across R_1 and R_4 must be equal:

$$V_1 = V_4. \tag{5.6}$$

But the voltages can be written in terms of current and resistance:

$$i_1 R_1 = i_2 R_4. \tag{5.7}$$

Similarly, the voltage drops across R_2 and R_3 must be equal, so

$$V_2 = V_3, \tag{5.8}$$

or, in terms of current and resistance,

$$i_1 R_2 = i_2 R_3. \tag{5.9}$$

Solving one equation for i_1 and substituting into the other yields an equation for the unknown resistance, R_3 in terms of the known resistance values:

$$R_3 = R_4 \frac{R_2}{R_1}. \tag{5.10}$$

1. Given the known resistances and battery potential listed in Table 5.2, what is the resistance of R_3 when the potentiometer (R_4) has been adjusted to 24 ohms to balance the bridge?
2. Use Kirchhoff's laws to find currents i_1 through i_3.

To solve this problem in Excel, we first enter the specified values into a worksheet, as shown in Figure 5.38.

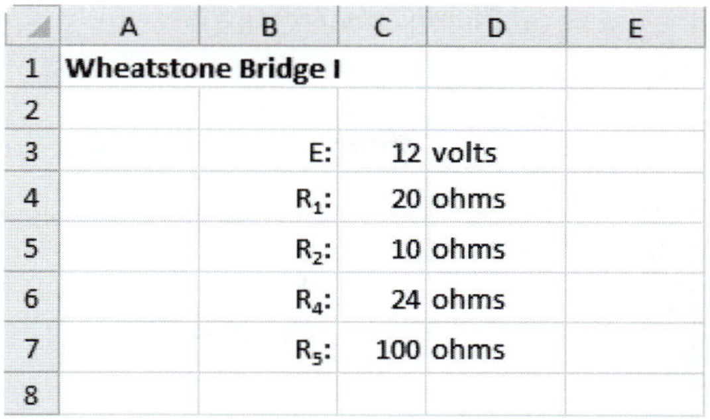

◢	A	B	C	D	E
1	Wheatstone Bridge I				
2					
3			E:	12	volts
4			R_1:	20	ohms
5			R_2:	10	ohms
6			R_4:	24	ohms
7			R_5:	100	ohms
8					

Figure 5.38
Entering the known values into the worksheet.

Table 5.2 Specified values

E	12 volts
R_1	20 ohms
R_2	10 ohms
R_3	24 ohms
R_5	100 ohms

Then the unknown resistance, R_3, is computed using equation 5.10 (Figure 5.39).

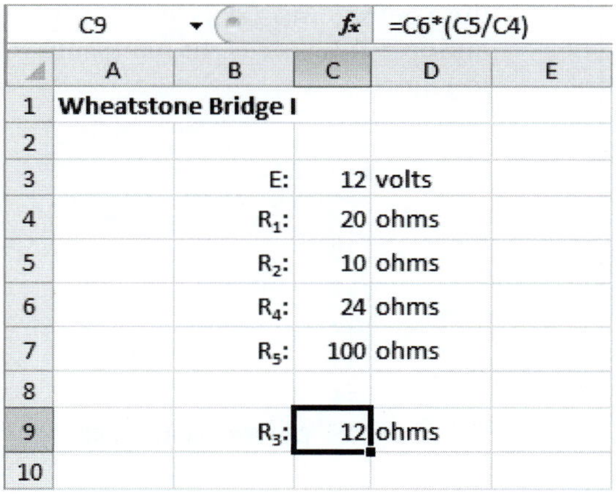

C9			f_x	=C6*(C5/C4)	
◢	A	B	C	D	E
1	Wheatstone Bridge I				
2					
3			E:	12	volts
4			R_1:	20	ohms
5			R_2:	10	ohms
6			R_4:	24	ohms
7			R_5:	100	ohms
8					
9			R_3:	12	ohms
10					

Figure 5.39
Determine the unknown resistance, R_3.

To find the currents, we need three equations to solve for the three unknowns. First, Kirchhoff's current law can be applied at point d:

$$i_1 + i_2 = i_3.$$

Then, the voltage law can be applied to the outer loop so that

$$E - V_5 - V_1 - V_2 = 0,$$

or, in terms of current and resistance,

$$E - i_3 R_5 - i_1 R_1 - i_1 R_2 = 0.$$

Finally, the voltage law can be applied to the inner loop, so

$$E - V_5 - V_4 - V_3 = 0,$$

or

$$E - i_3 R_5 - i_2 R_4 - i_2 R_3 = 0.$$

In matrix form, these equations become

$$\text{Coeff} = \begin{bmatrix} 1 & 1 & -1 \\ R_1 + R_2 & 0 & R_5 \\ 0 & R_4 + R_3 & R_5 \end{bmatrix} \quad \text{rhs} = \begin{bmatrix} 0 \\ E \\ E \end{bmatrix}. \tag{5.11}$$

In Excel, the coefficient matrix and right-hand-side vector are entered as shown in Figure 5.40.

	A	B	C	D	E	F	G	H
10								
11	[Coeff]	1	1	-1		[rhs]	0	
12		30	0	100			12	
13		0	36	100			12	
14								

Figure 5.40
Entering the coefficient matrix and right-hand-side vector.

After checking the determinant (see Figure 5.41), a solution is possible so the coefficient matrix is then inverted. The inverted coefficient matrix is then multiplied by the right-hand-side vector to find the currents. The result is shown in Figure 5.41.

G17		f_x	{=MMULT(CoeffInv,rhs)}					
	A	B	C	D	E	F	G	H
10								
11	[Coeff]	1	1	-1		[rhs]	0	
12		30	0	100			12	
13		0	36	100			12	
14								
15	Determinant:	-7680						
16								
17	[Coeff-inv]	0.47	0.02	-0.01		[i]	0.056	
18		0.39	-0.01	0.02			0.047	
19		-0.14	0.00	0.00			0.103	
20								

Figure 5.41
Completing the matrix problem for the currents.

The currents were found to be:

$$i_1 = 0.056 \text{ amp},$$
$$i_2 = 0.047 \text{ amp},$$
$$i_3 = 0.103 \text{ amp}. \tag{5.12}$$

CONDUCTION HEAT TRANSFER

The differential equation

$$\frac{\partial T}{\partial t} = \frac{k}{\rho C_p}\left[\frac{\partial^2 T}{\partial x^2} + \frac{\partial^2 T}{\partial y^2}\right]$$

describes energy conduction through a two-dimensional region and can be applied to a surface exposed to various boundary temperatures. For example, we might want to know the temperature distribution in a 50-cm × 40-cm metal plate exposed to boiling water (100°C) along two edges, ice water (0°C) on one edge, and room temperature (25°C) along another, as represented in Figure 5.42.

Figure 5.42
Boundary temperatures on the two-dimensional region.

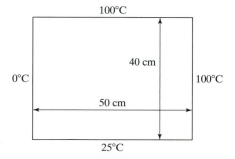

At steady state, $\frac{\partial T}{\partial t} = 0$, the equation simplifies considerably, to a form known as *Laplace's equation*:

$$0 = \frac{\partial^2 T}{\partial x^2} + \frac{\partial^2 T}{\partial y^2}$$

The partial derivatives can be approximated by using finite differences to produce the algebraic equation

$$0 = \left[\frac{T_{i+1,j} - 2T_{i,j} + T_{i-1,j}}{(\Delta x)^2}\right] + \left[\frac{T_{i,j+1} - 2T_{i,j} + T_{i,j-1}}{(\Delta y)^2}\right],$$

where subscript i,j represents a point on the plate at which the temperature is $T_{i,j}$. Then $T_{i-1,j}$ represents the temperature at a point to the left of point i,j; and $T_{i,j-1}$ represents the temperature at a point above point i,j, as indicated in Figure 5.43.

Figure 5.43
Defining grid point locations.

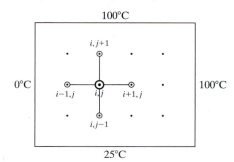

And, if we choose to make $\Delta x = \Delta y$, we get a particularly simple result:

$$0 = \left[T_{i+1,j} - 2T_{i,j} + T_{i-1,j}\right] + \left[T_{i,j+1} - 2T_{i,j} + T_{i,j-1}\right],$$

or

$$4T_{i,j} = \left[T_{i+1,j} + T_{i-1,j} \right] + \left[T_{i,j+1} + T_{i,j-1} \right] \quad \text{(general equation)}.$$

This equation says that the sum of the temperatures at the four points around any central point (i.e., any i,j) is equal to four times the temperature at the central point. (Remember, this is true only at steady state and only when $\Delta x = \Delta y$.) We can apply this equation at each interior point to develop a system of equations that, when solved simultaneously, will yield the temperatures at each point.

To help see how this is done, let's assign each interior point a letter designation and show the four points surrounding point A with a circle, as shown in Figure 5.44.

Figure 5.44
Assigning letter names to grid points and identifying point A's neighboring points.

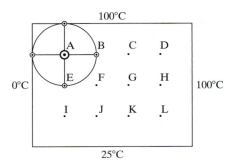

Applying the general equation at point A (see Figure 5.43) yields

$$4T_A = 0 + 100 + T_B + T_E$$

Then, we move the circle to point B, as shown in Figure 5.45.

Figure 5.45
The circle used to identify neighboring points has been moved to point B.

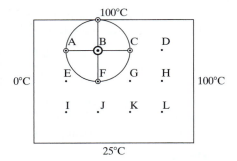

Applying the general equation at point B yields

$$4T_B = T_A + 100 + T_C + T_F$$

By continuing to apply the general equation at each interior point (points A through L), we generate 12 equations, one for each interior point:

$$4T_A = 0 + 100 + T_B + T_E$$
$$4T_B = T_A + 100 + T_C + T_F$$
$$4T_C = T_B + 100 + T_D + T_G$$
$$4T_D = T_C + 100 + 100 + T_H$$
$$4T_E = 0 + T_A + T_F + T_I$$
$$4T_F = T_E + T_B + T_G + T_J$$

$$4T_G = T_F + T_C + T_H + T_K$$
$$4T_H = T_G + T_D + 100 + T_L$$
$$4T_I = 0 + T_E + T_J + 25$$
$$4T_J = T_1 + T_F + T_K + 25$$
$$4T_K = T_J + T_G + T_L + 25$$
$$4T_L = T_K + T_H + 100 + 25$$

After rearranging (putting all constants on right-hand side), these equations can be written in matrix form as

$$[\text{Coeff}]\,[\text{T}] = [\text{rhs}]$$

where

$$[\text{Coeff}] = \begin{bmatrix} -4 & 1 & 0 & 0 & 1 & 0 & 0 & 0 & 0 & 0 & 0 & 0 \\ 1 & -4 & 1 & 0 & 0 & 1 & 0 & 0 & 0 & 0 & 0 & 0 \\ 0 & 1 & -4 & 1 & 0 & 0 & 1 & 0 & 0 & 0 & 0 & 0 \\ 0 & 0 & 1 & -4 & 0 & 0 & 0 & 1 & 0 & 0 & 0 & 0 \\ 1 & 0 & 0 & 0 & -4 & 1 & 0 & 0 & 1 & 0 & 0 & 0 \\ 0 & 1 & 0 & 0 & 1 & -4 & 1 & 0 & 0 & 1 & 0 & 0 \\ 0 & 0 & 1 & 0 & 0 & 1 & -4 & 1 & 0 & 0 & 1 & 0 \\ 0 & 0 & 0 & 1 & 0 & 0 & 1 & -4 & 0 & 0 & 0 & 1 \\ 0 & 0 & 0 & 0 & 1 & 0 & 0 & 0 & -4 & 1 & 0 & 0 \\ 0 & 0 & 0 & 0 & 0 & 1 & 0 & 0 & 1 & -4 & 1 & 0 \\ 0 & 0 & 0 & 0 & 0 & 0 & 1 & 0 & 0 & 1 & -4 & 1 \\ 0 & 0 & 0 & 0 & 0 & 0 & 0 & 1 & 0 & 0 & 1 & -4 \end{bmatrix} \quad [\text{T}] = \begin{bmatrix} T_A \\ T_B \\ T_C \\ T_D \\ T_E \\ T_F \\ T_G \\ T_H \\ T_I \\ T_J \\ T_K \\ T_L \end{bmatrix} \quad [\text{rhs}] = \begin{bmatrix} -100 \\ -100 \\ -100 \\ -200 \\ 0 \\ 0 \\ 0 \\ -100 \\ -25 \\ -25 \\ -25 \\ -125 \end{bmatrix}$$

The coefficient and right-hand-side (constant-value) matrices can be entered into Excel, as shown in Figure 5.46.

▲	A	B	C	D	E	F	G	H	I	J	K	L	M	N	O	P	Q	R
1	Heat Conduction																	
2																		
3			A	B	C	D	E	F	G	H	I	J	K	L				
4	[Coeff]	A	-4	1	0	0	1	0	0	0	0	0	0	0		[rhs]	-100	
5		B	1	-4	1	0	0	1	0	0	0	0	0	0			-100	
6		C	0	1	-4	1	0	0	1	0	0	0	0	0			-100	
7		D	0	0	1	-4	0	0	0	1	0	0	0	0			-200	
8		E	1	0	0	0	-4	1	0	0	1	0	0	0			0	
9		F	0	1	0	0	1	-4	1	0	0	1	0	0			0	
10		G	0	0	1	0	0	1	-4	1	0	0	1	0			0	
11		H	0	0	0	1	0	0	1	-4	0	0	0	1			-100	
12		I	0	0	0	0	1	0	0	0	-4	1	0	0			-25	
13		J	0	0	0	0	0	1	0	0	1	-4	1	0			-25	
14		K	0	0	0	0	0	0	1	0	0	1	-4	1			-25	
15		L	0	0	0	0	0	0	0	1	0	0	1	-4			-125	
16																		

Figure 5.46
The coefficient matrix and right-hand-side vector.

Now the coefficient matrix can be inverted and then multiplied by the $[r]$ vector to find the temperatures at the interior points. The result is shown in Figure 5.47.

	A	B	C	D	E	F	G	H	I	J	K	L	M	N	O	P	Q	R
17																		
18	[Coeff-inv]		-0.3	-0.1	-0.04	-0.01	-0.1	-0.07	-0.03	-0.01	-0.03	-0.03	-0.02	-0.01		[T]	50.4	=T$_A$
19				-0.1	-0.34	-0.11	-0.04	-0.07	-0.13	-0.08	-0.03	-0.03	-0.05	-0.04	-0.02	(°C)	70.6	=T$_A$
20				-0.04	-0.11	-0.34	-0.1	-0.03	-0.08	-0.13	-0.07	-0.02	-0.04	-0.05	-0.03		81.4	=T$_B$
21				-0.01	-0.04	-0.1	-0.3	-0.01	-0.03	-0.07	-0.1	-0.01	-0.02	-0.03	-0.03		90.2	=T$_C$
22				-0.1	-0.07	-0.03	-0.01	-0.33	-0.13	-0.05	-0.02	-0.1	-0.07	-0.03	-0.01		31.1	=T$_D$
23				-0.07	-0.13	-0.08	-0.03	-0.13	-0.39	-0.15	-0.05	-0.07	-0.13	-0.08	-0.03		50.8	=T$_E$
24				-0.03	-0.08	-0.13	-0.07	-0.05	-0.15	-0.39	-0.13	-0.03	-0.08	-0.13	-0.07		64.7	=T$_F$
25				-0.01	-0.03	-0.07	-0.1	-0.02	-0.05	-0.13	-0.33	-0.01	-0.03	-0.07	-0.1		79.5	=T$_G$
26				-0.03	-0.03	-0.02	-0.01	-0.1	-0.07	-0.03	-0.01	-0.3	-0.1	-0.04	-0.01		23.2	=T$_H$
27				-0.03	-0.05	-0.04	-0.02	-0.07	-0.13	-0.08	-0.03	-0.1	-0.34	-0.11	-0.04		36.6	=T$_I$
28				-0.02	-0.04	-0.05	-0.03	-0.03	-0.08	-0.13	-0.07	-0.04	-0.11	-0.34	-0.1		47.3	=T$_J$
29				-0.01	-0.02	-0.03	-0.03	-0.01	-0.03	-0.07	-0.1	-0.01	-0.04	-0.1	-0.3		62.9	=T$_K$
30																		

Figure 5.47
The calculated interior temperatures are shown in column Q.

KEY TERMS

Array	Matrices	Scalar multiplication
Array functions	Matrix	Singular
Array math	Matrix addition	Systems of linear
Automatic recalculation	Matrix inversion	algebraic equations
Coefficient matrix	Matrix multiplication	Transpose
[Ctrl-Shift-Enter]	Named array	Unknown vector
Determinant	Nonsingular	Vector
Invert	Right-hand-side vector	
Linear combination	Scalar	

SUMMARY

Naming a Matrix

1. Select the cells containing the array.
2. Enter the desired name in the Name box at the left side of the Formula Bar.

Process for Using Array Math in Excel

1. Enter the needed matrix or matrices.
2. Name the matrices (optional).
3. Select the cells that will hold the result.
4. Enter the array math formula into one cell (typically top-left).
5. Press [Ctrl-Shift-Enter] to enter the array formula into all selected cells.

Basic Matrix Math Operations

Adding Two Matrices

Requirement: The matrices to be added must be the same size.
Options: Use basic cell arithmetic or array math.
Process: Add element by element.

Matrix Multiplication by a Scalar

Requirement: Any matrix can be multiplied by a scalar value.
Options: Use basic cell arithmetic or array math.
Process: Multiply each element of the array by the scalar.

Multiplying Two Matrices

Requirement: The number of columns in the first matrix must equal the number of rows in the second matrix.

Size of result: The size of the product matrix is determined from the number of rows in the first matrix and the number of columns in the second matrix.

Process: Multiply across columns and down rows (this is handled by the **MMULT** function). To multiply two matrices:

1. Enter the matrices to be multiplied.
2. Give the matrices names (optional).
3. Determine the size of the product matrix.
4. Select the cells that will hold the product matrix.
5. Enter the matrix multiplication formula =MMULT(`first_matrix, second_matrix`).
6. Complete the multiplication by pressing [Ctrl-Shift-Enter].

Transposing a Matrix Using Edit/Paste (does not automatically recalculate)

Requirement: Any matrix can be multiplied by a scalar value.
Process: Interchange rows and columns by following these steps:

1. Select the array to be transposed and copy to the Windows clipboard using Ribbon options **Home/Clipboard/Copy**.
2. Indicate the cell that will contain the top-left corner of the result matrix.
3. Open the Paste Special dialog using Ribbon options **Home/Paste (menu)/ Paste Special. . . .**
4. Select Values in the Paste section and check the Transpose checkbox near the bottom of the dialog.
5. Click the [OK] button to close the dialog box and create the transposed matrix.

Transposing a Matrix Using Array Function TRANSPOSE (automatically recalculates)

Requirement: Any matrix can be multiplied by a scalar value.
Process: Interchange rows and columns by following these steps:

1. Enter the original matrix.
2. Indicate where the result should be placed, showing the exact size of the transposed matrix.
3. Enter the =TRANSPOSE(`matrix`) array function.
4. Complete the transpose operation by pressing [Ctrl-Shift-Enter] to transpose the matrix and place the formula in every cell in the result matrix.

Invert a Matrix

Requirement: Only a square, nonsingular matrix can be inverted. (A nonzero determinant indicates a nonsingular matrix.)

Process: Complex; fortunately it is handled by the **MINVERSE** function. To invert a matrix:

1. Enter the matrix to be inverted and name it (optional).
2. Indicate where the inverted matrix should be placed and the correct size (same size as original matrix).
3. Enter the =MINVERSE(matrix) array function. The matrix to be inverted can be indicated by name or cell range.
4. Press [Ctrl-Shift-Enter] to enter the array function in all the cells making up the result matrix.

Matrix Determinant

Requirement: The determinant is calculated only for a square matrix.

Process: Complex; fortunately it is handled by the **MDETERM** function. To calculate the determinant of a matrix:

1. Enter the matrix.
2. Use the formula =MDETERM(matrix) to calculate the determinant of the matrix.

Solving Systems of Linear Equations

Requirement: Must be as many equations as unknowns.

Process:

1. Write the equations in matrix form (coefficient matrix multiplying an unknown vector, equal to a right-hand-side vector).
2. Invert the coefficient matrix.
3. Multiply both sides of the equation by the inverted coefficient matrix.

PROBLEMS

5.1 Simultaneous Equations I

Use Excel's **MDETERM** array function to see whether there is a solution to the simultaneous equations represented by each of the coefficient and right-hand-side matrices shown.

$$\text{a) } C = \begin{bmatrix} 3 & 0 & 5 \\ 8 & 7 & 8 \\ 0 & 3 & 7 \end{bmatrix} \quad r = \begin{bmatrix} 3 \\ 8 \\ 2 \end{bmatrix},$$

$$\text{b) } C = \begin{bmatrix} 1 & 2 & 0 \\ 0 & 1 & 1 \\ 2 & 5 & 1 \end{bmatrix} \quad r = \begin{bmatrix} 4 \\ 3 \\ 3 \end{bmatrix},$$

$$\text{c) } C = \begin{bmatrix} 4 & 3 & 3 \\ 3 & 8 & 4 \\ 3 & 2 & 8 \end{bmatrix} \quad r = \begin{bmatrix} 7 \\ 3 \\ 2 \end{bmatrix},$$

If the solution exists, solve the equations by using matrix methods.

5.2 Simultaneous Equations II

Write the following sets of simultaneous equations in matrix form and check the determinant to see whether there is a solution:

$$\text{a)} \quad \begin{aligned} 0x_1 + 7x_2 + 1x_3 &= 3, \\ 3x_1 + 6x_2 + 3x_3 &= 8, \\ -3x_1 + 8x_2 - 1x_3 &= 2. \end{aligned}$$

$$\text{b)} \quad \begin{aligned} 1x_1 + 8x_2 + 4x_3 &= 0, \\ -1x_1 + 1x_2 + 7x_3 &= 7, \\ 6x_1 + 7x_2 - 2x_3 &= 3. \end{aligned}$$

$$\text{c)} \quad \begin{aligned} 6x_1 + 5x_2 - 1x_3 + 6x_4 &= 0, \\ -2x_1 + 2x_2 + 2x_3 + 2x_4 &= 1, \\ 1x_1 - 1x_2 - 1x_3 - 2x_4 &= 1, \\ 7x_1 - 3x_2 + 8x_3 + 4x_4 &= 4. \end{aligned}$$

If a solution exists, solve the equations using matrix methods.

5.3 Simultaneous Equations III

Write the following sets of simultaneous equations in matrix form and solve the new equations (if possible):

$$\text{a)} \quad \begin{aligned} 3x_1 + 1x_2 + 5x_3 &= 20, \\ 2x_1 + 3x_2 - 1x_3 &= 5, \\ -1x_1 + 4x_2 &= 7. \end{aligned}$$

$$\text{b)} \quad \begin{aligned} 6x_1 + 2x_2 + 8x_3 &= 14, \\ x_1 + 3x_2 + 4x_3 &= 5, \\ 5x_1 + 6x_2 + 2x_3 &= 7. \end{aligned}$$

$$\text{c)} \quad \begin{aligned} 4y_1 + 2y_2 + y_3 + 5y_4 &= 52.9, \\ 3y_1 + y_2 + 4y_3 + 7y_4 &= 74.2, \\ 2y_1 + 3y_2 + y_3 + 6y_4 &= 58.3, \\ 3y_1 + y_2 + y_3 + 3y_4 &= 34.2. \end{aligned}$$

5.4 Multiloop Circuits II

Find the currents i_1 through i_3 in the circuit diagrammed in Figure 5.48.

Figure 5.48

A multiloop circuit.

Known Values	
E	12 volts
R_1	10 ohms
R_2	20 ohms
R_5	50 ohms

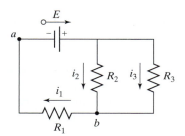

The resistances are listed in the accompanying table.

Material Balances on a Gas Absorber

The following equations are material balances for CO_2, SO_2, and N_2 around the gas absorber shown in Figure 5.49.

Figure 5.49
Gas absorber.

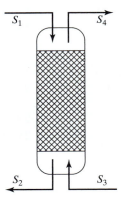

Stream S_1 is known to contain 99 mole% monoethanolamine (MEA) and 1 mole% CO_2. The flow rate in S_1 is 100 moles per minute. The compositions (mole fractions) used in the material balances are tabulated as

	S_1	S_2	S_3	S_4
CO_2	0.01000	0.07522	0.08000	0.00880
SO_2	0	0.01651	0.02000	0.00220
N_2	0	0	0.90000	0.98900
MEA	0.99000	0.90800	0	0
CO_2 Balance:	CO_2 in S_1 + CO_2 in S_3 = CO_2 in S_2 + CO_2 in S_4			
	1 mole + $0.08000 \cdot S_3 = 0.07522 \cdot S_2 + 0.00880 \cdot S_4$			
SO_2 Balance:	SO_2 in S_1 + SO_2 in S_3 = SO_2 in S_2 + SO_2 in S_4			
	$0 + 0.02000 \cdot S_3 = 0.01651 \cdot S_2 + 0.00220 \cdot S_4$			
N_2 Balance:	N_2 in S_1 + N_2 in S_3 = N_2 in S_2 + N_2 in S_4			
	$0 + 0.90000 \cdot S_3 = 0 + 0.98900 \cdot S_4$			

Solve the material balances for the unknown flow rates, S_2 through S_4.

5.6 Material Balances on an Extractor

This problem focuses on a low-cost, high-performance, chemical extraction unit: a drip coffee maker. The ingredients are water, coffee solubles (CS), and coffee grounds (CG). Stream S_1 is water only, and the coffee maker is designed to hold 1 liter of it. Stream S_2 is the dry coffee placed in the filter and contains 99% grounds and 1% soluble ingredients. The product coffee (S_3) contains 0.4% CS and 99.6% water. Finally, the waste product (S_4) contains 80% CG, 19.6% water, and 0.4% CS. (All percentages are on a volume basis.) Figure 5.50 is a diagram of the unit:

Write material balances on water, CS, and CG. Then solve the material balances for the volumes S_2 through S_4.

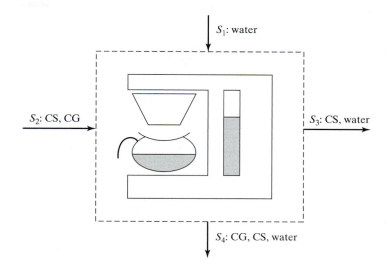

5.7 Flash Distillation

When a hot, pressurized liquid is pumped into a tank (flash unit) at a lower pressure, the liquid boils rapidly. This rapid boiling is called a flash. If the liquid contains a mixture of chemicals, the vapor and liquid leaving the flash unit will have different compositions, and the flash unit can be used as a separator. The physical principle involved is vapor-liquid equilibrium; the vapor and liquid leaving the flash unit are in equilibrium. This allows the composition of the outlet streams to be determined from the operating temperature and pressure of the flash unit. Multiple flash units can be used together to separate multicomponent mixtures, as illustrated in Figure 5.51.

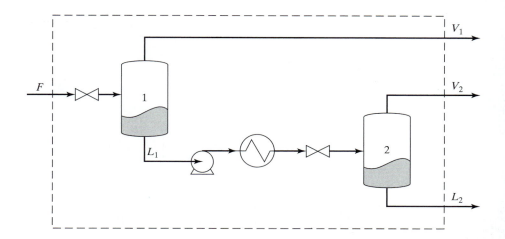

A mixture of methanol, butanol, and ethylene glycol is fed to a flash unit (Unit 1 in Figure 5.51) operating at 165°C and 7 atm. The liquid from the first flash unit is recompressed, reheated, and sent to a second flash unit (Unit 2) operating at 105°C and 1 atm. The mixture is fed to the process at a rate of 10,000 kg/h. Write material balances for each chemical and solve for the mass flow rate of each product stream (V_1, V_2, L_2). A material balance is simply a mathematical statement that all of the methanol (for example) going into the process has to come out again. (This assumes a steady state and no chemical reactions.) The following methanol balance is shown as an example:

$$\text{methanol in } F = \text{methanol in } V_1 + \text{methanol in } V_2 + \text{methanol in } L_2$$

$$0.300 \cdot (10,000 \, \text{kg/h}) = 0.716 \cdot V_1 + 0.533 \cdot V_2 + 0.086 \cdot L_2$$

The compositions of the feed stream F and of the three product streams are listed here:

Component	Mass Fraction in Stream			
	F	V_1	V_2	L_2
Methanol	0.300	0.716	0.533	0.086
Butanol	0.400	0.268	0.443	0.388
Ethylene Glycol	0.300	0.016	0.024	0.526

5.8 Wheatstone Bridge II

Because the resistances of metals vary with temperature, measuring the resistance of a metal sensor inserted in a material is one way to determine the temperature of the material. Resistance temperature detectors (RTDs) are devices commonly used to make temperature measurements.

To compute the resistance of the RTD, it can be built into a Wheatstone bridge as the unknown resistance R_3 as shown in Figure 5.52.

Figure 5.52
Wheatstone Bridge.

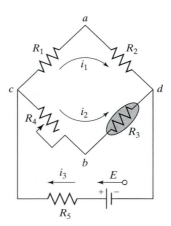

Once the bridge has been balanced, the unknown resistance can be computed from the known resistances, R_1 and R_2, and the setting on the potentiometer, R_4:

$$R_3 = R_4 \frac{R_2}{R_1}. \tag{5.13}$$

The temperature can be determined from the resistance as

$$R_T = R_0[1 + \alpha T], \tag{5.19}$$

where

R_T is the resistance at the unknown temperature, T;
R_0 is the resistance at $0°C$ (known, one of the RTD specifications); and
α is the linear temperature coefficient (known, one of the RTD specifications).

a) For the known resistances and battery potential listed below, what is the resistance of the RTD when the potentiometer (R_4) has been adjusted to 24 ohms to balance the bridge?

b) Use Kirchhoff's laws to find currents i_1 through i_3.

Known Values	
E	24 volts
R_1	15 ohms
R_2	15 ohms
R_5	40 ohms

6

Linear Regression in Excel

Objectives

After reading this chapter, you will know

- What linear regression is, and how it can be used.
- How to use Excel's regression functions to calculate slopes, intercepts, and R^2 values for data sets.
- How to perform regression analyses directly from a graph of your data, using trendlines.
  - Linear (slope, intercept)
  - Other Two-Coefficient Regression Models (Logarithmic, Exponential, Power)
  - Polynomial Regression Models
- How to use Excel's Regression Analysis tool with any linear regression model.

6.1 INTRODUCTION

When most people think of *linear regression,* they think about finding the slope and intercept of the best-fit straight line through some data points. While that is linear regression, it's only the beginning.

The "linear" in linear regression means that the equation used to fit the data points must be linear in the coefficients; it does not imply that the curve through the data points must be a straight line or that the equation can have only two coefficients (i.e., slope and intercept). All of the following equations are linear in the coefficients (the b's):

$$y_p = b_0 + b_1 x \qquad \text{slope and intercept model}$$
$$y_p = b_0 + b_1 x + b_2 x^2 + b_3 x^3 + b_4 x^4 \qquad \text{fourth-order polynomial model}$$
$$y_p = b_0 + b_1 \ln(x) \qquad \text{logarithm model}$$

And there are many number of other linear models that can be used in a regression analysis.

The "regression" part of linear regression involves finding the coefficients that produce the "best fit" of the regression model to a data set, where "best fit" means minimizing the "sum of the squared error" (SSE) between the actual y values in the data set and the y values predicted by the regression equation (y_p).

The process of finding the coefficients that minimize SSE is handled by Excel. Excel provides several methods for performing linear regressions, making it possible to either perform simple and quick analyses or apply highly detailed advanced regression models. Three approaches are covered in this chapter:

- Using regression functions
- Adding trendlines to graphs
- Using Excel's Regression Analysis Package

6.2 LINEAR REGRESSION USING EXCEL FUNCTIONS

When all you need is the *slope* and *intercept* of the best-fit straight line through your data points, Excel's **SLOPE** and **INTERCEPT** functions are very handy. Include the **RSQ** function to calculate the *coefficient of determination* or R^2 value, and a lot of simple data analysis problems are covered.

6.2.1 A Simple Example

The temperature and time values shown in Figure 6.1 are a simple data set that we can use to investigate the basics of regression analysis in Excel.

Figure 6.1
Temperature and time data set.

▲	A	B	C	D	E	F	G
1	Linear Regression with Excel's SLOPE() and INTERCEPT() Functions						
2							
3	Time (min.)	Temp. (K)					
4	0	298					
5	1	299					
6	2	301					
7	3	304					
8	4	306					
9	5	309					
10	6	312					
11	7	316					
12	8	319					
13	9	322					
14							

The **SLOPE** function can be used to determine the slope of the best-fit straight line through these data. The **SLOPE** function takes two arguments: the cell range containing the y values (dependent variable values) and the cell range containing the x values (independent variable values)—in that order.

In this example, temperature depends on time, not the other way around, so temperature (cells B4:B13) is the dependent variable, and time (A4:A13) is the independent variable. So, the **SLOPE** function should be used as:

```
=SLOPE(temperatures, times) or =SLOPE(B4:B13, A4:A13)
```

This is illustrated in Figure 6.2.

Figure 6.2
Using the **SLOPE** function.

	A	B	C	D	E	F	G
1	Linear Regression with Excel's SLOPE() and INTERCEPT() Functions						
2							
3	Time (min.)	Temp. (K)					
4	0	298		Slope:	2.78		
5	1	299		Intercept:			
6	2	301		R^2:			
7	3	304					
8	4	306					
9	5	309					
10	6	312					
11	7	316					
12	8	319					
13	9	322					
14							

Similarly, the intercept can be obtained by using the **INTERCEPT** function with the same arguments, as =INTERCEPT(B4:B13, A4:A13). The coefficient of determination (R^2) can be computed by using the **RSQ** function with the same arguments. The result is shown in Figure 6.3.

Figure 6.3
Using the **INTERCEPT** and
RSQ functions.

E6		f_x	=RSQ(B4:B13,A4:A13)				

	A	B	C	D	E	F	G
1	Linear Regression with Excel's SLOPE() and INTERCEPT() Functions						
2							
3	Time (min.)	Temp. (K)					
4	0	298		Slope:	2.78		
5	1	299		Intercept:	296.1		
6	2	301		R^2:	0.9864		
7	3	304					
8	4	306					
9	5	309					
10	6	312					
11	7	316					
12	8	319					
13	9	322					
14							

This tells us that the best-fit line through the data has slope (b_1) 2.7758 K/min (the units were inferred from the data) and intercept (b_0) 296.1 K. The regression model can be written as:

$$\text{Temp}_\text{p} = 296.1\,\text{K} + 2.7758\frac{\text{K}}{\text{min}}\,\text{time}$$

The R^2 value is 0.9864. Since $R^2 = 1.0$ is a perfect fit (the regression line goes right through every data point), $R^2 = 0.9864$ looks like a pretty good fit—but it is always a good idea to plot the data and the regression line to verify the fit visually. To do this, you first calculate predicted temperatures using the regression model with the best-fit coefficients, and then plot the original data and the predicted values on the same graph. This has been done in Figure 6.4.

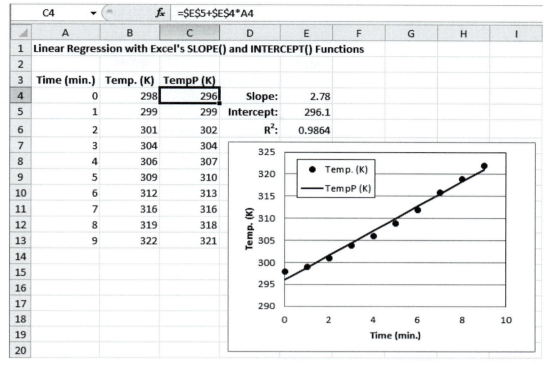

| C4 | | f_x | =$E$5+$E$4*A4 | | | | | |

Figure 6.4
Plotting the regression line with the original data to check the fit.

The graph in Figure 6.4 shows that the regression line is a reasonable approximation of the general upward slope in the data values, but there is clearly some curvature in the temperature data that the regression model is not accounting for. The moral here is to be careful about assuming the regression model "fits" the data just because the R^2 value is pretty close to 1. Taking time to graph the data is always a good idea—but, if you are going to graph your data, then Excel provides an even easier way to perform a linear regression: it is called a *trendline* and is the subject of the next section.

6.3 LINEAR REGRESSION USING EXCEL'S TRENDLINES

In Excel, a *trendline* is a calculated curve through a data set. The name implies that the trendline is intended to highlight the trends in the data. But because most of Excel's trendline options are calculated best-fit curves through the data values, trendlines are also an easy way to perform a regression analysis.

Using an Excel *trendline* to perform a regression is incredibly simple. Once the data have been graphed, regression takes just a few mouse clicks. And, when the regression is done with a trendline, the fitted curve (the trendline) is automatically added to the graph of the original data so that it is easy to visually check the "fit" with the data.

Excel provides the following types of trendlines:

- *Linear* Excel uses the term "linear" to mean a simple slope–intercept regression model.
- *Exponential*
- *Logarithmic*
- *Power*
- *Polynomial*

Note: Excel does provide one type of trendline that is not a regression line: the *moving average*. With a moving average, nearby points in a data set are averaged to provide some smoothing, but no equation is fit to the data set. No regression is performed. The moving average trendline is not covered in this chapter.

6.3.1 Simple Slope–Intercept Linear Regression

The process of performing a linear regression for a slope and intercept requires the computation of various sums involving the x (independent) and y (dependent) values in your data set. With these sums, you could use the following equations to calculate the slope b_1 and intercept b_0 of the straight line that best fits your data (but Excel will do it for you):

$$b_1 = \frac{\sum_i x_i y_i - \dfrac{1}{N_{\text{data}}} \sum_i x_i \sum_i y_i}{\sum_i (x_i^2) - \dfrac{1}{N_{\text{data}}} \left(\sum_i x_i\right)^2}$$

$$b_0 = \frac{\sum_i y_i - b_1 \sum_i x_i}{N_{\text{data}}} \tag{6.1}$$

In these equations, $\sum_i$ implies summation over all data points, $i = 1$ to N_{data}.

When you add a trendline to a graph, Excel calculates all of the required summations and then calculates the slope b_1, intercept b_0, and coefficient of determination R^2. The trendline is added to the graph so that you can see how well the fitted equation matches the data points. By default, the equation of the trendline and the R^2 value are not shown on the graph. You have to request them by checking boxes on the Trendline Options panel of the Format Trendline dialog box (see Figure 6.7).

To see how to use trendlines, we will again use the temperature vs. time data from Figure 6.1. The first step in using a trendline to obtain a regression equation is to plot the data using an XY (Scatter) plot with the data points indicated by data markers. This has been done in Figure 6.5.

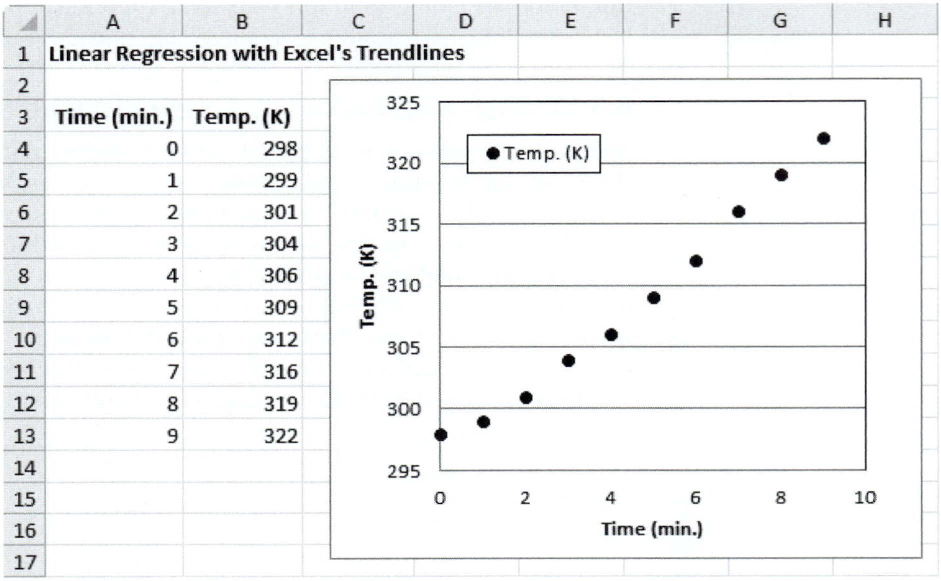

Figure 6.5
Plotting the temperature vs. time data.

	A	B	C	D	E	F	G	H
1	Linear Regression with Excel's Trendlines							
2								
3	Time (min.)	Temp. (K)						
4	0	298						
5	1	299						
6	2	301						
7	3	304						
8	4	306						
9	5	309						
10	6	312						
11	7	316						
12	8	319						
13	9	322						
14								
15								
16								
17								

Note: It is customary to plot data values with markers and fitted results with curves so that it is easy to see whether the fitted line actually goes through the data points.

To add the trendline, right-click on any data point on the graph, and select **Add Trendline**... from the pop-up menu (illustrated in Figure 6.6).

Figure 6.6
Adding a trendline via the data series' pop-up menu.

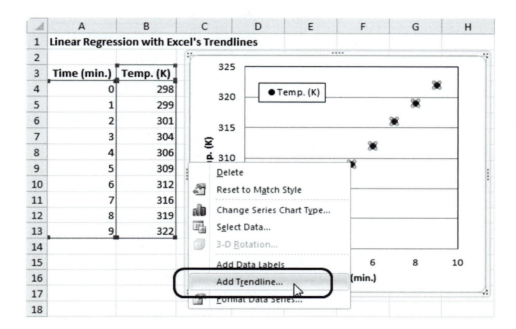

Alternatively, you can insert a trendline from the Ribbon when the graph is selected. Use Ribbon options **Chart Tools/Layout/Analysis/Trendline,** then select **More Trendline Options**.... Either way, the Format Trendline dialog shown in Figure 6.7 is opened.

From the Format Trendline dialog, you can select from six different types of trendline (but the Moving Average type is not a regression line).

Near the bottom of the Format Trendline dialog are three check boxes. Use these to:

- Set the intercept to a specified value (e.g., force the regression line through the graph origin)
- Show the equation of the regression line on the graph
- Display the R^2 value on the graph

These check boxes can be very useful when analyzing data.

As an example of using the Format Trendline dialog, select a **Linear** (slope–intercept) trendline, and select options to cause the equation and R^2 value to be displayed on the graph. These options were selected in Figure 6.7. When you click **Close** to close the Format Trendline dialog, the trendline is added to the graph as shown in Figure 6.8.

From the graph, we see that the regression equation is

$$y = 2.775x + 296.1, \tag{6.2}$$

Figure 6.7
The Format Trendline
dialog.

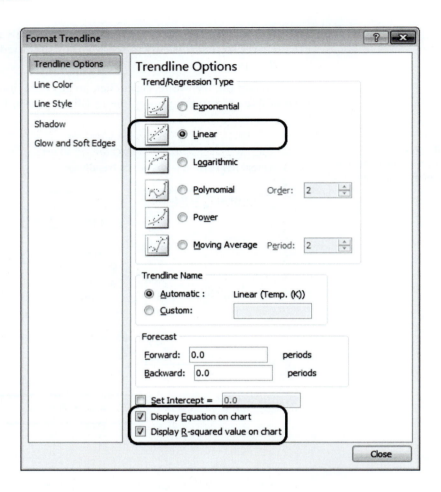

Figure 6.8
Applying the Linear
(slope–intercept) trendline.

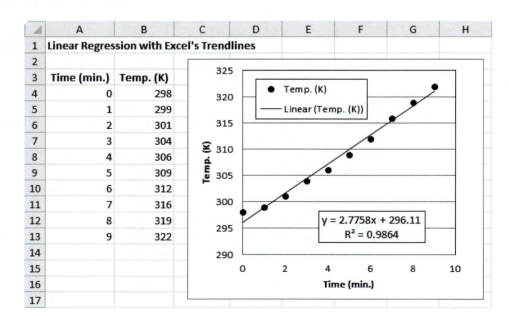

with the R^2 value 0.986. These are the same results we obtained by using the **SLOPE, INTERCEPT,** and **RSQ** functions, but it only took six mouse clicks to obtain these results:

1. Right-click on the data series to open the series pop-up menu
2. Click on **Add Trendline...** to open the Format Trendline dialog
3. Select **Linear** type
4. Check **Display Equation on chart**
5. Check **Display R-Squared Value on chart**
6. Click **Close**

The plotted trendline allows you to verify visually that the best-fit line really does (or doesn't) fit the data. In this case, the straight trendline is not fitting the data well. We need to try to find a regression equation that allows for some curvature.

6.4 OTHER TWO-COEFFICIENT LINEAR REGRESSION MODELS

Any equation relating x values to y values that is linear in the coefficients can be used in regression analysis, but there are a number of two-coefficient models that are commonly used for linear regression, as listed in Table 6.1.

Table 6.1 Two-Coefficient Linear Regression Models

Name	Equation	Linear Form	Data Manipulation
Linear fit	$y = k_0 + k_1 x$	$y = k_0 + k_1 x$ $b_0 = k_0$ $b_1 = k_1$	None required
Exponential fit	$y = k_0 e^{k_1 x}$	$\begin{aligned}\ln(y) &= \ln(k_0) + k_1 x \\ &= b_0 + b_1 x\end{aligned}$ $b_0 = \ln(k_0)$ $b_1 = k_1$	Take the natural log of all y values prior to performing regression. Regression returns b_0 and b_1, which can be related to k_0 and k_1 through the equations on the left.
Logarithmic fit	$y = k_0 + k_1 \ln(x)$	$y = k_0 + k_1 \ln(x)$ $b_0 = k_0$ $b_1 = k_1$	Take the natural log of all x values prior to performing regression.
Power fit	$y = k_0 x^{k_1}$	$\begin{aligned}\ln(y) &= \ln(k_0) + k_1 \ln(x) \\ &= b_0 + b_1 \ln(x)\end{aligned}$ $b_0 = \ln(k_0)$ $b_1 = k_1$	Take the natural log of all x and y values prior to performing regression.

Note: Excel will manipulate the data as needed to create these trendlines, perform the regression, and report the k values. You simply need to select the type of trendline you want to use with your x and y data values.

Each of the two-coefficient models is available as a regressed trendline in Excel. Select the type of regression that you want Excel to perform using the Format Trendline dialog (see Figure 6.7).

The slight upward bending visible in the data (see Figure 6.8) suggests an exponential fit might work. It's easy to give a try; just right-click on the existing trendline to display the trendline's pop-up menu (Figure 6.9), and select **Format Trendline…** to open the Format Trendline dialog.

Figure 6.9
The trendline's pop-up menu.

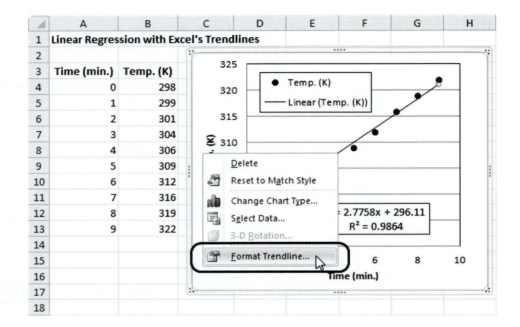

Then, select **Exponential** type on the Format Trendline dialog (see Figure 6.10). When you click the **Close** button, Excel will perform the regression using an exponential fit and show the result on the graph, as shown in Figure 6.11.

Figure 6.10
Selecting **Exponential** type on the Format Trendline dialog.

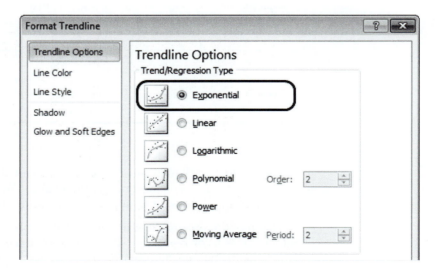

Figure 6.11
The Exponential trendline on the graph.

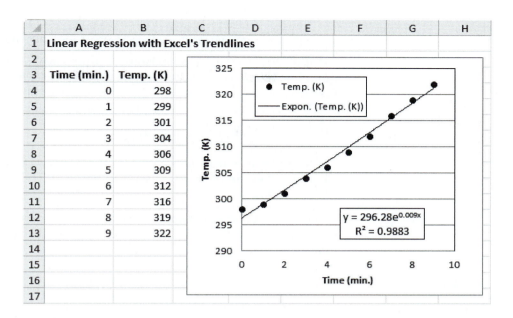

Note: The Logarithmic and Power types are not available for this data set, because it contains $x = 0$ (i.e., time $= 0$). Both of those regression equations take the natural log of x. Because $\ln(0)$ is not defined, these regression equations cannot be used with this data set. If you try to use them, Excel will display an error message.

From Figure 6.11, it is clear that using an exponential regression model doesn't improve the fit much, and the R^2 value is about the same as what we obtained with the linear model. There is only one more regression trendline type available for this data set: Polynomial.

6.5 POLYNOMIAL REGRESSION

Excel will automatically perform linear regression with polynomial regression models of orders 2 through 6. These models are listed in Table 6.2.

Table 6.2 Polynomial Regression Models

Order	Regression Model	Number of Coefficients
2	$y_p = b_0 + b_1 x + b_2 x^2$	3
3	$y_p = b_0 + b_1 x + b_2 x^2 + b_3 x^3$	4
4	$y_p = b_0 + b_1 x + b_2 x^2 + b_3 x^3 + b_4 x^4$	5
5	$y_p = b_0 + b_1 x + b_2 x^2 + b_3 x^3 + b_4 x^4 + b_5 x^5$	6
6	$y_p = b_0 + b_1 x + b_2 x^2 + b_3 x^3 + b_4 x^4 + b_5 x^5 + b_6 x^6$	7

Polynomial regression is still a linear regression, because the regression polynomials are linear in the coefficients (the b values). Since, during a regression analysis, all of the x values are known, each of the polynomial equations listed in Table 6.2 is a linear equation, regression using any of these equations is a linear regression. Polynomial regression differs from the two-coefficient regression models in that there are more than two coefficients.

Note: Technically, simple slope–intercept regression is also a first order polynomial regression, but the term is usually reserved for polynomials of order two or higher.

It is fairly standard nomenclature to write polynomials with high powers on the right; however, Excel reports polynomial trendline equations with the high powers first. It's not a big deal, but read the equations carefully.

To request a second-order polynomial regression for the temperature vs. time data:

1. Right-click on the existing trendline to display the trendline's pop-up menu (Figure 6.9)
2. Select **Format Trendline…** to open the Format Trendline dialog
3. Select **Polynomial** type, and set the **Order** to 2
4. Click **Close**

The second-order polynomial trendline shown in Figure 6.12 appears to fit the data well, and the R^2 value is much closer to unity than with the other trendlines we have used.

Figure 6.12
The second-order polynomial trendline.

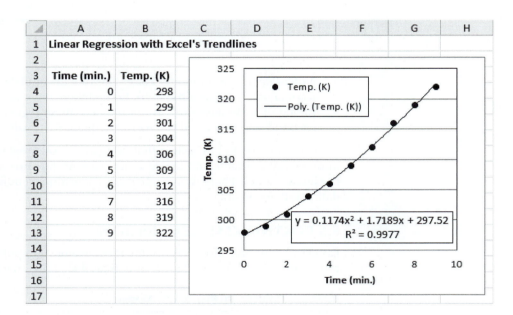

Polynomial regression using trendlines is as easy and simple as slope–intercept regression. But what happens if you want a seventh-order fit, and Excel's trendline capability can't be used? You can still use Excel's *Regression Analysis Package*, which is described in the next section.

6.6 LINEAR REGRESSION USING EXCEL'S REGRESSION ANALYSIS PACKAGE

Excel's Regression Analysis Package is fairly easy to use, but is more involved than simply asking for a trendline on a graph. There are two reasons why you might want to use the more complex approach:

1. You want to use a regression model that is not available as a trendline. The Regression Analysis Package can handle any linear-regression model.

2. You want more details about the regression process, or more detailed regression results than are provided by using trendlines.

The Regression Analysis Package is a powerful tool available with Excel but, by default, it is not activated when Excel is installed. Before we can use it, we need to see if it is installed and activated.

6.6.1 Activating the Regression Analysis Package

Excel's Regression Analysis Package, is part of the Excel's Analysis Toolpak Add-In, which is installed, but not activated when Excel is installed. You activate Add-Ins using the Excel Options dialog, which is opened as:

- Excel 2010: **File tab/Options**
- Excel 2007: **Office/Excel Options**

Then, select the **Add-Ins** panel shown in Figure 6.13. The available Excel Add-Ins will vary greatly from one computer to the next, so your screen may look quite different.

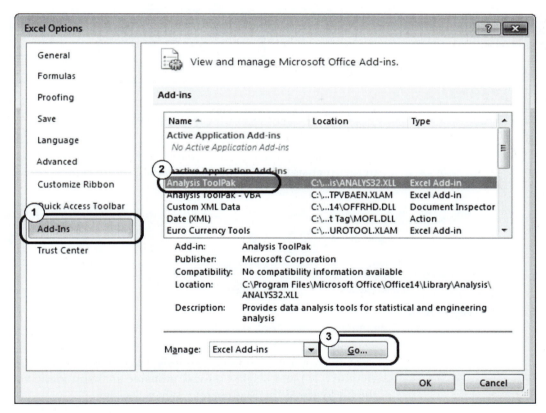

Figure 6.13
Activating the Analysis ToolPak Add-In.

The Add-Ins panel lists Active Application Add-Ins and Inactive Application Add-Ins.

- If the Analysis Toolpak is in the active list, the Regression Analysis Package is ready to go.
- If the Analysis Toolpak is not on any list, it wasn't installed; you will have to install the Analysis Toolpak.

- If it is in the inactive list, you need to activate it. To activate the Analysis ToolPak:

1. Click on the **Analysis ToolPak** list item (the name in the list)
2. Click the **Go…** button

These steps will cause the Add-Ins dialog to open, as shown in Figure 6.14.

Figure 6.14
The Add-Ins dialog.

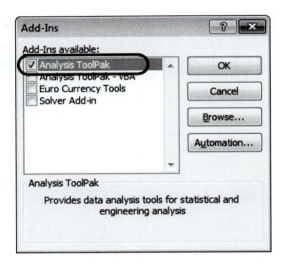

3. Check the box labeled **Analysis ToolPak**
4. Click the **OK** button

The Ribbon will then be updated, and a new button will appear on the **Data** tab in the Analysis group as the **Data Analysis** button. This is shown in Figure 6.15.

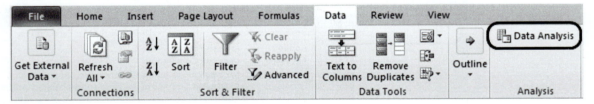

Figure 6.15
Ribbon options **Data/Analysis/Data Analysis** access the Analysis ToolPak.

Clicking on the **Data Analysis** button opens the Data Analysis dialog (Figure 6.16) and displays a list of available analysis tools. Choose **Regression** from the list to access the Regression Analysis Package.

6.6.2 A Simple Linear Regression for Slope and Intercept

As a first example of working with the Regression Analysis Package, we will once again regress the temperature vs. time data (Figure 6.1) with a simple slope–intercept regression model. The steps involved will include:

1. Open the Data Analysis list box
2. Select **Regression** to access the Regression Analysis Package
3. Indicate the cell addresses of the y values (dependent values)

Figure 6.16
The Analysis Tools available through the Data Analysis dialog.

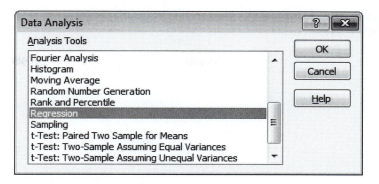

4. Indicate the cell addresses of the *x* values (independent values)
5. Indicate that the *x* and *y* cell ranges included labels (column headings)
6. Choose a location for the results
7. Indicate you want the results to be graphed
8. Tell Excel to perform the regression

Step 1. Open the Data Analysis list box. Use the following Ribbon options to open the Data Analysis list box shown in Figure 6.17.

Data/Analysis/Data Analysis

Figure 6.17
The Data Analysis dialog.

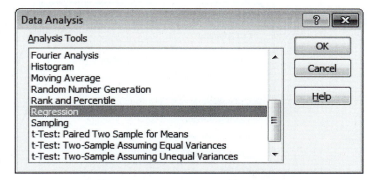

Step 2. Select Regression from the Data Analysis list box. Click **Regression** in the list, then click the **OK** button. This will open the Regression dialog box shown in Figure 6.18.

The Regression dialog is used to tell Excel:

- where to find the input values, and whether or not the input cell ranges include labels,
- where to put the output, and
- what features you want to include in the output.

Step 3. Tell Excel where to find the *y* values for the regression. Three fields on the Regression dialog have small buttons at the right edge of the fields (see Figure 6.19). These buttons are used to jump back to the worksheet (temporarily) so that cell locations can be identified.

Figure 6.18
The Regression dialog.

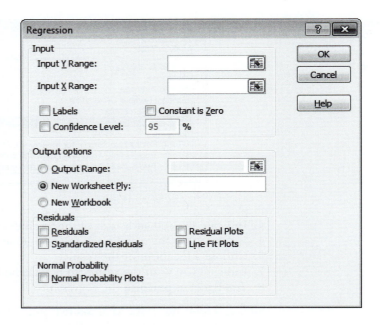

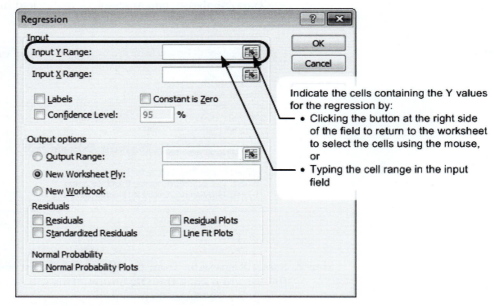

Figure 6.19
Identifying the *y* values using the Regression dialog.

One of these fields is labeled **Input Y Range:** (see Figure 6.19). Use the button at the right edge of the field to jump back to the worksheet, and select the cells containing the *y* values (temperature values in this example). Figure 6.19 shows this selection.

Note that the selected cell range in Figure 6.20 includes the column heading, "Temp. (K)," and the selection is indicated in the input field in the mini-Regression dialog.

Figure 6.20
Selecting the *y* values for the regression.

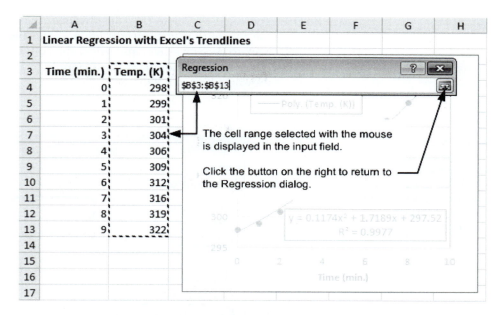

When you have selected the *y* values (temperature values), click the button at the right side of the input field in the mini-Regression dialog to jump back to the full Regression dialog.

Note: Common linear regression assumes that all of the uncertainty in the data is in the *y* values and that the *x* values are known precisely. It is important, therefore, to call the imprecise values the *y* values for regression analysis.

When the full Regression dialog is displayed, the selected cell range containing the *y* values is shown in the **Input Y Range:** field, as shown in Figure 6.21.

Figure 6.21
The Regression dialog after selecting the cell range containing the *y* values.

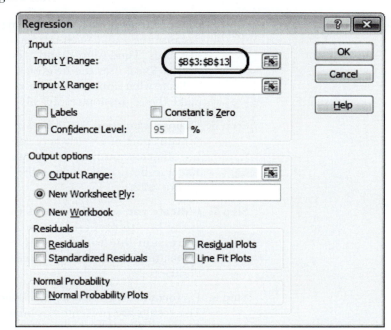

Step 4. Tell Excel where to find the *x* values. Similarly, indicate the *x* values, use the button at the right edge of the **Input X Range:** field to jump back to the worksheet and select the cells containing the *x* values (time values in this example). Then, return to the full Regression Dialog, as shown in Figure 6.22.

Figure 6.22
The Regression dialog after selecting the cell range containing the *y* values and indicating that the selected *x* and *y* cell ranges included labels.

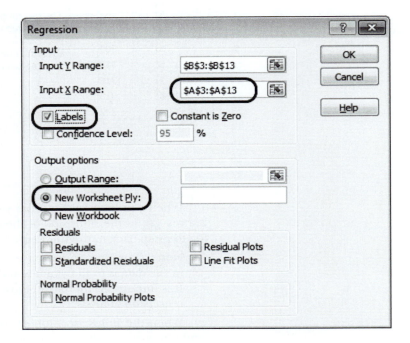

Step 5. Indicate that the *x* and *y* cell ranges included labels (column headings). Because the *x* and *y* cell ranges included the cells containing the column headings "Time (min.)" and "Temp. (K)," be sure to check the **Labels** box so that Excel uses the headings in the regression output. Using labels to indicate what *x* and *y* mean for your data set (time and temperature) helps make the regression results table easier to read and understand.

Step 6. Choose a location for the results. The regression results can fill a significant part of a worksheet screen. Because the results take up quite a bit of space, putting the output on a new worksheet is common. Tell Excel to do this by clicking the button next to **New Worksheet Ply:** (see Figure 6.22.)

Step 7. Indicate you want the results to be graphed. It is a good idea to look at the line fit plot and the residual plot to visually check the fit of the regression line with the data. This is done by checking the **Line Fit Plots** and **Residual Plots** boxes on the Regression dialog as shown in Figure 6.23.

Step 8. Perform the regression. Click the **OK** button to perform the regression. The results are presented as tables and graphs (if you requested graphs). A portion of the output is shown in Figure 6.24.

Figure 6.23
Selecting line fit and
residual plots.

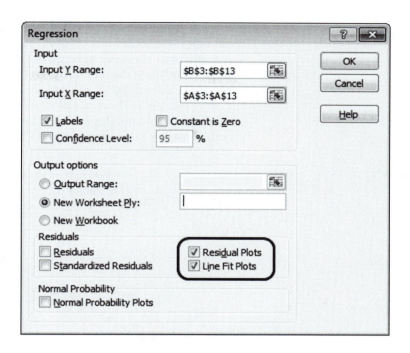

Figure 6.24
Regression results (partial).

The Coefficients column in the output table tells us that the best-fit line through the data has slope (b_1) 2.775 K/min (the units were inferred from the data) and intercept (b_0) 296.1 K, so the regression model is

$$\text{Temp}_p = 296.1\,\text{K} + 2.775\,\frac{\text{K}}{\text{min}}\,\text{time}.$$

The R^2 value is 0.9864. These are the same results we obtained with the other methods.

The plots allow you to verify visually that the best-fit line really does (or doesn't) fit the data. The Line Fit Plot (Figure 6.25) compares the temperature values in the data set to those predicted by the regression model. This is essentially the same plot using the linear trendline in Figure 6.8, although Excel initially presented the Line Fit Plot as a Column graph; the graph type has been changed to XY Scatter.

Figure 6.25

The Line Fit Plot.

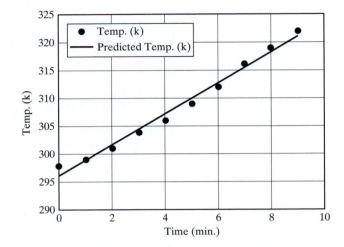

A Residual Plot shown in Figure 6.26 highlights the differences between the actual temperature data and the regression model's predicted values. A poor regression fit becomes very apparent when you look at the residual plot.

Figure 6.26

The Residual Plot.

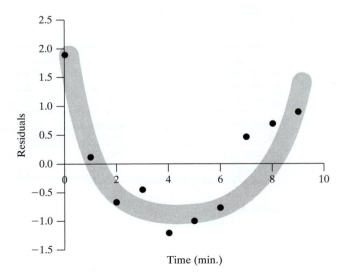

The *residual* is the difference between the data point y value and the regression line y value (predicted y value) at each x value. A residual plot highlights poor fits and makes them easy to spot. Strong patterns in a residual plot, such as the "U" shape shown in Figure 6.26, indicate that there is something happening in your data that the regression model is not accounting for. It's a strong hint that you should look for a better regression model.

6.6.3 Polynomial Regression Using the Regression Analysis Package

The best fit to the temperature–time data using trendlines was obtained by using the second-order polynomial

$$y_p = b_0 + b_1 x + b_2 x^2 \tag{6.3}$$

or, in terms of temperature, T, and time, t,

$$T_p = b_0 + b_1 t + b_2 t^2. \tag{6.4}$$

We will use this polynomial to demonstrate how to solve for more than two coefficients when using Excel's Regression Analysis Package.

The Regression Analysis Package in Excel allows you to have only one column of y values (dependent variables), but you can have multiple columns of x values (independent variables). These can be completely independent variables [e.g., enthalpy (dependent) as a function of temperature (independent) and pressure (independent)], or they can be the same variable in multiple forms [e.g., x and x^2 or z and $\ln(z)$.]

For this second-order polynomial, we'll need two columns of independent values, one for t values and another for t^2 values. The dependent values (T values) have been moved to column A in Figure 6.27 to allow the t and t^2 values to be placed next to each other in columns B and C.

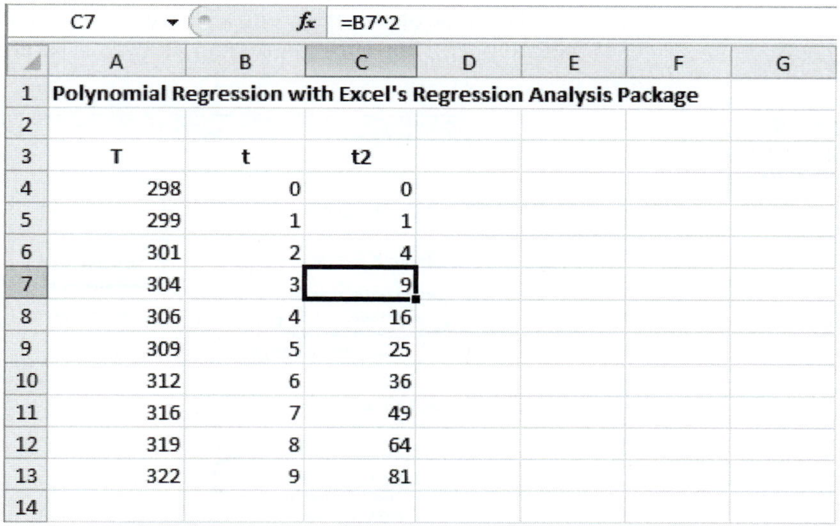

Figure 6.27
Preparing for polynomial regression.

The procedure for getting Excel to regress this data is almost the same as that used in the previous example.

1. Open the Data Analysis list box
2. Select **Regression** to access the Regression Analysis Package
3. Indicate the cell addresses of the y values (dependent values)
4. Indicate the cell addresses of the x values (independent values)

5. Indicate that the *x* and *y* cell ranges included labels (column headings)
6. Choose a location for the results
7. Indicate you want the results to be graphed
8. Tell Excel to perform the regression

The big difference is in step 4, when you tell Excel where to find the *x* values. In this case, you need to indicate the *t* and t^2 values in columns B and C. This step is illustrated in Figure 6.28.

Figure 6.28
Step 4 (select *x* values) for the polynomial regression.

	A	B	C	D	E	F	G
1	Polynomial Regression with Excel's Regression Analysis Package						
2							
3	T	t	t2				
4	298	0	0	Regression			
5	299	1	1	$B$3:$C$13			
6	301	2	4				
7	304	3	9				
8	306	4	16				
9	309	5	25				
10	312	6	36				
11	316	7	49				
12	319	8	64				
13	322	9	81				
14							

The rest of the regression process is unchanged. The Regression dialog just before performing the regression analysis is shown in Figure 6.29.

Figure 6.29
The Regression dialog for polynomial regression.

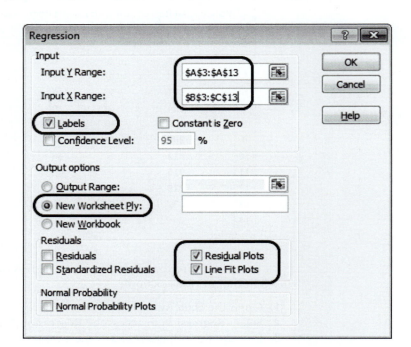

The regression output table (Figure 6.30) shows the results (again, only part of it is shown).

◢	A	B	C	D	E	F	G
1	SUMMARY OUTPUT						
2							
3	*Regression Statistics*						
4	Multiple R	0.9989					
5	R Square	0.9977					
6	Adjusted R Square	0.9971					
7	Standard Error	0.4584					
8	Observations	10					
9							
10	ANOVA						
11		*df*	*SS*	*MS*	*F*	*Significance F*	
12	Regression	2	642.93	321.46	1529.52	5.69E-10	
13	Residual	7	1.47	0.21			
14	Total	9	644.40				
15							
16		*Coefficients*	*Standard Error*	*t Stat*	*P-value*	*Lower 95%*	*Upper 95%*
17	Intercept	297.52	0.36	825.40	1.01E-18	296.67	298.37
18	t	1.72	0.19	9.22	3.66E-05	1.28	2.16
19	t2	0.12	0.02	5.89	6.08E-04	0.07	0.16
20							

Figure 6.30
Regression results (partial).

Again, the regression coefficients obtained by using the Regression Analysis Package are the same as those obtained by using the polynomial trendline.

$$T_p = 297.5 + 1.72t + 0.12t^2.$$

One piece of information that is available only from the Regression Analysis Package is the residual plot, as shown in Figure 6.31.

The residual plot shows no obvious trends. This suggests that the model is doing about as good a job of fitting these data as can be done.

6.6.4 Other Linear Models

The models used in the preceding examples,

$$T_p = b + b_1 t \quad \text{and}$$
$$T_p = b_0 + b_1 t + b_2 t^2, \tag{6.5}$$

are both linear models (linear in the coefficients, not in time). Excel's Regression Analysis Package works with any linear model, so you could try fitting equations such as

$$T_p = b_0 + b_1 \sin h(t) + b_2 \text{atan}(t^2) \tag{6.6}$$

Figure 6.31
The Residual Plot.

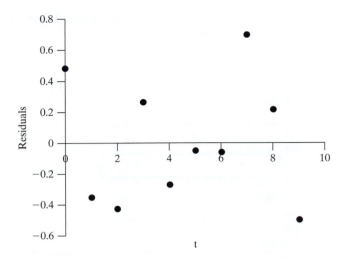

or

$$T_p = b_0 \exp(t^{0.5}) + b_1 \ln(t^3). \tag{6.7}$$

There is no reason to suspect that either of these last two models would be a good fit to the temperature vs. time data, but both equations are linear in the coefficients (the b's) and are compatible with Excel's Regression Analysis Package. (Actually, there is one problem: the natural logarithm in the last model won't work when $t = 0$ appears in the data set.)

In general, you try to choose a linear regression model either from some theory that suggests a relationship between your variables or from looking at a plot of the data set.

So far, we have looked at polynomial regression where the dependent variable is a function of one independent variable, t, even though t appeared in two forms: t and t^2. Remember that you can also use a regression equation that has multiple independent variables, such as

$$V_p = b_0 + b_1 P + b_2 T, \tag{6.8}$$

which says, for example, that the volume of a gas depends on pressure and temperature. In this equation there is one dependent variable, V_p, and two independent variables, P and T. This equation is also compatible with Excel's Regression Analysis Package.

6.6.5 Forcing the Regression Line through the Origin (0, 0)

If you do not want Excel to compute an intercept (i.e., if you want to force the curve to go through $(0, 0)$ by setting $b_0 = 0$), there is a **Constant is Zero** check box you can select on the Regression dialog box, indicated in Figure 6.32.

There are times when theory predicts that the curve should go through $(0, 0)$. In such situation, you should force the regression line through the origin by checking the **Constant is Zero** box.

Figure 6.32
Forcing the regression
curve through the intercept.

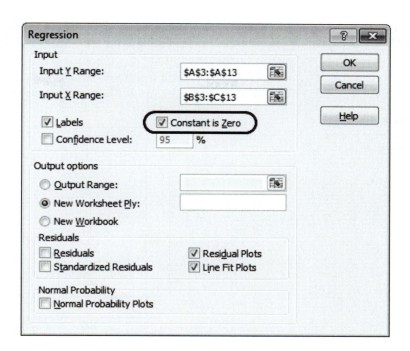

PRACTICE!

In Table 6.3, two sets of y values are shown. The noisy data were calculated from the clean data by adding random values. Try linear regression on these data sets. What is the impact of noisy data on the calculated slope, intercept, and R^2 value?

Table 6.3 Clean and Noisy Data

X	Y_{CLEAN}	Y_{NOISY}
0	1	0.85
1	2	1.91
2	3	3.03
3	4	3.96
4	5	5.10
5	6	5.90

[ANSWERS: Clean Data: $y_p = 1 + 1x$, $R^2 = 1$; Noisy Data: $y_p = 0.904 + 1.021\ x$, $R^2 = 0.998$.]

EXAMPLE 6.1

Thermocouple output voltages are commonly considered to be directly proportional to temperature. But, over large temperature ranges, a power fit can do a better job. Compare the simple linear fit and the exponential fit for

(*continued*)

copper–copper/nickel (Type T) output voltages (reference junction at 0°C) between 10 and 400°C.[1]

First, we perform a simple linear regression for a slope and intercept using a Linear trendline. The result is shown in Figure 6.33.

Figure 6.33
Using a Linear (slope–intercept) trendline.

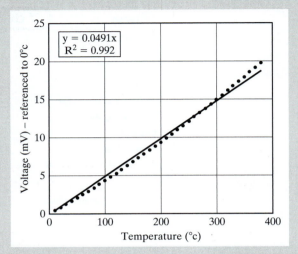

The R^2 value 0.992, looks OK, but there is clearly some difference between the data values and the regression line.

Next, we modify the trendline to see how a power fit looks. The result is shown in Figure 6.34.

Figure 6.34
Using a Power trendline.

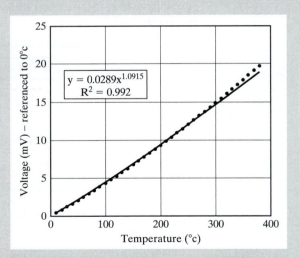

The R^2 value is closer to unity, and although there is still some difference between the actual data and the predicted values, especially at higher temperatures, the power fit looks a lot better at low temperatures.

----

[1]Data from Omega Engineering online reference tables at http://www.omega.com/temperature/Z/pdf/z207.pdf. Type T was selected for this example because this alloy is known to have a nonlinear relationship between temperature and voltage (over large temperature ranges).

RECALIBRATING A FLOW METER

Flow meters in industrial situations come with calibration charts for standard fluids (typically air or water), but for critical applications, the calibration must be checked periodically to see whether the instrument still provides a reliable reading.

When purchased, the calibration sheet for a turbine flow meter provided the following equation relating the meter output (frequency, Hz) to water flow velocity (m/s):

$$v = 0.0023 \; + \; 0.0674f$$

After the meter had been in use for a period of 1 year, it was removed from service for recalibration. During a preliminary test, the following data were obtained from a test system:

Velocity (m/s)	Frequency (Hz)
0.05	0.8
0.27	4.2
0.53	8.2
0.71	10.9
0.86	13.1
1.10	16.8
1.34	20.5
1.50	23.0
1.74	26.6
1.85	28.4
2.15	32.9
2.33	36.6
2.52	38.5
2.75	42.1

Does the meter need to be recalibrated? If so, what is the new calibration equation?

First, we can use the original calibration equation to calculate predicted velocity values and plot the results to see whether the original calibration equation is still working. The result is shown in Figure 6.35.

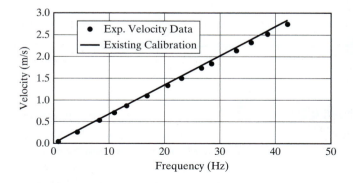

Figure 6.35

Existing calibration equation with experimental data.

The data values do not seem to agree with the original calibration line at higher velocities, so it is time to recalibrate. To do so, simply add a linear trendline to the graph and ask Excel to display the equation of the line. The result (with the new calibration equation) is shown in Figure 6.36.

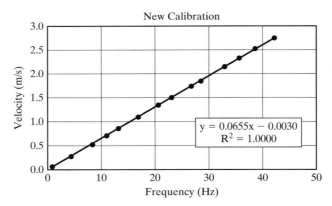

Figure 6.36
A Linear trendline added to the experimental data to obtain a new calibration equation.

The new calibration equation is

$$v = -0.0030 + 0.0655f$$

KEY TERMS

Best fit
Coefficient of
   determination (R^2)
Dependent variable
   (y-axis)
Exponential trendline
Independent variable
   (x-axis)
Intercept

Linear regression
Linear trendline
Origin
Polynomial trendline
Power trendline
Regression
Regression analysis
   package
Regression coefficients

Regression equation
Regression model
Residual
Residual plot
Slope
Trendline
XY Scatter plot

SUMMARY

Using Excel's Regression Functions

For simple slope–intercept calculations, Excel's built-in regression functions are useful. The drawback is they don't show you the regression line superimposed on the data values, and verifying the fit visually is always a good idea. If you plot the data, it is faster to use a trendline to find the slope and intercept, but, although the trendline displays the values, it does not make them available for use in the rest of the worksheet.

SLOPE(*y*, *x*)	Returns the slope of the straight line through the *x* and *y* values.
INTERCEPT(*y*, *x*)	Returns the intercept of the straight line through the *x* and *y* values.
RSQ(*y*, *x*)	Returns the coefficient of determination (R^2) for the straight line through the *x* and *y* values.

Using Trendlines

If you don't need a lot of information about the regression results—just the regression equation and the R^2 value—then, trendlines are fast and easy. To add a trendline, do the following:

1. Graph your data.
2. Right-click on any data point, and select Trendline from the pop-up menu.
3. Use the Type tab to select the type of trendline. (Set the order of the polynomial, if needed.)
4. Use the options on the Format Trendline dialog to force the intercept through the origin or to have Excel to display the equation of the line and the R^2 value, if desired.

Using the Regression Analysis Package

General Procedure

1. Choose a linear regression model: the model must be linear in the coefficients.
2. Set up the required columns of *x* and *y* values in the worksheet:
   - You may have only one column of *y* values, but multiple columns of *x* values (e.g., x, x^2, x^3 as required for the regression model you want to fit).
3. Have Excel to perform the regression analysis.
4. Find your regression results in the output table created by Excel. As a minimum, check for
   - **The coefficients** listed as Intercept (if you asked Excel to compute one) and the Coefficients for *X* Variable 1, *X* Variable 2, and so on (as many as are needed for your model).
   - **The R^2 value**; the value 1.0 indicates a perfect fit.
5. Check the line fit plot and residual plots (if you requested them) to verify visually that your model does (or does not) fit the data.

Using the Regression Analysis Package

Open the Data Analysis List Box using Ribbon options **Data/Analysis/Data Analysis** [Excel 2003: Tools/Data Analysis…]

1. Select **Regression** from the list.
2. Show Excel where to find the *y* values.
3. Show Excel where to find the *x* values.
4. Choose a location for the regression results; on a new worksheet ply is the most common location.
5. Indicate whether you want the line fit plot and residual plot prepared (generally recommended).
6. Click **OK** on the Regression dialog box to have Excel to perform the regression and produce the output summary tables and any graphs you requested.

PROBLEMS

6.1 Graphing Functions

Graph the following common regression functions, using the specified coefficients and ranges:

a. $y = a + bx$, $a = 2, b = 0.3, 0 \leq x \leq 5$.
b. $y = a + b/x$ $a = 2, b = 0.3, 1 \leq x \leq 5$.
c. $y = ae^{bx}$, $a = 2, b = 0.3, 0 \leq x \leq 5$.
d. $y = ae^{-bx}$, $a = 2, b = 0.3, 0 \leq x \leq 5$.

Compare the general shape of the curves produced by the regression functions with the plot shown in Figure 6.37.

Figure 6.37
Data for regression.

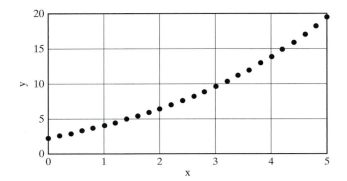

Which regression function(s) is(are) most likely to provide a good fit to the data?

6.2 Simple Linear Regression

Plot each of the three data sets presented in the accompanying table to see whether a straight line through each set of points seems reasonable. If so, add a linear trend-line to the graph and have Excel to display the equation of the line and the R^2 value on the graph.

The same x data have been used in each example to minimize typing.

X	Y₁	Y₂	Y₃
0	2	0.4	10.2
1	5	3.6	4.2
2	8	10.0	12.6
3	11	9.5	11.7
4	14	12.0	28.5
5	17	17.1	42.3
6	20	20.4	73.6
7	23	21.7	112.1

6.3 Thermocouple Calibration Curve

Thermocouples are made by joining two dissimilar metal wires. The contact of the two metals results in a small, but measurable, voltage drop across the junction. This voltage drop changes as the temperature of the junction changes;

thus, the thermocouple can be used to measure temperature if you know the relationship between temperature and voltage. Equations for common types of thermocouples are available, or you can simply take a few data points and prepare a calibration curve. This is especially easy for thermocouples because for small temperature ranges, the relationship between temperature and voltage is nearly linear.

Use a linear trendline to find the coefficients of a straight line through the data shown here:

T (°C)	V (mV)
10	0.397
20	0.798
30	1.204
40	1.612
50	2.023
60	2.436
70	2.851
80	3.267
90	3.682

Note: The thermocouple voltage changes because the temperature changes—that is, the voltage depends on the temperature. For regression, the independent variable (temperature) should always be on the x axis, and the dependent variable (voltage) should be on the y axis.

6.4 Heat Transfer by Conduction

Thermal conductivity is a property of a material related to the material's ability to transfer energy by conduction. Materials with high thermal conductivities, such as copper and aluminum, are good conductors. Materials with low thermal conductivities are used as insulating materials.

Figure 6.38 depicts a device that could be used to measure the thermal conductivity of a material. A rod of the material to be tested is placed between a resistance heater (on the right) and a block containing a cooling coil (on the left). Five thermocouples are inserted into the rod at evenly spaced intervals. The entire apparatus is placed under a bell jar, and the space around the rod is evacuated to reduce energy losses.

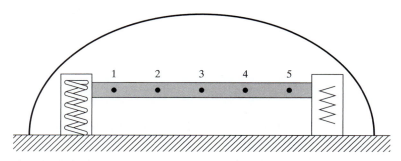

Figure 6.38
Heat conduction experiment.

To run the experiment, a known amount of power is sent to the heater, and the system is allowed to reach steady state. Once the temperatures are steady, the power level and the temperatures are recorded. A data sheet from an experiment with the device is reproduced next.

Thermal Conductivity Experiment	
Rod diameter	2 cm
Thermocouple spacing	5 cm
Power	100 watts

Thermocouple	Temperature (K)
1	348
2	387
3	425
4	464
5	503

The thermal conductivity can be determined by using Fourier's law,

$$\frac{q}{A} = -k\frac{dT}{dx} \tag{6.9}$$

where

q is the power applied to the heater
A is the cross-sectional area of the rod

The q/A is the energy flux and is a vector quantity, having both magnitude and direction. With the energy source on the right, the energy moves to the left or in the negative x direction, so the flux in this problem is negative.

    a. From the information supplied on the data sheet, prepare a table of position x, and temperature T, values, in a worksheet, with the position of thermocouple 1 set as $x = 0$.

    b. Graph the temperature vs. position data, and add a linear trendline to the data. Ask Excel to display the equation of the line and the R^2 value.

    c. Use Fourier's law and the slope of the regression line to compute the material's thermal conductivity.

Thermal conductivity is usually a function of temperature. Does your graph indicate that the thermal conductivity of this material changes significantly over the temperature range in this problem? Explain your reasoning.

6.5 Calculating Heat Capacity

The heat capacity at constant pressure is defined as

$$C_p = \left(\frac{\partial \hat{H}}{T}\right)_p \tag{6.10}$$

where

$\hat{H}$ is specific enthalpy.
T is absolute temperature.

If enthalpy data are available as a function of temperature at constant pressure, the heat capacity can be computed. For steam, these data are readily available.[2] For

[2]The data are from a steam table in *Elementary Principles of Chemical Engineering*, 3rd ed., by Felder, R. M. and R. W. Rousseau, New York: Wiley, 2000. These data are also available at http://www.coe.montana. edu/che/Excel.

example, the following table shows the specific enthalpy for various absolute temperatures at a pressure of 5 bars:

T (°C)	Ĥ(kJ/kg)
200	2855
250	2961
300	3065
350	3168
400	3272
450	3379
500	3484
550	3592
600	3702
650	3813
700	3926
750	4040

a. Plot the specific enthalpy on the y axis against absolute temperature on the x axis.

b. Add a linear trendline to the data, and ask Excel to display the equation of the line and the R^2 value.

c. Compute the heat capacity of steam from the slope of the trendline.

d. The ideal gas (assumed) heat capacity for steam is approximately 2.10 kJ/kg K at 450°C. How does your result compare with this value?

e. Does it appear that the heat capacity of steam is constant over this temperature range? Why or why not?

6.6 Vapor–Liquid Equilibrium

When a liquid mixture is boiled, the vapor that leaves the vessel is enriched in the more volatile component of the mixture. The vapor and liquid in a boiling vessel are in equilibrium, and vapor–liquid equilibrium (VLE) data are available for many mixtures. The data are usually presented in tabular form, as in the table at the end of this problem statement, which represents VLE data for mixtures of methanol (MeOH) and ethanol (EtOH) boiling at 1 atm. From the graph of the VLE data, we see that, when 50:50 liquid mixture of the alcohol is boiled, the vapor will contain about 60% methanol (Figure 6.39).

Figure 6.39

Vapor-Liquid equilibrium curve for methanol and ethanol.

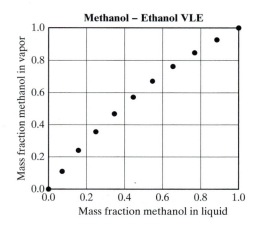

VLE data are commonly used in designing distillation columns, but an equation relating vapor mass fraction to liquid mass fraction is a lot handier than tabulated values.

Use trendlines on the tabulated VLE data to obtain an equation relating the mass fraction of methanol in the vapor (y) to the mass fraction of vapor in the liquid (x). Test several different linear models (e.g., polynomials) to see which model gives a good fit to the experimental data.

Data: The VLE values shown next were generated by using Excel, with the assumption that these similar alcohols form an ideal solution. The **X** column represents the mass fraction of methanol in the boiling mixture. (Mass fraction of ethanol in the liquid is calculated as $1 - x$ for any mixture.) The mass fraction of methanol in the vapor leaving the solution is shown in the **Y** column.

These data are available in electronic form at the text's website http://www.coe.montana.edu/che/Excel.

X_{MeOH}	Y_{MeOH}
1.000	1.000
0.882	0.929
0.765	0.849
0.653	0.764
0.545	0.673
0.443	0.575
0.344	0.471
0.250	0.359
0.159	0.241
0.072	0.114
0.000	0.000

6.7 Calculating Latent Heat of Vaporization

The Clausius–Clapeyron equation can be used to determine the latent heat (or enthalpy change) of vaporization, $\Delta \hat{H}_v$. The equation is

$$\ln(p_{vapor}) = -\frac{\Delta \hat{H}_v}{RT} + k, \tag{6.11}$$

where

p_{vapor} is the vapor pressure of the material,
R is the ideal gas constant,
T is the absolute temperature,
k is a constant of integration.

There are a few assumptions built into this equation:
1. The molar volume of the liquid must be much smaller than the molar volume of the gas (not true at high pressures).
2. The gas behaves as an ideal gas.
3. The latent heat of vaporization is not a function of temperature.

Water vapor data from the website of an Honors Chemistry class[3] is reported in the following table. A plot of the natural log of vapor pressure vs. T^{-1} should, if the assumptions are valid, produce a straight line on the graph.

a. Create a plot of natural log of vapor pressure of water on the y axis against T^{-1} on the x axis.

b. Add a linear trendline, and have Excel to show the equation of the line and the R^2 value.

c. What is the latent heat of vaporization of water, as computed from the slope of the trendline?

d. Does it look as if the assumptions built into the Clausius–Clapeyron equation are valid for this data set? Why, or why not?

T (°C)	p_{vapor}(mm Hg)	$\ln(p_{vapor})$
90	526.8	6.265
92	567.0	6.340
94	610.9	6.415
96	657.6	6.489
98	707.3	6.561
100	760.0	6.633
102	816.9	6.704
104	876.1	6.774
106	937.9	6.844
108	1004.4	6.912
110	1074.6	6.980

Note: What do you do with the units on vapor pressure inside that natural logarithm? In this problem, it doesn't matter; the units you choose for vapor pressure change the value of k, but not the slope of the trendline.

6.8 Orifice Meter Calibration

Orifice meters are commonly used to measure flow rates, but they are highly nonlinear devices. Because of this, special care must be taken when preparing calibration curves for these meters. The equation relating volumetric flow rate V, to the measured pressure drop ΔP, across the orifice is

$$\dot{V} = \frac{A_0 C_0}{\sqrt{1-\beta^4}} \sqrt{\frac{2 g_c \Delta P}{\rho}}. \tag{6.12}$$

For purposes of creating a calibration curve, the details of the equation are unimportant (as long as the other terms stay constant). It is necessary that we see the theoretical relationship between flow rate and pressure drop

$$\dot{V} \propto \sqrt{\Delta P}. \tag{6.13}$$

---

[3]Dr. Tom Bitterwolf's Honors Chemistry class at the University of Idaho. Data used with the permission of Dr. Bitterwolf.

Also, the pressure drop across the orifice plate depends on the flow rate, not the other way around. So, $\sqrt{\Delta P}$ should be regressed as the dependent variable (y values) and the volumetric flow rate as the independent variable (x values):

$\dot{V}(ft^3 \cdot MIN)$	$\Delta P(psi)$
3.9	0.13
7.9	0.52
11.8	1.18
16.7	2.09
19.6	3.27
23.6	4.71
27.5	6.41
31.4	8.37
36.3	10.59
39.3	13.08

a. Calculate $\sqrt{\Delta P}$ values at each flow rate from the tabulated data.
b. Regress $\dot{V}$ and $\sqrt{\Delta P}$, using Excel's Regression Analysis Package (found by selecting Tools/Data Analysis...) to create a calibration curve for this orifice meter.
c. Check the line fit and residual plots to make sure that your calibration curve really fits the data.

7

Excel's Statistics Functions

Objectives

After reading this chapter, you will know

- How to distinguish samples and populations and how to select a representative sample
- How to use Excel for common statistical calculations,

including means, standard deviations, and variances
- How to create histograms with Excel's Data Analysis package
- How to calculate confidence intervals about sample mean values

7.1 OVERVIEW

Data analysis is a standard part of an engineer's day-to-day work, aimed at understanding a process better and making better products. Whenever you try to glean information from data sets, you soon find yourself needing some *statistics*.

A statistical analysis is performed on a data set, and the rows and columns of values fit naturally into an Excel worksheet. Perhaps it is no surprise, then, that Excel has a variety of built-in functions to help out with statistical calculations. Because of these features, Excel has become a commonly used tool for statistical analysis.

7.2 POPULATIONS AND SAMPLES

If a dairy producer has 238 cows that are milked every day, those 238 cows form a *population*—the herd. If the dairy operator keeps track of the daily milk production for each cow, the data set represents the daily milk production for a population.

If, one day, the dairy is running behind schedule and the operator decides to record the milk production for only 24 cows and use the values from those 24 cows to estimate the total milk production for the entire herd, then the data set from the 24 cows represents data from a *sample* of the total population.

Whenever a sample is used to predict something about a population, the sampling method must be carefully considered. For example, if a hired hand, not understanding what the dairy operator planned to do with the data, recorded data for the 24 cows that usually gave the most milk, the data set is not much good for predicting the total milk production for the herd. The dairy operator needs a *representative sample*. The usual way to try to achieve this is to choose a *random sample*. The key to getting a representative random sample is to try to make sure that every cow has an equal chance of being a part of the sample.

7.2.1 Alternative Example

If cows aren't your cup of tea, consider mice—computer mice. If every mouse coming off the assembly line during the month of November was tested (for ease of rolling, button functionality, cursor tracking, etc.), then the data set represents a population: mice produced in November. (It could also represent a sample of mice produced in the fourth quarter.) If the company decided to test only part of the mice coming off the assembly line, the mice that were tested would represent a sample of the total population. If the sample is to be considered a representative sample, some care must be taken in designing the sampling protocol. For example, if mice are collected for testing only during the day shift, defects introduced by sleepy employees on the night shift will never be observed. Again, the usual way to try to get a representative sample is to choose a random sample. The key to getting a representative random sample is to make sure that every mouse produced has an equal chance of being a part of the sample.

7.2.2 Multiple Instrument Readings

Instruments, such as digital thermometers and scales, are imperfect measuring devices. It is common, for example, to weigh something several times to try to get a "better" value than would be obtained from a single measurement. Any time you record multiple readings from an instrument, you are taking a sample from an infinite number of possible readings. The multiple readings should be treated as a sample, not a population. The impact of this will become apparent when the equations for standard deviations and variances are presented.

PRACTICE!

The SO_2 content in the stack gas from a coal-fired boiler is measured halfway up the stack at six different positions across the diameter, and the concentration values are averaged.

- Do these data represent a sample or a population?
- If it's a sample, is it a representative sample? Is it a random sample? Why, or why not?

[ANSWERS: These data represent a sample, since not every possible bit of gas is tested. If the flow in the sack is turbulent (likely), then the tumbling eddies of flow up the stack would give all portions of the gas the same chance of passing a sensor, so the sampling should be random. It's a representative random sample.]

7.2.3 Arithmetic Mean or Average

A commonly calculated statistical quantity is the *arithmetic mean* or *average*. The formulas for the *population mean*, μ (mu), and *sample mean*, $\bar{x}$, are

$$\mu = \frac{\sum_{i=1}^{N_{\text{pop}}} x_i}{N_{\text{pop.}}},$$

$$\bar{x} = \frac{\sum_{i=1}^{N_{\text{sample}}} x_i}{N_{\text{sample}}}. \tag{7.1}$$

Both equations add up all of the values in their respective data set and then divide by the number of values in their data set. The same Excel function, **AVERAGE,** is used for computing the population mean and the sample mean. Different symbols are used for these means as a reminder that they have different meanings:

- The population mean is the arithmetic average value for the data set.
- The sample mean should be interpreted as the best estimate of the population mean.

Excel provides the **AVERAGE(range)** function for computing arithmetic means or averages. The range in the function represents a range of cells in a worksheet. The average of the values in a small data set consisting of six random integers is computed in the worksheet shown in Figure 7.1.

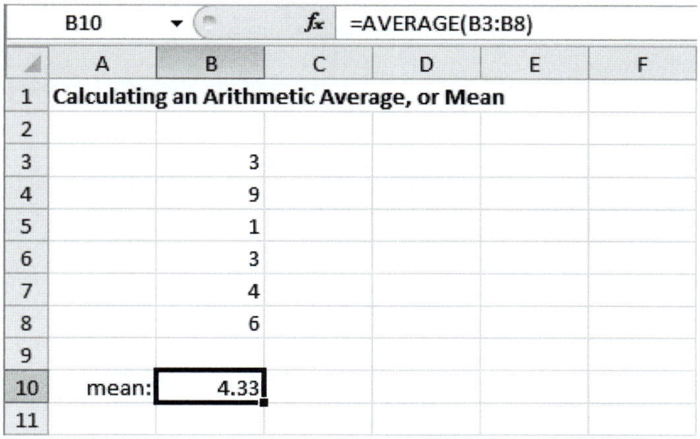

Figure 7.1
Using the **AVERAGE** function.

	B10	▾		f_x	=AVERAGE(B3:B8)	
	A	B	C	D	E	F
1	Calculating an Arithmetic Average, or Mean					
2						
3		3				
4		9				
5		1				
6		3				
7		4				
8		6				
9						
10	mean:	4.33				
11						

The AVERAGE function was used in cell B10 as

$$\texttt{=AVERAGE(B3:B8)}$$

To determine the average, or mean, of the values in cell range B3:B8.

7.3 STANDARD DEVIATIONS AND VARIANCES

The *variance* and *standard deviation* of a data set both provide information about the spread of the values (or *dispersion*) about the mean. A small variance or standard deviation suggests that *all* of the data values are clustered closely around the mean value; a large variance or standard deviation tells you that there is a wide range of values in the data set. The variance and standard deviation are related; the standard deviation is simply the square root of the variance. The standard deviation is more commonly reported in engineering work and has the advantage that it has the same units as the values in the data set. The mean and standard deviation of a data set is commonly reported as *mean* $\pm$ *standard deviation.* For example, if a data set containing pressure values has a mean of 138 kPa, and a standard deviation of 16 kPa, these would be reported as 138 $\pm$ 16 kPa.

While the **AVERAGE** function can be used to calculate mean values for both samples and populations, separate standard deviation and variance functions are needed for samples and populations because the computational formulas are slightly different for populations and samples. The differences in the formulas are accounted for in the worksheet functions listed in Table 7.1.

Table 7.1 Excel functions for averages, standard deviations, and variances

	Population	Sample
Arithmetic Average, or Mean	AVERAGE	AVERAGE
Standard Deviation	STDEVP	STDEV
Variance	VARP	VAR

To illustrate how the standard deviation can be used to indicate how widely distributed data set values are around the mean, consider the data sets shown in Figure 7.2. These data sets shown in the following worksheet have the same mean value, but quite different standard deviations.

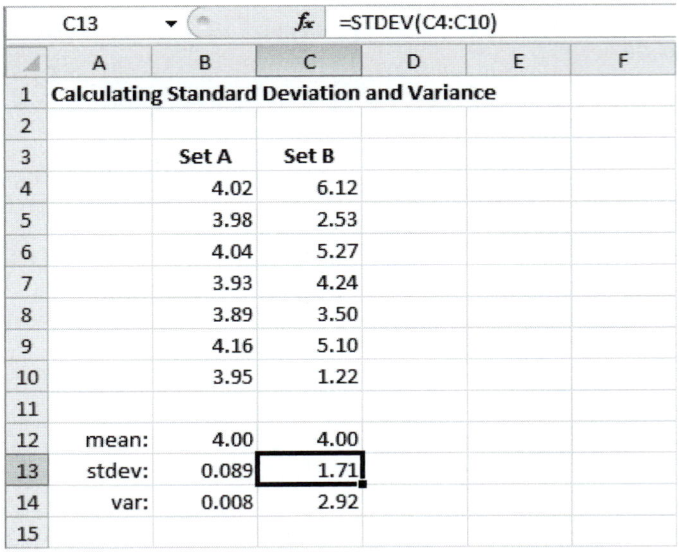

C13 =STDEV(C4:C10)

	A	B	C	D	E	F
1	Calculating Standard Deviation and Variance					
2						
3		Set A	Set B			
4		4.02	6.12			
5		3.98	2.53			
6		4.04	5.27			
7		3.93	4.24			
8		3.89	3.50			
9		4.16	5.10			
10		3.95	1.22			
11						
12	mean:	4.00	4.00			
13	stdev:	0.089	1.71			
14	var:	0.008	2.92			
15						

Figure 7.2
Data sets with the same mean, but different standard deviations.

The greater spread around the mean value (4.00) in Set B is easily seen if the data are plotted, as shown in Figure 7.3.

Figure 7.3
Data sets A and B plotted
to show variability
in values.

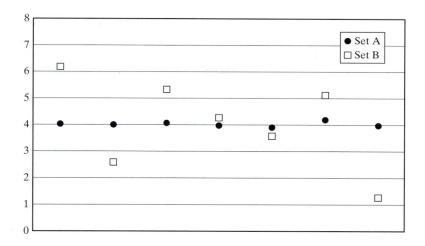

7.3.1 Population Statistics

The Greek symbol σ (sigma) is used for the standard deviation of a population. The population variance is indicated as σ^2:

$$\sigma^2 = \frac{\sum_{i=1}^{N_{pop}} (x_i - \mu)^2}{N_{pop}}, \tag{7.2}$$

$$\sigma = \sqrt{\sigma^2}.$$

The Excel functions for computing the variance and standard deviation of a population are

$$VARP(range)$$
$$STDEVP(range)$$

In the worksheet shown in Figure 7.4, these two functions have been used to calculate the standard deviation and variance in a small data set (assumed here to be a population). The formulas used in column B are shown in column C.

Figure 7.4
Calculating standard
deviation and variance
of a population.

	B12		f_x	=VARP(B3:B8)		
	A	B	C	D	E	F
1	**Population Standard Deviation and Variance**					
2						
3		3				
4		9				
5		1				
6		3				
7		4				
8		6				
9						
10	mean:	4.33	=AVERAGE(B3:B8)			
11	stdev:	2.56	=STDEVP(B3:B8)			
12	var:	6.56	=VARP(B3:B8)			
13						

7.3.2 Samples of a Population

The denominator of the variance equation changes to $N_{sample} - 1$ when your data set represents a sample. The symbol s is used for the standard deviation of a sample and the sample variance is indicated as s^2:

$$s^2 = \frac{\sum_{i=1}^{N_{sample}} (x_i - \bar{x})^2}{N_{sample} - 1}, \tag{7.3}$$

$$s = \sqrt{s^2}.$$

Excel provides functions for computing the variance and standard deviation of a sample:

$$\text{VAR(range)}$$
$$\text{STDEV(range)}$$

Applying these formulas to the same data set used in Figure 7.4, but not treated as a sample, produces the results shown in Figure 7.5.

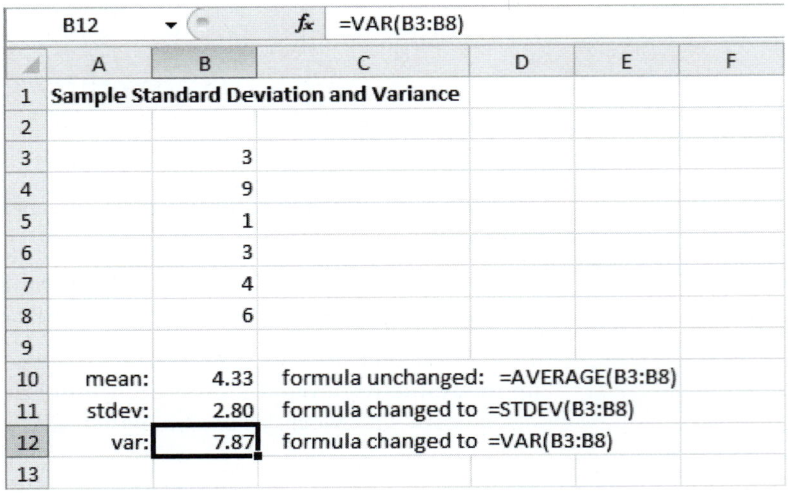

Figure 7.5

Calculating standard deviation and variance of a sample.

B12		f_x	=VAR(B3:B8)		
A	B	C	D	E	F
1 Sample Standard Deviation and Variance					
2					
3	3				
4	9				
5	1				
6	3				
7	4				
8	6				
9					
10 mean:	4.33	formula unchanged: =AVERAGE(B3:B8)			
11 stdev:	2.80	formula changed to =STDEV(B3:B8)			
12 var:	7.87	formula changed to =VAR(B3:B8)			
13					

The data values are the same in Figures 7.4 and 7.5, but the sample standard deviation and variance are a little larger than the population standard deviation and variance, because of the $N_{sample} - 1$ in the denominator of equation 7.3.

PRACTICE!

Two thermocouples are placed in boiling water, and the output voltages are measured five times. Calculate the standard deviations for each thermocouple to see which is giving the most reliable output. Is the voltage data presented here a sample or a population? Why?

TC$_A$ (mV)	TC$_B$ (mV)
3.029	2.999
3.179	3.002
3.170	3.007
3.022	3.004
2.928	3.013

[ANSWERS: TC$_A$ has a standard deviation of 0.107 while TC$_B$'s is 0.005. The smaller standard deviation with TC$_B$ is desirable in instrument readings (assuming they are both working correctly). The measurements are a sample of all possible measurements from the instrument.]

7.4 ERRORS, DEVIATIONS, AND DISTRIBUTIONS

A *deviation* is the difference between a measured value in a sample and the sample mean:

$$\text{dev}_i = x_i - \bar{x}. \tag{7.4}$$

A related quantity, *error*, is computed for populations:

$$\text{error}_i = x_i - \mu. \tag{7.5}$$

A *deviation plot* is sometimes used to look at the distribution of deviation (or error) values around the mean. In a quality-control setting, for example, patterns in a deviation plot can offer hints that something is wrong.

Example: Tools made by two machines are supposed to be identical, and tolerances are measured on each tool as it comes off the line. When the deviations were plotted against time (or sample number if the samples were taken at regular intervals), the graph shown in Figure 7.6 was obtained.

Figure 7.6
Deviations plotted over time.

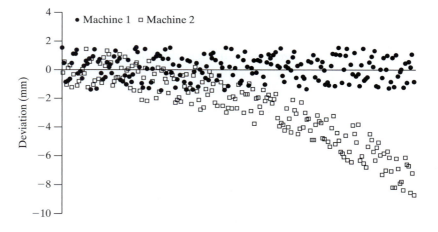

It looks as if something is going wrong with Machine 2. This type of trend could be seen in the tolerance measurements directly, but problems can often be hidden in the scale used to graph the original measurements, as shown in Figure 7.7.

Figure 7.7
Tolerance values plotted
over time.

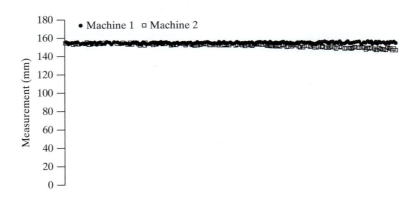

7.4.1 Frequency Distributions

Producing a *frequency distribution*, or *histogram*, can help you understand how the error in your data is distributed. Excel's Data Analysis package will automatically create frequency distributions (and perform several other statistical analyses, such as *t*-tests and ANOVA). However, the Data Analysis package is an Add-In that is typically installed but not activated when Excel is installed on a computer. First, we need to check to see if the Data Analysis package is already available and activate it if necessary.

Checking to see if the Data Analysis package is already activated
Look for a button labeled **Data Analysis** on the Ribbon's **Data** tab, in the Analysis group.

> **Data/Analysis/Data Analysis**

If the button is there, the Data Analysis package is activated and ready to use. You can skip the next few steps.

Checking to see if the Data Analysis package can be activated
Activating an Add-In is done via the Excel Options dialog which is accessed as:

- Excel 2010: File tab/Options
- Excel 2007: Office/Excel Options

Once the Excel Options dialog is open, select the **Add-Ins** panel and look for **Analysis ToolPak** in the Add-In List (see Figure 7.8).

The Add-Ins panel lists Active Application Add-Ins and Inactive Application Add-Ins.

- If the Analysis Toolpak is in the active list (as shown in Figure 7.8), the Regression Analysis package is ready to go.
- If the Analysis Toolpak is not on any list, it wasn't installed; you will have to get out the CDs to install the Analysis Toolpak.
- If it is in the inactive list, you need to activate it. To activate the Analysis ToolPak:
  1. Click on the **Analysis ToolPak** list item (the name in the list, as in Figure 7.8).
  2. Click the **Go...** button
  These steps will cause the Add-Ins dialog to open, as shown in Figure 7.9.
  3. Check the box labeled **Analysis ToolPak**.
  4. Click the **OK** button.
  The Ribbon will then be updated and a new button will appear on the Data tab in the Analysis group as the **Data Analysis** button. This is shown in Figure 7.10.

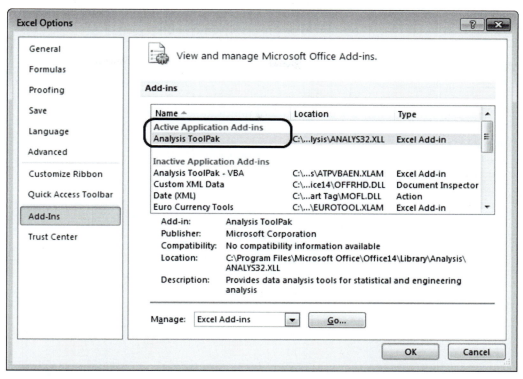

Figure 7.8
Managing Add-Ins.

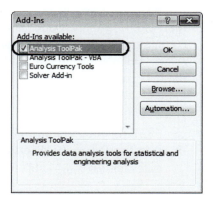

Figure 7.9
The Add-Ins dialog.

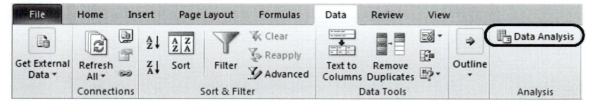

Figure 7.10
Ribbon options **Data/Analysis/Data Analysis** to access the Analysis ToolPak.

Once the Data Analysis package is available for use, we can use it to create a histogram or frequency distribution.

EXAMPLE 7.1

On a good day, when the tools made by the machines 1 and 2 are nearly identical, the deviation plot might look like Figure 7.11.

Figure 7.11
Deviation plot when machines 1 and 2 are working better.

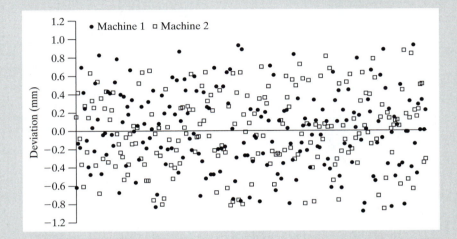

Notice that the magnitudes of the deviations are generally smaller than in the previous deviation plot. The machines are working better.

The basic idea of a frequency distribution is to separate the values into bins and then plot how many values fall into each bin. For deviations, if the bins near zero contain more values than the bins far from zero, then your deviations (sample) or errors (population) might follow a *normal distribution* (the classic *bell curve*). If all of the bins contain approximately the same number of values, then you might have a *uniform distribution*. Frequency distributions can be created by using deviations, as in this example, or directly from the original data values.

To create a frequency distribution in Excel, use the following steps:

1. Decide how many bins you want to use. The number of bins is arbitrary, but the following list is workable for the deviations plotted in Figure 7.11:

Bin	Lower Limit	Upper Limit
1	−1.0	−0.8
2	−0.8	−0.6
3	−0.6	−0.4
4	−0.4	−0.2
5	−0.2	0.0
6	0.0	0.2
7	0.2	0.4
8	0.4	0.6
9	0.6	0.8
10	0.8	1.0

Figure 7.12
Excel worksheet showing machine tolerance values, calculated deviation values, and a set of bin limits for the frequency distribution.

	A	B	C	D	E	F	G
1	Machine 1 Tolerance Distribution						
2							
3	Tolerance	Deviation	Bins				
4	153.38	-0.62	-1.0				
5	153.87	-0.13	-0.8				
6	153.81	-0.19	-0.6				
7	154.69	0.69	-0.4				
8	154.43	0.43	-0.2				
9	154.24	0.24	0.0				
10	153.89	-0.11	0.2				
11	153.60	-0.40	0.4				
12	153.52	-0.48	0.6				
13	154.04	0.04	0.8				
14	153.78	-0.22	1.0				
15	154.53	0.53					
16	154.12	0.12					
201	154.00	0.00					
202	154.34	0.34					
203	154.00	0.00					
204	154.22	0.22					
205							

2. Create a column of bin limits (cells C2:C9 in Figure 7.12).

Note the following:

- If you do not create a column of bin limits and do not indicate any bin limits in Step 6, Excel will automatically create them for you.
- The data set shown in Figure 7.12 contains a lot of data, so most of the rows were hidden to save space. To hide rows
  o Select the rows to be hidden
  o Right-click on the selected rows and choose Hide from the pop-up menu.

When you select a cell range across hidden rows, the hidden rows will be included in the selected cell range.

3. Open the Data Analysis dialog (Figure 7.13) by using Ribbon options **Data/ Analysis/Data Analysis** [Excel 2003: Tools/Data Analysis...].

Figure 7.13
The Data Analysis Dialog.

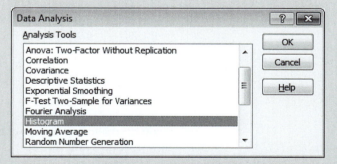

4. Excel offers a wide variety of data analysis options. Choose **Histogram** from the selection list as indicated in Figure 7.13. The Histogram dialog will be displayed, as shown in Figure 7.14.

5. Now indicate the cell range containing the deviation values in the **Input Range:** field. You can either type in the cell range or click on the button at the right side of the field to jump to the worksheet and indicate the range by using the mouse, as illustrated in Figure 7.15.

(continued)

Figure 7.14
The Histogram dialog.

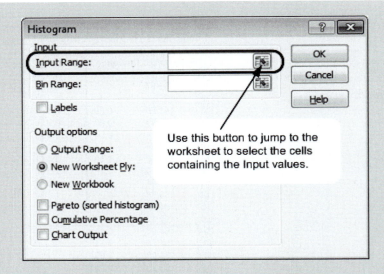

Figure 7.15
Selecting the Input Range.

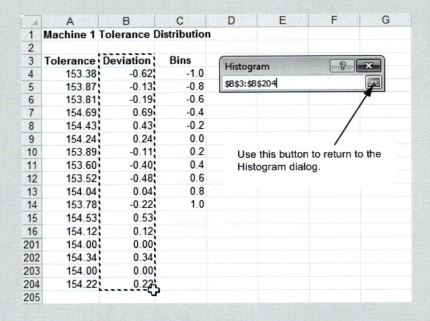

When you have selected the input range (with column heading (optional)), click the button at the right side of the input field on the mini-Histogram dialog (see Figure 7.15) to return to the full Histogram dialog. The selected input range is displayed on the dialog (Figure 7.16).

6. In the same manner, indicate the cell range containing the bin limit values in the **Bin Range:** field as illustrated in Figure 7.17. This step is optional; if you do not indicate bin limits, Excel will use default bin limits.

7. If the cell range selected as the input range contained a column heading (as in Figure 7.15), be sure to check the **Labels** checkbox on the Histogram dialog.

8. Now tell Excel how to handle the output data. Typically, you will want to put the results on a **New Worksheet Ply,** and you will want **Chart Output**.

Figure 7.16
Back at the full Histogram dialog.

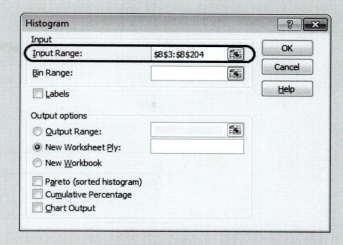

Figure 7.17
Indicating the cell range containing the bin limits (optional).

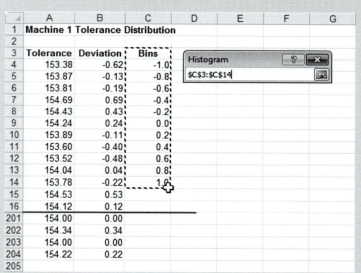

Figure 7.18
Indicating that the input range contained a column heading.

(*continued*)

Figure 7.19
Setting the Output options.

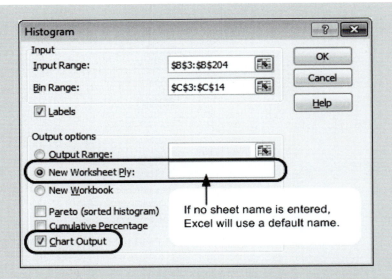

Both of these options have been indicated on the dialog box shown in Figure 7.19.

9. Click the **OK** button to create the histogram plot.

The histogram is shown in Figure 7.20. This data set approximates a "bell curve" shape suggesting the deviation values may be normally distributed.

Figure 7.20

The completed histogram (frequency distribution).

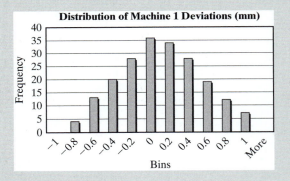

7.5 CONFIDENCE INTERVALS

When your data set is a sample of a larger population, it is easy to compute sample statistics like $\bar{x}$ and s, but you frequently want to use the sample information to say something about the population. A *confidence interval* allows you to use your sample results ($\bar{x}$ and s) to put numerical bounds on the range of possible values for the population mean, μ.

Using a sample to learn something about an entire population is common. You are probably familiar with public opinion polls, and if not, just wait until the next election.

News organizations, political parties, and special interest groups all use public opinion polls to find out how the general population is thinking about particular issues. The polls are carried out by asking a small sample, typically only a few hundred people, to provide data. With such a small sample, there is always a possibility that the sample results do not accurately reflect the opinions of the entire population, so the poll results usually indicate an uncertainty in the result (plus or minus four percentage points is very common).

There is also the question of how the sample was obtained. If you choose random telephone listings, your sample will exclude people whose names aren't in telephone directories. If you use voting records, you exclude nonvoters. Coming up with a representative sample is sometimes difficult.

The public opinion polls point out a couple of things you should keep in mind when developing your own data collection strategies:

- Samples can be used to learn something about the population.
- Any time you use a sample, you introduce some uncertainty into the result.
- If you use a sample, you need to take care how you choose the sample.

A pollster who wants to be 100% confident (zero uncertainty) would have to poll everyone in the population. If you are not going to poll everyone, you have to accept a level of uncertainty in your result—you choose a *confidence level.*

7.5.1 Confidence Levels and the Level of Significance

When you cannot be 100% confident, you must decide how much uncertainty you are willing to accept, or—put another way—how often you are willing to be wrong. It is very common to use a confidence level of 95%, implying that you are willing to be wrong 5% of the time, or one time out of every 20. There are other times when 95% confidence is totally unrealistic. No one would design a banked curve on a highway, for example, such that 19 out of every 20 cars would successfully negotiate the curve, with one out of 20 going over the edge.

A confidence level of 95% will be used for the remainder of this chapter, but remember that the choice of confidence level is up to you and should be chosen appropriately for the problem you are working on.

The *level of significance,* α, is closely related to the confidence level. The level of significance corresponding to a confidence level of 95% is calculated as

$$\alpha = 1 - 0.95 = 0.05. \tag{7.6}$$

7.5.2 Confidence Intervals: Bounding the Extent of Uncertainty

As an example of using a small sample to say something about the population, consider the small data set used earlier in the chapter (Figure 7.1). The data were collected as a sample and has sample mean $\bar{x} = 4.33$; standard deviation, $s = 2.80$; and variance, $s^2 = 7.87$ as shown in Figure 7.21.

So, given the sample results shown in Figure 7.21, what is the population mean, μ?

Who knows? If the sample was well chosen, the population mean is probably close to 4.33. Sometimes "probably close to" is not good enough, and we need to say that the population mean will be between two numeric limits; we need to determine the confidence interval for the population mean. In setting the limits for the confidence interval, there are a few things to keep in mind:

- The most likely value of the population mean, μ, is the sample mean we calculated from the sample data, $\bar{x}$, so the range of possible population means will always be centered on $\bar{x}$.
- The size of the range will depend on how willing you are to be wrong. If you want to avoid ever being wrong, then set the limits to $\pm\infty$. It's not a very

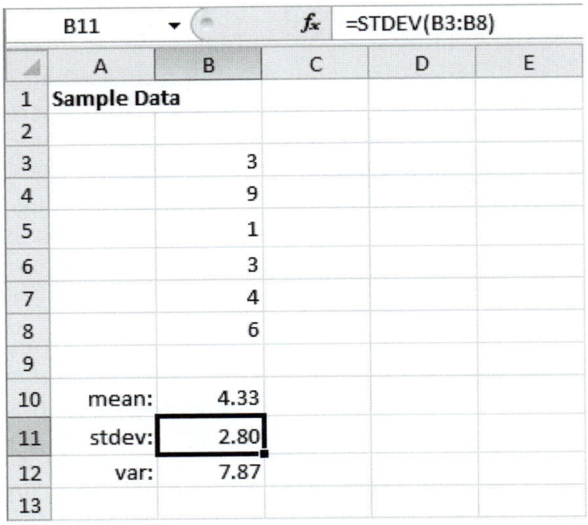

	A	B	C	D	E
1	Sample Data				
2					
3		3			
4		9			
5		1			
6		3			
7		4			
8		6			
9					
10	mean:	4.33			
11	stdev:	2.80			
12	var:	7.87			
13					

Figure 7.21 Calculating the sample mean, standard deviation, and variance.

useful result, but you can be sure the population mean will always fall in that range.

- The more uncertainty you are willing to accept (i.e., the lower your confidence level), the narrower the range of possible population mean values becomes. So lower confidence levels produce narrower confidence intervals.

The confidence interval uses sample information ($\bar{x}$, s) to say something about the population mean, μ. Specifically, the confidence interval indicates the range of likely population mean values.

With only six values in the small data set and a wide range of values in the set (ranging from 1 to 9), you should expect a pretty wide confidence interval (unless you are willing to be wrong a high percentage of the time). We'll calculate the 95% confidence interval ($\alpha = 0.05$) for this data set. The procedure is summarized as follows:

1. Compute the mean and standard deviation of the sample.
2. Determine the number of values in the sample.
3. Decide on a confidence level.
4. Compute the level of significance from the confidence level.
5. Compute the *t*-value (a statistical quantity, described below).
6. Calculate the upper and lower bounds of the confidence interval.

We now apply this procedure to determine the 95% confidence interval for the data shown in Figure 7.21.

Step 1. Compute the Mean and Standard Deviation of the Sample This has already been done; the results are shown in Figure 7.21.

$$\bar{x} = 4.33$$
$$s = 2.80$$

Step 2. Determine the Number of Values in the Sample With the small data set used here, the number of values is obviously six. For larger data sets, it is convenient to use the worksheet function **COUNT** to determine the number of values.

```
=COUNT(range)
```

Step 3. Decide on a Confidence Level We have opted for a confidence level of 95%, or 0.95, in this example.

Step 4. Compute the Level of Significance from the Confidence Level This was done earlier as well:

$$\alpha = 1 - 0.95 = 0.05. \tag{7.7}$$

Step 5. Compute the *t*-value The *t*-value is a statistical quantity calculated using a *t* distribution and requires knowledge of the number of values in the sample, N_{sample}, and the calculated level of significance, α.

The *t* distribution is similar to a normal distribution (a.k.a. bell curve, or *z* distribution), but it is designed to handle small data sets. The "bell" of a *t* distribution gets wider and shorter as the sample size gets smaller. For large samples (30 or more values), there is little difference between the *t* and *z* distributions.

In the histogram of Figure 7.20, we saw how the deviation values for Machine 1 were distributed. The *t* distribution is simply a mathematical description of how values in a small sample distribute around the mean. The *t* distribution is used as a probability map when calculating a confidence interval. We want to know how wide a range is needed to account for 95% of the area under the *t* distribution (see Figure 7.22), with the other 5% split between the two tails of the *t* distribution.

Figure 7.22
Using a distribution as a probability map.

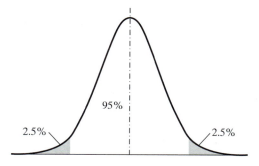

The *t*-values provide part of the information we need to set the limits of the confidence interval; we also need information about our sample ($\bar{x}$, s, N_{sample}).

Excel's **TINV** function can be used to determine the *t*-values we need (*t*-values for a two-tailed *t* distribution). This function takes two parameters:

$$\texttt{=TINV(a, DOF)}$$

α level of significance
DOF   degrees of freedom $= N_{sample} - 1$

For this example, we need a *t*-value corresponding to a level of significance of 0.05, with DOF $= 5$, since $N_{sample} = 6$. Excel's **TINV** function (used as $\texttt{=TINV(0.05,5)}$) returns the value $t = 2.571$, as is seen in Figure 7.23.

The upper and lower bounds on the confidence interval are found by using the formulas

$$B_L = \bar{x} - t\,\frac{s}{\sqrt{N_{sample}}},$$

and

$$B_U = \bar{x} + t\,\frac{s}{\sqrt{N_{sample}}}. \tag{7.8}$$

Figure 7.23
Calculating the *t*-value.

	E8			f_x	=TINV(E4,E6)			
	A	B	C	D	E	F	G	H
1	Sample Data							
2							Formulas in Column E	
3		3		Conf Int:	95%			
4		9		α:	0.05		=1-E3	
5		1		N$_{sample}$:	6		=COUNT(B3:B8)	
6		3		DOF:	5		=E5-1	
7		4						
8		6		t-value:	2.571		=-TINV(E4,E6)	
9								
10	mean:	4.33						
11	stdev:	2.80						
12	var:	7.87						
13								

The results are shown in Figure 7.24.

The confidence interval has been determined and we can now state that, with 95% confidence, the population mean, μ, for the data shown in Figure 7.24 lies between 1.39 and 7.28. That is a very large range, but this was a very small data set with a relatively large standard deviation.

Figure 7.24
The upper and lower bounds (limits) on the confidence interval.

	E10			f_x	=B10-E8*B11/SQRT(E5)			
	A	B	C	D	E	F	G	H
1	Sample Data							
2							Formulas in Column E	
3		3		Conf Int:	95%			
4		9		α:	0.05		=1-E3	
5		1		N$_{sample}$:	6		=COUNT(B3:B8)	
6		3		DOF:	5		=E5-1	
7		4						
8		6		t-value:	2.571		=-TINV(E4,E6)	
9								
10	mean:	4.33		B$_L$:	1.39		=B10-E8*B11/SQRT(E5)	
11	stdev:	2.80		B$_U$:	7.28		=B10+E8*B11/SQRT(E5)	
12	var:	7.87						
13								

7.5.3 Standard Error of the Sample

The quantity $\dfrac{s}{\sqrt{N_{sample}}}$ has a name: It is called the *standard error of the sample*, or *SES*. For the data in our example,

$$SES = \frac{s}{\sqrt{N_{sample}}} = \frac{2.80}{\sqrt{6}} = 1.145.$$

Once SES is known, multiply it by the *t*-value to obtain the quantity that is added to and subtracted from the sample mean to determine the upper and lower bounds of the confidence interval.

In this example, the *t*-value is 2.571, so $t \cdot SES = 2.571 \cdot 1.145 = 2.943$.

The confidence interval can be written as

$$(\bar{x} - t \cdot \text{SES}) \leq \mu \leq (\bar{x} + t \cdot \text{SES}).$$

which corresponds to

$$B_L \leq \mu \leq B_U. \tag{7.9}$$

For the data in this example, the 95% confidence interval can be written as

$$1.39 \leq \mu \leq 7.28. \tag{7.10}$$

This is the same result found in Figure 7.24.

7.5.4 Confidence Intervals Using Excel's Descriptive Statistics Package

One of the tools in Excel's Data Analysis package is designed to calculate descriptive statistics for a data set. To use the Descriptive Statistics package

1. Use Ribbon options **Data/Analysis/Data Analysis** [Excel 2003: Tools/Data Analysis...] to open the Data Analysis dialog.
2. Select Descriptive Statistics from the list of analysis tools shown in the Data Analysis dialog (Figure 7.25).

Figure 7.25
Using the Descriptive Statistics package.

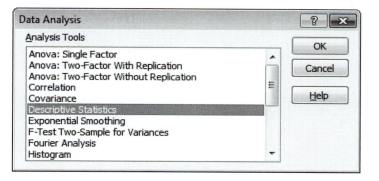

The Descriptive Statistics dialog will open (Figure 7.26).

3. Indicate the range containing the data set (B3:B8) in the **Input Range:** field.
4. Let Excel know where you want the results placed by specifying the top-left cell of the **Output Range:** or indicating that the results should be placed on a **New Worksheet Ply**.
5. Choose the type of results that should be calculated including **Summary statistics**.

As shown in Figure 7.26, **Summary statistics** (mean, standard deviation, variance, SES, etc.) and the desired **Confidence Level for [the] Mean** has been set to 95%.

6. Click the **OK** button to close the Descriptive Statistics dialog. The results are calculated and presented on the worksheet as shown in Figure 7.27.

The descriptive statistics are the same as those calculated earlier in the chapter by using Excel's functions, so the Descriptive Statistics package is simply an alternative method for calculating these quantities. But notice the value 2.943 in cell C25. The label is "Confidence Level (95.0%)," but that value is actually $t \cdot \text{SES}$. This is a very handy way of calculating the values needed to determine a confidence interval, because the confidence interval is simply

$$(\bar{x} - t \cdot \text{SES}) \leq \mu \leq (\bar{x} + t \cdot \text{SES}).$$

So the confidence interval is $(4.333 - 2.943) \leq \mu \leq (4.333 + 2.943)$, or $1.39 \leq \mu \leq 7.28$. (same as before).

Figure 7.26
The Descriptive Statistics dialog.

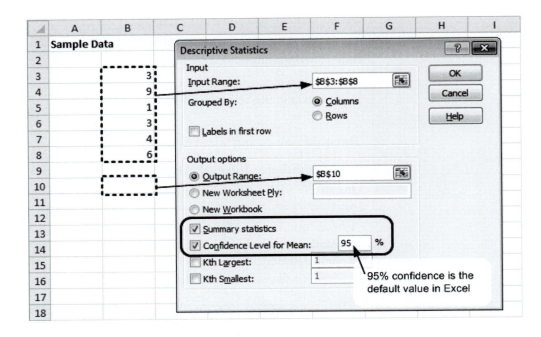

Figure 7.27
Descriptive Statistics Output.

	A	B	C	D
1	**Sample Data**			
2				
3			3	
4			9	
5			1	
6			3	
7			4	
8			6	
9				
10		*Column1*		
11				
12		Mean	4.333333	
13		Standard Error	1.145038	
14		Median	3.5	
15		Mode	3	
16		Standard Deviation	2.804758	
17		Sample Variance	7.866667	
18		Kurtosis	0.670066	
19		Skewness	0.876235	
20		Range	8	
21		Minimum	1	
22		Maximum	9	
23		Sum	26	
24		Count	6	
25		Confidence Level(95.0%)	2.943413	
26				

PRACTICE!

Two thermocouples are placed in boiling water and the output voltages are measured five times. Calculate the sample mean and standard deviation for each thermocouple. Then compute the confidence interval for the population mean for each thermocouple, assuming a 95% confidence level. The following table shows the measured values:

TC_A (mV)	TC_B (mV)
3.029	2.999
3.179	3.002
3.170	3.007
3.022	3.004
2.928	3.013

ANSWERS:

	$\bar{x}$	s	Confidence Interval
TC_A	3.07	0.107	$2.933 \leq \mu \leq 3.199$
TC_B	3.01	0.005	$2.998 \leq \mu \leq 3.012$

A More Realistic Example

To generate a more realistic data set, I searched the lab for the oldest, most beat-up weight set I could find. Then I chose the worst-looking weight in the set (the 2-g weight) and weighed it 51 times on a very sensitive balance, but I left the scale doors open with air moving through the room to generate some noise. (The scale is attached to a printer, and I printed the mass 51 times.) The results of the repeated sampling of the mass were a sample mean, $\bar{x}$, of 1.9997 grams and a standard deviation, s, of 0.0044 grams or

$$\bar{x} = 1.9997 \pm 0.0044 \text{ grams}. \tag{7.11}$$

A 95% confidence interval can be used to put bounds on the expected value of the true mass. The confidence interval for this example is computed in the worksheet shown in Figure 7.28. The formulas used in column C are shown in column F.

Figure 7.28
Confidence interval for a real data set.

	A	B	C	D	E	F	G
1	Two-Gram Weight Data: Confidence Interval						
2							
3						Formulas in Column C	
4		Mean:	1.9998	grams			
5		St. Dev.:	0.0044	grams			
6		N_{sample}:	51				
7							
8		a:	0.05				
9		DOF:	50			=C6-1	
10		t:	2.0086			=TINV(C8,C9)	
11							
12		B_L:	1.9986	grams		=C4-C10*(C5/SQRT(C6))	
13		B_U:	2.0010	grams		=C4+C10*(C5/SQRT(C6))	
14							

The 95% confidence interval for this example can be written as

$$1.9986 \, g \leq \mu \leq 2.0010 \, g. \tag{7.12}$$

For an old, beat-up weight, it's still pretty close to 2 grams (with 95% certainty).

KEY TERMS

Arithmetic mean	Frequency distribution	Standard deviation
Average	Histogram	Standard error of the
Bell curve	Level of significance	sample (SES)
Confidence interval	Normal distribution	Statistics
Confidence level	Population	t distribution
Data analysis	Population mean	t-value
Degrees of freedom	Random sample	Uniform distribution
Deviation	Representative sample	Variance
Deviation plot	Sample	z distribution
Dispersion	Sample mean	
Error	Sampling method	

SUMMARY

Terminology	
Population	When a data set includes information from each member of a group, or all possible measurements, it is said to represent the population.
Sample	When only a subset of the entire population is measured or only some of all possible measurements are taken, the data set is called a sample. Whenever a sample is used, care should be taken to ensure that the sample is representative of the population. This generally means that a random sample should be taken.
Random Sampling	Each member of a population should have an equal chance of being part of the sample.
Error	The difference between a value and the true value—typically, the population mean.
Deviation	The difference between a value and another value—typically, the sample mean.
Arithmetic Mean	The average value in the data set. Describes the central value in the data: **AVERAGE(range)**
Standard Deviation	A value that gives a sense of the range of scatter of the values in the data set around the mean: **STDEV(range)** sample **STDEVP(range)** population
Variance	The square of the standard deviation: **VAR(range)** sample **VARP(range)** population

Confidence Level	Statistical tests are based on probabilities. For each test you choose a confidence level—the percentage of the time you want to be right. A confidence level of 95%, or 0.95, is very common.
Level of Significance	One minus the (fractional) confidence level.
Confidence Interval	If you know the mean and standard deviation of your sample and choose a confidence level, you can calculate the range of values (centered on the sample mean) in which the population mean will probably lie. The range of possible values is called the confidence interval for the population mean.

Frequency Distributions

To create a frequency distribution in Excel, use the following steps:

1. Decide how many bins you want to use.
2. Create a column of bin limits (or skip this step to let Excel assign bins).
3. Open the Data Analysis dialog by using Ribbon options **Data/Analysis/Data Analysis** [Excel 2003: Tools/Data Analysis…].
4. Choose "Histogram" from the selection list.
5. Indicate the cell range containing the values to be sorted into bins for the histogram in the Input Range: field.
6. Indicate the cell range containing the bin limit values (skip this step to let Excel assign bins).
7. Check the Labels checkbox on the Histogram dialog if labels were included in the Input Range.
8. Tell Excel how to handle the output data. (Typically, on a "New Worksheet Ply.")
9. Click the [OK] button to create the histogram plot.

Confidence Intervals

1. Compute the mean and standard deviation of the sample.
2. Determine the number of values in the sample.
3. Decide on a confidence level.
4. Compute the level of significance from the confidence level.
5. Compute the t-value.
6. Calculate the upper and lower bounds of the confidence interval.

$$B_L = \bar{x} - t\,\frac{s}{\sqrt{N_{\text{sample}}}}$$

$$B_U = \bar{x} + t\,\frac{s}{\sqrt{N_{\text{sample}}}}$$

Alternatively, Use Excel's Descriptive Statistics Package

1. Use Ribbon options **Data/Analysis/Data Analysis** [Excel 2003: Tools/Data Analysis…] to open the Data Analysis dialog.
2. Select **Descriptive Statistics** from the list of analysis tools shown in the Data Analysis dialog.
3. Indicate the range containing the data set.
4. Let Excel know where you want the results placed.
5. Choose the type of results that should be calculated.
6. Click the **OK** button to close the Descriptive Statistics dialog and calculate the results.

The values labeled as "Confidence Level (95.0%)" are added to and subtracted from the sample mean to determine the upper and lower bounds of the confidence interval, respectively.

PROBLEMS

7.1 Exam Scores

Calculate the average and standard deviation for the set of exam scores shown in the accompanying table. Using a 90% = A, 80% = B, and so on, grading scale, what is the average grade on the exam?

Scores
92
81
72
67
93
89
82
98
75
84
66
90
55
90
91

7.2 Thermocouple Reliability I

Two thermocouples are supposed to be identical, but they don't seem to be giving the same results when they are both placed in a beaker of water at room temperature. Calculate the mean and standard deviation for each thermocouple. Which one would you use in an experiment?

TC_1 (mV)	TC_2 (mV)
0.3002	0.2345
0.2998	0.1991
0.3001	0.2905
0.2992	0.3006
0.2989	0.1559
0.2996	0.1605
0.2993	0.5106
0.3006	0.3637
0.2980	0.4526
0.3014	0.4458
0.2992	0.3666
0.3001	0.3663
0.2992	0.2648
0.2976	0.2202
0.2998	0.2889

7.3 Thermocouple Reliability II

The two thermocouples mentioned in the previous problem are supposed to be identical, but they don't seem to be giving the same results when they are both placed in a beaker of water at room temperature. What is the confidence interval for each set of readings ($\alpha = 0.05$)? Which thermocouple would you use in an experiment?

7.4 Sample Size and Confidence Interval

Some PRACTICE! Problems in this chapter have used five readings from a thermocouple. Those are actually the first five readings from a larger data set, which is reproduced here and available on the text's website, www.chbe.montana.edu/Excel:

TC_A (mV)					
3.029	2.838	3.039	2.842	3.083	3.013
3.179	2.889	2.934	2.962	3.076	2.921
3.170	3.086	2.919	2.893	3.009	2.813
3.022	2.792	2.972	3.163	2.997	2.955
2.928	2.937	2.946	3.127	3.004	2.932

   a. Calculate the confidence interval for the population mean (95% confidence), using the five data points in the left column, the 10 points in the left two columns, and all 30 data points.

   b. How significant is the number of data points for the size of the confidence interval?

7.5 Thermocouple Calibration Curve

A thermocouple was calibrated by using the system illustrated. The thermocouple and a thermometer were dipped in a beaker of water on a hot plate. The power was set at a preset level (known only as 1, 2, 3 . . . , on the dial) and the thermocouple readings were monitored on the computer screen. When steady state had been reached, the thermometer was read, and 10 thermocouple readings were recorded. Then the power level was increased and the process repeated.

Figure 7.5P
Calibrating a
thermocouple.

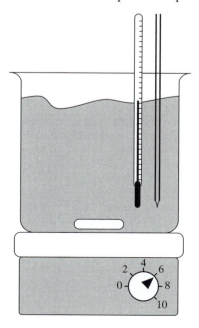

The accumulated calibration data (steady-state data only) are tabulated as follows and are available electronically at www.chbe.montana.edu/Excel.

Power Setting	Thermometer (°C)	Thermocouple Readings (mV)									
0	24.6	1.253	1.175	1.460	1.306	1.117	1.243	1.272	1.371	1.271	1.173
1	38.2	1.969	1.904	2.041	1.804	2.010	1.841	1.657	1.711	1.805	1.670
2	50.1	2.601	2.506	2.684	2.872	3.013	2.832	2.644	2.399	2.276	2.355
3	60.2	3.141	2.920	2.931	2.795	2.858	2.640	2.870	2.753	2.911	3.180
4	69.7	3.651	3.767	3.596	3.386	3.624	3.511	3.243	3.027	3.181	3.084
5	79.1	4.157	4.322	4.424	4.361	4.115	4.065	4.169	4.376	4.538	4.809
6	86.3	4.546	4.384	4.376	4.548	4.654	4.808	4.786	4.535	4.270	4.156
7	96.3	5.087	5.197	5.390	5.624	5.634	5.335	5.525	5.264	4.993	5.267
8	99.8	5.277	5.177	4.991	5.190	5.228	5.274	4.990	5.010	4.916	4.784

Note: This is very noisy thermocouple data, so that the error bars will be visible when graphed. Actual thermocouples are much more precise than the values reported here.

 a. Calculate the average and standard deviation of the thermocouple readings at each power setting.
 b. Plot the thermocouple calibration curve with temperature on the x axis and average thermocouple reading on the y axis.
 c. Add a linear trendline to the graph and have Excel display the equation for the trendline and the R^2 value.
 d. Use the standard deviation values to add error bars (± 1 std. dev.) to the graph.

7.6 Bottling Plant Recalibration

A bottling plant tries to put just a bit over 2 L in each bottle. If the filling equipment is delivering <2.00 L per bottle, it must be shut down and recalibrated.
The last 10 sample volumes are listed here:

Volume (L)
1.92
2.07
2.03
2.04
2.11
1.94
2.01
2.03
2.13
1.99

a. Design a random sampling protocol for a bottling plant filling 20,000 bottles per day. How many samples should you take? How often? How would you choose which bottles to sample?

b. Calculate the mean and standard deviation of the last 10 sample volumes.

c. Should the plant be shut down for recalibration?

7.7 Extracting Vanilla

When you want to extract vanilla from vanilla beans, you usually grind the bean to speed the extraction process, but you want very uniformly sized particles so that the vanilla is extracted from all of the particles at the same rate. If you have a lot of different-sized particles, the small particles extract faster (removing the vanilla and some poorer flavored components), while the larger particles could leave the process incompletely extracted.

A sample was collected from the vanilla-bean grinder and split into 50 parts. The average particle size of each part was determined. The following data were collected, and they are available on the text's website:

Average Particle Size (μm)				
91	147	150	165	130
157	117	114	139	131
105	64	137	115	97
116	94	99	163	144
135	112	116	138	129
138	115	106	139	76
147	116	115	101	107
129	122	120	131	104
88	98	157	110	138
152	113	128	130	125

a. Calculate the standard deviation of the particle sizes in the sample. Is the standard deviation less than 15 μm? (If not, the grinder should be repaired.)

b. How would you determine the average particle size of the ground vanilla beans?

8

Excel's Financial Functions

Objectives

After reading this chapter, you will know

- The principles behind the time value of money and making engineering economic decisions
- The terminology used with time-value-of-money calculations
- How to move money through time at specified interest rates

- How to use built-in Excel functions for present values (PVs), future values (FVs), and regular payments
- How to use Excel to calculate an internal rate of return
- How to use time-value-of-money concepts to make economic decisions
- How to use Excel's depreciation function

8.1 TIME, MONEY, AND CASH FLOWS

Making decisions that are based upon economic considerations is a common task for an engineer. The decision may be as simple as deciding which vendor to use for the parts needed for a construction project. The cost of the parts will be a large part of the decision-making process, but other nonmonetary considerations also come into play, such as the availability of the parts and the reliability of the vendor at delivering the parts on time. These nonmonetary questions can be significant parts of the decision-making process; they are termed *intangibles*. Here we will consider only the monetary aspects of making decisions.

Other common decisions include

1. Choosing between competing technologies.
2. Making the best use of limited resources.

Examples are as follows:

1. An electronic switching device could use a mechanical relay or a solid-state relay. The mechanical relay would be less expensive, but slower to open and close, and it would wear out and have to be replaced periodically. The solid-state relay would be more reliable and operate more quickly, but would be more expensive than a mechanical relay. Which type should be specified?

2. Two engineers working for the same company each have an idea that will save money. Ben's idea will cost $10,000 to implement, but will save the company $200 a day after the project is implemented. Anna's idea will cost $2.3 million to build and will actually increase operating costs by $360,000 per month during a 6-month startup period. However, after the startup period, the savings will be $520,000 per month forever. Where should the company put its money?

In the first example, the speed and reliability of the solid-state relays might make them the better choice if the additional cost is not excessive. If they are a lot more expensive, the costs and hassle associated with replacing the mechanical relays would need to be considered.

In the second example, the company might decide to implement both ideas. Or, the company might not have $2.3 million available or be unable to handle the increased operating costs for the 6-month startup period, so Anna's project might not get funded, even if it is a great idea.

It is important to know how to make these economic decisions.

8.1.1 Cash Flows

A big return on an investment in 2 years will not help you pay your rent next month. For your personal finances or your company's finances, the cash must be available when it is needed. To keep track of the availability of cash over time, we look at the *cash flow*. In the second example, Anna's idea required a large initial outlay of cash ($2.3 million) and regular cash payments for 6 months ($360,000 per month) to take care of the increased operating costs. The company would need a cash flow sufficient to handle all of these costs; otherwise, the company would go bankrupt before it began reaping the benefit of the $520,000 per month cost savings.

Cash flows can be diagrammed in a couple of ways. A common method of illustrating expense and income payments over time is to use arrows on a *timeline*. Arrows pointing toward the timeline indicate out-of-pocket *expenses* (from you, or from your company). Arrows pointing away indicate *incomes*. The timeline for the first 24 months of Anna's idea is shown in Figure 8.1.

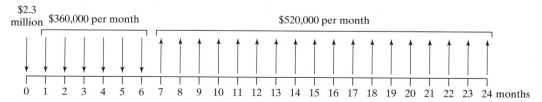

Figure 8.1
Cash flow timeline.

An alternative is to scale the arrows to the size of the expense or income to keep track of the total funds available to the company over time. For example, if the company had $3.5 million in their capital fund when they began implementing Anna's idea, the capital fund balance as a cash flow diagram (neglecting the time value of money) might look like Figure 8.2.

Figure 8.2
Capital fund balance over time.

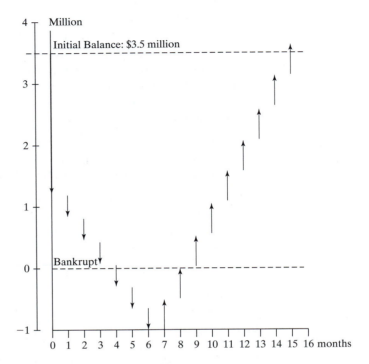

Trying to implement Anna's idea with $3.5 million in the capital fund would bankrupt the fund in 4 months. The diagram shows that (ignoring interest rates for a moment) nearly another million dollars is required to cover the operating costs in the first 6 months. If the company had $4.5 million available for the project (at time zero), then Anna's idea would pay back the capital fund in 15 months, and afterward the $520,000 per month savings would be pure profit.

8.1.2 Types of Expenses

Expenses are categorized by when they occur and by whether they recur. Single payments, such as the $2.3 million expense at time zero in the last example, occur frequently in time-value-of-money problems. When these expenses occur at time zero, they are called *PVs (Present Values)* and are given the symbol PV. Single payments at known later times are called *future values*, or FV.

Regularly recurring, constant-amount expenses, such as the $360,000 operating cost in the last example, are also common, and time-value-of-money formulas have been developed to handle this type of payment. A classic type of regularly occurring expense is an annual payment into an investment. This annual payment is called an *annuity payment*, and the symbol A (for annuity) is sometimes used for any regularly recurring payment, even if the interval between payments is not 1 year. In this text, *regular payments* will be labeled with the symbol Pmt. Regularly occurring expenses with differing amounts are handled as a series of single payments.

Expense payments may be made at the beginning of the time period (common for rents and tuition) or at the end of the period (common for provided services, such as utility and telephone bills). Excel's time-value-of-money functions allow you to specify whether the payments are made at the beginning or end of the time periods.

8.1.3 Types of Income

Types of income are also categorized by when they occur and by whether they recur. Single payments at time zero are called *PVs* or, in the context of a loan, the *principal*. Income resulting from the sale of used equipment at the end of a time period (the *service life* of the equipment) is called a *salvage value* and is a type of FV. Incomes could also be received at regular intervals and are included in Excel's time-value-of-money functions as regular payments (to you) (Pmt). If the amount of the periodic income is not constant, the incomes are handled as a series of single payments.

8.2 INTEREST RATES AND COMPOUNDING

Money left in a bank grows with time because of interest payments. However, if the nation's *inflation rate* is greater than the bank's *interest rate*, the buying power, or value, of your banked money decreases with time. The inflation rate is effectively a negative interest rate.

The value of money over time depends on the interest rate. There is no single interest rate used. Banks give their customers one rate for saving accounts, but often charge a different rate for car loans and yet another rate for home loans. Your company might use an effective interest rate when deciding which projects to fund. Before you can work with money over time, you need to know the applicable rate(s).

The *compounding period* is also significant. Bank loans typically require monthly payments and calculate the interest you owe on a monthly basis (monthly compounding). Many banks compound daily on savings accounts, although many credit unions use quarterly compounding. Excel's time-value-of-money functions use *periodic interest rates* to handle any of these options.

There are several ways of expressing interest rates:

- **Periodic interest rates** are the interest rates used over a single compounding period. If a bank pays interest daily, the interest rate they use to compute the amount of the interest payment is the daily (periodic) interest rate.
- **Nominal interest rates** are the most common, and least useful, way of expressing interest rates. A nominal interest rate is an annual rate that ignores the effect of compounding. If you have a periodic interest rate of 1% per month, the nominal annual rate would be 12% compounded monthly. The "compounded monthly" is an important part of the nominal-interest-rate specification; you must know the compounding interval in order to use a nominal annual interest rate, and you use a nominal rate by converting it into a periodic interest rate.
- **Effective annual interest rates** are calculated annual interest rates that yield the same results as periodic interest rates with nonannual compounding.
- **Annual Percentage Rates (APRs)** are the way lending institutions in the United States are required to present interest rate information on funds you borrow. They are not the same as an effective annual interest rate for two reasons. First, any up-front fees that are part of the total cost of borrowing are included when

calculating the APR. Second, the APR tells you what the total cost of borrowing would be with annual compounding, but that does not mean that annual compounding will be used. Most loans that quote an APR rate use monthly compounding. The actual interest cost is therefore greater than what you calculate using the APR and a compounding period of 1 year. If there are no up-front fees, the APR can be used as a nominal interest rate if the compounding period is known.

- **Annual Percentage Yields (APYs)** are the way savings institutions in the United States are required to present interest rate information on interest-bearing accounts, such as savings accounts or certificates of deposit (CDs). The APY is a nominal interest rate calculated by assuming daily compounding.

8.2.1 Periodic Interest Rates (i_p)

Nonannual compounding periods are very common. The interest rate per compounding period (e.g., percent per day) is called a *periodic interest rate*. Excel's time-value-of-money functions all use periodic interest rates.

> **Note:** You must be careful that the periodic interest rate and the number of payments are on the same time basis. If your interest rate is expressed as percent per month, then the number of compounding periods used in the function must also be in months.

8.2.2 Nominal Interest Rates (i_{nom})

Nominal interest rates are a type of annual interest rate, but usually with nonannual compounding. You must know the compounding period in order to compute a periodic interest rate from the nominal rate. Banks are required to report APR and APY, but most still report nominal interest rates, as well as APR and APY. Table 8.1 is typical of how a bank might present their CD rates:

Table 8.1 Certificate of deposit

Compounded Daily: 365 Basis	Rate	APY
6 Month	5.85%	6.02%
1 Year	5.20%	5.34%
1 1/2 Year	5.30%	5.44%
3 Year	5.60%	5.76%
5 Year	5.70%	5.87%

The rate column shows a nominal interest rate. The compounding period is stated in the table, "Compounded Daily: 365 Basis." This means that the bank is assuming a 365-day year when calculating the APY from the nominal rate. (Banks are required to use a 365-day basis when calculating APY.) We can use this example to learn how to convert the nominal interest rate to a periodic rate and an effective annual rate.

Converting a Nominal Interest Rate (i_{nom}) to a Periodic Interest Rate (i_p)

To convert a nominal interest rate to a periodic interest rate, simply divide the nominal rate by the number of compounding periods per year. For the 6-month CD shown in the preceding table, the nominal rate is

$$i_{nom} = 5.85\%, \text{ or } 0.0585. \tag{8.1}$$

The table states that compounding is daily and uses a 365-day-year basis. The periodic rate (daily interest rate) is found by dividing the nominal rate by 365 days per year:

$$i_p = i_{nom}/365$$
$$= 5.85\%/365 = 0.0160\% \text{ per day}$$
$$= 0.0585/365 = 0.000160 \text{ fractional interest rate per day.} \tag{8.2}$$

8.2.3 Effective Annual Interest Rates (i_{nom})

It is sometimes convenient to compute the *effective annual interest rate* for nonannual periodic interest payments. The effective annual interest rate is simply the rate that produces the same end-of-year value with a single interest payment (at the end of the year) as would be produced by more frequent periodic interest payments.

For example, $100.00 invested at a monthly interest rate of 1% would be worth $112.68 12 months later (see Figure 8.3).

Figure 8.3
Value of $100 invested at a monthly interest rate of 1% for 12 months.

	A	B	C	D	E
1	Value of $100 Invested at 1% monthly				
2					
3	Month	At Start	Interest	End Value	
4	1	$ 100.00	$ 1.00	$ 101.00	
5	2	$ 101.00	$ 1.01	$ 102.01	
6	3	$ 102.01	$ 1.02	$ 103.03	
7	4	$ 103.03	$ 1.03	$ 104.06	
8	5	$ 104.06	$ 1.04	$ 105.10	
9	6	$ 105.10	$ 1.05	$ 106.15	
10	7	$ 106.15	$ 1.06	$ 107.21	
11	8	$ 107.21	$ 1.07	$ 108.29	
12	9	$ 108.29	$ 1.08	$ 109.37	
13	10	$ 109.37	$ 1.09	$ 110.46	
14	11	$ 110.46	$ 1.10	$ 111.57	
15	12	$ 111.57	$ 1.12	$ 112.68	
16					

The interest rate required to generate the same result with a single interest payment is 12.68%. So, the effective annual interest rate equivalent to a 1% monthly rate is 12.68%.

Effective Annual Interest Rate (i_e) from Periodic Interest Rate (i_p)

Because the effective annual interest rate is designed to produce the same future (end-of-year) value as the periodic interest rate, we begin by finding out how much an investment made at time zero would be worth 1 year later using the periodic interest rate (i_p). To do this, we use the following equation for calculating an FV when given a PV and a periodic interest rate (i_p):

$$\text{FV} = \text{PV} \cdot \left(1 + i_p\right)^{N_{PPY}}. \tag{8.3}$$

Here,

i_p is the periodic interest rate,
N_{PPY} is the number of compounding periods per year,
FV is the future value, and
PV is the present value.

We need to get the same FV by using the effective annual interest rate (i_e), and a compounding period of 1 year. With the effective annual interest rate, the FV equation becomes

$$FV = PV \cdot \left(i + i_e\right)^1. \tag{8.4}$$

Because the FV must be the same regardless of whether the periodic interest rate or the effective annual interest rate is used, we can set the FV equations equal to each other to solve for i_e:

$$i_e = \left(1 + i_p\right)^{N_{PPY}} - 1. \tag{8.5}$$

At the beginning of this section, it was stated that the effective annual interest rate equivalent to a periodic rate of 1% per month was 12.68%. We can use this fact to test the conversion equation we just derived:

$$i_e = \left(1 + 0.01\right)^{12} - 1 = 0.1268 \tag{8.6}$$

or

$$12.68\%.$$

Effective Annual Interest Rates from Nominal Rates

The periodic interest rate is simply the nominal rate divided by the number of periods per year, so the equation for (i_e) in terms of the nominal interest rate can be written as

$$i_e = \left(1 + \frac{i_{nom}}{N_P}\right)^{N_{PPY}} - 1. \tag{8.7}$$

EXAMPLE 8.1

If the nominal rate for the 6-month CD described in the Table 8.1 is 5.85% and the number of periods per year is 365, then

$$i_e = \left(1 + \frac{5.85\%}{365}\right)^{365} - 1 = 6.02\%. \tag{8.8}$$

The effective annual interest rate (and APY) is 6.02%, as reported in the bank's table of CD yields (Table 8.1).

Excel provides a function for calculating effective annual interest rates from nominal rates. The **EFFECT** function has the syntax

```
=EFFECT(Nominal_Rate, NPPY)
```

where

Nominal_Rate is the nominal (annual) interest rate (required) and
NPPY is the number of periods per year (required)

The nominal rate may be entered as a percentage (e.g., 5.85%) or as a fractional rate (0.0585). An example of its use is shown in Figure 8.4.

Figure 8.4
Using the **EFFECT** function to calculate an effective annual interest rate.

C6			f_x	=EFFECT(C3,C4)	
	A	B	C	D	E
1	Calculating an Effective Annual interest Rate				
2					
3	Nominal Rate:		5.85%		
4	Payments per Year:		365		
5					
6	Effective Annual Rate:		6.02%		
7					

The corresponding function to convert an effective annual interest rate to a nominal annual interest rate is the **NOMINAL** function:
The syntax of the nominal function is

$$=NOMINAL(Eff\_Ann\_Rate, NPPY)$$

where

Eff_Ann_Rate   is the effective annual interest rate (required) and
NPPY   is the number of periods per year (required)

The effective annual rate may be entered as a percentage (e.g., 6.02%) or as a fractional rate (0.0602). The use of the **NOMINAL** function is illustrated in Figure 8.5.

Figure 8.5
Using the **NOMINAL** function to calculate a nominal interest rate.

C6			f_x	=NOMINAL(C3,C4)	
	A	B	C	D	E
1	Calculating a Nominal interest Rate				
2					
3	Effective Annual Rate:		6.02%		
4	Payments per Year:		365		
5					
6	Nominal Rate:		5.85%		
7					

8.3 MOVING AMOUNTS THROUGH TIME

Excel's *time-value-of-money functions* allow you to calculate the effect of interest rates on money as it moves through time. It is a standard part of an engineer's job to calculate what a proposed project will cost, how much a possible improvement might be worth, or how long it will take for a new piece of equipment to pay for itself. It is also part of nearly everyone's life to think about borrowing for an education, a vehicle, or a home; or savings toward a goal. All of these are time-value-of-money problems.

8.3.1 Present and Future Values

One hundred dollars invested for 1 year at 1% per month will be worth $112.68 (see Figure 8.3). There is an equality relating those two amounts at different points in time; it is the equation for calculating an FV when given a PV, periodic interest rate, and number of compounding periods:

$$\text{FV} = \text{PV} \cdot \left(1 + i_p\right)^{N_{per}} \tag{8.9}$$

In this equation,

i_p	is the periodic interest rate,
N_{per}	is the number of compounding periods between the present and future,
FV	is the FV, and
PV	is the present value.

Note: The N_{per} (number of compounding periods over the term of the loan or investment) in this equation is not the same as the N_{PPY} (number of compounding periods per year) that is used in calculating the effective annual interest rate.

One hundred dollars today and the $112.68 1 year in the future are considered equivalent when making economic decisions (if the interest rate is 1% per month). The numerical values are not equivalent, but the two values at their respective points in time are equivalent. This is an example of the *time value of money* and illustrates an important point: To compare numerical values, the values must be at the same point in time. Moving values through time (at specified interest rates) is the subject here.

Calculating FVs from PVs

If you have $1000 today, what will it be worth in 20 years if:

1. You put it in a savings account at 5% APR?
2. You invest it in the stock market with an average annual rate of 12.3%?

The "$1000 today" is a PV. The value at the end of the specified period (20 years) is an FV. Excel provides a function, called **FV,** which can be used to calculate FVs when given PVs, periodic interest rates, and the number of compounding periods. The syntax of the function is

```
=FV(Rate, Nper, Pmt, PV, Type)
```

where

Rate	is the interest rate per compounding period (required),
Nper	is the number of periods (required),
Pmt	is the amount of any regular payment each period (required—if there is no regular payment, include a 0 for this argument),
PV	is the amount of any PV (optional), and
Type	indicates whether payments occur at the beginning of the period (Type = 1) or at the end of the period (Type = 0). If Type is omitted, payments at the end of the period are assumed.

Note: Excel, like most financial software programs, uses a *sign convention* to indicate which direction money is moving. Out-of-pocket expenses are negative, and incomes are positive.

We can use the **FV** function to answer questions like "If you have $1000 today, what will it be worth in 20 years?"

Question 1 If you have $1000 today, what will it be worth in 20 years if you put it in a savings account at 5% APR?

$1000 invested at 5% APR for 20 years would be worth $2653.30 (assuming annual compounding). The use of the **FV** function to calculate this result is illustrated in the worksheet shown in Figure 8.6.

Figure 8.6
Calculating the FV of a savings investment.

C10		f_x	=FV(C4,C5,C6,C7,C8)		

	A	B	C	D	E	F
1	**Future Value of an Investment**					
2	$1000 invested at 5% APR for 20 years.					
3						
4		Rate:	5.0%	per year		
5		N_{per}:	20	years		
6		Pmt:	0	(no regular payments)		
7		PV:	-$1,000	(out of pocket expense)		
8		Type:	0	(end of year payments)		
9						
10		FV:	$2,653.30			
11						

Note the following:

1. In Excel, 5% and 0.05 are equivalent. You can enter interest rates either as percentages (including the percent sign) or as decimal. In this example, the rate was entered as a percentage (5%).
2. The $1000 investment is an expense and was entered into the worksheet as a negative value. The $2653.30 is income, so the result was positive.

Question 2 If you have $1000 today, what will it be worth in 20 years if you put it in the stock market with an average annual rate of return of 12.3%?

If the $1000 were invested in the stock market with an average annual rate of return of 12.3%, it would be worth $10,176, as shown in Figure 8.7. (But stock market rates are not guaranteed. Historical rates make no promises for the future, and sometimes stocks can lose a great deal of value quickly.)

Figure 8.7
Calculating the potential FV of a stock investment.

	A	B	C	D	E	F
1	**Future Value of an Investment**					
2	$1000 invested at 12.3% APR for 20 years.					
3						
4		Rate:	12.3%	per year		
5		N_{per}:	20	years		
6		Pmt:	0	(no regular payments)		
7		PV:	-$1,000	(out of pocket expense)		
8		Type:	0	(end of year payments)		
9						
10		FV:	$10,176.42			
11						

Calculating PVs from FVs

Excel also provides a function for calculating PVs from FVs, the **PV** function. Its syntax is similar to that of the **FV** function, namely,

$$=PV(Rate, Nper, Pmt, FV, Type)$$

where

Rate	is the interest rate per period (required),
Nper	is the number of periods (required),
Pmt	is the amount of any regular payment each period (required—if there is no regular payment, include a 0 for this argument),
FV	is the amount of any FV (optional), and
Type	indicates whether payments occur at the beginning of the period (Type = 1) or end of the period (Type = 0). If Type is omitted, payments at the end of the period are assumed.

Again, out-of-pocket expenses should be entered as negative values, and incomes are treated as positive.

The **PV** function can be used to answer questions such as the following:

1. How much do I need to invest today at 5% APR in order to have $1000 available in 1 year?
2. How much do we need to reserve from this year's budget surplus to buy a $30,000 piece of equipment when it becomes available in 15 months? (Assume that your company makes an effective annual rate of 7% on reserved monies.)

A worksheet capable of answering the first question is shown in Figure 8.8.

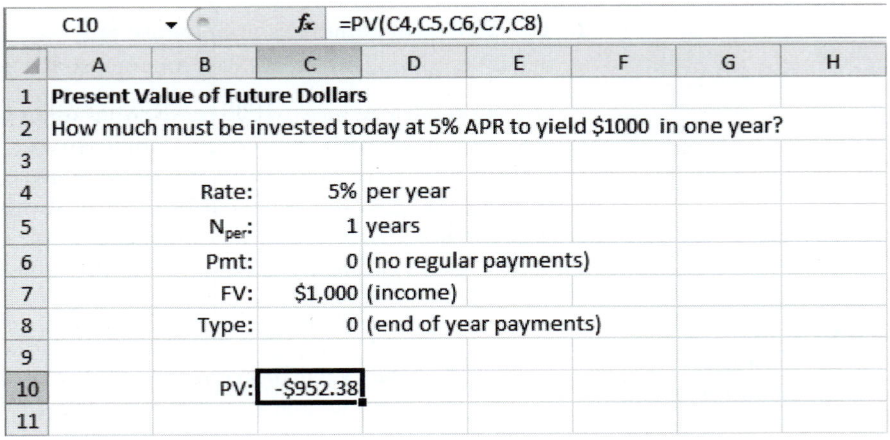

Figure 8.8
Determining the investment needed to reach a goal.

An investment of $952.38 (shown in the worksheet as a negative value since it is an expense) at 5% APR would be worth $1000 1 year later.

An additional calculation is required in order to answer the second question, because the number of periods is expressed in months, while the interest rate is on an annual basis. We must convert the 15 months to years. This has been done in cell C5 in the worksheet shown in Figure 8.9.

Figure 8.9

Determining the needed investment for a future purchase.

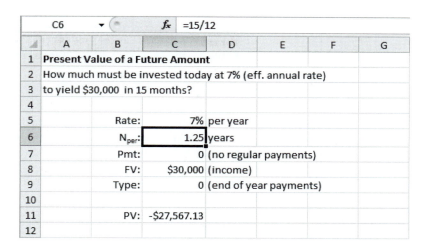

ECONOMICS

Preparing for Retirement: Part 1

It is fairly common to try to predict the amount of money you will need to live comfortably after retirement. Some of the first questions to be considered are

- How much income do I need to live comfortably today?
- With inflation, how much income will I need when I retire?

All economic predictions are full of assumptions. In this example, we will have to assume an inflation rate, the number of years you have left before retirement, and the amount of income required today to live comfortably. A common assumption made when deciding what income level is required to live comfortably is that, by the time you retire, most of your major bills will be paid off, such as your home mortgage and the education expenses for your children.[1] Here are the assumptions we'll use in this calculation:

1. Income required to live comfortably today (without major bill payments): $25,000 per year.
2. Inflation rate: 4% per year.
3. Years left before retirement: 25.

With inflation, how much income will I need when I retire?

To answer this question, we need to find the FV that is equivalent to $25,000 today. We've already developed a worksheet for calculating FVs from PVs. It is presented again in Figure 8.10, modified for this application.

You will need an annual income of over $66,000 to have the same buying power as $25,000 today. This, of course, assumes that the inflation rate will be 4% for the next 25 years. If the inflation rate is higher during some of those years, the necessary future income would also be higher, but the interest rates on the investments used to generate that income tend to go up when inflation goes up. So, you may need more income after periods of high inflation, but your investments also tend to yield higher returns during those high-inflation years.

---

[1]The biggest uncertainty tied to this assumption is the cost of health care during your retirement years. This analysis leaves it out. Including several hundred dollars a month (in today's dollars) for health insurance changes the required annual income considerably.

	A	B	C	D	E	F	G
1	Future Value of a Current Amount						
2	What is the equivalent value, 25 years in the future, of $25,000 today?						
3							
4		Rate:	4%	per year			
5		N_{per}:	25	years			
6		Pmt:	0	(no regular payments)			
7		PV:	-$25,000	(out of pocket expense)			
8		Type:	0	(end of year payments)			
9							
10		FV:	$66,646				
11							

Figure 8.10
Calculating FV.

In another application, we will look at how much money you will need to have available to generate an income of over $66,000 per year.

8.3.2 Regular Payments

Loan payments, periodic maintenance payments, rental payments, and even salaries are regular, periodic payments. Because *regular payments* are very common, Excel provides functions for working with these payments over time. The **FV** and **PV** functions described before can be used with payments, and there is a function **PMT** that calculates payment amounts from PVs and FVs.

The **PMT, FV,** and **PV** functions can be used to answer such questions as the following:

1. What will be the monthly payment on a 4-year car loan at 7% APR interest, compounded monthly, if the loan amount is $22,000?
2. If your company earns 5% (effective annual rate) on unspent balances, how much money do you need to have in the annual budget to cover the $600 per month rental rate (payments at the end of each month) on a piece of lab equipment?
3. If a relative gives you $10,000 and you invest it at 7% nominal rate, compounded monthly, how much can you spend each month and have the money last 5 years?

The **PMT** function uses the syntax

$$=PMT(Rate, Nper, PV, FV, Type)$$

where

Rate	is the interest rate per period (required),
Nper	is the number of periods (required),
PV	is the amount of any FV (required—if there is no PV, include a 0 for this argument),
FV	is the amount of any FV (optional), and
Type	indicates whether payments occur at the beginning of the period (Type = 1) or end of the period (Type = 0). If Type is omitted, payments at the end of the period are assumed.

Again, out-of-pocket expenses should be entered as negative values, incomes as positive values.

Question 1 To calculate the required payment on a 4-year, $22,000 car loan at 7% APR, compounded monthly, the **PMT** function is used, as illustrated in Figure 8.11.

Figure 8.11
Calculating Car Loan
Payments.

	C11	▾	f_x =PMT(C5,C6,C7,C8,C9)			
	A	B	C	D	E	F
1	**Calculating Payments**					
2	What is the payment on a 4-year, $22,000 loan					
3	at 7% APR (monthly compounding)?					
4						
5		Rate:	0.583%	per month		
6		N_{per}:	48	months		
7		PV:	$22,000	(loan principal)		
8		FV:	$0			
9		Type:	0			
10						
11		Pmt:	-$526.82			
12						

Because the interest rate on the loan was an APR, treated as a nominal annual rate, the periodic interest rate corresponding to the compounding period (1 month) had to be computed. This was done in cell C4, with the following formula:

$$C4: = 7\%/12$$

The total number of compounding periods was computed in cell C5 as

$$C5: = 4*12 \quad (4 \text{ years, } 12 \text{ months per year})$$

The $526.82 payment is shown as a negative number because it represents an out-of-pocket expense.

Question 2 If your company earns 5% (effective annual rate) on unspent balances, how much money do you need to have in the annual budget to cover the $600 per month rental rate (payments at the end of each month) on a piece of lab equipment?

In this question, the amount of the regular payment is stated, and you are asked to work out how much needs to be allocated in the budget at time zero to cover all of the rent payments. The **PV** function will be used to answer this question.

The effective annual rate, 5%, needs to be converted to a periodic (monthly) interest rate to correspond to the monthly rent payments. The equation for converting an effective annual rate to a periodic rate is

$$i_p = \left[\sqrt[N_{PPY}]{i_e + 1} \right] - 1, \tag{8.10}$$

where

i_e is the effective annual interest rate,
i_p is the periodic interest rate, and
N_{PPY} is the number of compounding periods per year.

An effective annual interest rate of 5% is equivalent to a monthly interest rate of 0.4074%:

$$i_p = \left[\sqrt[12]{0.05 + 1} \right] - 1 = 0.004074, \tag{8.11}$$

or

$$0.4074\%, \text{ per month.}$$

The amount required in the annual budget to cover the rent payments is found by using the **PV** function, as shown in Figure 8.12.

Figure 8.12
Calculating the present value needed to cover a year of rent payments.

C10			f_x	=PV(C4,C5,C6,C7,C8)		
	A	B	C	D	E	F	G
1	Present Value of Future Payments						
2	How much must be available today to cover a year of rent payments?						
3							
4		Rate:	0.4074%	per month			
5		N_{per}:	12	months			
6		Pmt:	$600	per month			
7		FV:	0				
8		Type:	0	(end of year payments)			
9							
10		PV:	-$7,012.91				
11							

Because the unspent funds are invested, slightly over $7000 is enough to pay all 12 rental payments for the year.

Question 3 If a relative gives you $10,000 and you invest it at 7% APR, compounded monthly, how much can you spend each month and have the money last 5 years?

This time you are looking for a monthly payment, so you would use the **PMT** function to answer this question.

First, the nominal rate of 7% compounded monthly needs to be converted to a periodic rate. This was done in cell C4 of the worksheet shown in Figure 8.13.

$$C4: = 7\%/12$$

The number of monthly payments over the 5-year period is computed in cell C5 as

$$C5: = 5*12$$

If you assume payments at the end of the month (Type = 0), the result is $198.01, as shown in Figure 8.13.

Figure 8.13
Determine the monthly payment that will make $10,000 last for 5 years.

	A	B	C	D	E	F
1	Calculating Payments					
2	What monthly payment will make $10,000					
3	invested at 7% APR last for 5 years?					
4						
5		Rate:	0.583%	per month		
6		N_{per}:	60	months		
7		PV:	-$10,000	(investment expense)		
8		FV:	$0			
9		Type:	0			
10						
11		Pmt:	$198.01	(positive, an income)		
12						

What if you want the income to last forever? You can't put infinity into the **PMT** function's *N*per argument. You can put large numbers into cell C5 and keep increasing the number of months until the calculated payment stops changing. But what you are doing is calculating the monthly interest payment on a $10,000 investment at 0.583% per month. If you spend only the interest, the monthly payments will last forever.

The interest payment on a $10,000 investment at 0.583% per month is $58.30:

$$0.00583 \times \$10,000 = \$58.30. \tag{8.12}$$

ECONOMICS

Preparing for Retirement: Part 2

In the first part of this application example, we found that an income of $66,000 per year would be required 25 years from now to provide buying power equivalent to $25,000 today. If you stop working 25 years from now, you will need enough money in a retirement fund to generate that $66,000 per year.

Question 1 How much money must be in the retirement fund (at 7% effective annual rate) to generate an annual interest payment of $66,000? (Remember that, if you spend only the interest, the income can continue forever.)

We have

$$\$66,000 = 0.07 \times \text{PV},$$

$$\text{PV} = \frac{\$66,000}{0.07} = \$943,000. \tag{8.13}$$

You need to have accumulated $943,000 in your retirement account to be able to live off the interest and have $66,000/year. But, if you don't plan to live forever, you can reduce the amount required by spending down the principal over time.

Question 2 How much money must be in your retirement fund (at 7% effective annual rate) to generate payments of $66,000 per year for 20 years? (The principal will be gone after 20 years.)

To answer this question, we need to find the PV (in an account at 7%) that is equivalent to a series of $66,000 payments over 20 years. This can be calculated using Excel's **PV** function, as shown in Figure 8.14.

	C11	▼	f_x	=PV(C5,C6,C7,C8,C9)			
	A	B	C	D	E	F	G
1	Present Value of Future Payments						
2	How much must be available at retirement to cover 20 years						
3	of annual payments?						
4							
5		Rate:	7%	per year			
6		N$_{per}$:	20	years			
7		Pmt:	$66,000	per year (desired income)			
8		FV:	0				
9		Type:	0	(end of year payments)			
10							
11		PV:	-$699,205				
12							

Figure 8.14

Determining the present value needed to generate annual payments of $66,000 for 20 years.

You will need $699,000 available at retirement to generate 20 years of annual payments of $66,000 per year.

Question 3 How much do you need to put into your retirement account (at 12% APY compounded monthly) each month for 25 years to generate a fund of $699,000? You can use Excel's **PMT** function to answer this question, as shown in Figure 8.15.

C11			f_x	=PMT(C5,C6,C7,C8,C9)	
	A	B	C	D	E	F
1	Calculating Payments					
2	What monthly payment will create a retirement fund					
3	of $699,000 after 25 years?					
4						
5		Rate:	1.00%	per month		
6		N_{per}:	300	months		
7		PV:	$0			
8		FV:	$699,999			
9		Type:	0			
10						
11		Pmt:	-$373			
12						

Figure 8.15
Monthly payments needed to fund the retirement account in 25 years.

The answer: It only takes a monthly investment of $373 to create the (modest) retirement account (if you can get 12% APY).

What happens if you wait 15 years and give yourself only 10 years to create the account? What monthly payment is needed to fund the same retirement account in 10 years? Figure 8.16 shows the calculation.

	A	B	C	D	E	F
1	Calculating Payments					
2	What monthly payment will create a retirement fund					
3	of $699,000 after 10 years?					
4						
5		Rate:	1.00%	per month		
6		N_{per}:	120	months		
7		PV:	$0			
8		FV:	$699,999			
9		Type:	0			
10						
11		Pmt:	-$3,043			
12						

Figure 8.16
Monthly payments needed to fund the retirement account in 10 years.

The moral: It is easier on your lifestyle if you start saving for retirement early.

8.4 NET PRESENT VALUE

When a company invests in developing a new technology, the project's cash flow typically has a big development and startup expense up-front, followed by annual incomes (hopefully) for many years. When the expenses and incomes occur at various points in time, it is hard to pin down what the total value of the project really is. Moving all of the expenses and incomes to the present can help make it clear whether the project is a money maker or loser. Moving all of the expenses and incomes to the present and adding the PVs is called determining the *net present value* (NPV).

EXAMPLE 8.2

If a friend asks to borrow $1000 and promises to pay you $250 after 1 year, then $200 more every year for 4 more years. Is that a good deal? Is that a better deal than you could get by putting the $1000 in a 5% APR CD?

To help answer these questions, you might want to find the NPV of the series of payments. You can't use the Pmt argument in the **PV** function, because the payments are not all equal, but you can calculate the PV of each payment, and then add them. This is illustrated in Figure 8.17.

Figure 8.17
Calculating the NPV of a series of payments.

	C4	▼		f_x	=SUM(C7:C11)	
	A	B	C	D	E	
1	Net Present Value of a Series of Payments					
2						
3		Rate:	5%	per year		
4		NPV:	$913.51			
5						
6	Payment	Amount	PV			
7	1	-$250	$238.10			
8	2	-$200	$181.41			
9	3	-$200	$172.77			
10	4	-$200	$164.54			
11	5	-$200	$156.71			
12						

The **PV** function was used to determine the PV of every payment. For example, the $250 paid at the end of the first year was found to have a PV of $238.10 as

C7: `=PV(5%,1 year,0,-$250,0)`   or   `=PV($C$3,A7,0,B7,0)`

The 5% interest rate was used to compare how well the proposed payment plan compares to simply putting the $1000 in a CD at 5% APR. The result is not good; the net present worth of all of the payments is <$1000. You are getting less back by loaning the money that you would by investing in the CD.

(continued)

Excel has a function, **NPV,** which calculates NPVs directly from a stream of incomes. The syntax of the **NPV** function is

$$=NPV(iP, Incomes1, Incomes2, ...)$$

where

iP	is the interest rate per period for the cash flow (required),
Incomes1	is a value, cell address, or range of cells containing one or more end-of-period incomes, with expenses as negative incomes (required), and
Incomes2, ...	are used to include a series of individual end-of-period incomes (up to 29) if the incomes are not stored as a range of cells.

The worksheet shown in Figure 8.18 illustrates the use of the **NPV** function.

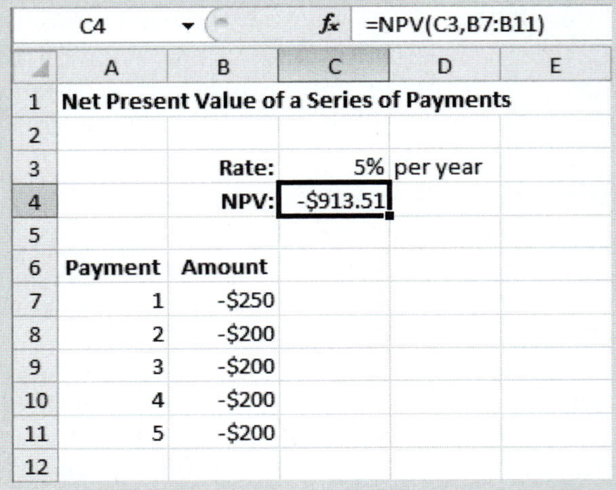

Figure 8.18
Calculating the NPV of a series of payments using the **NPV** function.

	C4			f_x	=NPV(C3,B7:B11)	
	A	B	C	D	E	
1	Net Present Value of a Series of Payments					
2						
3		Rate:	5%	per year		
4		NPV:	-$913.51			
5						
6	Payment	Amount				
7	1	-$250				
8	2	-$200				
9	3	-$200				
10	4	-$200				
11	5	-$200				
12						

The NPV came out the same when it was calculated by adding a bunch of PVs (Figure 8.17) and when the **NPV** function was used.

EXAMPLE 8.3

A new heat exchanger will cost $30,000, but is expected to generate an income of $7000 per year for the first 2 years and then $5500 per year for another 5 years. However a maintenance expense of $6000 is expected in the fourth year. If the interest rate used by your company is 6.5% per year, what is the NPV?

A cash flow timeline for this heat exchanger is shown in Figure 8.19. A worksheet that can be used to determine the NPV is shown in Figure 8.20.

Figure 8.19
Expenses and incomes
for the heat exchanger.

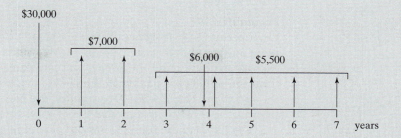

Figure 8.20
Calculating the NPV of a
series of expenses and
incomes.

C5		f_x	=NPV(C3,D8:D14)+C4			

	A	B	C	D	E	F	G
1	Net Present Value						
2							
3		Rate:	6.50%	per year			
4		Initial Cost:	-$30,000	(negative because it is an expense)			
5		NPV:	-$1,768				
6							
7	Year	Expense	Income	Net Income			
8	1		$7,000	$7,000			
9	2		$7,000	$7,000			
10	3		$5,500	$5,500			
11	4	$6,000	$5,500	-$500			
12	5		$5,500	$5,500			
13	6		$5,500	$5,500			
14	7		$5,500	$5,500			
15							

Because the initial expense is not at the end of a period, it was not included in the cell range sent to the **NPV** function. But, since the $30,000 expense is already at time zero, it can simply be added on to the result returned by **NPV**. This was done in cell C5 in Figure 8.20, using the function call:

$$\text{C5: =NPV(C3,D8:D14)+C4}$$

Notes:

- The **NPV** function works with the incomes, not the PVs of the incomes. It calculates the PVs internally.
- The incomes sent to the **NPV** function must be equally spaced in time and occur at the end of each interval. If you have both an expense and an income at the same point in time, send NPV the net income at that time.
- Do not include a payment (or income) at time zero in the values sent to the **NPV** function, because time zero is not the end of an interval. If there is an amount at time zero, it should be added on to the result returned by **NPV** function.

Notice, in Figure 8.20, that the NPV was found to be −$1768. At 6.5% interest, the investment in the heat exchanger would lose money. (This isn't really surprising. Heat exchangers are not typically expected to generate income; they are purchased because they are needed to get a job done.)

ECONOMICS

Payout Period

Engineering projects generally require a source of cash up-front to purchase equipment and get things started. Then, income starts coming in, and, eventually, income exceeds expenses, and the project starts making money. The moment when income from the project has paid back the startup expenses and initial operating costs marks the end of the *payout period*. Determining the payout period is often done without considering the time value of money. This is equivalent to setting the interest rate to zero.

The NPV can be used to find the payout period for a cash flow. You simply start at the beginning of the cash flow and calculate the NPV at time zero. Then you include income and expenses for one time period and recalculate the NPV. You continue to include more time intervals until the NPV becomes positive. That's the end of the payout period.

Considering the heat-exchanger example again, the $30,000 expense at time zero creates a NPV of $-\$30,000$ at time zero. At the end of the first year, the income of $7000 changes the NPV to $-\$33,000$ (if the time value of money is ignored by setting $i_p = 0$). Building the incomes over time into a worksheet (see Figure 8.21), it is easy to calculate the NPV year by year using the **NPV** function.

K6				f_x	=NPV($C$3,K8:K14)+$C$4							
	A	B	C	D	E	F	G	H	I	J	K	L	M
1	Payout Period												
2													
3		Rate:	0.00%	per year									
4		Initial Cost:	-$30,000	(negative because it is an expense)									
5					Year:	1	2	3	4	5	6	7	
6					NPV:	-$23,000	-$16,000	-$10,500	-$11,000	-$5,500	$0	$5,500	
7	Year	Expense	Income	Net Income									
8	1		$7,000	$7,000		$7,000	$7,000	$7,000	$7,000	$7,000	$7,000	$7,000	
9	2		$7,000	$7,000			$7,000	$7,000	$7,000	$7,000	$7,000	$7,000	
10	3		$5,500	$5,500				$5,500	$5,500	$5,500	$5,500	$5,500	
11	4	$6,000	$5,500	-$500					-$500	-$500	-$500	-$500	
12	5		$5,500	$5,500						$5,500	$5,500	$5,500	
13	6		$5,500	$5,500							$5,500	$5,500	
14	7		$5,500	$5,500								$5,500	
15													

Figure 8.21
Using the NPV to determine the payout period.

The heat-exchanger project (with $i_p = 0$) has a payout period of 6 years.

Note: It is possible to account for the time value of money when computing the payout period. Simply use an interest rate greater than zero when calculating the NPV.

8.5 INTERNAL RATE OF RETURN

The *internal rate of return* is defined as the interest rate that causes the NPV to be zero. You can determine the internal rate of return by changing the interest rate used in a worksheet like that shown in Figure 8.20 until the calculated NPV equals zero.

In Figure 8.22, we see that the internal rate of return for the heat-exchanger project is 4.68%. (Excel's Goal Seek feature was used to get this value quickly.)

	C5		▾	●		f_x	=NPV(C3,D8:D14)+C4			
◢	A		B		C	D	E	F	G
1	Internal Rate of Return								
2									
3			Rate:		4.68%	per year -- determined by iteration			
4			Initial Cost:		-$30,000	(negative because it is an expense)			
5			NPV:		$0				
6									
7		Year	Expense		Income	Net Income			
8		1			$7,000	$7,000			
9		2			$7,000	$7,000			
10		3			$5,500	$5,500			
11		4	$6,000		$5,500	-$500			
12		5			$5,500	$5,500			
13		6			$5,500	$5,500			
14		7			$5,500	$5,500			
15									

If you don't want to use Goal Seek to find the interest rate that makes the NPV zero, Excel provides the **IRR** function. The syntax of the **IRR** function is

$$=IRR(Incomes, irrGuess)$$

where

Incomes   is a range of cells containing the incomes, with expenses as negative incomes (required), including an income or expense at time zero, and

irrGuess   is a starting value, or guess value, for the internal rate of return. (This is optional, and Excel uses 10% if it is omitted.)

The use of the **IRR** function is illustrated in Figure 8.23.

Notes:

- The **IRR** function works with the incomes, not the PVs of the incomes. It calculates the PVs internally.
- The incomes sent to the **IRR** function must be equally spaced in time and, with the exception of the first value (an amount at time zero), must occur at the end of the interval. If you have both an expense and an income at the same point in time, send **IRR** the net income at that time.
- The rate returned by the **IRR** function is the rate per time interval (% per year, % per month, etc.)
- The **IRR** function uses an iterative approach to solving for IRR. If the solution does not converge on a value within 20 iterations, **IRR** returns a **#NUM! error**. A closer *irrGuess* value may help in finding the correct internal rate of return.

Figure 8.23
Determining the internal rate of return using the **IRR** function.

	C3			f_x	=IRR(D6:D13)		
	A	B	C	D	E	F	G
1	Internal Rate of Return						
2							
3			IRR:	4.68%	per year -- calculated by Excel		
4							
5	Year	Expense	Income	Net Income			
6	0	$30,000		-$30,000			
7	1		$7,000	$7,000			
8	2		$7,000	$7,000			
9	3		$5,500	$5,500			
10	4	$6,000	$5,500	-$500			
11	5		$5,500	$5,500			
12	6		$5,500	$5,500			
13	7		$5,500	$5,500			
14							

PRACTICE!

Would the internal rate of return increase or decrease if the exchanger could be sold for $20,000 at the end of 7 years? Why?

[ANSWER: The internal rate of return increases to 14.5%.]

8.6 ECONOMIC ALTERNATIVES: MAKING DECISIONS

One of the most common reasons for doing time-value-of-money calculations is to make a decision between two or more economic alternatives. Choosing between alternatives is a routine part of an engineer's job. Here are a couple of examples of the types of economic decisions engineers must make:

- Which is the better deal: The expensive compressor with lower annual maintenance costs and longer life, or the cheaper compressor with higher annual maintenance costs and shorter life?
- Which is the better deal when financing a new truck: 1.5% APR on the loan, or $1500 cash back to apply toward the downpayment?

To answer questions such as these, it is necessary to move all of the money to the same point in time, usually a PV.

EVALUATING ALTERNATIVES

Choosing the Best Compressor

Two compressor options are available for a new process design:

- Option A is a $60,000 compressor with a 10-year life expectancy, annual maintenance costs of $2200, and a salvage value of $12,000 when it is removed from service.
- Option B is a $32,000 compressor with a 5-year life expectancy, annual maintenance costs of $2600, and no salvage value.

Which is the better deal?

APPLICATION

Before you can make a decision, you need to get all of the dollars moved to the same point in time (usually a PV) by using a specified interest rate. Here, we will assume an effective annual rate of 7%. You also need to get the total time span for each option to be the same, in this case by purchasing a second compressor in year 5 for Option B. (Alternatively, you can calculate an equivalent cost per year for each option to make the decision.)

The cash flow diagrams for each option are shown in Figure 8.24.

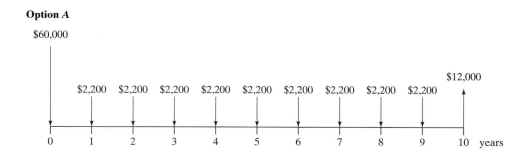

Option A

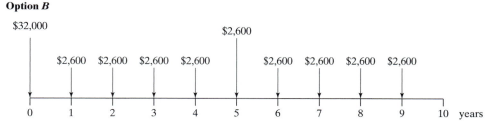

Option B

Figure 8.24
Cash flow diagrams for compressor options A and B.

The PV of Option A is approximately $68,000, as calculated in the worksheet shown in Figure 8.25.

	A	B	C	D	E	F	G	H	I	J
1	**Present Value of Option A**									
2										
3		Item	PV		iP	Nper	Pmt	FV	Type	
4					(per year)	(years)				
5		Initial purchase cost (at time zero):	$60,000							
6		Annual maintenance costs:	$14,334		7%	9	-$2,200	0	0	
7		Salvage value:	-$6,100		7%	10	0	$12,000	0	
8										
9		**TOTAL PV, OPTION A:**	$68,233							
10										

Figure 8.25
Present value of Option A.

Cells E6 through I7 contain the values shown. The values or formulas in column C are as follows:

C5:	`$60,000`	(a value)
C6:	`=PV(E6,F6,G6,H6,I6)`	(The "H6" and "I6" are optional and could be omitted.)
C7:	`=PV(E7,F7,G7,H7,I7)`	(The "I7" could be omitted.)

Calculating the PV of Option B requires a few more steps. One way to simplify the calculations is to take $2600 from the purchase price of the second pump in year 5 ($32,000 − $2,600 = $29,400) and use it to create a uniform series of $2600 payments in years 1 through 9, as shown on the cash flow diagram in Figure 8.26.

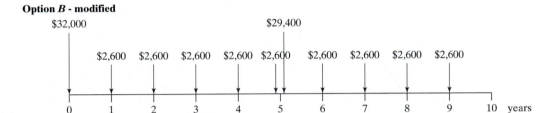

Figure 8.26
Modified (equivalent) cash flow diagram for Option B.

Now, the PV of Option B can be calculated as shown in Figure 8.27.

	A	B	C	D	E	F	G	H	I	J
11		**Present Value of Option B**								
12										
13		Item	PV		iP	Nper	Pmt	FV	Type	
14					(per year)	(years)				
15		Initial purchase cost, Comp. 1 (at time zero):	$32,000							
16		Initial purchase cost, Comp. 2 (year 5):	$20,962		7%	5	0	-$29,400	0	
17		Annual maintenance costs:	$16,940		7%	9	-$2,600	0	0	
18		Salvage value:	$0		7%	10	0	$0	0	
19										
20		**TOTAL PV, OPTION B:**	$69,901							
21										

Figure 8.27
Present value of Option B.

The values or formulas in column C are as follows:

C15:	`$32,000`
C16:	`=PV(E16,F16,G16,H16,I16)`
C17:	`=PV(E17,F17,G17,H17,I17)`

The PV of Option B is approximately $70,000, almost $2000 higher than Option A. Option A is the way to go (at least, if you think the compressor will be needed for more than 5 years).

EXAMPLE 8.4

SELECTING THE BEST CAR-FINANCING DEAL

For years, advertisements for new cars have offered low interest rates as incentives. Those ads typically offer either a low interest rate or cash back (usually applied to the downpayment), but not both. The cash-back option is provided for individuals who can pay cash for the new car, since they will have no interest in a low APR on a loan.

If you will finance the purchase, you get to choose between the following options:

1. Pay the full purchase price, but get a reduced interest rate on the loan.
2. Pay a standard interest rate (e.g., 8% APR, compounded monthly), but get "cash back" to lower the effective purchase price.

Which is the better deal? You can use time-value-of-money calculations to find out. Specifically, you can use Excel's **PMT** function to calculate which option results in the lower monthly payment.

A new pickup truck sells for $27,000. You plan to offer a downpayment of $5000 and finance the rest. The company's latest advertisement offers either 1.5% APR for up to 48 months or $1500 cash back.

If you take the cash back, it would effectively reduce the purchase price to $25,500. After your $5000 downpayment, the remaining $20,500 would be financed with a 8% APR loan for 48 months. Which is the better deal for you?

Figure 8.28 shows the payment calculation for the 1.5% APR loan.

Figure 8.28
Calculating the loan payment for the reduced APR loan.

◢	A	B	C	D	E	F	G
1	**Calculating Payments - Low APR Loan**						
2	What is the monthly payment on a 4-year, $25,000 loan at 1.5% APR?						
3							
4		Rate:	0.125%	per month			
5		N_{per}:	48	months			
6		PV:	$25,000	(loan principal)			
7		FV:	$0				
8		Type:	0				
9							
10		Pmt:	-$536.94				
11							

The lower principal, but higher rate loan is shown in Figure 8.29.

Figure 8.29
Calculating the loan payment for the higher APR, reduced principal loan.

◢	A	B	C	D	E	F	G
1	**Calculating Payments - Reduced Principal Loan**						
2	What is the monthly payment on a 4-year, $20,500 loan at 8% APR?						
3							
4		Rate:	0.667%	per month			
5		N_{per}:	48	months			
6		PV:	$20,500	(loan principal)			
7		FV:	$0				
8		Type:	0				
9							
10		Pmt:	-$500.46				
11							

(continued)

The cash back offers the lowest payments. For this example, that's the better deal; but the situation can change if you need to provide a smaller downpayment, or if you can't get an 8% loan.

8.7 DEPRECIATION OF ASSETS

Taxes are a very significant cost of doing business and must be considered when estimating the profit or loss from a proposed business venture. *Depreciation* is an important factor in tax planning, and numerous methods have been used over the years to calculate depreciation amounts. Only a few of the more common methods will be presented here. Excel provides a built-in function for the most common depreciation method.

8.7.1 Methods of Depreciation

Tax codes subject most capital expenditures to depreciation. *Capital expenditures* are substantial expenditures for *assets* (things like equipment purchases) that will be used for more than 1 year. (That is, the impact of the equipment purchase is felt over multiple taxation periods.) Expenses can be used to offset income before the calculation of taxes, but capital expenditures impact multiple tax years, so the amount of the capital expenditure that may be used to offset income is also spread over multiple taxation periods. The allowed amount each year is the depreciation.

The amount of the capital expenditure that may be deducted each year may be computed by several depreciation methods. US tax laws require businesses to use a depreciation system called *Modified Accelerated Cost Recovery System (MACRS)*, which includes multiple depreciation methods.

Straight-Line Depreciation Method

With straight-line depreciation, you simply deduct the same percentage of the asset's net value each year of the asset's service life. The net value is the difference between the *initial cost* and the *salvage value* (if any) and is termed the *depreciation basis* of the asset.

If you buy a heat exchanger for $40,000, it has an expected service life of 10 years, and an expected salvage value of $2000, how much can you deduct each year?

Using the *straight-line depreciation* method, the amount that can be deducted each year, *D,* is the depreciation basis (initial cost minus the salvage value) divided by the expected service life:

$$D = \frac{C_{\text{init}} - S}{N_{\text{SL}}} = \frac{\$40{,}000 - \$2{,}000}{10} = \$3{,}800, \tag{8.14}$$

where

D	is the *depreciation amount* (same each year using the straight-line method),
C_{init}	is the initial cost of the asset,
S	is the salvage value at the end of the service life, and
N_{SL}	is the service life (also called the *recovery period*) of the asset.

Because the capital expense of an asset must be depreciated over its service life, you must keep track of the current value of the asset throughout its service life. The current value of an asset is called its *book value, B.* For straight-line depreciation, the book value in the current year is last year's book value minus the depreciation amount, so

$$B_j = B_{j-1} - D, \qquad (8.15)$$

where j represents the current year and $j-1$ represents the previous year. B_0 is equivalent to the initial cost of the asset.

Note: Depreciation is an accounting practice used to spread the cost of assets over several tax years. The book value used for tax purposes simply keeps track of the amount of the original purchase cost that has yet to be deducted. It is not necessarily related to the actual resale value of a piece of equipment.

The book value of an asset subject to straight-line depreciation decreases linearly over the service life of the asset (that's why it is called straight-line depreciation), as is shown in Figure 8.30.

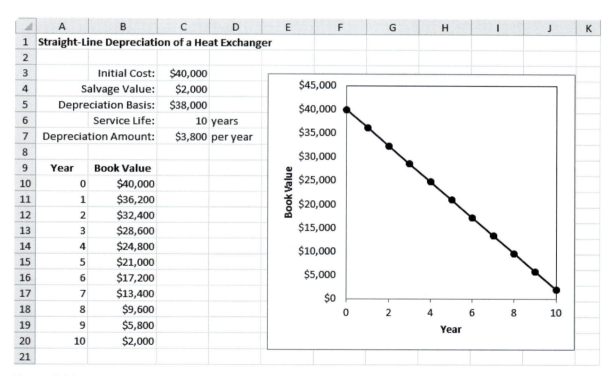

Figure 8.30
Book value of a $40,000 heat exchanger with a service life of 10 years and expected salvage value of $2000.

200% or Double-Declining-Balance Depreciation
Declining-balance methods calculate the allowable depreciation amount, based on:

- Current book value of an asset
- Declining depreciation factor, F_{DB},
- Depreciation rate R_{DB},
- Salvage value S of the asset

The depreciation amount in a given year is calculated as

$$D_j = (B_{j-1} - S)F_{DB}, \tag{8.16}$$

or, if the salvage value is not considered (as is the case with MACRS),

$$D_j = B_{j-1}F_{DB}, \tag{8.17}$$

where

D_j	is the allowable depreciation amount in the current year (year j),
B_{j-1}	is the book value at the end of the previous year ($B_0 = C_{init}$),
N_{SL}	is the service life of the asset,
F_{DB}	is the declining-balance depreciation factor, and
R_{DB}	is the declining-balance percentage. For the double-declining-balance method, $R_{DB} = 200\%$, or 2.

The declining-balance depreciation factor is computed for the double-declining method as

$$F_{DB} = \frac{200\%}{N_{SL}}. \tag{8.18}$$

Figure 8.31 shows a worksheet designed to calculate depreciation for the heat-exchanger example ($C_{init} = \$40,000$, $S = \$2,000$, $N_{SL} = 10$ years), using the double-declining-balance method of depreciation.

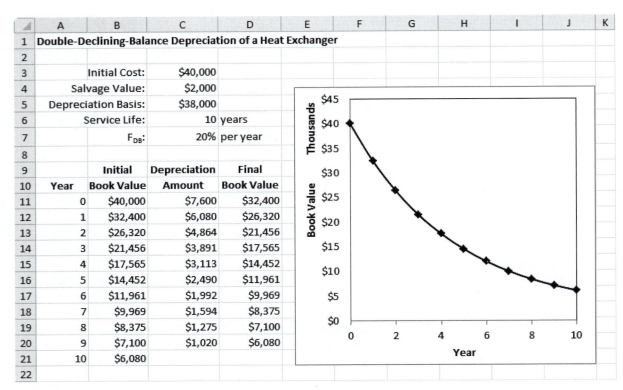

Figure 8.31
Double-declining-balance depreciation for the heat-exchanger example.

Notice that the declining-balance depreciation factor, F_{DB}, is 20%. That means that you can depreciate 20% of the net value of the asset each year. This creates depreciation amounts that are much larger in the first few years than are possible with the straight-line depreciation method. This allows significantly larger tax deductions in the first years after capital expenditures.

MACRS Depreciation System

The MACRS system is designed to depreciate assets as quickly as possible.

- In the early years of an asset's service life, MACRS uses declining-balance methods for most capital expenditures. These methods allow larger depreciation amounts.
- In the later years of an asset's service life, the straight-line depreciation method allows larger depreciation amounts, so the MACRS system switches to straight-line depreciation in the later years to maximize depreciation.
- MACRS does not consider salvage value when calculating depreciation amounts.

Sounds complicated? It is, but Excel provides the **VDB** function to handle the details.

When the MACRS system is used, the tax codes define the service life of various types of assets with service lives of 3, 5, 7, 10, 15, or 20 years. 200% declining-balance methods are used for service lives of 10 years or less, 150% declining-balance methods for service lives of 15 or 20 years.

MACRS also uses a half-year convention that assumes that assets are placed in service in the middle of the tax year and calculates a half-year of depreciation in the first and last years of service. When the remaining service life falls below 1 year, the straight-line depreciation factor for that tax year is 1.00 (100%), which means that the entire remaining book value may be deducted.

Excel provides the **VDB** function to handle MACRS depreciation. The syntax of the **VDB** function is

```
=VDB(Cinit, S, NSL, Perstart, Perend, FDB, NoSwitch)
```

where

Cinit	is the initial cost of the asset (required),
S	is the salvage value (set to zero for MACRS depreciation) (required),
NSL	is the service life of the asset (required),
Perstart	is the start of the period over which the depreciation amount will be calculated (required),
Perstop	is the end of the period over which the depreciation amount will be calculated (required), and
FDB	is the declining-balance percentage (optional; Excel uses 200% if omitted).
NoSwitch	tells Excel whether to switch to straight-line depreciation when the straight-line depreciation factor is larger than the declining-balance depreciation factor (optional). (Set this to FALSE to switch to straight line, to TRUE not to switch; default is FALSE):

To see how the **VDB** function is used, again consider the heat-exchanger example. MACRS depreciation is illustrated in Figure 8.32.

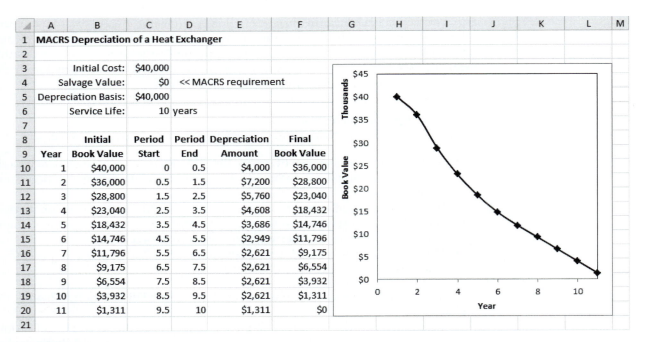

Figure 8.32
MACRS depreciation for the heat-exchanger example.

The **VDB** function in cell E10 tells Excel:

$$=\text{VDB}(\$C\$3,\$C\$4,\$C\$6,C10,D10,200\%,\text{FALSE})$$

- the initial cost is $40,000 (always in cell C3—hence the dollar signs)
- no salvage value should be used
- the service life is 10 years
- the start period and end period for the depreciation calculation are found in cells C10 and D10
- double-declining-balance depreciation should be used ($R_{DB} = 200\%$),
- NoSwitch is set to FALSE to cause the calculation to switch to straight-line depreciation when that method gives a larger depreciation factor.

The half-period at the beginning and end of the service life are necessitated by the half-year convention (required by tax code) that assumes that all assets are placed into service halfway through a tax year.

KEY TERMS

Annual Percentage Rate (APR)
Annual Percentage Yield (APY)
Annuity payment
Assets
Book value
Capital expenditures
Cash flow
Compounding period
Declining-balance depreciation
Depreciation
Depreciation amount
Depreciation basis
Effective annual interest rates
Expenses
Future value (FV)
Incomes
Inflation rate
Initial cost
Intangibles
Internal rate of return
Interest rate

Modified Accelerated Cost Recovery System (MACRS)	Principal	Straight-line depreciation
	Recovery period	
	Regular payment (Pmt)	Taxes
Net present value	Salvage value	Time value of money
Nominal interest rates	Service life	
Payout period	Sign convention	Time value of money functions
Periodic interest rates	(incomes +,	
Present value (PV)	expenses −)	Timeline

SUMMARY

There are several ways of expressing interest rates:

- **Periodic interest rates** are the interest rates used over a single compounding period.
- **Nominal interest rates** are annual rates that ignore the effect of compounding. You use a nominal rate by converting it into a periodic interest rate.
- **Effective annual interest rates** are calculated annual interest rates that yield the same results as periodic interest rates with nonannual compounding.
- **APRs** include any up-front fees that are part of the total cost of borrowing. If there are no up-front fees, the APR can be used as a nominal interest rate if the compounding period is known.
- **APY** are nominal interest rates calculated on the assumption of daily compounding.

Excel's Time-Value-of-Money Functions

Nomenclature

Rate	is the interest rate per period.
Nper	is the number of periods.
Pmt	is the amount of any regular payment each period.
PV	is the amount of any present value.
Type	indicates whether payments occur at the beginning of the period (Type = 1) or end of the period (Type = 0). If Type is omitted, payments at the end of the period are assumed.

PV Function

PV (Rate, Nper, Pmt, FV, Type)

FV Function

FV (Rate, Nper, Pmt, PV, Type)

Payment Function

PMT (Rate, Nper, PV, FV, Type)

Net Present Value (NPV)

To calculate the NPV, we use

=NPV(iP, Incomes1, Incomes2, ...)

where

iP	is the interest rate per period for the cash flow (required),
Incomes1	is a value, cell address, or range of cells containing one or more incomes, with expenses as negative incomes, (required), and
Incomes2, ...	is used to include a series of individual incomes (up to 29) if the incomes are not stored as a range of cells.

Internal Rate of Return

This is the periodic interest rate that produces the NPV zero for a set of incomes and expenses, so

$$=\texttt{IRR(Incomes, irrGuess)}$$

where

Incomes	is a range of cells containing the incomes, with expenses as negative incomes (required), and
irrGuess	is a starting value, or guess value, for the internal rate of return. (This is optional, so Excel uses 10% if it is omitted.)

Evaluating Options

Principles

- Compare values at the same point in time (PVs, FVs, or equivalent annual cost).
- Projects should have the same duration before comparison, or use equivalent annual cost.

Depreciation of Assets

Straight-Line Depreciation

$$D = \frac{C_{\text{init}} - S}{N_{\text{SL}}}, \tag{8.19}$$

where

D	is the depreciation amount,
C_{init}	is the initial cost of the asset,
S	is the salvage value at the end of the service life, and
N_{SL}	is the service life of the asset.

Double-Declining-Balance Depreciation

$$D_j = (B_{j-1} - S)\,F_{\text{DB}},$$

and

$$F_{\text{DB}} = \frac{200\%}{N_{\text{SL}}}, \tag{8.20}$$

where

D_j	is the allowable depreciation amount in the current year (year j),
B_{j-1}	is the book value at the end of the previous year ($B_0 = C_{\text{init}}$),
N_{SL}	is the service life of the asset,
F_{DB}	is the declining-balance depreciation factor, and
R_{DB}	is the declining-balance percentage. For the double-declining-balance method, $R_{\text{DB}} = 200\%$.

MACRS Depreciation

A rapid depreciation method based on the double-declining-balance and straight-line depreciation methods. The depreciation amount used each year is the larger of the double-declining-balance and straight-line depreciation amounts. MACRS depreciation factors normally are used to determine annual depreciation amounts with this method. Excel's **VDB** function can also handle MACRS depreciation:

```
=VDB(Cinit, S, NSL, Perstart, Perend, FDB, NoSwitch)
```

where,

Cinit	is the initial cost of the asset (required),
S	is the salvage value, required (set to zero for MACRS depreciation),
NSL	is the service life of the asset (required),
Perstart	is the start of the period over which the depreciation amount will be calculated (required),
Perstop	is the end of the period over which the depreciation amount will be calculated (required), and
FDB	is the declining-balance percentage (optional; Excel uses 200% if it is omitted).
NoSwitch	tells Excel whether to switch to straight-line depreciation when the straight-line depreciation factor is larger than the declining-balance depreciation factor. (This is optional; set it to FALSE to switch to straight line, to TRUE not to switch. The default is FALSE.)

PROBLEMS

8.1 Loan Calculations I

Terry wants to purchase a used car for $8000. The dealer has offered to finance the deal at 8.5% APR for 4 years with a 10% downpayment. Terry's bank will provide a loan at 7.8% for 3 years, but wants a 20% downpayment.

a. Calculate the required downpayment and monthly payment for each loan option. (Assume monthly compounding and monthly loan payments.)

b. The PV for both options (including the downpayments) is the same, $8000. What additional factors should Terry consider when deciding which is the best deal for her?

8.2 Loan Calculations II

It is not uncommon to borrow $100,000 or more to purchase a home. Banks typically offer 15-, 20-, or 30-year loans.

a. Calculate the monthly loan payment to repay a $100,000 home loan at 7% APR over 30 years.

b. How much more must be paid each month to pay off the loan in 15 years?

8.3 Loan Calculations III

John has estimated that trading in his current car and adding the money he has saved will allow him to pay $8000 in cash when he buys a new car. He also feels that his budget can handle a $350 payment each month, but doesn't want to make payments for more than 3 years. His credit union will lend him money at 6.8% APR (monthly compounding). What is the maximum-priced car he can afford?

8.4 Loan Calculations IV

Some of the loans college students get for school start charging interest immediately, even if repayment is not required until the student leaves school. The unpaid interest while the student is in school is added to the principal.

If a student borrows $2300 at 8% APR compounded monthly when he or she starts school, how much will be owed 5.5 years later when loan repayment begins? (The 5.5-year figure assumes 5 years to graduate plus 6 months after graduation to begin loan payments.)

8.5 Loan Calculations V

Patricia and Bill want to buy a home, and they feel that they can handle a $1000-per-month payment. If they make $1000 payments for 30 years on a 7% APR loan, how much can they borrow?

8.6 Time-Value-of-Money Problems I

Fill in the missing values in the following table:

Present Value	FV	Reg. Payment	Interest Rate	Compounding	Years
$50,000		—	5% APR	Annual	12
$50,000		—	5% APR	Weekly	12
	$250,000	—	17.3% (effective annual rate)	Annual	7
—	$250,000		12.4% (effective annual rate)	Annual	30
$50,000	—		5% APR	Weekly	12

8.7 Time-Value-of-Money Problems II

Compute the PV of the cash flows shown in Figure 8.33 if the interest rate is 4% APR compounded monthly.

8.8 Evaluating Options I

Consumers today have a choice of buying or leasing new cars. Compare the NPV of the following options:

3-Year-Life Calculations

- Option A: Lease a new $26,000 sport utility vehicle for 36 months with $380 monthly payments and $1750 required up-front. At the end of the lease, you return the car and get no cash back.
- Option B: Purchase a new $26,000 sport utility vehicle with a 60-month loan at 9% APR and $1750 downpayment. At the end of 3 years, sell the vehicle for $15,000 and use the money to pay off the rest of the loan and generate some cash.

Figure 8.33
Cash flow timelines.

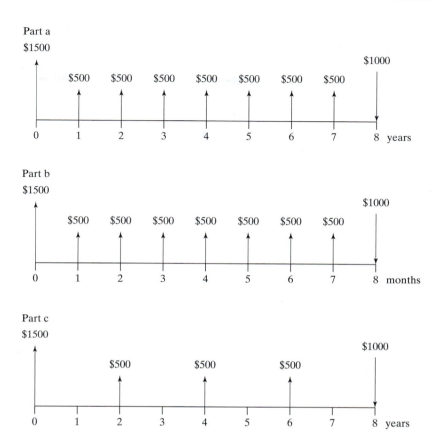

9-Year-Life Calculations

- Option C: Use three consecutive 3-year leases to cover the 9-year time span. Assume 4% annual inflation rate for the monthly payments and up-front costs each time a new vehicle is leased.
- Option D: Purchase a new $26,000 sport utility vehicle with a 60-month loan at 9% APR and $1750 downpayment. Keep the vehicle for 9 years. Assume no maintenance costs the first 3 years (warranty coverage) and $400 maintenance in the fourth year, with maintenance costs increasing by $200 each year, thereafter. After 9 years, sell the vehicle for $5000.

Sketch the cash flow diagram for each option and calculate the NPV for each option, using an interest rate of 6% APR. Which seems like the best deal?

Note: Why a 6% rate when the bank loan is 9%? The 9% is what the bank owners want to get for their money. The 6% is a rate that the person buying the car could readily get for his or her own money if it was not tied up in the vehicle.

8.9 Internal Rate of Return I

Find out the internal rate of return for the cash flows illustrated in Figure 8.34.

Figure 8.34
Cash flow timelines.

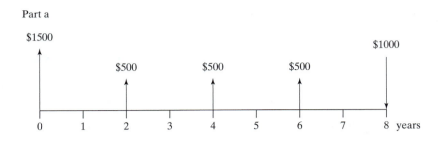

Part a

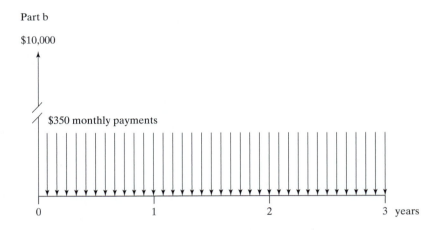

Part b

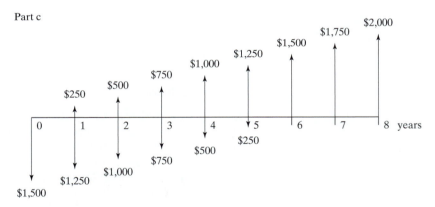

Part c

8.10 Internal Rate of Return II

Archway Industries requires projects to provide an internal rate of return of at least 4% to be considered for funding. Sara's proposed project requires an initial investment of $350,000 for new equipment plus $24,000 annually for operating costs (years 1 through 12). The project is expected to generate revenues of $14,000 the first year, $170,000 for years 2 through 6, and $100,000 for years 7 through 12. At that time, the equipment will be scrapped at an estimated cost of $400,000.

Will Sara's project be considered for funding?

8.11 Internal Rate of Return III

When major expenses come at the end of a project (e.g., nuclear power-plant cleanup costs) as illustrated in Figure 8.35, something interesting can happen to the internal rate of return.

Figure 8.35
Cash flow with a major expense at the end of the project.

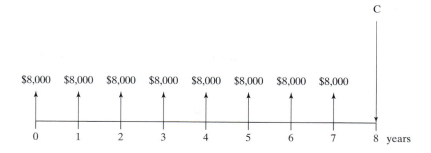

a. Calculate the internal rate of return for the preceding cash flow, diagrammed when the end-of-product cost, *C*, is
   i. $70,000. (The internal rate of return is near 1% for this case, and the **IRR** function might need that value as an initial guess for its iteration.)
   ii. $80,000.
   iii. $90,000.
b. Calculate the NPV of the cash flow if the interest rate is 7% when *C* is
   i. $70,000.
   ii. $80,000.
   iii. $90,000.
c. Which case—(i), (ii), or (iii)—produces the highest internal rate of return?
d. Which case—(i), (ii), or (iii)—produces the highest NPV?
e. Which is the best project in which to invest?

8.12 Depreciation I

Create a depreciation schedule showing annual depreciation amounts and end-of-year book values for a $26,000 asset with a 5-year service life and a $5000 salvage value, using the straight-line depreciation method.

8.13 Depreciation II

Create a depreciation schedule showing annual depreciation amounts and end-of-year book values for a $26,000 asset with a 5-year service life, using the MACRS depreciation method.

8.14 Time-Value-of-Money Problems III

Find the PV of the cash flows shown in Figure 8.36 if the interest rate is 9% APR compounded monthly.

8.15 Evaluating Options II

Your company is expanding into a new manufacturing area, but there are two possible ways to do this:

- Option A: Computerized manufacturing equipment requiring an initial outlay of $2.3 million for equipment with operating costs (excluding materials, which will be the same for either option) of $20,000 per year for general costs and $80,000 per year in wages for equipment operators. Products from a new computerized line are expected to generate revenues of $650,000 per year over the 20-year life expectancy of the equipment.
- Option B: Manual manufacturing equipment requiring an outlay of $0.94 million for equipment with operating costs (excluding materials) of $70,000

per year of general costs, plus $240,000 per year for wages to the skilled technicians required to run the manually operated equipment. Products from a new manually operated line are expected to generate revenues of $530,000 per year over the 20-year life expectancy of the equipment.

a. Which option provides the higher internal rate of return?
b. If the company uses a rate of 6% (effective annual rate) for comparing investment options, what is the NPV of each option?
c. What nonfinancial factors should be considered in making the decision on which option to implement?

Figure 8.36
Cash flow timelines.

Part a

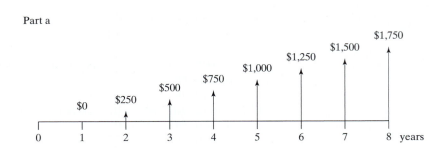

Part b

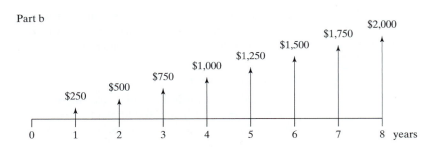

Part c

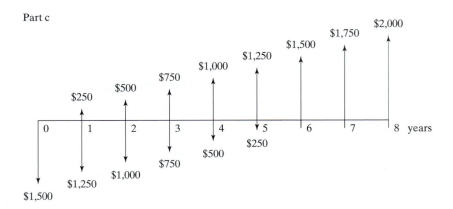

9 Iterative Solutions Using Excel

Objectives

After reading this chapter, you will know how to

- Rewrite equations in two standard forms to facilitate iterative solutions
- Use a graph to find solutions, or roots, of equations
- Find roots of equations using several iterative solution methods, including
  - "Guess and Check" iteration
  - Direct substitution and in-cell iteration

- Excel's Goal Seek to solve for roots of equations
- Excel's Solver to solve for roots of equations
- Solve optimization problems by using Excel's Solver
- Set up and solve nonlinear regression problems in Excel
- Solve linear programming problems in Excel

9.1 INTRODUCTION

Some equations are easy to rearrange and solve directly. For example,

$$PV = nRT \tag{9.1}$$

is easily rearranged to solve for the volume of an ideal gas:

$$V = \frac{nRT}{P}. \tag{9.2}$$

Sometimes, things are a little tougher, such as when you need to solve for the superficial velocity ($\overline{V}_0$) or porosity (ε) in the Ergun equation[1]:

$$\frac{\Delta p}{L} = \frac{150 \, \overline{V}_0 \mu}{\Phi_s^2 D_p^2} \cdot \frac{(1 - \varepsilon)^2}{\varepsilon^3} + \frac{1.75 \, \rho \overline{V}_0^2}{\Phi_s D_p} \cdot \frac{1 - \varepsilon}{\varepsilon^3}, \tag{9.3}$$

Here,

$\dfrac{\Delta p}{L}$ is the pressure drop per unit length of packing in a packed bed,

$\overline{V}_0$ is the superficial velocity of the fluid through the bed,

Φ_s is the sphericity of the particles in the bed,

D_P is the particle diameter,

ε is the porosity of the packing,

μ is the absolute viscosity of the fluid flowing through the bed, and

ρ is the density of the fluid flowing through the bed.

There are times when it is easier to find a *solution,* or *root,* by using *iterative methods,* and sometimes they are the only way. Iterative techniques can be used to solve a variety of complex equations.

9.2 ITERATIVE SOLUTIONS

A common way of describing an *iterative solution technique* is to say it is a "guess and check" method. Nearly all iterative solution techniques require an *initial guess* to be provided by the user. Then the equation is solved, using the guessed value, and a result is calculated. A test is performed to see whether the guess is "close enough" to the correct answer. If not, a new guess value is used; the repeating process is called *iteration.* The key to efficient iterative solutions is a method that is good at coming up with guess values. We will look at several methods:

1. Using a plot to search for roots
2. Guessing and checking
3. Direct-substitution method
4. In-cell iteration
5. Excel's Solver

With Excel 2010, the Solver has been improved and now includes multiple solution methods. Excel attempts to automatically choose the most appropriate method based on the characteristics of the problems being solved. Excel's Solver is a good, general-purpose iterative solver with many applications.

9.2.1 Standard Forms

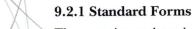

The equation to be solved should be put into a *standard form* if it is to be solved by iterative methods. Two standard forms will be used in this chapter, shown in Table 9.1. For example, consider the equation $x^3 + 12 = 17x$:

Table 9.1 Standard equation forms for iterative solutions

Form	Example	Convenient For...
(1) Set equation equal to 0	$x^3 - 17x + 12 = 0$	Plot method, Excel's Solver
(2) Get an x by itself on the left side	$x = \dfrac{x^3 + 12}{17}$	Direct-substitution method and in-cell iteration

[1]*Unit Operations of Chemical Engineering,* 5th ed., McCabe, W. L., J. C. Smith, and P. Harriott: New York: McGraw-Hill, 1993, p. 154.

The direct-substitution method and in-cell iteration in Excel require form 2. Form 1 is usually more convenient for finding solutions when using graphs, and a version of form 1 is required when using the Solver. (The right-hand side may be any constant, not just 0.) Either form can be used if you just "guess and check" by hand until you find a solution.

9.3 USING A PLOT TO SEARCH FOR ROOTS

If you have a fair idea that the root or roots of the equation should lie within a finite range of values, you can run a series of guesses in the range through your equation, solve for the computed values, plot the results, and look for the roots. The first standard form, with all terms on one side of the equation, works well when using a plot to search for roots. The equation listed previously can be written as

$$x^3 - 17x + 12 = 0 \tag{9.4}$$

in standard form 1. However, when we try guess values of x in the equation, the result will usually not equal 0 (unless we have found a root). The equation might be written as

$$f(x) = x^3 - 17x + 12. \tag{9.5}$$

To search for roots, we will try various values of x (i.e., various guesses), solve for $f(x)$, and graph $f(x)$ vs. x. When $f(x) = 0$, we have found a root. First, we create a column of x values in the range $0 \le x \le 5$ as shown in Figure 9.1. We're guessing that there might be a root in that range.

Figure 9.1

Creating a column of guess values.

	A	B	C	D
1	Using a Plot to Search for Roots			
2				
3	x			
4	0.0			
5	0.4			
6	0.8			
7	1.2			
8	1.6			
9	2.0			
10	2.4			
11	2.8			
12	3.2			
13	3.6			
14	4.0			
15	4.4			
16	4.8			
17	5.0			
18				

Then we compute $f(x)$ at each x value, as illustrated in Figure 9.2.

Notice that the signs of the computed values change twice in the column of $f(x)$ values; every time the sign of the computed value changes sign, you're close to a solution. Already we know there are roots between $x = 0.4$ and 0.8, and $x = 3.6$ and 4.0.

Next, we graph the $f(x)$ and x values, using an XY scatter plot as shown in Figure 9.3.

Figure 9.2

Calculate computed values for each guess value.

	B4		f_x	=A4^3-17*A4+12	
	A	B	C	D	E
1	Using a Plot to Search for Roots				
2					
3	x	f(x)			
4	0.0	12.0			
5	0.4	5.3			
6	0.8	-1.1			
7	1.2	-6.7			
8	1.6	-11.1			
9	2.0	-14.0			
10	2.4	-15.0			
11	2.8	-13.6			
12	3.2	-9.6			
13	3.6	-2.5			
14	4.0	8.0			
15	4.4	22.4			
16	4.8	41.0			
17	5.0	52.0			
18					

Figure 9.3

Plotting the f(x) and x values as an XY scatter graph.

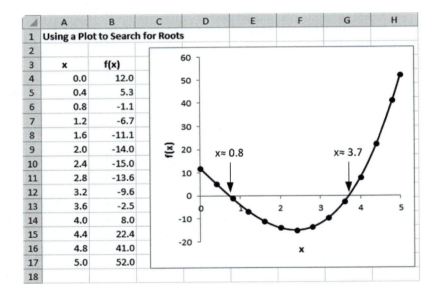

From the graph, it looks as if there are roots near $x = 0.8$ and $x = 3.7$. To get more accurate results, narrow the range of x values used. For example, we might use 20 x values between 0.6 and 1.0 to try to find out the value of the root near $x = 0.8$ more accurately.

The x^3 in the original equation suggests that there are three roots, but this graph shows only two. We should expand the range of x values to search for the third root. In Figure 9.4, x values between −5 and +4 were investigated.

From the new plot, it looks as if the additional root is near $x = -4.5$.

A plot is a fairly easy way to get approximate values for the roots, but, if you want more precise values, then numerical iteration techniques can be useful. But most numerical iteration techniques require initial guesses, and a plot is a great way to get a good initial guess.

Figure 9.4
Searching for a third root.

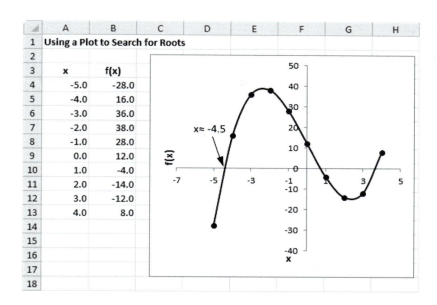

9.4 SIMPLE "GUESS AND CHECK" ITERATION

If you know that you have a root near $x = 0.8$, one of the easiest ways to find the root is simply to create a worksheet with a place to enter guess values and a formula that evaluates $f(x)$, like the worksheet shown in Figure 9.5.

Figure 9.5
Simple "guess and check" iteration in a worksheet.

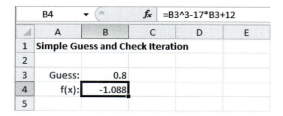

Then keep entering guesses until the $f(x)$ value is nearly 0 (because the equation was written in form 1). This is illustrated in Figure 9.6.

Figure 9.6
Guess and check iteration, getting close to a root.

	A	B	C	D	E
1	Simple Guess and Check Iteration				
2					
3	Guess:	0.728			
4	f(x):	0.009828			
5					

In the worksheet in Figure 9.6, $f(x)$ is getting close to, but is not, 0. Iterative methods can go on forever if you try to get $f(x)$ exactly to 0. Instead, you search for a root value that is close enough for your needs. Later, we will see how a *tolerance* value is used to judge whether the value is close enough. For now, realize that the guess value shown here is good to two decimal places; there is still inaccuracy in the third decimal place, the digit that is still being "guessed."

Sometimes, it is easier to keep track of the iteration process if you leave a trail of previous guesses and calculated values, as shown in Figure 9.7.

Figure 9.7
Leaving a trail of guess
values can facilitate the
guess and check process.

B11		f_x	=A11^3-17*A11+12	
	A	B	C	D	E
1	Simple Guess and Check Iteration				
2					
3	Guess	f(x)			
4	0.8	-1.0880			
5	0.7	0.4430			
6	0.75	-0.3281			
7	0.72	0.1332			
8	0.73	-0.0210			
9	0.725	0.0561			
10	0.727	0.0252			
11	0.728	0.0098			
12					

9.4.1 Writing Your Equation in Standard Form 2 ("Guess and Check" Form)

When your equation is in standard form 1, you keep trying guess values of x until $f(x)$ is close enough to 0. The alternative is to put your equation into standard form 2, sometimes called "guess and check" form. This simply requires getting one of the variables to be solved for by itself on the left side of the equation. The example equation can be rewritten in form 2 as

$$x = \frac{x^3 + 12}{17} \tag{9.6}$$

or

$$x = \sqrt[3]{17x - 12}. \tag{9.7}$$

Either version will work. The first version has been used in these examples.

The isolated variable is called the computed value and is renamed x_C. The x value in the expression on the right side of the equation is called the guess value and is renamed x_G:

$$x_C = \frac{x_G^3 + 12}{17}. \tag{9.8}$$

During the solution process, x_G will be assigned values, and x_C will be computed. When x_C and x_G are equal, you have found a root. This equation actually has three roots ($x = -4.439$, 0.729, and 3.710), but only one can be found at a time. You can try to find all roots by starting the iteration process with different guess values, but not all iterative solution methods can find all roots for all equations.

With your equation in form 2, you keep trying x_G values until x_C is equal (actually, "close enough") to x_G. The worksheet shown in Figure 9.8 shows the initial guess $x_G = 0.8$ and the initial computed value $x_C = 0.736$:

Figure 9.8
Guess and check iteration
using standard form 2.

B4		f_x	=(B3^3+12)/17	
	A	B	C	D	E
1	Guess and Check Iteration: Form 2				
2					
3	x_G:	0.8			
4	x_C:	0.736			
5					

After a few iterations, we are getting close to a root value as shown in Figure 9.9.

Figure 9.9

Converging on a root using standard form 2.

	A	B	C	D	E
1	Guess and Check Iteration: Form 2				
2					
3	x_G:	0.728			
4	x_C:	0.728578			
5					

We can quantify how close we are getting to a root by calculating the difference between the guessed and computed values:

$$\Delta x = x_C - x_G. \tag{9.9}$$

Often a precision setting, or tolerance (e.g., TOL = 0.001), is used with the calculated difference to specify stopping criteria such as "Stop when **ABS**(Δx) < TOL". This is illustrated in Figure 9.10.

Figure 9.10

Using a tolerance value to decide when the guess is "close enough."

B7		f_x	=ABS(B4-B3)		
	A	B	C	D	E
1	Guess and Check Iteration: Form 2				
2					
3	x_G:	0.728			
4	x_C:	0.728578			
5					
6	TOL:	0.001	<< chosen		
7	ABS(Δx):	0.000578			
8					
9	Test:	Stop: close enough			
10					

The test in cell B9 uses an **IF** function to test if the absolute difference is less than or equal to the chosen tolerance value.

B9:   `=IF(B7<=B6,"Stop: close enough","Keep guessing")`

When you get tired of guessing values yourself, you might want to come up with an automatic guessing mechanism—that's what the remaining methods are all about.

9.5 DIRECT-SUBSTITUTION TECHNIQUE

The simplest automatic guessing mechanism is to use the previous computed value as the next guess value. Standard form 2 must be used to make this possible. This is called *direct substitution* of the computed value. Using the direct-substitution method, all you need to provide is the initial guess; the method takes it from there. The chief virtue of this method is simplicity, but there are roots that this method simply cannot find.

When the direct-substitution method is used in Excel, the original guess is still entered by hand. In Figure 9.11, it has been entered into cell A4.

Figure 9.11
Entering the initial guess
for the direct-substitution
method.

	A	B	C	D	E
1	Direct Substitution Method				
2					
3	x_G	x_C			
4	0.8				
5					

Then, a computed value is generated using the guess value. The first computed value is shown in cell B4 in Figure 9.12.

Figure 9.12
Finding the first
computed value.

B4			f_x	=(A4^3+12)/17	

	A	B	C	D	E
1	Direct Substitution Method				
2					
3	x_G	x_C			
4	0.8	0.7360			
5					

The key to the direct-substitution method is using the previous computed value as the next guess value, as illustrated in Figure 9.13.

Figure 9.13
Using the previous
computed value as the
next guess value.

A5			f_x	=B4	

	A	B	C	D	E
1	Direct Substitution Method				
2					
3	x_G	x_C			
4	0.8	0.7360			
5	0.7360				
6					

The "formula" for the new guess value in cell A5 is just =B4. To complete row 5, the formula in cell B4 is copied to cell B5 to calculate the new computed value (see Figure 9.14).

Figure 9.14
Calculating the second
computed value.

B5			f_x	=(A5^3+12)/17	

	A	B	C	D	E
1	Direct Substitution Method				
2					
3	x_G	x_C			
4	0.8	0.7360			
5	0.7360	0.7293			
6					

For more iterations, simply copy the cells in row 5 down the worksheet as far as necessary, as shown in Figure 9.15.

Figure 9.15
Copy the formulas in rows
5 down the worksheet to
iterate toward a solution.

	A	B	C	D	E
1	Direct Substitution Method				
2					
3	x_G	x_C			
4	0.8	0.7360			
5	0.7360	0.7293			
6	0.7293	0.7287			
7	0.7287	0.7286			
8	0.7286	0.7286			
9	0.7286	0.7286			
10	0.7286	0.7286			
11					

As you can see in Figure 9.15, the direct-substitution method did find a root: $x = 0.7286$. Adding the difference calculation (Figure 9.16) allows us to see how quickly this method found this root.

Figure 9.16
Displaying the difference,
$\Delta x = x_C - x_G$.

D10		▼	f_x	=B10-A10	
	A	B	C	D	E
1	Direct Substitution Method				
2					
3	x_G	x_C		Δx	
4	0.8	0.7360		-0.064000	
5	0.7360	0.7293		-0.006665	
6	0.7293	0.7287		-0.000631	
7	0.7287	0.7286		-0.000059	
8	0.7286	0.7286		-0.000006	
9	0.7286	0.7286		-0.000001	
10	0.7286	0.7286		0.000000	
11					

It took just a few steps to find this solution.

To try to find the root near $x = 3.7$, let's try an initial guess slightly larger than the expected root, say, $x_G = 3.8$: The iteration process with this starting value is shown in Figure 9.17.

Figure 9.17
Trying to find the root near
$x = 3.7$ using an initial
guess of $x = 3.8$.

	A	B	C	D	E
1	Direct Substitution Method				
2					
3	x_G	x_C		Δx	
4	3.8	3.9336		0.133647	
5	3.9336	4.2863		0.352682	
6	4.2863	5.3383		1.051971	
7	5.3383	9.6546		4.316283	
8	9.6546	53.6419		########	
9	53.6419	########		########	
10	########	########		########	
11					

The pound symbols indicate that Excel cannot fit the number into the format specified for the cell. In this case, the numbers are getting too large: The direct-substitution method is *diverging*. To try to get around this, try an initial guess on the other side of the root, say, $x_G = 3.6$: The iteration process with this initial guess is shown in Figure 9.18.

Figure 9.18

Trying to find the root near $x = 3.7$ using an initial guess of $x = 3.6$.

	A	B	C	D	E
1	Direct Substitution Method				
2					
3	x_G	x_C		Δx	
4	3.6	3.4504		-0.149647	
5	3.4504	3.1221		-0.328222	
6	3.1221	2.4961		-0.626037	
7	2.4961	1.6207		-0.875395	
8	1.6207	0.9563		-0.664403	
9	0.9563	0.7573		-0.198971	
10	0.7573	0.7314		-0.025893	
11	0.7314	0.7289		-0.002532	
12	0.7289	0.7287		-0.000238	
13	0.7287	0.7286		-0.000022	
14	0.7286	0.7286		-0.000002	
15	0.7286	0.7286		0.000000	
16					

This time the direct-substitution method did find a root ($x = 0.7286$), but not the root near $x = 3.7$. Direct substitution simply cannot find the root at 3.71 from this equation (unless you are lucky enough to enter 3.7102 as the initial guess).

9.5.1 In-Cell Iteration

In the preceding method, we caused Excel to repeatedly calculate (iterate) by copying the formulas in row 5 down the worksheet. Excel can perform this iteration within a cell. To tell Excel to iterate within a cell, you use the value displayed in a cell in a formula in the *same* cell. Excel calls this a *circular reference*. Generally, you try to avoid circular references in worksheets (they are usually unintended errors), but if you want Excel to iterate within a cell, you must intentionally create a circular reference.

In-cell iteration requires your equation to be written in standard form 2, the guess-and-check form. Excel displays the result of the calculation (i.e., the computed value x_C). If we put the right side of our equation, $(x_G^3 + 12)/17$, into cell A4 and refer to this cell wherever the equation calls for x_G, we create the circular reference. This is illustrated in Figure 9.19.

Figure 9.19

Creating a circular reference to allow for in-cell iteration.

A4			f_x	=(A4^3+12)/17		
	A	B	C	D	E	F
1	In-Cell Iteration					
2	Cell A4 contains the formula and displays the final result					
3						
4	0.0000					
5						

If you try this in Excel, you will get the error message shown in Figure 9.20 saying there is a circular reference.

Figure 9.20

Excel warning when you create a circular reference.

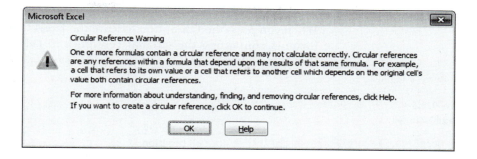

Clicking the **OK** button on the warning box allowed the formula containing the circular reference to be created, but Excel still did not perform in-cell iteration. In-cell iteration is not activated by default in Excel, because iteration slows down worksheet performance, and, most of the time, a circular reference indicates a data entry error, not an intentional request for an iterative solution. To activate iteration, use the Excel Options dialog, accessed as follows:

- Excel 2010: **File tab/Options**
- Excel 2007: **Office/Excel Options**
- Excel 2003: **File/Options**

On the Excel Options dialog, choose the **Formulas** panel.

Activate iteration by checking the box labeled **Enable iterative calculation** as illustrated in Figure 9.21.

Figure 9.21

Enabling in-cell iteration.

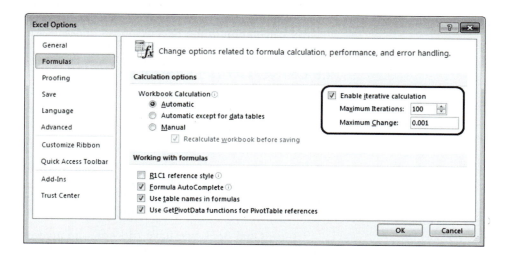

[Excel 2003: Tools/Options…, then use the Calculation tab, check "Iteration"].

By default, Excel will iterate for no more than 100 cycles and will stop iterating when the computed value differs from the guess by <0.001 (this is the default tolerance value). You can change these values, and solutions using this direct-substitution method could well require more than 100 iteration cycles.

Note: If the root is not found within the specified number of iterations, this method provides no feedback, but simply leaves the last result displayed in the cell. You should always check the result to make sure the displayed value actually satisfies your equation.

When you press the **OK** button to leave the Excel Options dialog, Excel iterates the formula containing the circular reference in cell A4 and displays the result, as shown in Figure 9.22.

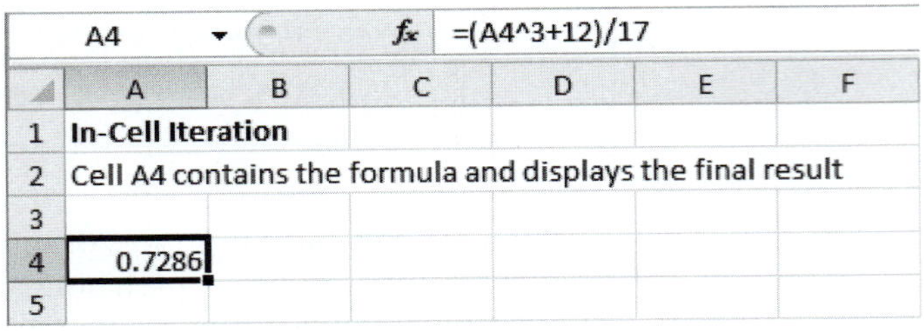

Figure 9.22

The result of in-cell iteration.

	A	B	C	D	E	F
1	In-Cell Iteration					
2	Cell A4 contains the formula and displays the final result					
3						
4	0.7286					
5						

A4 f_x =(A4^3+12)/17

This approach has another very serious limitation: It always uses 0 as the initial guess. That means it can find no more than one of the equation's roots. You cannot use this method to find multiple roots. Fortunately, Excel provides two alternative methods:

- Goal Seek
- The Solver

We will look at using Goal Seek next.

Note: If you are using Excel and following along in the text, you probably want to turn in-cell iteration back off by clearing the checkbox next to **Enable iterative calculation** in the Excel Options dialog, Formulas panel.

9.6 USING GOAL SEEK IN EXCEL

The *Goal Seek* feature in Excel allows you to solve problems backward: to find the input values needed to get the answers you want. This is fairly common in engineering calculations. Here are a couple of examples where Goal Seek might be used:

- I know the flow velocity should not exceed 3 m/s; what pipe diameter is needed to keep the flow to this velocity?
- The output voltage has to be 5V for compatibility with the rest of the device; what size resistor is needed to produce an output voltage of 5V?

To use Goal Seek, we have to know the answer we want. If we tried to use standard form 2 (guess and check form), we would need to know the root in order to find the root. That won't work; but if we use form 1, we know that we want to find the x values that cause $f(x)$ to be 0. With form 1, we know the answer we want; it's 0.

Figure 9.23 shows a worksheet set up for using Goal Seek; it is essentially the same worksheet used for the simplest guess and check approach (Figure 9.5).

Figure 9.23
Getting ready to use Goal
Seek to find a root.

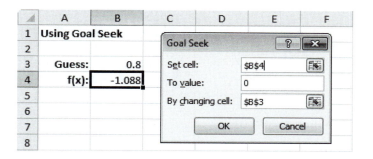

To use Goal Seek, we need a cell to hold the guess value (cell B3 in Figure 9.23) and a cell (B4) containing the formula that computes the result (the $f(x)$ value).

To open the Goal Seek dialog, use Ribbon options

Data/Data Tools/What-If Analysis/Goal Seek…

[Excel 2003: Tools/Goal Seek…]

The Goal Seek dialog is shown in Figure 9.24.

Figure 9.24
The Goal Seek dialog.

The goal is to get $f(x)$ in cell B4 to be 0, so we enter B4 in the **Set cell:** field (see Figure 9.24) and indicate that the goal is to get the value to zero by entering 0 in the **To value:** field. We want the guess value in cell B3 changed until the goal is met, so we enter B3 in the **By changing cell:** field. Once the Goal Seek input fields have been filled in, clicking on the **OK** button causes Excel to change the guess values until the goal is met (or the iteration fails). The successful result is shown in Figure 9.25.

Figure 9.25
The result of using Goal
Seek to find a root.

Goal Seek not only leaves the root in the guess cell (cell B3), but it pops up a box to let you know how close it came to meeting the goal. The Goal Seek Status box shown in Figure 9.25 shows that the goal was met very closely.

Can Goal Seek find all three roots? Since Goal Seek starts with the initial guess you provide, and uses a considerably more advanced technique than direct substitution, Goal Seek is pretty good at finding solutions. In Figures 9.26 and 9.27, the other two roots for the example equation are displayed.

Figure 9.26
Finding the root at 3.71
(initial guess, $x = 3.6$).

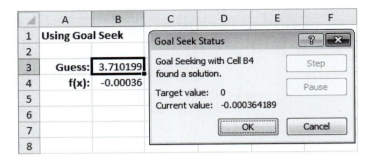

Figure 9.27
Finding the root at -4.44
(initial guess, $x = -4$).

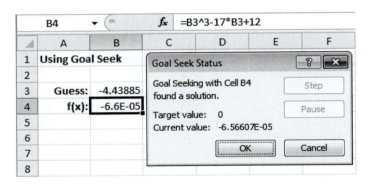

FLUID MECHANICS

Finding the Right Pipe Diameter

High fluid velocities cause high pressure losses due to fluid friction, which means high pumping costs. To control pressure losses, fluid velocities in water transport pipes are often kept to about 3 m/s or less.

A design engineer for a new housing complex needs to provide 0.20 m³/s of water to the unit. What pipe diameter should be used to keep the fluid velocity at 3 m/s?

A worksheet to solve this problem with Goal Seek is shown in Figure 9.28. The initial guess for the pipe diameter is 1 m (arbitrary), in cell C4.

	A	B	C	D	E	F
1	**APPLICATION: FLUID MECHANICS**					
2					**Formulas in Column B**	
3	Volumetric Flow Rate:	0.200	m³/s			
4	Pipe Diameter:	1.000	m/s			
5						
6	Area for Flow:	0.785	m²		=PI()/4*B4^2	
7	Fluid Velocity:	0.255	m/s		=B3/B6	
8						

Figure 9.28
The worksheet, ready for Goal Seek.

APPLICATION

Next, the Goal Seek dialog is opened and necessary inputs are provided as shown in Figure 9.29.

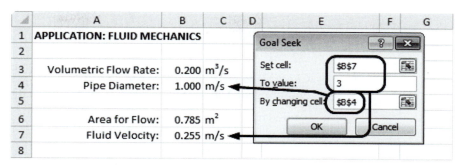

Figure 9.29
Setting the Goal Seek parameters.

Clicking the **OK** button on the Goal Seek dialog causes Excel to iterate for the necessary pipe diameter. The result is shown in Figure 9.30.

	A	B	C	D	E	F	G
1	APPLICATION: FLUID MECHANICS						
2							
3	Volumetric Flow Rate:	0.200	m³/s		Goal Seek Status		
4	Pipe Diameter:	0.291	m/s		Goal Seeking with Cell B7		Step
5					found a solution.		
6	Area for Flow:	0.067	m²		Target value: 3		Pause
7	Fluid Velocity:	3.000	m/s		Current value: 3.000		
8					OK		Cancel

Figure 9.30
The result after iteration.

A pipe with an inside diameter of 0.291 m, or 291 mm, is required. The engineer would then specify the next larger standard size pipe. A 300 mm (nominal size) schedule 40 steel pipe has an inside diameter of 303 mm. Since this is just a little larger than needed, the pipe will work and the actual fluid velocity will be a little less than 3 m/s.

Goal Seek is easy to use and powerful, but Excel provides an even more powerful option, the Solver.

9.7 INTRODUCTION TO EXCEL'S SOLVER

Excel provides a very powerful iteration tool that it calls the Solver; it will do everything that Goal Seek will do, and more.

9.7.1 Activating the Solver

The *Solver* is an Excel Add-In that is installed but not activated when Excel is installed. Before we can use the Solver, we need to see if it is already available and activate it if necessary.

Checking to see if the Solver is already activated

Look for a button labeled **Solver** on the Ribbon's **Data** tab, in the Analysis group.

Data/Data Analysis/Solver

If the button is there, the Solver is activated and ready to use. You can skip the next few steps.

Checking to see if the Solver can be activated

Activating an Add-In is done via the Excel Options dialog which is accessed as follows:

- Excel 2010: **File tab/Options**
- Excel 2007: **Office/Excel Options**
- Excel 2003: **File/Options**

On the Excel Options dialog:

1. Choose the **Add-Ins** Panel.
2. Look for **Solver Add-In** in the Add-Ins List (see Figure 9.31).

Figure 9.31
Managing Add-Ins.

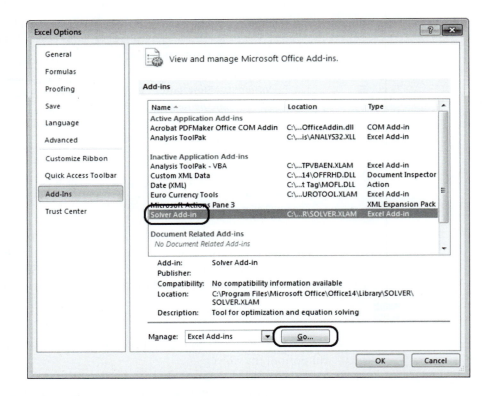

The **Add-Ins** panel lists Active Application Add-Ins and Inactive Application Add-Ins.

- If the Solver Add-In is in the active list, the Solver is ready to go.
- If the Solver Add-In is not on any list, it wasn't installed; you will have to get out the CD's to install the Solver Add-In.
- If it is in the inactive list, you need to activate it. To activate the Solver Add-In:
  1. Click on the **Solver Add-In** list item (the name in the list, as in Figure 9.31).
  2. Click the **Go...** button.

These steps will cause the Add-Ins dialog to open, as shown in Figure 9.32.

Figure 9.32
The Add-Ins dialog.

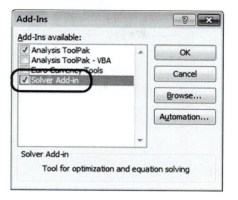

3. Check the box labeled **Solver Add-In**.
4. Click the **OK** button.

When the Solver Add-In is activated, the Ribbon will be updated and a **Solver** button will appear on the **Data** tab in the Analysis group, as shown in Figure 9.33.

Figure 9.33
Ribbon options **Data/Analysis/Solver** access the Solver. Once the Solver is available for use, we can use it for iterative calculations.

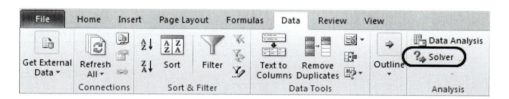

9.7.2 Using the Solver

Like Goal Seek, the Solver requires that the equation be written in a variation of standard form 1. The sample equation in form 1 is $x^3 - 17x + 12 = 0$ and is ready to be used with the Solver.

Using the Solver to find roots requires a worksheet just like the one we used with Goal Seek (Figure 9.23); only the title has been changed in Figure 9.34.

Figure 9.34
A worksheet to find roots using the Solver.

	B4		f_x	=B3^3-17*B3+12	
	A	B	C	D	E
1	**Using the Solver**				
2					
3	Guess:	0.8			
4	f(x):	-1.088			
5					

The Solver requires an initial guess value (cell B3) and a cell containing the formula being evaluated (cell B4). Excel evaluates the formula in cell B4 with the initial guess, but it does not iterate until you use the Solver. Start the Solver using Ribbon options **Data/Analysis/Solver** [Excel 2003: Tools/Solver…].

The Solver Parameters dialog will be displayed as shown in Figure 9.35.

The cell references and value that must be changed have been indicated in Figure 9.35.

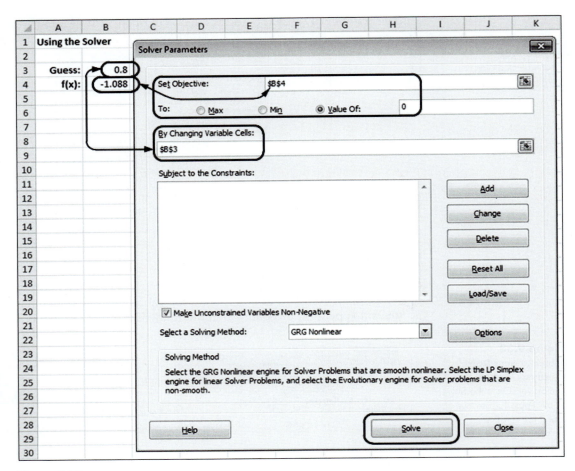

Figure 9.35
The Solver Parameters dialog.

- **Set Objective:** The objective cell is the cell containing the formula: cell B4. When you use the mouse to indicate the target cell, Excel automatically adds dollar signs to indicate an absolute cell address. This doesn't hurt anything, but absolute addressing is not required for simple problems.
- **To:** We want the Solver to keep guessing until the target formula produces the result 0 (because our equation was set equal to 0). We tell the Solver this by setting the **To:** to **Value of:** 0.
- **By Changing Variable Cells:** We tell the Solver where the guess value is located with the **By Changing Variable Cells:** field. The value in this cell will be changed by the Solver during the solution process.

Note: The Solver can find roots, as we are using it here, and it can also be used to find maximum and minimum values, with and without constraints. Using the Solver to find minimum and maximum values is presented later in the chapter.

Once the required information has been set in the Solver Parameters dialog box, click on the **Solve** button to iterate for a solution. Once the iteration process is complete, Excel displays the Solver Results dialog with some options on how to handle the results, as shown in Figure 9.36.

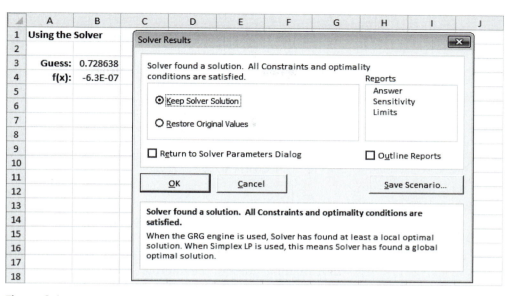

Figure 9.36
The Solver Results dialog.

Here the Solver Results dialog indicates that a solution was found, so we can press **OK** to keep the solution. The option to **Restore Original Values** (i.e., the initial guess) can be useful when the Solver fails to find a solution, but that happens rarely. Excel will also prepare short reports (in a new worksheet) describing the answer, sensitivity, and limits (for problems with constraints).

With an initial guess of 0.8, the Solver found the root at 0.729. To search for other roots, change the initial guess and run the Solver again. The Solver has no trouble finding all of the roots with this simple example.

ENGINEERING ECONOMICS

Internal Rate of Return on a Project

The Solver can be used to find the internal rate of return on a series of incomes and expenses (listed as negative incomes). The internal rate of return is the interest rate (actually, the time value of money) that causes the incomes and expenses associated with a project to balance out, generating a net present value of 0. Internal rates of return are a fairly common way of looking at the value of a proposed project:

- A negative internal rate of return on a proposed project indicates that the project will lose money and is not a good investment for a company.
- A small positive rate of return indicates the project will not make very much money, so the company might want to put its money somewhere else.
- A large positive internal rate of return is a good investment.

Common internal rate of return functions, like Excel's **IRR** function, assume that incomes and expenses occur at regular intervals. When the incomes and expenses do not occur at regular intervals, you need to build the timing of the incomes and expenses into the calculation of the internal rate of return. The Solver and Excel's present value function, **PV,** can be used to find the internal rate of return on non-uniformly spaced incomes and expenses, such as of the 10-year series of cash flows illustrated in Figure 9.37.

APPLICATION

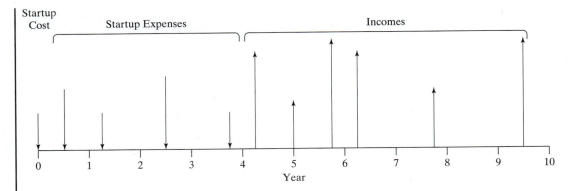

Figure 9.37
Anticipated incomes and expenses over a 10-year period.

First, the times, incomes, and expenses are entered into a worksheet, along with an initial guess of the internal rate of return. This is illustrated in Figure 9.38.

Figure 9.38
Anticipated incomes and expenses over a 10-year period, as an Excel worksheet.

	A	B	C	D	E	F
1	**Internal Rate of Return with Non-Uniform Time Intervals**					
2						
3		IRR:	2.0%	per year (guess value)		
4		Net PV:				
5						
6	**Time (years)**	**Income**				
7	0.00	-$375,000				
8	0.50	-$625,000				
9	1.25	-$375,000				
10	2.50	-$750,000				
11	3.75	-$375,000				
12	4.25	$1,000,000				
13	5.00	$500,000				
14	5.75	$1,125,000				
15	6.25	$1,000,000				
16	7.75	$625,000				
17	9.50	$1,125,000				
18						

The present value of each income or expense (indicated as a negative income) is calculated by using Excel's **PV** function, as shown in Figure 9.39.

The arguments on the **PV** function in cell C7 are the periodic interest rate (cell $C$3), the length of time between the present and the time of the income (cell A7), the periodic payment (0 in this example), and the amount of the future income (cell B7).

Note: The **PV** function changes the sign of the result because it assumes that you will invest (an expense) at time zero to get a return (an income) in the future. In this case, we are using the **PV** function just to move a future income to the present, so a minus sign is included on the future income in the formula to undo the sign change that the **PV** function will introduce.

Figure 9.39
Calculating the present
value of each income
or expense using Excel's
PV function.

	C7	▼		f_x	=PV($C$3,A7,0,-B7)		
	A	B	C	D	E	F	
1	Internal Rate of Return with Non-Uniform Time Intervals						
2							
3			IRR:	2.0%	per year (guess value)		
4			Net PV:				
5							
6	Time (years)	Income	PV				
7	0.00	-$375,000	-$375,000				
8	0.50	-$625,000	-$618,842				
9	1.25	-$375,000	-$365,831				
10	2.50	-$750,000	-$713,774				
11	3.75	-$375,000	-$348,161				
12	4.25	$1,000,000	$919,283				
13	5.00	$500,000	$452,865				
14	5.75	$1,125,000	$1,003,926				
15	6.25	$1,000,000	$883,586				
16	7.75	$625,000	$536,079				
17	9.50	$1,125,000	$932,075				
18							

The net present value of the series of incomes and expenses is simply the sum of the PV values in cells C7 through C17. Excel's **SUM** function was used in cell C4 to determine the net present value, as shown in Figure 9.40.

Figure 9.40
Calculating the net
present value (at the
guessed rate of return).

	C4	▼		f_x	=SUM(C7:C17)		
	A	B	C	D	E	F	
1	Internal Rate of Return with Non-Uniform Time Intervals						
2							
3			IRR:	2.0%	per year (guess value)		
4			Net PV:	$2,306,205			
5							
6	Time (years)	Income	PV				
7	0.00	-$375,000	-$375,000				
8	0.50	-$625,000	-$618,842				
9	1.25	-$375,000	-$365,831				
10	2.50	-$750,000	-$713,774				
11	3.75	-$375,000	-$348,161				
12	4.25	$1,000,000	$919,283				
13	5.00	$500,000	$452,865				
14	5.75	$1,125,000	$1,003,926				
15	6.25	$1,000,000	$883,586				
16	7.75	$625,000	$536,079				
17	9.50	$1,125,000	$932,075				
18							

The internal rate of return is the interest rate that causes the present value of the incomes to just offset the present value of the expenses. That is, it is the interest rate that makes the net present value zero. We can ask the Solver to adjust the interest rate (cell C3) until the net present value (cell C4) is 0, as shown in Figure 9.41.

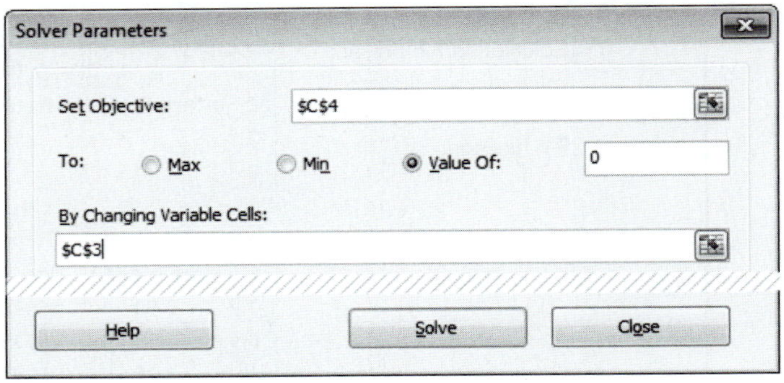

Figure 9.41
The Solver Parameters dialog with necessary inputs shown.

After clicking the **Solve** button, we see the computed internal rate of return in Figure 9.42.

Figure 9.42
The computed internal rate of return.

This cash flow has an internal rate of return of 17.4%, so it is not a bad investment.

9.8 OPTIMIZATION USING THE SOLVER

The Solver can also be used for *optimization* problems. As a simple example, consider the equation

$$y = 10 + 8x - x^2. \tag{9.10}$$

This equation is graphed in Figure 9.43 and has the maximum value of 26 at $x = 4$.

To find the maximum by using the Solver, the right side of the equation is entered in one cell and a guess for x in another, as shown in Figure 9.44.

Figure 9.43
A plot of
$y = 10 + 8x - x^2$.

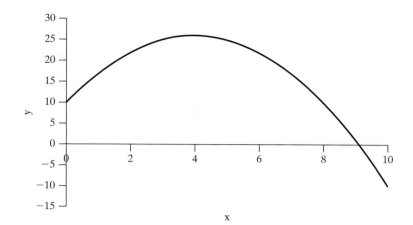

Figure 9.44
Preparing to find the
maximum value.

	B4		f_x	=10+8*B3-B3^2	
	A	B	C	D	E
1	Optimizing Using the Solver				
2					
3	Guess:	2			
4	Result:	22			
5					

Then the Solver is asked to find the maximum value by varying the value of *x* in cell B3, as shown in Figure 9.45.

Figure 9.45
The Solver Parameters
dialog with necessary
inputs.

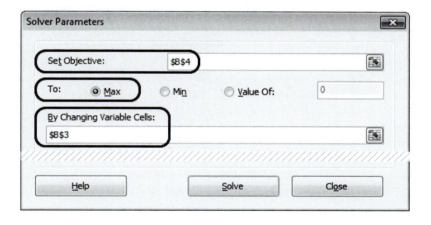

The result is shown in Figure 9.46.

Figure 9.46
The result of finding the
maximum value using
the Solver.

	B3		f_x	3.99999999136139	
	A	B	C	D	E
1	Optimizing Using the Solver				
2					
3	Guess:	4			
4	Result:	26			
5					

The Solver's result was not quite 4, (3.99999999136139). The Solver stopped when the guessed and computed values were within a predefined precision. The desired precision can be changed by using the Options button in the Solver Parameters dialog.

Adding Constraints

If you want the maximum value of y with $x \leq 3$, a *constraint* must be added to the system. This is done by using the constraints box on the Solver Parameters dialog, as shown in Figure 9.47.

Figure 9.47
The Solver Parameters dialog with an added constraint.

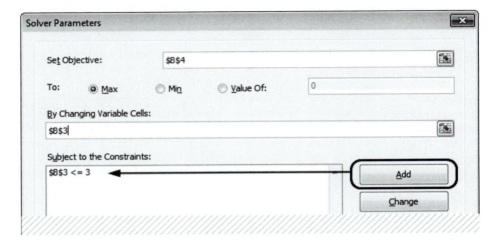

To include this constraint, click on the Add button at the right of the constraints box. This brings up the Add Constraint dialog shown in Figure 9.48.

Figure 9.48
The Add Constraint dialog.

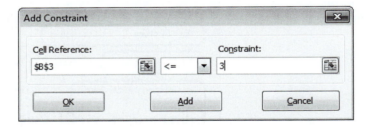

After running the Solver, we see that, with the constraint, the maximum y value is 25 at the x value 3 (Figure 9.49).

Figure 9.49
The optimized result with the constraint.

◢	A	B	C	D	E
1	Optimizing Using the Solver				
2					
3	Guess:	3		x=3 generates the maximum value of f(x) when x is constrained to be <= 3.	
4	Result:	25			
5					

Solving for Multiple Values

The Solver will change values in multiple cells to try to find your requested result. For example, the function $f(x, y) = \sin(x) \cdot \cos(y)$ (shown in Figure 9.50) has a maximum at $x = 1.5708$ (or $\pi/2$) and $y = 0$ in the following region ($-1 \leq x \leq 2$, $-1 \leq y \leq 2$).

Figure 9.50
A surface plot of
$f(x, y) = \sin(x) \cdot \cos(y)$.

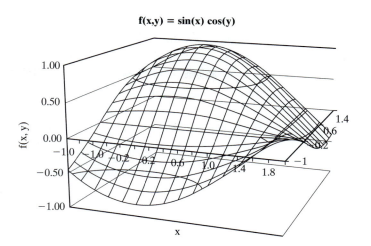

$$f(x,y) = \sin(x)\,\cos(y)$$

To use the Solver to find the values of x and y that maximize $f(x,y)$, enter guesses for x and y and the equation to be solved, as shown in Figure 9.51.

Figure 9.51
Maximizing two variables.

B6			f_x	=SIN(B4)*COS(C4)	
	A	B	C	D	E
1	Optimizing Two Variables Using the Solver				
2					
3		**x**	**y**		
4	**Guesses:**	0.000	0.000		
5					
6	**Result:**	0.000			
7					

The function evaluation was entered into cell B6 as

$$=\text{SIN(B4)}*\text{COS(C4)}$$

We then ask the Solver to find the values of x and y that maximize the value of the function. The Solver Parameters dialog is shown in Figure 9.52.

Figure 9.52
The Solver Parameters
dialog.

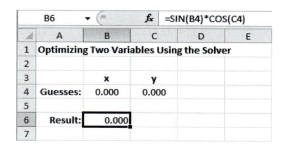

We have asked the Solver to:

- Maximize the value in cell B6 (which contains the function)
- Change cells B4:C4 (containing values for x and y)
- Search in a constrained area

The constraints shown in Figure 9.52 limit the search region to $-1 \leq x \leq 2$, $-1 \leq y \leq 2$; otherwise, there are an infinite number of solutions (the Solver would find only one). The result is shown in Figure 9.53.

Figure 9.53

The Solver's result.

▲	A	B	C	D	E
1	Optimizing Two Variables Using the Solver				
2					
3		x	y		
4	Guesses:	1.571	0.000		
5					
6	Result:	1.000			
7					

The solution that the Solver finds is the true maximum at $x = 1.5708$ (or $\pi/2$) and $y = 0$:

This function also has a local maximum at $x = -1$, $y = 2$ as shown in Figure 9.54.

Figure 9.54

The local maximum at $x = -1$, $y = 2$.

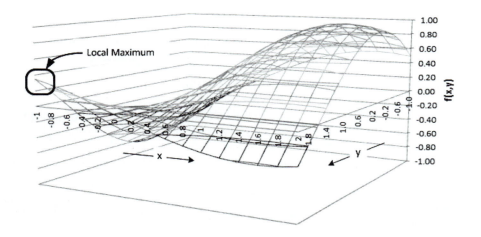

We can try guess values near this local maximum to see whether the Solver will find the *local maximum* rather than the true maximum over the constraint region. Guesses of $x = -0.9$ and $y = 1.8$ caused the Solver to find the solution shown in Figure 9.55.

Figure 9.55

The Solver's solution using initial guesses near the local minimum.

▲	A	B	C	D	E
1	Optimizing Two Variables Using the Solver				
2					
3		x	y		
4	Guesses:	-1.000	2.000		
5					
6	Result:	0.350			
7					

The Solver did find the local maximum! Remember to test your solutions; visualization (such as the three-dimensional graph used here) is a good idea as well.

9.9 NONLINEAR REGRESSION

The idea behind *regression analysis* is to find coefficient values that minimize the total, unsigned (squared) error between a regression model (a function) and the values in a data set. Linear regression models can be solved directly for the coefficients. When the regression model is nonlinear in the coefficients, an iterative minimization approach can be used to find the coefficients. Excel's Solver can be used to solve the minimization problem.

The topic of *nonlinear regression* will be presented with an example: fitting experimental vapor pressure data to Antoine's equation. Data on the vapor pressure of water at various temperatures are easy to find. The data graphed in Figure 9.56 came from a website (unfortunately, no longer available) prepared by an honors chemistry class at the University of Idaho and have been used with the permission of the instructor, Dr. Tom Bitterwolf. The units on vapor pressure are millimeters of mercury, and temperature is in °C.

Figure 9.56

Vapor pressure data.

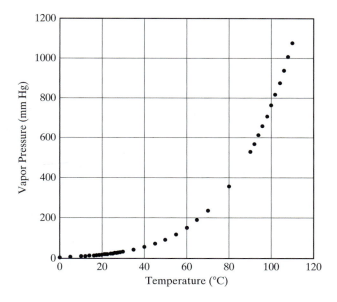

There are several forms of Antoine's equation for vapor pressure in use. One common form is

$$\log(p^*) = A - \frac{B}{C + T}, \tag{9.11}$$

where A, B, and C are the coefficients used to fit the equation to vapor pressure (p^*) data over various temperature (T) ranges. This version of Antoine's equation uses a base-10 logarithm. Antoine's equation is a *dimensional equation*; you must use specific units with Antoine's equation. The units come from the data used in the regression analysis used to find the coefficients. In this example, T has units of °C and p^* has units of mm Hg.

The first step in solving for the Antoine equation coefficients is getting the data and initial guesses for the coefficients into a worksheet. In Figure 9.57, arbitrary values of 1 for each coefficient were used. If you have estimates for the size of each coefficient, you can use them. Better guesses will speed up the iteration, but the arbitrary guesses provide a good test of the Solver's abilities.

Figure 9.57

Setting initial guesses for coefficients A, B, and C.

	A	B	C	D	E	F	G	H
1	Vapor Pressure of Water							
2		Guesses						
3	A:	1						
4	B:	1						
5	C:	1						
6								
7	T (°C)	p* (mm Hg)						
8	0	4.58						
9	5	6.54						
10	10	9.21						
11	12	10.52						
12	14	11.99						
13	16	13.63						
14	17	14.53						

Note: This data set contains 40 data points; only a few will be shown in the figures.

Regression analysis is designed to minimize the *sum of the squared error* (SSE) between the data values, y, and the predicted values calculated from the regression equation, y_p:

$$SSE = \sum_{i=1}^{N} (y_i - y_{p_i})^2. \qquad (9.12)$$

In this example, we want to minimize the SSE between the logarithm of the experimental vapor pressure values and the value of $\log(p^*)$, predicted by the Antoine equation.

The next step is to take the logarithm of the experimental vapor pressures to generate the y values for the regression. This is illustrated in Figure 9.58.

Figure 9.58

Generating the y values for the regression.

C8			f_x	=LOG10(B8)			
	A	B	C	D	E	F	G	H
1	Vapor Pressure of Water							
2		Guesses						
3	A:	1						
4	B:	1						
5	C:	1						
6								
7	T (°C)	p* (mm Hg)	y					
8	0	4.58	0.661					
9	5	6.54	0.816					
10	10	9.21	0.964					
11	12	10.52	1.022					
12	14	11.99	1.079					
13	16	13.63	1.134					
14	17	14.53	1.162					

Next, we calculate $\log(p^*)$, using the guesses in the Antoine equation, to create a column of y_P values (predicted values). These are shown in Figure 9.59.

Figure 9.59
Using the regression equation (the Antoine equation) with the guessed coefficients to compute predicted values.

	D8		f_x	=$B$3-$B$4/($B$5+A8)			
	A	B	C	D	E	F	G	H
1	Vapor Pressure of Water							
2		Guesses						
3	A:	1						
4	B:	1						
5	C:	1						
6								
7	T (°C)	p* (mm Hg)	y	y_P				
8	0	4.58	0.661	0.000				
9	5	6.54	0.816	0.833				
10	10	9.21	0.964	0.909				
11	12	10.52	1.022	0.923				
12	14	11.99	1.079	0.933				
13	16	13.63	1.134	0.941				
14	17	14.53	1.162	0.944				

Then, the error associated with each data point is defined as the difference between the y values from the data and the predicted y value:

$$\text{error}_i = y_i - y_{p_i} \tag{9.13}$$

The calculation of error values is illustrated in Figure 9.60.

Figure 9.60
Error value for each data point.

	E8		f_x	=C8-D8			
	A	B	C	D	E	F	G	H
1	Vapor Pressure of Water							
2		Guesses						
3	A:	1						
4	B:	1						
5	C:	1						
6								
7	T (°C)	p* (mm Hg)	y	y_P	error			
8	0	4.58	0.661	0.000	0.661			
9	5	6.54	0.816	0.833	-0.018			
10	10	9.21	0.964	0.909	0.055			
11	12	10.52	1.022	0.923	0.099			
12	14	11.99	1.079	0.933	0.145			
13	16	13.63	1.134	0.941	0.193			
14	17	14.53	1.162	0.944	0.218			

If we just summed the individual errors, many of the positive errors would cancel out negative errors. To prevent this, the SSE is calculated. This has been done by using Excel's **SUM** function in cell D3 in Figure 9.61.

Finally, we ask the Solver to find the values of A, B, and C that minimize the SSE. The completed Solver Parameters dialog is shown in Figure 9.62.

Figure 9.61
Computing the squared errors and the sum of the squared errors (SSEs).

	D3			f_x	=SUM(F8:F49)			
	A	B	C	D	E	F	G	H
1	**Vapor Pressure of Water**							
2		**Guesses**		**SSE (Sum of Squared Error)**				
3	A:	1		69.347				
4	B:	1						
5	C:	1						
6								
7	T ($^\circ$C)	p* (mm Hg)	y	y_P	error	error^2		
8	0	4.58	0.661	0.000	0.661	0.437		
9	5	6.54	0.816	0.833	-0.018	0.000		
10	10	9.21	0.964	0.909	0.055	0.003		
11	12	10.52	1.022	0.923	0.099	0.010		
12	14	11.99	1.079	0.933	0.145	0.021		
13	16	13.63	1.134	0.941	0.193	0.037		
14	17	14.53	1.162	0.944	0.218	0.047		

Figure 9.62
The completed Solver Parameters dialog.

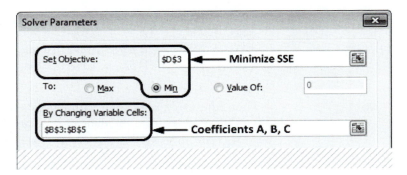

The Solver varies the values of A, B, and C (in cells B3 through B5) to find the minimum value of the SSE (in D3). The result is shown in Figure 9.63.

Figure 9.63
The Solver's results.

	A	B	C	D	E	F	G	H
1	**Vapor Pressure of Water**							
2		Guesses		**SSE (Sum of Squared Error)**				
3	A:	7.994		0.000				
4	B:	1688						
5	C:	230.1		Computed				
6				Coefficients				
7	T ($^\circ$C)	p* (mm Hg)	y	y_P	error	error^2		
8	0	4.58	0.661	0.658	0.003	0.000		
9	5	6.54	0.816	0.814	0.002	0.000		
10	10	9.21	0.964	0.963	0.001	0.000		
11	12	10.52	1.022	1.021	0.001	0.000		
12	14	11.99	1.079	1.078	0.000	0.000		
13	16	13.63	1.134	1.135	0.000	0.000		
14	17	14.53	1.162	1.162	0.000	0.000		

The minimum value of SSE was 0 to the four significant figures displayed in cell D3, and the coefficient values were found to be

$$A = 7.994$$
$$B = 1688 \tag{9.14}$$
$$C = 230.1$$

Reported values for Antoine coefficients for water are the following[2]:

$A = 7.96681$ 0.34% difference

$B = 1669.21$ 1.13% difference

$C = 229.00$ 0.48% difference

However, the reported coefficients are for temperatures between 60 and 150°C; the data set contains values for temperatures between 0 and 120°C. One of the homework problems for this chapter asks you to reanalyze the portion of the data set between 60 and 150°C to allow a better comparison of the Solver's regression results and the published regression coefficients.

Summary of Regression Analysis Using the Solver

The basic steps for using the Solver to perform a regression are the following:

1. Select a regression model (Antoine's equation in this example).
2. Enter the data set and initial guesses for the regression model coefficients (A, B, and C).
3. Calculate the y values for the regression model if needed. (They are frequently part of the data set.)
4. Calculate the y_P values, using the regression model and the x values from the data set.
5. Calculate the error values (error $= y - y_P$).
6. Calculate the squared errors.
7. Calculate the SSE by using the **SUM** function.
8. Ask the Solver to minimize the SSE by changing the values of the coefficients in the regression model.
9. Test your result.

9.9.1 Testing the Regression Result

The last step, testing the result, is always a good idea after any regression analysis. The coefficients that give the "best" fit of the regression model to the data might still do a pretty poor job. The usual ways to test the result are to calculate a *coefficient of determination* and then to plot the original data with the predicted values.

The coefficient of determination, usually called the R^2 value, is computed from the SSE and the *total sum of squares (SSTo)*. SSTo is defined as

$$\text{SSTo} = \sum_{i=1}^{N} (y_i - \bar{y})^2,$$ (9.15)

where

y_i is the y value for the ith data point in the data set, with i ranging from 1 to N,

$\bar{y}$ is the arithmetic average of the y values in the data set and

N is the number of values in the data set.

The R^2 value is calculated as

$$R^2 = 1 - \frac{\text{SSE}}{\text{SSTo}}.$$ (9.16)

For a perfect fit, every predicted value is identical to the data set value, and SSE $= 0$. So, for a perfect fit, $R^2 = 1$.

---

[2]*Lange's Handbook of Chemistry*, 9th ed., Sandusky, OH, Handbook Publishers, Inc., 1959.

In the worksheet, the average of the y values is computed by the **AVERAGE** function, as shown in Figure 9.64.

Figure 9.64
Finding the average of the y values for the SSTo calculation.

	E3		f_x	=AVERAGE(C8:C49)			
	A	B	C	D	E	F	G	H
1	Vapor Pressure of Water						
2		Guesses		SSE	AVG(y)		
3	A:	7.994		0.000	1.959		
4	B:	1688					
5	C:	230.1					
6							
7	T (°C)	p* (mm Hg)	y	yₚ	error	error²	
8	0	4.58	0.661	0.658	0.003	0.000	
9	5	6.54	0.816	0.814	0.002	0.000	
10	10	9.21	0.964	0.963	0.001	0.000	
11	12	10.52	1.022	1.021	0.001	0.000	
12	14	11.99	1.079	1.078	0.000	0.000	
13	16	13.63	1.134	1.135	0.000	0.000	
14	17	14.53	1.162	1.162	0.000	0.000	

Then, the squared differences between the y values and the average y value are computed (column G), and the total sum of squares, SSTo, is computed (cell F3) as shown in Figure 9.65.

Figure 9.65
Calculating SSTo.

	G8		f_x	=(C8-$E$3)^2			
	A	B	C	D	E	F	G	H
1	Vapor Pressure of Water						
2		Guesses		SSE	AVG(y)	SSTo	
3	A:	7.994		0.000	1.959	29.16	
4	B:	1688					
5	C:	230.1					
6							
7	T (°C)	p* (mm Hg)	y	yₚ	error	error²	(y-yₐᵥ)²
8	0	4.58	0.661	0.658	0.003	0.000	1.686
9	5	6.54	0.816	0.814	0.002	0.000	1.308
10	10	9.21	0.964	0.963	0.001	0.000	0.990
11	12	10.52	1.022	1.021	0.001	0.000	0.878
12	14	11.99	1.079	1.078	0.000	0.000	0.775
13	16	13.63	1.134	1.135	0.000	0.000	0.680
14	17	14.53	1.162	1.162	0.000	0.000	0.635

The SSTo value in cell F3 of Figure 9.65 was computed by using the **SUM** function the values in column G.

$$F3: \ = SUM(G8:G49)$$

Finally, the R^2 value is computed from the SSE and SSTo values as shown in Figure 9.66.

The R^2 value of 0.9999986 is very close to one, suggesting a good fit.

Figure 9.66
Calculating R^2.

G3			f_x	=1-D3/F3			
	A	**B**	**C**	**D**	**E**	**F**	**G**	**H**

	A	B	C	D	E	F	G	H
1	Vapor Pressure of Water							
2		Guesses		SSE	AVG(y)	SSTo	R^2	
3	A:	7.994		0.000	1.959	29.16	0.9999986	
4	B:	1688						
5	C:	230.1						
6								
7	T ($^\circ$C)	p* (mm Hg)	y	y_p	error	error^2	$(y-y_{avg})^2$	
8	0	4.58	0.661	0.658	0.003	0.000	1.686	
9	5	6.54	0.816	0.814	0.002	0.000	1.308	
10	10	9.21	0.964	0.963	0.001	0.000	0.990	
11	12	10.52	1.022	1.021	0.001	0.000	0.878	
12	14	11.99	1.079	1.078	0.000	0.000	0.775	
13	16	13.63	1.134	1.135	0.000	0.000	0.680	
14	17	14.53	1.162	1.162	0.000	0.000	0.635	

Using Plots to Validate Your Regression Results

Test the fit by using graphs; there are several that may be used.

1. **Plot the original data and the predicted values on the same graph.** The common practice is to graph data values by using markers and predicted values as a curve. This allows you to see whether the regression results go through the data points. This has been done in Figure 9.67.

Figure 9.67

Vapor pressure data as a function of temperature.

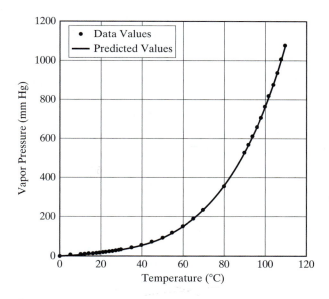

When the original vapor pressures and the vapor pressures calculated by using the Antoine equation with the regressed coefficients are plotted together, the values appear to agree. It looks as if the regression worked.

2. **Plot the original data against the predicted values.** In the plot shown in Figure 9.68, vapor pressure (data) is plotted against vapor pressure (predicted). Temperature has not been plotted.

Figure 9.68
Plot of vapor pressure
(data) against vapor
pressure (predicted).

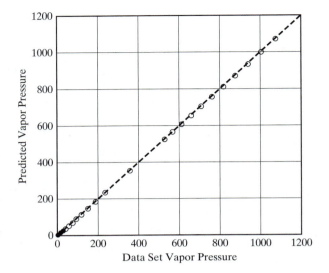

When plotted like this, the data points should ideally follow a diagonal line across the graph. The diagonal has been added so that it is easy to see if the regression follows the diagonal.

- The plot of p_p^* against p^* is shown by the data markers.
- The diagonal line was created by plotting p^* against p^*.

A poor regression fit will show up as scatter around the diagonal line, or patterns. If there were patterns in the scatter (e.g., more scatter at one end of the line, or curvature), you might want to try a different regression model to see whether you could get a better fit.

3. **Deviation plot.** In the field of regression analysis, it is common to refer to the difference between the data value (p^*) and the predicted value (p_p^*) as error, but, in other fields, this difference is referred to as a *deviation*. (The term error is often reserved for the difference between the data value and a known or true value.) When these differences are plotted, the graph typically is called a *deviation plot*, as shown in Figure 9.69.

Figure 9.69
Deviation plot.

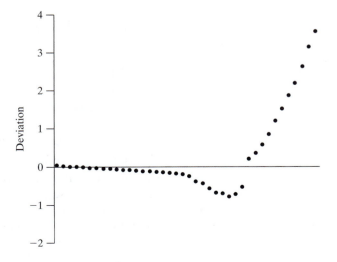

In this line plot of the deviations $[p^* - p^*{}_P]$, there is a clear pattern in the deviation values. The magnitude of the deviations is small, much less than percent of the p^* values, but if a high-precision fit is needed, you might want to look for a better regression model.

THERMODYNAMICS

Bubble Point Temperature I

When a liquid mixture is heated at constant pressure, the temperature at which the first vapor bubble forms (incipient boiling) is called the *bubble point temperature*. At the bubble point, the liquid and vapor are in equilibrium, and the total pressure on the liquid is equal to the sum of the partial pressures of the components in the mixture, so

$$P = \sum_{i=1}^{N_{comp}} p_i \tag{9.17}$$

where

P is the total pressure on the liquid,
p_i is the partial pressure of the ith component in the mixture, and
N_{comp} is the number of components in the mixture.

If the solution can be considered ideal, the partial pressures can be related to the vapor pressures of the components by

$$P = \sum_{i=1}^{N_{comp}} x_i p_i^* \tag{9.18}$$

where

x_i is the mole fraction of component i in the mixture and
p_i^* is the vapor pressure of the ith component in the mixture.

There are good correlating equations, such as Antoine's equation, for vapor pressure as a function of temperature:

$$\log(p_i^*) = A_i - \frac{B_i}{T_{bp} + C_i} \tag{9.19}$$

Solving for the vapor pressure and substituting into the preceding summation yields the fairly involved equation

$$P = \sum_{i=1}^{N_{comp}} x_i \cdot 10 \left[A_i - \frac{B_i}{T_{bp} + C_i} \right] \tag{9.20}$$

where

A_i, B_i, C_i are the Antoine equation coefficients for component i and
T_{bp} is the bubble point temperature of the mixture

This equation can be used to find the bubble point temperature if the composition (x values) and total pressure (P) are known. This is a fairly unpleasant summation to solve by hand, but it is a good candidate for Excel's Solver. We will ask the Solver to keep trying T_{bp} values until the summation is equal to the total pressure.

Consider the following mixture of similar components (i.e., ideal solution is a reasonable assumption)[3]:

Mole	Component Fraction	Antoine Coefficients		
		A	**B**	**C**
1-Propanol	0.3	7.74416	1437.686	199.463
1-Butanol	0.2	7.36366	1305.198	173.427
1-Pentanol	0.5	7.18246	1287.625	161.330

Use Excel's Solver to estimate the bubble point temperature for this mixture at a total pressure equal to 2 atmospheres (1520 mm Hg).

First, the known values are entered into a worksheet, as shown in Figure 9.70.

▲	A	B	C	D	E	F	G	H
1	Bubble Point Temperature I							
2								
3		Pressure:						
4		T$_{bp}$:						
5								
6								
7			Mole Fraction	A	B	C		
8		Propanol:	0.3	7.74416	1437.686	198.463		
9		Butanol:	0.2	7.36366	1305.198	173.427		
10		Pentanol:	0.5	7.18246	1287.625	161.330		
11								

Figure 9.70
Preparing the worksheet with known values.

Next, a guess for the bubble point temperature is entered and, from that, the component vapor pressures are calculated. This is illustrated in Figure 9.71.

G8			f_x	=10^(D8-E8/($C$4+F8))				

▲	A	B	C	D	E	F	G	H
1	Bubble Point Temperature I							
2								
3		Pressure:						
4		T$_{bp}$:		100	°C (guess)			
5								
6								p*
7			Mole Fraction	A	B	C	(mm Hg)	
8		Propanol:	0.3	7.74416	1437.686	198.463	845.7	
9		Butanol:	0.2	7.36366	1305.198	173.427	389.2	
10		Pentanol:	0.5	7.18246	1287.625	161.330	180.0	
11								

Figure 9.71
Using a guess for bubble point temperature and solving for component vapor pressures.

The total pressure can then be calculated using the vapor pressures and the mole fractions in the equation:

$$P = \sum_{i=1}^{N_{comp}} x_i p_i^* \tag{9.21}$$

---

[3]Data are from *Elementary Principles of Chemical Processes*, 3rd ed., by Felder, R. M. and R. W. Rousseau, New York: Wiley, 2000.

This calculation is shown in Figure 9.72.

	C3		f_x	=(C8*G8)+(C9*G9)+(C10*G10)				
	A	B	C	D	E	F	G	H
1	Bubble Point Temperature I							
2								
3		Pressure:	421.5	mm Hg				
4		T_{bp}:	100	°C (guess)				
5								
6							p*	
7			Mole Fraction	A	B	C	(mm Hg)	
8		Propanol:	0.3	7.74416	1437.686	198.463	845.7	
9		Butanol:	0.2	7.36366	1305.198	173.427	389.2	
10		Pentanol:	0.5	7.18246	1287.625	161.330	180.0	
11								

Figure 9.72
Calculating the total pressure (based on the guessed bubble point temperature).

The calculated total pressure is nowhere near the 1520 mm Hg we want, so we will ask Excel's Solver to try other temperature values until the desired total pressure is found. The completed Solver Parameters dialog is shown in Figure 9.73.

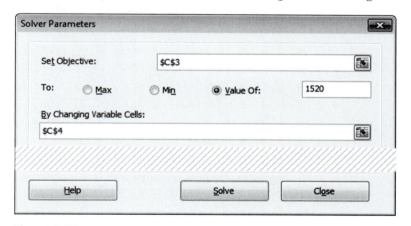

Figure 9.73
The Solver Parameters dialog.

The final result is shown in Figure 9.74. According to our result, this mixture is expected to produce a vapor pressure equal to 1520 mm Hg (2 atm) at 137°C.

	A	B	C	D	E	F	G	H
1	Bubble Point Temperature I							
2								
3		Pressure:	1520.0	mm Hg				
4		T_{bp}:	137.0	°C (calculated)				
5								
6							p*	
7			Mole Fraction	A	B	C	(mm Hg)	
8		Propanol:	0.3	7.74416	1437.686	198.463	2877.1	
9		Butanol:	0.2	7.36366	1305.198	173.427	1444.2	
10		Pentanol:	0.5	7.18246	1287.625	161.330	736.0	
11								

Figure 9.74
The calculated result, a bubble point temperature of 137°C.

9.10 LINEAR PROGRAMMING

An engineering problem encountered fairly commonly is finding the maximum or minimum of some function, subject to one or more constraints. As a simple example that can be graphed to help visualize the problem, consider the *minimization problem*

$$\text{Minimize:} \quad 3x_1 - 5x_2$$

subject to the following constraints that place the solution within the two-dimensional region illustrated in Figure 9.75.

$$x_1 \geq 4$$
$$x_1 \leq 8$$
$$x_2 \geq 2$$
$$x_2 \leq 5$$

Figure 9.75
The solution domain.

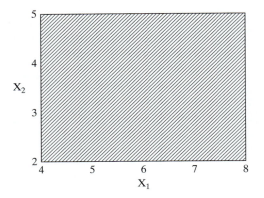

The following additional constraints further limit the *solution domain* to a band across the region, as illustrated in Figure 9.76.

$$x_2 \leq \tfrac{1}{2}x_1 + 1$$
$$x_2 \geq \tfrac{1}{4}x_1 + 1$$

Figure 9.76
Additional constraints reduce the domain of the solution. (The dot shows the initial feasible solution that was chosen used for the iteration process.)

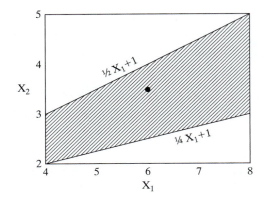

Minimization of a linear function subject to a series of linear constraints is a *linear programming* problem.

Linear programming problems are frequently written in matrix form, as:

Minimize: $[c]^T[x]$ (objective function)

Subject to: $[A][x] \leq [b]$ (general constraints)

$[x] \geq 0$ (nonnegativity constraint)

In this form, the function to be optimized (called the *objective function*) is typically written as a minimization function. Additionally, each constraint must be written as a less-than-or-equal-to statement.

While it is convenient to write the linear programming problem in matrix form, it is not necessary in order to solve the problem in Excel. For this example, we will drop the matrix notation, but we will use it in an application example at the end of the chapter.

The problem can be summarized as follows:

Minimize: $3x_1 - 5x_2$ (objective function)

Subject to: $x_1 \geq 4$ (constraints)

$$x_1 \leq 8$$

$$x_2 \geq 2$$

$$x_2 \leq 5$$

$$\tfrac{1}{2}x_1 - x_2 \geq -1$$

$$\tfrac{1}{4}x_1 - x_2 \leq -1$$

This can be entered into an Excel worksheet as shown in Figure 9.77.

Figure 9.77
Worksheet with initial values (nonoptimized).

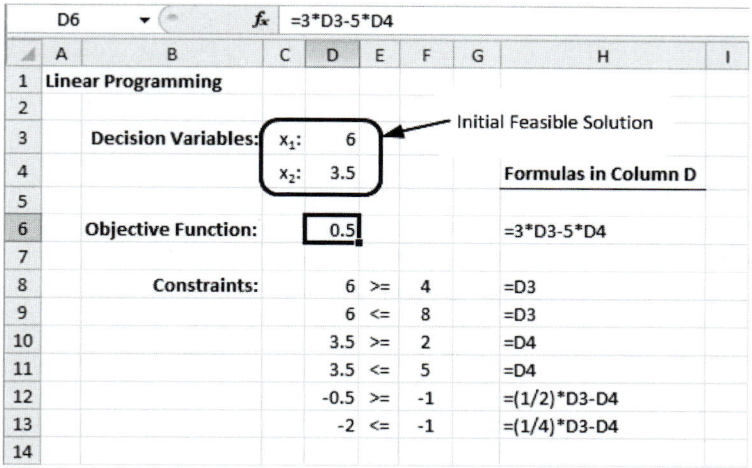

D6 | f_x =3*D3-5*D4

	A	B	C	D	E	F	G	H	I
1	**Linear Programming**								
2									
3		**Decision Variables:**	x_1:	6			Initial Feasible Solution		
4			x_2:	3.5				**Formulas in Column D**	
5									
6		**Objective Function:**		0.5				=3*D3-5*D4	
7									
8		**Constraints:**		6	>=	4		=D3	
9				6	<=	8		=D3	
10				3.5	>=	2		=D4	
11				3.5	<=	5		=D4	
12				-0.5	>=	-1		=(1/2)*D3-D4	
13				-2	<=	-1		=(1/4)*D3-D4	
14									

Initial feasible values for each decision variable (i.e., each x value) must be provided. Providing starting values is standard practice any time an iterative solution process is used. The cells containing the values of each decision variable are used in the objective function `=3*D3-5*D4` in cell D6.

Cells D8:D13 also use the values of the decision variables to evaluate each constraint. The formulas used in cells D8:D13 are listed in cells H8:H13. The operators listed in column E and the values listed in column F are not used in the solution, but are included to make the constraints easier to read. The actual constraints are entered into the Solver as part of the solution process.

Once the problem has been set up with an initial feasible solution (Figure 9.77), then we start the Solver, using Ribbon options **Data/Analysis/Solver** [Excel 2003:

Tools/Solver...] and begin entering parameters in the Solver Parameters dialog, shown in Figure 9.78.

- Set the target cell to the cell containing the evaluated objective function, cell D9.
- Select **Min** to minimize the objective function.
- Indicate the cells containing the decision variable values in the **By Changing Variable Cells:** field (cells D3:D4)

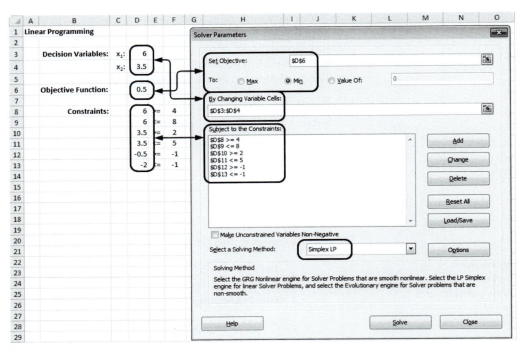

Figure 9.78
The Solver Parameters dialog.

Then, begin to add constraints by clicking the **Add** button on the Solver Parameters dialog.
The process of entering the first constraint is illustrated in Figure 9.79.

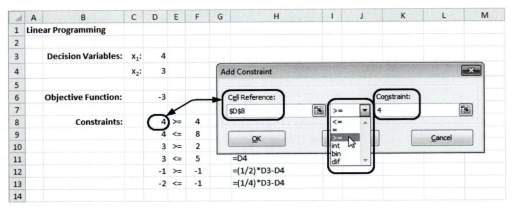

Figure 9.79
Enter the first constraint using the Add Constraint dialog.

The first constraint is entered by indicating the cell reference that contains the formula that uses the decision variables (cells D3:D4) to evaluate the constraint. The first constraint cell reference is cell D8, which contains the formula =D3, that is, it contains the value of x_1. The constraint is $x_1 \geq 4$. The greater-than-or-equal-to operator is selected from the drop-down list of operators on the Add Constraint dialog, and the constraint value, 4, is entered into the **Constraint** field. (Or, you can reference cell F8 in the **Constraint** field.) To enter the first constraint and add another, click the **Add** button on the Add Constraint dialog.

The second constraint is added in the same manner (see Figure 9.80) by:

- Indicating the **Cell Reference** that contains the formula used to evaluate that constraint, cell D9.
- Selecting the less-than-or-equal-to operator from the drop-down list.
- Entering the constraint value, 8, in the **Constraint** field.

Figure 9.80
Entering the second constraint using the Add Constraint dialog.

Each of the six constraints is added in the same way. After entering all of the constraints, click the **OK** button to return to the Solver Parameters dialog. The six constraints are shown in the dialog, as shown in Figure 9.81.

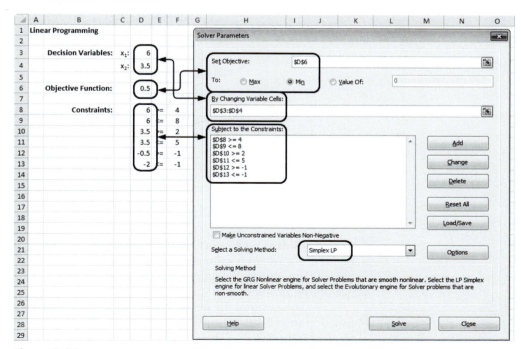

Figure 9.81
The Solver Parameters dialog with constraints added.

Notice that Excel has automatically selected the **Simplex LP** (linear programming) solution method for this problem. With the new version of the Solver (with Excel 2010), Excel automatically selects an appropriate solution method based on the characteristics of the problem. You can override Excel's selection using the **Select a Solving Method** drop-down menu.

You can also enforce nonnegativity constraints by checking the **Make Unconstrained Variables Non-Negative** box on the Solver dialog. In Excel 2010, this is done on the Solver Parameters dialog. In previous versions of the Solver, this was done using the Options dialog (click the **Options** button on the Solver Parameters dialog to access the Options dialog). The Solver Options dialog is shown in Figure 9.82.

Figure 9.82
The Solver Options dialog.

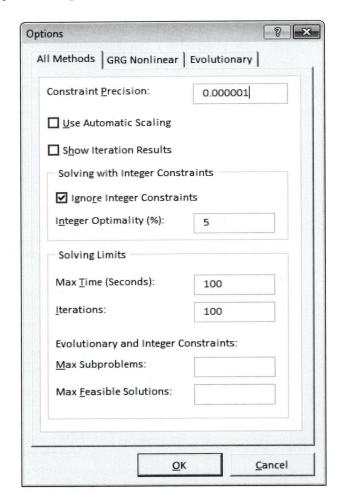

The Solver's Options dialog allows you to adjust the iteration process, if needed. The default values work well for most problems. Click **OK** to return to the Solver Parameters dialog and then click the **Solve** button to solve the linear programming problem.

The Solver varies the decision values in cells D3 and D4 to minimize the objective function in cell D9. The results are displayed in the worksheet, and a Solver results window is displayed as shown in Figure 9.83. Because iterative solutions can fail, the Solver Results window gives you an opportunity to restore the original values if needed.

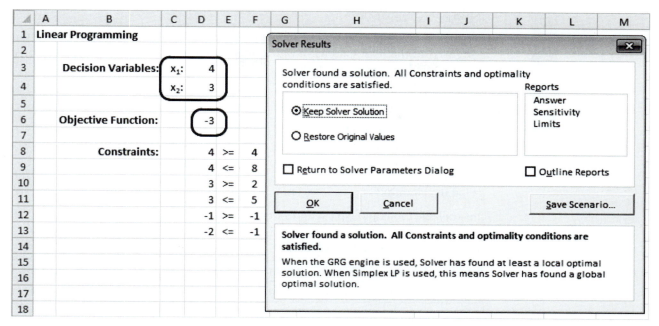

Figure 9.83
The Solver results.

In this example, the optimal solution is at the top-left corner of the solution domain, at $x_1 = 4$ and $x_2 = 3$, as illustrated in Figure 9.84.

Figure 9.84
The optimal solution.

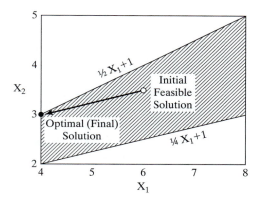

The minimum value of the objective function, shown in cell D6, was found to be -3. Also, in cells D8 through D13, each constraint has been evaluated with the optimal values of x_1 and x_2, and each constraint has been satisfied.

Many linear programming problems are easy to set up and solve by using Excel's Solver; however, there are a couple of limitations. The Solver that comes with Excel

- Is limited to 200 decision variables (we used 2 in this example)
- Has a limited ability to handle strictly integer decision variables.

Enhanced Solvers that can handle much larger problems and integer problems can be purchased from the company that created the Solver, if needed.

LINEAR PROGRAMMING

As a slightly more complex linear programming example, consider the problem of trying to find the least expensive way to meet five dietary minimum requirements. We may state the problem as follows:

Minimize: cost of food
Subject to: At least

- 1.5 mg thiamin (each day)
- 1.7 mg riboflavin
- 20 mg niacin
- 400 μg folic acid
- 60 mg vitamin C

The available foods and typical nutrient values (per serving) are listed in the Table 9.2.

Table 9.2 Nutritional values of food

Food #	Name	Thiamin	Riboflavin	Niacin	Folic Acid	Vitamin C	Cost
		mg	mg	mg	μg	mg	$
1	Rice	0	0	4	11	0	0.25
2	Cabbage	0	0	0	23	124	0.17
3	Spinach	0	0.1	0	123	28	0.35
4	Peanuts	0.4	0	20	20	0	0.95
5	Or. Juice	0.2	0	0	109	64	0.75
6	Tuna	0	0.1	10	0	0	1.45
7	Turkey	0	0	10	0	0	0.85
8	Pork	0.8	0	6	0	0	1.85
9	Bread	0	0.8	0	0	0	0.15
10	Lima Beans	0.4	0	0	78	9	0.35

This problem has enough decision variables to prove that matrix methods are helpful. The decision variables are stored in vector $[x]$, where x_1 through x_{10} represent the number of servings of each food. The $[c]$ vector is the vector of cost per serving for each food in dollars. The constraint right-hand-side vector, $[r]$, represents the daily minimum requirement of each nutrient. Then,

$$[c] = \begin{bmatrix} 0.25 \\ 0.17 \\ 0.35 \\ 0.95 \\ 0.75 \\ 1.45 \\ 0.85 \\ 1.85 \\ 0.15 \\ 0.35 \end{bmatrix} \quad [r] = \begin{bmatrix} 1.5 \\ 1.7 \\ 20 \\ 400 \\ 60 \end{bmatrix}$$

Note: The units on the values in $[r]$ are mg per day (except for folic acid, which is in μg per day).

The $[A]$ matrix represents the constraint-coefficient matrix. Each of these matrices is shown in the following worksheet in Figure 9.85 (initial guesses of 1 for each x have been included in the decision variable cells, O7:O16).

	A	B	C	D	E	F	G	H	I	J	K	L	M	N	O	P
1	Linear Programming Application															
2																
3		Food:	1	2	3	4	5	6	7	8	9	10				
4																
5		$[c]^T$	0.25	0.17	0.35	0.95	0.75	1.45	0.85	1.85	0.15	0.35		$[c]^T [x]$	7.12	
6	Nutrient															
7	thiamin	[A]	0	0	0	0.4	0.2	0	0	0.8	0	0.4		[x]	1.00	
8	riboflavin		0	0	0.1	0	0	0.1	0	0	0.8	0			1.00	
9	niacin		4	0	0	20	0	10	10	6	0	0			1.00	
10	folic acid		11	23	123	20	109	0	0	0	0	78			1.00	
11	vitamin C		0	124	28	0	64	0	0	0	0	9			1.00	
12															1.00	
13		[A] [x]	1.8	>=	[b]	1.5									1.00	
14			1	>=		1.7									1.00	
15			50	>=		20		Initial Guesses for							1.00	
16			364	>=		400		Amount of Each Food							1.00	
17			225	>=		60										
18																

Figure 9.85
The worksheet with initial guesses [x] for the amount of each food.

In matrix form, the problem statement is

Minimize: $[c]^T[x]$ (objective function—minimize total cost)

Subject to: $[A][x] \geq [b]$ (constraints—must meet or exceed daily minimum for each nutrient)

In the worksheet shown in Figure 9.84, the objective function is in cell O5:

O5: =MMULT(C5:L5,O7:O16)

This is an array function, so the function was entered by pressing [Ctrl-Shift-Enter], and cell O5 contains the formula {=MMULT(C5:L5, O7:O16)}, where the braces indicate an array.

The constraints are evaluated in cells C13 through C17. This is accomplished by multiplying matrices $[A]$ and $[x]$. To do this, the five cells containing constraint evaluations (C13:C17) were selected, then the matrix multiplication function was entered as

=MMULT(C7:L11,O7:O16) this is =MMULT([A], [X])

followed by [Ctrl-Shift-Enter] to finish the array function. The array function {=MMULT (C7:L11,O7:O16)} is stored in each cell in the $[A]$ $[x]$ vector (cells C13:C17).

At this point, the worksheet is using the initial guesses for the amount of each food (each x) to evaluate the objective function in cell O5 and the constraints in cells C13:C17. The initial guesses indicate that the total cost is $7.12 each day and that the riboflavin and folic acid requirements are not being met—that is, the 1's in the $[x]$ vector do not represent an initial feasible solution. The Solver might be able to find a solution anyway.

The next step is to use the Solver to attempt to find optimal values for each x that minimize the total cost while satisfying each constraint. To start the Solver, use Ribbon options **Data/Analysis/Solver** [Excel 2003: Tools/Solver...]. The Solver Parameters dialog, shown in Figure 9.86, will open.

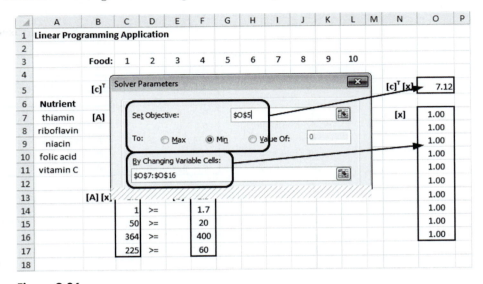

Figure 9.86
The Solver Parameters dialog.

The Set Target Cell: field points to the cell containing the objective function, cell O5. Minimization is requested, and the decision variables in cells O7:O16 are indicated in the **By Changing Variable Cells:** field.

Next, the constraints must be entered. All five constraints are greater-than-or-equal-to constraints, so all five constraints can be entered at once. To do this, click the **Add** button on the Solver Parameters dialog to open the Add Constraint dialog shown in Figure 9.87.

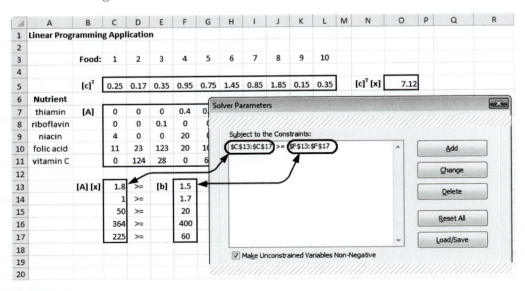

Figure 9.87
Adding the constraints to the Solver Parameters dialog.

Note: In Figures 9.86 and 9.87, only portions of the Solver Parameters dialog are shown. The complete dialog is illustrated in Figure 9.88.

All five cells that contain constraint evaluations, cells C13:C17, are indicated in the Cell Reference field, and all five constraint values (in cells F13:F17) are indicated in the Constraint field. The greater-than-or-equal-to operator is selected from the drop-down operator list. The **OK** button is used to return to the Solver Parameters dialog, shown in Figure 9.88.

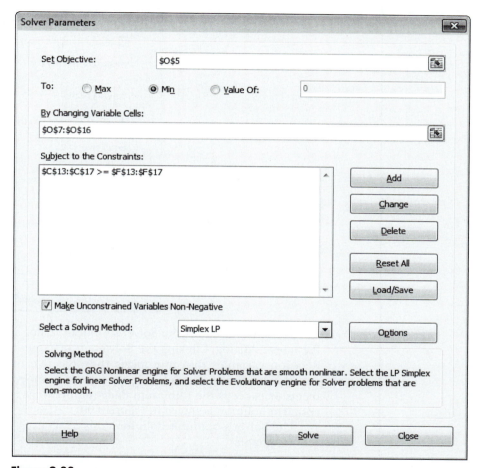

Figure 9.88
The Solver Parameters dialog, with constraints.

With versions of the Solver prior to Excel 2010, the Options dialog must be used to:

- Select a linear model solving method
- Restrict to nonnegative decision variables

However, both of these selections are made on the Solver Parameters dialog in Excel 2010.

Note: The non-negativity constraint prevents the Solver from attempting to minimize cost by using negative servings of food.

You can use the Solver's Options dialog to adjust the **Constraint Precision** if you want a faster or more precise solution. The Option dialog is shown in Figure 9.89.

Figure 9.89
The Solver Options dialog.

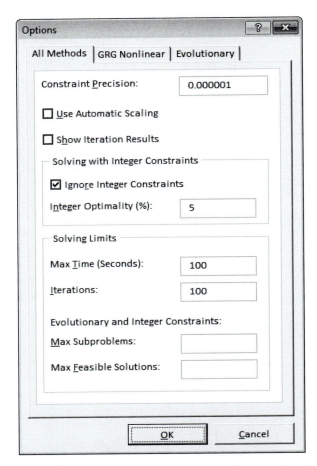

The **OK** button on the Solver's Options dialog is used to return to the Solver Parameters dialog. Clicking the **Solve** button causes the Solver to seek a new set of decision variables that minimizes the objective function while satisfying each constraint—that is, the Solver looks for the number of servings of each food that meets or exceeds the minimum daily requirement for each nutrient at minimum cost. The solution is shown in Figure 9.90.

The Solver's optimum solution costs $2.68 each day and includes the foods in following table:

Food #	Name	Servings
1	Rice	0
2	Cabbage	0
3	Spinach	1.35
4	Peanuts	1
5	Or. Juice	0
6	Tuna	0
7	Turkey	0
8	Pork	0
9	Bread	1.96
10	Lima Beans	2.75

▲	A	B	C	D	E	F	G	H	I	J	K	L	M	N	O	P
1	**Linear Programming Application**															
2																
3		Food:	1	2	3	4	5	6	7	8	9	10				
4																
5		$[c]^T$	0.25	0.17	0.35	0.95	0.75	1.45	0.85	1.85	0.15	0.35		$[c]^T[x]$	2.68	
6	**Nutrient**															
7	thiamin	**[A]**	0	0	0	0.4	0.2	0	0	0.8	0	0.4		**[x]**	0.00	
8	riboflavin		0	0	0.1	0	0	0.1	0	0	0.8	0			0.00	
9	niacin		4	0	0	20	0	10	10	6	0	0			1.35	
10	folic acid		11	23	123	20	109	0	0	0	0	78			1.00	
11	vitamin C		0	124	28	0	64	0	0	0	0	9			0.00	
12															0.00	
13		**[A] [x]**	1.5	>=	**[b]**	1.5									0.00	
14			1.7	>=		1.7									0.00	
15			20	>=		20									1.96	
16			400	>=		400									2.75	
17			62.4	>=		60										
18																

Figure 9.90
The optimized solution.

Notice that the Solver did not return integer numbers of servings—the version of the Solver that comes with Excel does not solve integer programming models—but noninteger portions of servings are certainly possible, so this noninteger solution is valid for this problem.

Only four of the foods are actually required to meet all of the minimum daily requirements. These four foods provide the following nutrients:

Nutrient	Amount From Food	Daily Requirement
	mg	mg
Thiamin	1.5	1.5
Riboflavin	1.7	1.7
Niacin	20	20
Folic Acid	0.400	0.400
Vitamin C	62.4	60

All of the constraints have been met minimally except for vitamin C, which is slightly in excess of the daily minimum.

This is the minimum cost solution that provides the required amounts of each nutrient; it might be acceptable, unless you really don't like one of the foods. If that is the case, you might want to consider optimizing on a different objective function—one related to taste rather than cost. That is the subject of one of the homework problems at the end of this chapter.

KEY TERMS

Circular reference
Coefficient of
  determination (R^2)
Computed value
Constraint
Deviation
Deviation plot
Dimensional equation
Direct substitution
Divergence
Diverging
Error
Goal Seek

Guess and check
Guessed value
In-cell iteration
Initial guess
Iteration
Iterative methods
Iterative solution
  technique
Linear programming
Local maximum
Nonlinear regression
Objective function
Optimization

Precision setting
Regression analysis
Root
Solution
Solution domain
Solver
Standard form
Sum of the squared error
  (SSE)
Target cell
Tolerance
Total sum of squares
  (SSTo)

SUMMARY

Equation Standard Forms for Iterative Solutions

Form 1 Rearrange the equation to get all of the terms on one side, set equal to 0:

$$x^3 - 17x + 12 = 0. \tag{9.22}$$

Form 2 Get one instance of the desired variable on the left side of the equation.

$$x = \frac{x^3 + 12}{17}. \tag{9.23}$$

Form 2 is often called "guess and check" form, with the calculated variable x_C by itself and the guessed variable x_G included in the right side of the equation:

$$x_G = \frac{x_G^3 + 12}{17}. \tag{9.24}$$

Using a Plot to Search for Roots of Equations

Use form 1 and replace the 0 with $f(x)$.
Plot $f(x)$ for a chosen range of x values.
The x values that cause $f(x)$ to be equal to 0 are the roots of your equation. These are easily seen as the x values where the $f(x)$ curve crosses the x axis.

Iterative Methods

"Guess and Check" Iteration

You can use either form 1 or form 2 for "guess and check" iteration. The use of form 1 is summarized here.

Put your guess value in one cell and the computed value (using the guess) in another cell. Keep trying guess values until the computed value is very nearly 0 (because form 1 was used).

Benefits: It is simple.
Drawbacks: It is tedious.

Direct Substitution

Put the equation into "guess and check" form (form 2) with a calculated value of the desired quantity on the left and a function of the desired quantity (the guess value) on the right:

$$x_C = f(x_G).$$

With the direct-substitution method, the calculated value at the end of one iteration becomes the guess value for the next iteration.

> **Benefits:** This is easy to do in a worksheet.
>
> **Drawbacks:** This method often fails to converge on a root and cannot find all roots.

In-Cell Iteration

Use "guess and check" form (form 2) for the equation. The $f(x_G)$ goes in the cell, and the x_C is the displayed result of the calculation.

In-cell iteration is not enabled by default. Enable iteration by selecting Tools/ Options/and Iterations from the calculation panel.

> **Benefits:** It is straightforward.
>
> **Drawbacks:** It gives no indication of whether it stopped iterating because it found a root or because it reached the maximum number of iterations. (Check your results!) And, it always uses 0 as the initial guess and therefore cannot find multiple roots.

Goal Seek

Use form 1 for the equation, replacing the 0 with $f(x)$. Put your guess value in one cell and compute $f(x)$ in another cell.

To open the Goal Seek dialog, use Ribbon options

Data/Data Tools/What-If Analysis/Goal Seek…

[Excel 2003: Tools/Goal Seek…]

- The Set cell: field should contain the address of the cell in which $f(x)$ is computed.
- The To value: field should be set to 0.
- The By changing cell: field should contain the address of the cell in which the guess value is stored.

> **Benefits:** It is straightforward and powerful.
>
> **Drawbacks:** Goal Seek is simple to use and works well. It has fewer options than the Solver, but you don't need all of the Solver's options just to find roots of equations.

Solver

Needs a constant on one side of the equation. (Basically, form 1 is used, but it can have any constant on the right, not just 0.)

This capability might not be installed or activated by default. Installation requires the Excel Installation CD. To activate it, select Tools/Add-Ins and check the Solver Add-In box in the list of available add-ins. (If Solver Add-In is not listed, it needs to be installed.)

Benefits: The Solver is easy to use, powerful, and fast, can handle constraints, and can maximize and minimize.

Optimization by Using the Solver: Select **Max** or **Min** on the **To:** line of the Solver Parameters dialog box.

Include constraints by clicking the **Add** button on the **Subject to Constraints** line. Constraints must be constructed by using the Add Constraint dialog box.

To optimize on multiple values, include multiple cell references in the **By Changing Variable Cells** field.

Nonlinear Regression Using The Solver

Nonlinear regression is an optimization problem that seeks to minimize the SSE between dependent values predicted by a regression model (y_P) and those from the data set (y):

$$\text{SSE} = \left[\sum_{i=1}^{N} (y_i - y_{p_i})^2 \right].$$

(9.25)

Here,

y_p is the value predicted by the regression model and

y is the data set value.

The model coefficients go in the **By Changing Variable Cells** field, and the cell containing the SSE is the target.

Linear Programming Using the Solver

In linear programming problems, a linear objective function is optimized subject to one or more linear constraints. Because all of the equations involved are linear, the problem can be written concisely in matrix form.

Minimize: $[c]^T[x]$ (objective function)

Subject to: $[A][x] \leq [b]$ (constraints)

Additionally, the decision variables are often constrained to be nonnegative values.

Linear programming problems are readily solved by using Excel's Solver with the following parameters:

Field	Contents
Target Cell	Objective Function, $[c]^T[x]$
Equal To	Min
By Changing Cells	Decision Variables, $[x]$ (Initial values for each decision variable must be specified in the worksheet)
Subject to the Constraints	Constraints, $[A][x] \leq [b]$ (Nonnegativity can be imposed on the Solver Parameters dialog in Excel 2010, or using the Options dialog in prior versions)

PROBLEMS

9.1 Iterative Solutions I

The three solutions to the following equation, $x = 1$, $x = 3$, and $x = 7$, are evident by inspection:

$$(x - 1)(x - 3)(x - 7) = 0.$$

(9.26)

The same equation can be written as a polynomial:

$$x^3 - 11x^2 + 31x - 21 = 0. \tag{9.27}$$

a. Use Excel's Solver to find the three solutions to this polynomial.

b. Use the Solver to find the solutions to the following polynomial:

$$x^3 - 10x^2 + 35x - 27 = 0. \tag{9.28}$$

9.2 Iterative Solutions II

The graph of a J_1 Bessel function between $x = 0$ and 10 shown in Figure 9.91 indicates that there are three roots (solutions) to the equation.

$$J_1(x) = 0. \tag{9.29}$$

The solutions are at $x = 0$ and near $x = 4$ and $x = 7$. Use the Solver with Excel's **BESSELJ** function to find more precise values for the roots near 4 and 7.

Figure 9.91
Plot of $J_1(x)$ between $x = 0$ and $x = 10$.

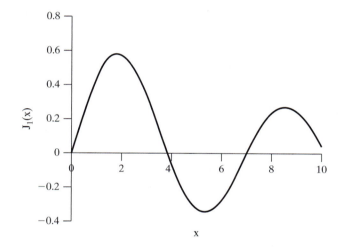

9.3 Nonideal Gas Equation

Determine the molar volume $\hat{V}$ of ammonia at 300°C and 1200 kPa by using

a. the ideal gas equation

$$P\hat{V} = RT; \tag{9.30}$$

b. the Soave–Redlich–Kwong (SRK) equation[4] and Excel's Solver:

$$P = \frac{RT}{\hat{V} - b} - \frac{\alpha a}{\hat{V}(\hat{V} + b)} \tag{9.31}$$

Here,

α, a, and b are coefficients for the SRK equation, defined as

$$a = 0.42747\frac{(RT_C)^2}{P_C},$$

$$b = 0.08664\frac{RT_C}{P_C},$$

$$\alpha = \left[1 + m(1 - \sqrt{T_r})\right]^2,$$

---

[4]From *Elementary Principles of Chemical Processes*, 3rd ed., Felder, R. M. and R. W. Rousseau, New York: Wiley, 2000.

$$m = 0.48508 + 1.55171\,\omega - 0.1561\omega^2,$$

$$T_r = \frac{T}{T_c},$$
(9.32)

P is the absolute pressure (1200 kPa),
P_C is the critical pressure of the gas (11,280 kPa for ammonia),
T is the absolute temperature ($300°C + 273 = 573$ K),
T_C is the critical temperature of the gas (405.5 K for ammonia),
ω is the Pitzer acentric factor for the gas (0.250 for ammonia),
$\hat{V}$ is the molar volume (liters per mole)
R is the ideal gas constant [9.314 (liter Pa)/(mole K)].
(9.33)

9.4 Internal Rate of Return

Use Excel's **NPV** function with an arbitrary (guess) interest rate to calculate the net present value of the cash flow illustrated in Figure 9.92. Then, use Excel's Solver to find the interest rate that makes the net present value 0.

Figure 9.92
Cash flow timeline.

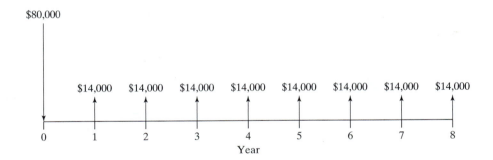

9.5 Bubble Point Temperature II

Compute the bubble point temperature for a mixture containing 10 mole % 1-propanol, 70 mole % 1-butanol, and 20 mole % 1-pentanol at a pressure equal to 1000 mm Hg.

9.6 Bubble Point Temperature III

Use Excel's Solver to estimate the bubble point temperature of a mixture of four hydrocarbons at 3 atmospheres pressure (2280 mm Hg). Composition and other required data for the mixture are tabulated as follows[5]:

Component	Mole Fraction	Antoine Coefficients		
		A	B	C
n-Octane	0.1	9.98174	1351.756	209.100
i-Octane	0.5	9.88814	1319.529	211.625
n-Nonane	0.3	9.93764	1430.459	201.808
n-Decane	0.1	9.95707	1503.568	194.738

[5] Data are from *Elementary Principles of Chemical Processes*, 3rd ed., Felder, R. M. and R. W. Rousseau, New York: Wiley, 2000.

9.7 Nonlinear Regression I

Rework the nonlinear regression example used in this chapter (fitting the Antoine equation to vapor pressure data) using only data in the temperature range $60 \leq T \leq 150°C$. (The complete data set is listed next, only a portion is needed for this problem.)

a. Use the Solver to find the best-fit values for the coefficients in Antoine's equation (A, B, and C).

b. Compare your results with the following published values[6]:

T (°C)	P* (mm Hg)	T(°C)	P* (mm Hg)	Coefficient	Published Value
0	4.58	40	55.3	A	7.96681
5	9.54	45	71.9	B	1668.21
10	9.21	50	92.5	C	228.000
12	10.52	55	118		
14	11.99	60	149.4		
16	13.63	65	187.5		
17	14.53	70	233.7		
18	15.48	80	355.1		
19	19.48	90	525.8		
20	17.54	92	567		
21	18.65	94	610.9		
22	19.83	96	657.6		
23	21.07	98	707.3		
24	22.38	100	760		
25	23.76	102	815.9		
26	25.21	104	875.1		
27	29.74	106	937.9		
28	28.35	108	1004.4		
29	30.04	110	1074.6		
30	31.82	150	3570.4		
35	42.2	200	11659.2		

9.8 Nonlinear Regression II

A decaying sine wave is a common waveform. It occurs in oscillating circuits, controlled process responses, and wave dynamics. A decaying sine wave may have the form

$$y_p = e^{-Ax} \sin(Bx)$$

where A and B are model coefficients, typically obtained by regression analysis. Now

• Perform a nonlinear regression on the data shown next to compute the values of coefficients A and B.

---

[6]*Lange's Handbook of Chemistry*, 9th ed., Sandusky, OH, Handbook Publishers, Inc., 1959.

- Plot the regression equation with the original data values to verify visually that the regression model fits the data.
- Calculate the R^2 value for the regression result.

The plotted data values shown in Figure 9.93 are listed in the table that follows, but they are also available at the website, http://www.chbe.montana.edu/excel
The values on the website can be imported into your Excel worksheet.

Figure 9.93

Decaying sine wave.

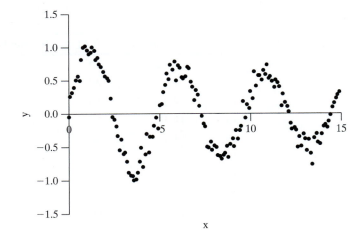

Data values for Problem 8

x	y	x	y	x	y	x	y	x	y
0.0	−0.05	3.0	−0.60	9.0	0.73	9.0	−0.38	12.0	0.10
0.1	0.26	3.1	−0.58	9.1	0.69	9.1	−0.49	12.1	0.02
0.2	0.32	3.2	−0.72	9.2	0.55	9.2	−0.26	12.2	−0.15
0.3	0.39	3.3	−0.88	9.3	0.53	9.3	−0.38	12.3	−0.24
0.4	0.50	3.4	−0.92	9.4	0.56	9.4	−0.26	12.4	−0.21
0.5	0.56	3.5	−0.93	9.5	0.70	9.5	−0.19	12.5	−0.25
0.6	0.49	3.6	−1.00	9.6	0.68	9.6	0.18	12.6	−0.50
0.7	0.81	3.7	−0.99	9.7	0.47	9.7	−0.07	12.7	−0.35
0.8	1.00	3.8	−0.88	9.8	0.40	9.8	0.13	12.8	−0.49
0.9	1.02	3.9	−0.72	9.9	0.20	9.9	0.07	12.9	−0.31
1.0	0.96	4.0	−0.51	7.0	0.35	10.0	0.33	13.0	−0.38
1.1	0.90	4.1	−0.80	7.1	0.28	10.1	0.23	13.1	−0.55
1.2	0.92	4.2	−0.62	7.2	0.11	10.2	0.63	13.2	−0.39
1.3	1.00	4.3	−0.31	7.3	−0.04	10.3	0.41	13.3	−0.59
1.4	0.95	4.4	−0.58	7.4	−0.15	10.4	0.57	13.4	−0.76
1.5	0.80	4.5	−0.35	7.5	−0.19	10.5	0.57	13.5	−0.37
1.6	0.84	4.6	−0.35	7.6	−0.50	10.6	0.50	13.6	−0.31
1.7	0.74	4.7	−0.18	7.7	−0.51	10.7	0.65	13.7	−0.43
1.8	0.70	4.8	−0.06	7.8	−0.43	10.8	0.59	13.8	−0.45

1.9	0.63	4.9	−0.22	7.9	−0.54	10.9	0.73	13.9	−0.31
2.0	0.56	5.0	0.12	8.0	−0.48	11.0	0.53	14.0	−0.19
2.1	0.54	5.1	0.15	8.1	−0.50	11.1	0.56	14.1	−0.17
2.2	0.49	5.2	0.32	8.2	−0.63	11.2	0.48	14.2	−0.20
2.3	0.23	5.3	0.43	8.3	−0.64	11.3	0.39	14.3	−0.12
2.4	−0.05	5.4	0.60	8.4	−0.68	11.4	0.50	14.4	−0.02
2.5	−0.09	5.5	0.51	8.5	−0.63	11.5	0.46	14.5	0.09
2.6	−0.19	5.6	0.73	8.6	−0.59	11.6	0.40	14.6	0.16
2.7	−0.34	5.7	0.66	8.7	−0.49	11.7	0.11	14.7	0.23
2.8	−0.54	5.8	0.78	8.8	−0.66	11.8	0.29	14.8	0.28
2.9	−0.39	5.9	0.49	8.9	−0.46	11.9	0.17	14.9	0.32

9.9 Linear Programming

Solve the following linear programming problem:

Maximize: $12 x_1 + 21 x_2 + 7 x_3 + 15 x_4$
Subject to:

$$4 x_1 + 3 x_2 + 2 x_3 + 5 x_4 \leq 400$$
$$2 x_1 + 1.5 x_2 + 3.3 x_3 + 2 x_4 \leq 150$$
$$x_1 \geq 45$$
$$x_2 \leq 100$$
$$x_1, x_2, x_3, x_4 \geq 0 \text{ (nonnegative)}$$

Rework the linear programming application example presented in this chapter with a new objective function. This time, the objective will be to minimize a "distaste factor" rather than cost. The distaste factor can have a value between 1 and 10, where

1—I don't mind this food at all; serve it up!
4—I can eat some of it, but tire of it quickly.
8—I would prefer not to eat this food.
10—I'll gag if I have to eat this food.

The distaste factors are obviously specific to each individual, but a set of distaste factors has been assigned to each food in the following table:

Food #		1	2	3	4	5	6	7	8	9	10
	Name	**Rice**	**Cabbage**	**Spinach**	**Peanuts**	**Or. Juice**	**Tuna**	**Turkey**	**Pork**	**Bread**	**Lima Beans**
Nutrient											
Thiamin	mg	0	0	0	0.4	0.2	0	0	0.8	0	0.4
Riboflavin	mg	0	0	0.1	0	0	0.1	0	0	0.8	0
Niacin	mg	4	0	0	20	0	10	10	6	0	0
Folic Acid	μg	11	23	123	20	109	0	0	0	0	78
Vitamin C	mg	0	124	28	0	64	0	0	0	0	9
Distaste Factor		1	4	3	1	1	2	1	3	1	7
Cost	$	0.25	0.17	0.35	0.95	0.75	1.45	0.85	1.85	0.15	0.35

a. Solve the linear programming problem to minimize distaste factor while still satisfying each of the nutritional constraints listed in the application example. How many servings of each type of food are needed, and what is the overall distaste level?

b. Create your own set of distaste factors for these foods and repeat step (a).

10

Sharing Excel Information with Other Programs

Objectives

After reading this chapter, you will know

- How to use the Windows Clipboard to move information between programs, such as:
  - Excel and Word
    - Using Excel data to create a table in Microsoft Word
    - Using data in a Word document in an Excel worksheet
    - Placing an Excel graph into a Word document
  - Excel and Mathcad
    - Using Excel data to create an array in Mathcad
    - Using Mathcad arrays to fill cell ranges in Excel
  - Excel and Matlab
    - Using Excel data to create an array in Matlab
    - Using Matlab arrays to fill cell ranges in Excel
  - Using Web data in Excel
- How to use an Excel component in Mathcad to share data with an Excel worksheet
- The difference between embedded and linked objects and the uses of each
- How to use external data sources with Excel

10.1 INTRODUCTION TO SHARING EXCEL INFORMATION WITH OTHER PROGRAMS

The original method for exchanging information between programs was to save one program's results in a file (usually a text file or database file) and have the second program read the file. This is still done, and it is very effective for many situations (e.g., very large data sets), but there are now other options. A common data transfer mechanism uses the *Windows Clipboard* as a temporary storage location and *copy-and-paste operations* to move information to and from the Clipboard. Information can be copied to the Clipboard from a variety of programs, and then pasted from the Clipboard back into the same program, or into a different program.

In many cases, a connection between the data source program and the destination program can be maintained; this is called a *link*. When a data link is used, each time the data in the source program is updated, the data in the linked program can be automatically updated as well.

The process used to move data within and between programs is called *Object Linking and Embedding (OLE)*. In computer terms, an *object* is a collection of data and the procedures and methods required to handle the data. Objects can contain other objects. For example, an Excel workbook is an object, but so is a graph within the workbook, and the legend displayed on the graph, and each data point presented on the graph.

Excel can also be configured to periodically read data from a source file (called an *external data source*) to automatically update the worksheet. Updating worksheets using external data sources is not a commonly used feature in Excel, but it is extremely important to some engineers who utilize the feature to monitor the performance of their equipment.

This chapter focuses on the various mechanisms for moving data to and from Excel.

10.2 USING COPY AND PASTE TO MOVE INFORMATION

The simplest way to move information is to copy it from the source program and paste it into the destination program using the Windows Clipboard as a temporary storage location. This is illustrated in Figure 10.1.

Figure 10.1
Using the Clipboard as a temporary storage location.

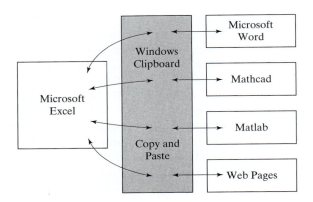

The copy-and-paste operations that use the Windows clipboard are available with nearly all Windows programs. Most current software packages try to assist the data transfer process by analyzing the contents of the copied content and acting in "appropriate" ways:

* Numbers are pasted into cells as values in Excel
* Excel cell contents become tables in Microsoft Word
* Images are displayed as a picture

Much of this is seamless and intuitive to the user, but some information on how the process works can help if you want to control how two programs share information.

Using the simplest copy-and-paste operation, you can transfer information (as *objects*, in computer terms) from one program to another, and the objects are not *linked*. The objects may or might not be *embedded* (it depends on the type of object).

- *Embedded Objects:* When an object is *embedded*, it means that it is stored with the file in which it is embedded (the destination) and that the connection, or link, to the program that created the object is severed. For example, an Excel graph object can be embedded in a Word document. After the Excel graph is embedded in the document file:
  - The graph will be stored as part of the Word document.
  - No connection to the Excel workbook used to create the graph is maintained.
  - The graph object is still recognizable as an Excel graph.

  Because the object is still understood to be an Excel graph, double-clicking on the graph object in Word will start Excel so that the object can be edited by using Excel. But changing a value in the original Excel workbook will not cause the embedded graph to be updated, because the link between the original Excel workbook and the embedded graph was severed.
- *Linked Objects:* When an object is *linked*, a connection between the object and the source file used to create the object is maintained. Any changes to the source file will cause the object to be changed in the destination file as well. For example, if an Excel graph object is linked to a Word document, any changes to the data values in the Excel workbook will cause the graph to be updated in both the Excel file and the Word file.

Information (technically, objects) that are moved from one program to another may or may not be linked; you can often choose whether or not to create the link as part of the paste process.

10.2.1 Moving Information between Excel and Word without Linking

The simpler copy-and-paste operations move information between programs without creating a data link. We will focus on nonlinked data transfer operations first. Using data links is presented in the next section.

Copying Cell Contents from Excel to Word

A range of cells in Excel becomes a table in Word. This is one of the easiest ways to create a table in Word.

To copy the range of cells in Excel to the clipboard,

1. Select the cells containing the information to be copied;
2. Copy the cell contents to the Windows clipboard, using Excel Ribbon options **Home/Clipboard/Copy** [Excel 2003: Edit/Copy].

To paste the information into Word as a table,

1. Position the cursor in Word at the location where the table should be created;
2. Paste the contents of the Windows clipboard using Word Ribbon options **Home/Clipboard/Paste** [Word 2003: Edit/Paste].

Example: Pasting a data set from Excel into a Word table.

The temperature vs. time data shown in Figure 10.2 can be pasted into Word as a table.

Figure 10.2

Temperature and time data from a regression analysis.

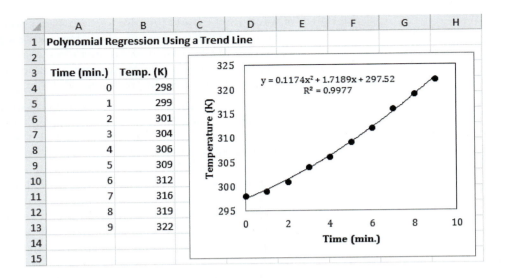

Step 1. **Select the Cells Containing the Information to be Copied** In Excel, select the cells to be copied (shown in Figure 10.3).

Figure 10.3

Copying the Excel data to the Windows clipboard.

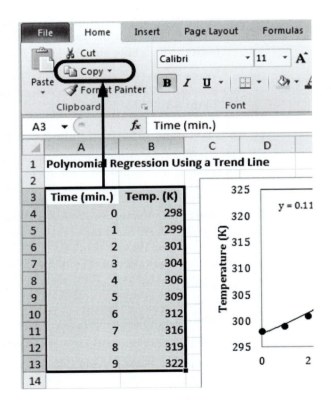

Step 2. **Copy the Selected Cells to the Windows Clipboard** Click the **Copy** button on the Ribbon's **Home** tab and Clipboard group. The **Copy** button is indicated in Figure 10.3.

Step 3. Position the Cursor in Word at the Location Desired for the Table In Word, move the cursor to the location where you want the table to be inserted. This is illustrated in Figure 10.4.

Step 4. Paste the Information into Word In Word, use Ribbon options **Home/ Clipboard/Paste** [Word 2003: Edit/Paste] to paste the data from the Windows clipboard into Word. The result is shown in Figure 10.5.

Figure 10.4
Positioning the cursor in Word where the table should be placed.

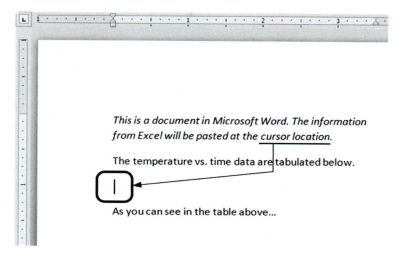

Figure 10.5
The Excel data pasted into Word as a table.

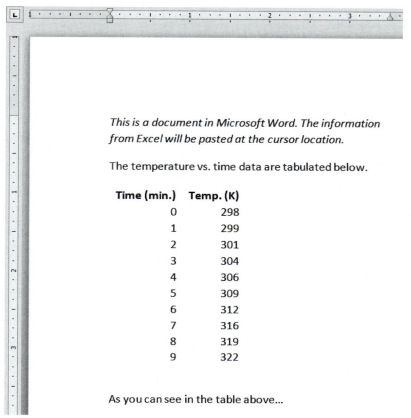

At this point, the information from Excel has been copied to, and is stored in, Word. There is no link between the table in Word and the data in Excel.

- If the Excel cells contained labels or values, the cells of the Word table will display text (words or numbers).
- If the Excel cells contained formulas, the cells of the Word table will display the results of the formulas as text.
- To format the table, use Word's formatting options.

Copying a Word Table into Excel

The process of copying the contents of a Word table to Excel is basically the same. Copy the table in Word to the clipboard:

1. Select the cells of the table containing the information to be copied.
2. Copy the cell contents to the Windows clipboard using Word Ribbon options **Home/Clipboard/Copy** [Word 2003: Edit/Copy].

Paste the information into Excel cells:

1. Click on the cell that will contain the top-left corner of the pasted range of cells.
2. Paste the contents of the Windows clipboard using Excel Ribbon options **Home/Clipboard/Paste** [Excel 2003: Edit/Paste].

The information from the clipboard has been copied to, and is stored in, Excel. There is no link between the table in Word and the data in Excel.

- If the cells of the Word table contained text, the Excel cells will contain labels.
- If the Word table contained numbers, the Excel cells will contain values.

Copying an Excel Graph to Word

Creating a graph in Excel and then pasting it into Word is a very easy way to get a graph into a report.

To copy a graph from Excel to Word:

1. In Excel, select the graph. (Click on the whitespace around the plot area to select the entire graph.) A heavy border and drag handles (small clusters of dots in Excel 2007) will appear on the corners and edges of the graph, as illustrated in Figure 10.6.

   Note: If you click on one of the objects that make up a graph (axis, series, titles, etc.) that object will be selected rather than the entire graph. By clicking in the whitespace around the plot area, you will select the entire graph, not just one element of the graph.

2. Copy it to the Windows clipboard using Excel Ribbon options **Home/Clipboard/Copy** [Excel 2003: Edit/Copy].
3. Position the Word cursor at the location where the table should be pasted.
4. Paste the contents of the Windows clipboard using Word Ribbon options **Home/Clipboard/Paste** [Word 2003: Edit/Paste].

The result is shown in Figure 10.7.

An Excel graph object is pasted into Word as an embedded object. No link is maintained with the original file, but double-clicking on the graph in Word allows you to edit the graph just as you would in Excel. In Figure 10.8, the trendline object has been selected for editing. The graph is in Word, but the pop-up menu is an Excel menu. Clicking the **Format Trendline . . .** option on the pop-up menu would open Excel's Format Trendline dialog.

Figure 10.6
The selected graph in Excel.

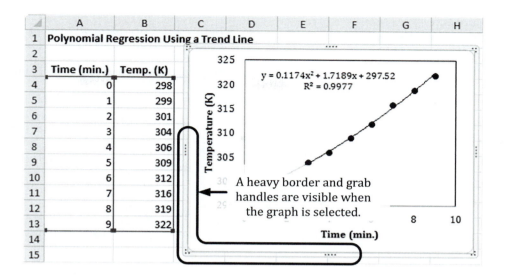

Figure 10.7
The graph pasted into Word.

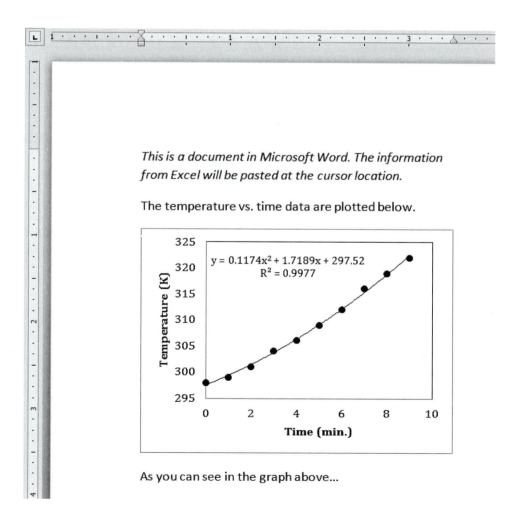

Figure 10.8
Editing an embedded Excel
graph from within Word.

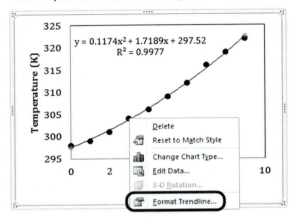

*This is a document in Microsoft Word. The information
from Excel will be pasted at the cursor location.*

The temperature vs. time data are plotted below.

As you can see in the graph above...

Any changes you make to the embedded graph in Word are not reflected in the graph in Excel. An embedded graph is a copy of the original, and there is no link between the embedded graph in Word and the original in Excel.

Note: It is possible to paste an Excel graph into Word as a linked object; just use **Paste Special** and select **Paste Link** as part of the paste process. With a linked Excel graph, the changes made to the Excel worksheet are automatically reflected in the linked graph in the Word document.

10.2.2 Moving Information between Excel and Mathcad

Excel and Mathcad® are both commonly used math packages, and each has particular strengths and weaknesses. It is very handy to be able to move information back and forth to make use of the best features of each program. Most objects copy to and from Mathcad as graphic images—just graphical "snapshots" of the Mathcad objects. An exception is an array of values, which can be moved to and from Excel.

Copying Excel Values to Mathcad
Mathcad can handle arrays of numbers, but it is easier to enter columns of values in Excel. Once you have a column of values in Excel, you can copy and paste them to Mathcad. The procedure is as follows:

1. Enter the values in Excel.
2. Select the range of cells containing the values.

   Note: If there are any labels or other nonnumbers in the column of values, Mathcad will interpret the information to be pasted as an image and place a picture of the Excel values on the Mathcad worksheet.

3. Copy the values to the clipboard from Excel (Ribbon options **Home/Clipboard/Copy**).
4. Enter the variable name that will be used with the matrix in Mathcad and then the *define as equal to* := (entered by pressing the [:(colon)] key).

5. Click on the placeholder that Mathcad creates on the right side of the *define as equal to* symbol.
6. Paste the values into the placeholder, using **Edit/Paste** from Mathcad.

EXAMPLE 10.1

COPYING THE TIME VALUES TO MATHCAD

Steps 1 and 2. Enter the Values in Excel, and Select the Range of Cells to Be Copied (see Figure 10.9)

Figure 10.9
Selecting the values to be copied.

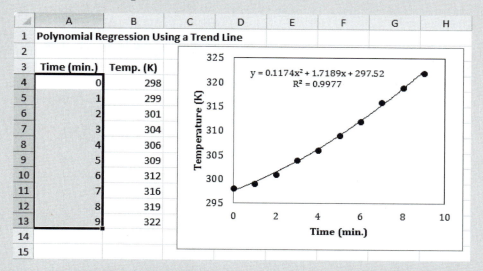

Step 3. Copy the Cell Values to the Clipboard   In Excel, copy the selected values to the Windows clipboard using Excel Ribbon options **Home/Clipboard/Copy**.

Steps 4 and 5. Create the Matrix Variable Definition in Mathcad that will Receive the Values   The variable name *time* has been used in the Mathcad worksheet shown in Figure 10.10.

Figure 10.10
A Mathcad worksheet with variable *time* declared.

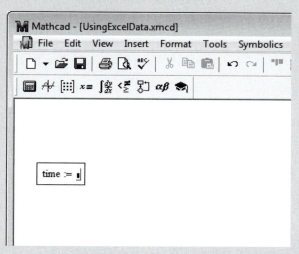

(continued)

Step 6. Paste the Contents of the Clipboard into Mathcad The empty place-holder after the time variable definition will receive the values from the clipboard. Click on the placeholder to select it (as shown in Figure 10.10), then paste the values from the clipboard using Mathcad's menu options **Edit/Paste**. The Excel values on the clipboard will be placed in the placeholder as a Mathcad array. The result is shown in Figure 10.11.

Figure 10.11
The time values from Excel are now the *time* array in Mathcad.

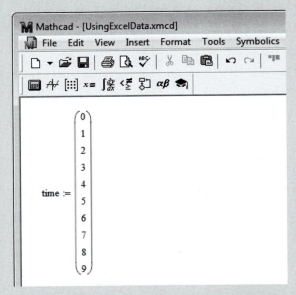

There is no link from the *time* matrix in Mathcad back to the Excel worksheet, and this is not an embedded Excel object on the Mathcad worksheet. This is simply a one-way transfer of information from Excel to Mathcad.

Copying an Array from Mathcad to Excel

To move an array of values the other direction, from Mathcad to Excel, the process is very similar, but you must be certain to select only the values in the Mathcad array, not the entire definition (the entire *equation region*, in Mathcad's terminology). When the values in the array have been selected, the selection bars surround the array, but not the "define as equal to" symbol or the variable name. The correct selection is illustrated in Figure 10.12.

Figure 10.12
Select only the values in Mathcad before copying them to the clipboard.

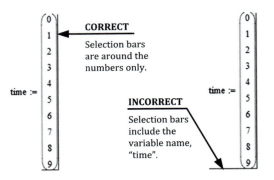

Once the array has been selected, use **Edit/Copy** from Mathcad to copy the values to the clipboard. Then use Excel Ribbon options **Home/Clipboard/Paste** to paste the values into Excel cells.

10.2.3 Moving Information between Excel and Matlab

Copying Excel Values to Matlab

Matlab is another popular math package, and it is very simple to move an array of values between Excel and Matlab using the Clipboard.

To move the values from a range of cells in Excel into a Matlab array, follow these steps:

1. Select the cells in Excel.
2. Copy the cell values to the Clipboard using Ribbon options **Home/Clipboard/ Copy**.
3. Define a new variable in Matlab
   a. Click on the New Variable button in the Matlab Workspace, as indicated in Figure 10.13. The new variable will initially be called "unnamed."
   b. Assign the variable a name. The name "Temp" was used in Figure 10.13.

Figure 10.13
Creating a new variable in Matlab.

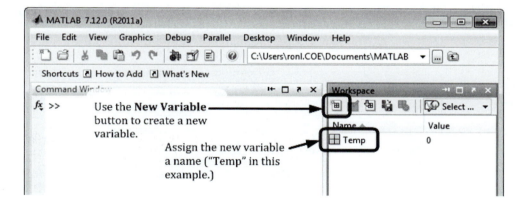

4. Double-click on the new variable to open the Variable editor, shown in Figure 10.14.

Figure 10.14
Matlab's Variable Editor.

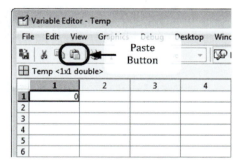

5. Click the **Paste** button (or **File/Paste**) in Matlab's Variable Editor. The Excel values on the Clipboard will be pasted into the array in the Variable Editor. The result is shown in Figure 10.15.

Figure 10.15
The Excel values pasted into Matlab's Variable Editor.

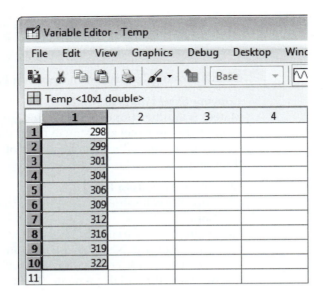

Once the variables have been pasted into Matlab, they can be used in any Matlab calculation. In Figure 10.16, Matlab variables Time and Temp are used in Matlab's polynomial regression function, **polyfit**.

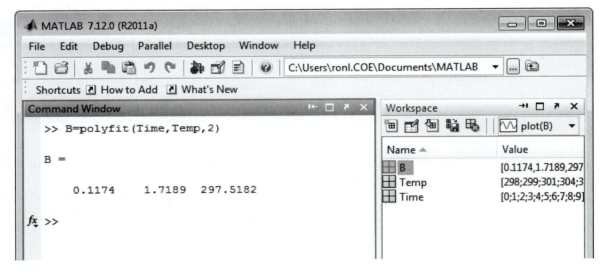

Figure 10.16
Using values pasted into Matlab for polynomial regression. The results of the regression (the coefficients) were assigned to a new Matlab variable, *B*.

Copying an Array from Matlab to Excel

The process to move values from a Matlab variable into Excel is similar:

1. Double-click on the variable name in the Matlab Workspace to open the variable in the Variable Editor, shown in Figure 10.17.
2. Select the values and copy them to the Clipboard using Matlab options **File/Copy**.

Figure 10.17
The *B* variable in the Matlab Variable Editor.

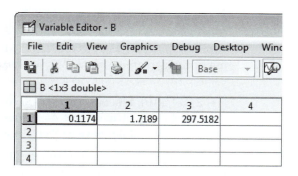

3. In Excel, select the top-left cell of the range that should receive the value from the Clipboard, then paste the values using **Home/Clipboard/Paste**. The result is shown in Figure 10.18.

Figure 10.18
The polynomial coefficients from Matlab pasted into Excel cells C4:E4.

	A	B	C	D	E	F
1	Polynomial Regression Using a Trend Line					
2						
3	Time (min.)	Temp. (K)	Coefficients			
4	0	298	0.117424242	1.718939	297.5182	
5	1	299				
6	2	301				
7	3	304				
8	4	306				
9	5	309				
10	6	312				
11	7	316				
12	8	319				
13	9	322				
14						

10.2.4 Moving Information between Excel and the World Wide Web

Copying Tabular Data from HTML Pages to Excel

Tabular data from HTML pages on the World Wide Web can be copied and pasted into Excel. To do so, select the cell values you want to copy (you must copy entire rows in HTML files) and copy them to the clipboard by selecting **Edit/Copy** from a browser. Then, in Excel, use one of the following approaches:

- Use Excel Ribbon options **Home/Clipboard/Paste** to paste the data into Excel.
- Use Excel Ribbon options **Home/Clipboard/Paste (menu)** to open the Paste menu. Then select **Paste Special . . .** to open the Paste Special dialog. Then select **Text** or **Unicode Text** from the **As** list.

An HTML page has quite a few formatting tags interspersed among the data values to tell the browser how to arrange the data on the web page. Sometimes Excel will interpret the HTML formatting as nonnumeric and try to paste the copied information as labels. To get around this, use the Paste Special dialog to paste as **Text** or **Unicode Text**. When pasting as text, the HTML tags are ignored which makes the numbers stand out.

Creating HTML Pages from Excel

This section is included here for the sake of completeness, but you do not use the copy-and-paste process to create HTML pages from Excel. Instead, use the following options to create an HTML version of your Excel worksheet. The menu options used to save a worksheet as a web page vary:

- Excel 2010: **File tab/Save As/Save As Type: Web Page**
- Excel 2007: **Office/Save As/Other Formats/Save As Type: Web Page**
- Excel 2003: **File/Save as Web Page . . .**

You will have an option during the save process to indicate whether you want the entire workbook saved as a Web page (with tabs for the various sheets), just the current worksheet, or the currently selected cell range.

When you choose to save as a web page, the Save As dialog will display a **Page Title**: field (probably empty) and a **Change Title . . .** button. It is considered good form to give all web pages titles; they are used by the browser and search engines.

Then click **Save** or **Publish** to create the web page. The **Publish** button opens the Publish as Web Page dialog to give you additional options on what to include in the web page. Typically, after clicking **Save** on the Save As dialog, an HTML document is prepared, along with a folder with the same file name. The folder is used to house any images that the page needs, such as the graph on the polynomial regression page.

10.3 EMBEDDED AND LINKED OBJECTS

Whenever information (actually, an object) is shared between two programs, there is a *source* program in which the object was created and a *destination* that receives either the object itself or a *link* to the object in the other program. When the destination program receives and stores a copy of the object, the object is said to be embedded in the destination program. When the object continues to be stored with the source program and the destination program only knows how to find the object, the object is linked.

- If you double-click on a **linked object** in a destination file, the original source file in which the object was created will be opened by the program used to create the object. Any changes you make to the linked object will be made to the source file and displayed in both the source and destination file.

  Note: The object to be linked must be in a saved file before you create the link. If OLE doesn't know where the object is stored, it cannot update the link.

- If you double-click on an **embedded object** in a destination file, the program used to create the object will be started—but only the copy of the object that was embedded will be displayed, not the original. Any changes you make to the embedded object will be made to the copy of the object stored with the destination file.

The copy-and-paste operations described above occasionally create linked objects (e.g., embedding an Excel graph in Word), but more often, you use the *paste special* operation to embed and link objects.

10.3.1 Embedded and Linked Objects in Word and Excel

Using the Paste Special dialog to Paste an Excel Graph into Word

If the polynomial regression graph was copied to the clipboard from Excel, we could use Word's Paste Special dialog to paste the graph into Word. You access the

Paste Special dialog using Word Ribbon options **Home/Clipboard/Paste (menu)** to open the Paste Options menu, and then select the **Paste Special** . . . option. This is illustrated in Figure 10.19.

[In Word 2003, use Edit/Paste Special . . .]

Figure 10.19
Opening Word's Paste
Special dialog.

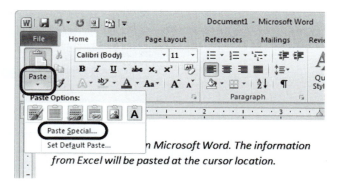

Notice in Figure 10.19 that the Paste Options: menu in Excel 2010, shown again in Figure 10.20, has six icons that provide quick access to various paste options.

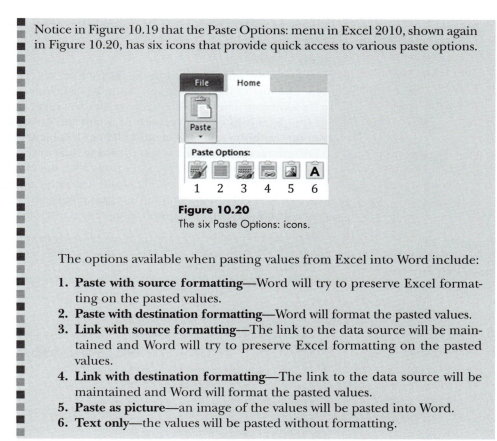

Figure 10.20
The six Paste Options: icons.

The options available when pasting values from Excel into Word include:

1. **Paste with source formatting**—Word will try to preserve Excel formatting on the pasted values.
2. **Paste with destination formatting**—Word will format the pasted values.
3. **Link with source formatting**—The link to the data source will be maintained and Word will try to preserve Excel formatting on the pasted values.
4. **Link with destination formatting**—The link to the data source will be maintained and Word will format the pasted values.
5. **Paste as picture**—an image of the values will be pasted into Word.
6. **Text only**—the values will be pasted without formatting.

The Paste Special dialog (shown in Figure 10.21) provides several options on how to paste the contents of the clipboard (the Excel graph) into Word.

Word's default for Excel chart objects causes the Excel graph to be pasted (embedded) as a **Microsoft Office Graphic Object**. This default was used earlier

Figure 10.21
Word's Paste Special
dialog when the clipboard
holds an Excel graph.

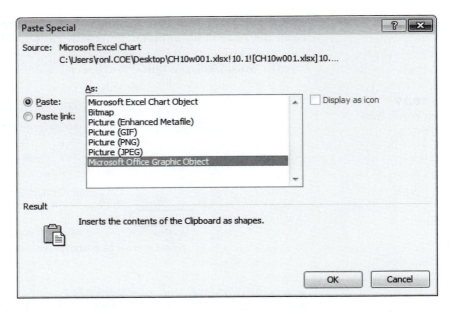

when we pasted the Excel graph into Word with the simple copy-and-paste operation (see Figure 10.7).

Most of the other options in the **As:** list paste the graph as an image. This would place a picture of the Excel graph in the Word document, but it would no longer be an Excel object. There is also the paste as **Microsoft Office Excel Chart Object**; this was the default in earlier versions of Microsoft Word. It still works, but editing is smoother with the newer **Microsoft Office Graphic Object** option.

Alternatively, you could select the **Paste link:** option, indicated in Figure 10.22.

This option is used to create a link from the Word document to the graph stored in the Excel worksheet. The usual reason for doing this is to allow the graph displayed in the Word document (but stored in the Excel worksheet) to be updated automatically, any time the graph in the Excel worksheet is changed.

Figure 10.22
Using the **Paste link:**
option when the clipboard
holds an Excel graph.

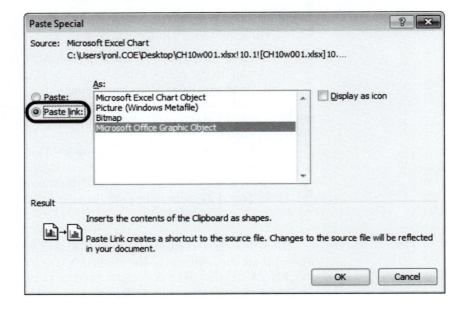

If the graph was pasted as a linked object, double-clicking on the graph in Word would cause the Excel worksheet used to create the graph to be opened.

Note: Before you create a link from Word to an Excel graph, be sure to save the Excel file. In order to create the link the file name must be known.

Using Paste Special . . . to Paste Excel Cell Values as a Table in Word

If you copy a range of Excel cell values to the clipboard, such as the table of temperature and time values used earlier, you have a number of paste options available under the Paste Special dialog box, as shown in Figure 10.23.

Figure 10.23
The Paste Special dialog box when the clipboard contains an Excel cell range.

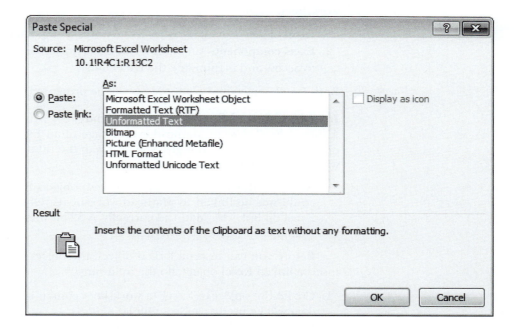

The possible ways to paste the cell values listed in the **As:** list are nearly the same for both the **Paste:** and **Paste link:** options. The difference is:

- The **Paste link:** option will cause the table displayed in Word to be updated any time the cell values in Excel are changed.
- The **Paste:** option will place the current clipboard contents into the Word document and not maintain a link to the Excel worksheet.

The various ways the clipboard contents can be pasted into the Word document are listed in the **As:** box shown in Figure 10.23.

- **HTML Format** is the default and allows the new table to be formatted by using Word's table-formatting options, which are quite extensive.
- **Formatted Text (RTF)** is also a good choice for most tables.
- **Unformatted Text** is a good choice when you want the values from the Clipboard, but you want to format the values in Word.
- **Unformatted Unicode Text** is just like the **Unformatted Text** option, but uses the newer Unicode character codes that provide support for international alphabets.

- **Picture** and **Bitmap** are graphic image formats and will not allow formatting of the table contents.
- The **Word Hyperlink** method (**Paste link**: only) creates a table of hyperlinks back to the original Excel data. The hyperlinks display the contents of the cells when the table is created and do not update if the Excel values are changed.

10.3.2 Using Excel with Mathcad

It is possible to insert an embedded or linked Excel object in Mathcad. This effectively makes it possible to use the capabilities of Excel with Mathcad.

There are two ways to insert a piece of an Excel worksheet into a Mathcad worksheet:

1. **Excel object**—an image of an Excel worksheet; no Mathcad access to Excel data.
2. **Excel component**—an image of an Excel worksheet, plus the ability for Mathcad to access and manipulate the Excel data.

Both Excel objects and components will be presented in the following sections.

Inserting a Linked Excel Object in Mathcad

A linked *Excel object* in Mathcad is an image of an Excel worksheet. Linked Excel objects in Mathcad do not give you access to the numbers on the Excel worksheet, which limits their usefulness.

Note: As this text was being prepared, objects from 64-bit versions of Excel could not be linked to Mathcad worksheets. Objects from 32-bit versions of Excel do link with Mathcad correctly. A 32-bit version of Excel was used for the examples in this section of the text.

Before you can insert a linked object, the object must exist in a saved file. To insert a linked Excel object, do the following:

1. Create the object in Excel (a worksheet containing the temperature and time data will be used as an example).
2. From Mathcad, use **Insert/Object** to open the Insert Object dialog, shown in Figure 10.24.

Figure 10.24
Mathcad's Insert Object dialog.

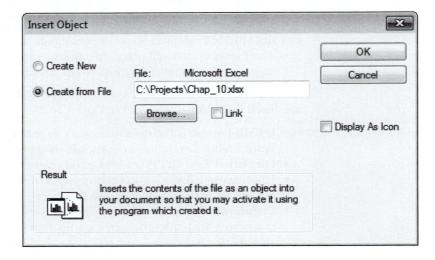

3. Select **Create from File**, check the **Link** box, and **Browse** for the file containing the Excel object (the Excel workbook).

> **Note:** If you create a new Excel workbook object, it will be embedded, not linked.

4. Click **OK** to close the Insert Object dialog and insert the linked Excel worksheet into the Mathcad worksheet. The result is illustrated in Figure 10.25.

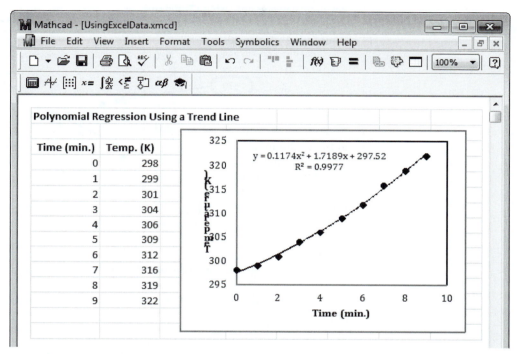

Figure 10.25
The linked Excel worksheet displayed in Mathcad.

Because it is a linked object, changes to the Excel worksheet will be shown on the Mathcad worksheet also. For example, if the temperature at 2 minutes is (arbitrarily) changed in the Excel worksheet to 310 K, that will be reflected in the image on the Mathcad worksheet, as shown in Figure 10.26.

An Excel object in Mathcad is not extremely useful because Mathcad cannot access the numbers in the Excel worksheet. But there is something called an *Excel component* in Mathcad that does provide access to the Excel data.

Using Excel Components in Mathcad

In this example, we will use Excel data in Mathcad to perform a spline fit, a feature missing from Excel. We access Excel data using Mathcad's Excel component.

> **Note:** As this text was being prepared, the Excel component in Mathcad 15 would not successfully connect to 64-bit Excel. Mathcad's Excel component worked fine with 32-bit versions of Excel. A 32-bit version of Excel was used for the examples in this section of the text.

> **Note:** Mathcad's Excel component embeds a copy of an Excel file into a Mathcad worksheet; no link to the original Excel file is maintained.

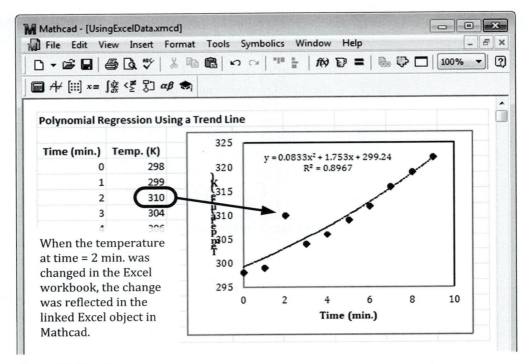

When the temperature at time = 2 min. was changed in the Excel workbook, the change was reflected in the linked Excel object in Mathcad.

Figure 10.26
The linked Excel worksheet automatically reflects changes made to the original.

The steps involved in getting Excel data into Mathcad using an Excel component are

1. Enter the data into an Excel worksheet and save it.
2. Start Mathcad and open Mathcad's Component Wizard (shown in Figure 10.27) using menu options **Insert/Component** . . .

Figure 10.27
The Mathcad Component Wizard.

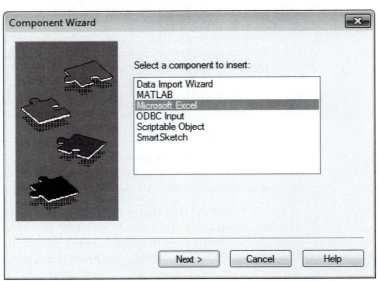

3. Select **Microsoft Excel** and click **Next >** to open the Excel Setup Wizard, shown in Figure 10.28.

Figure 10.28
The Excel Setup Wizard
(page 1).

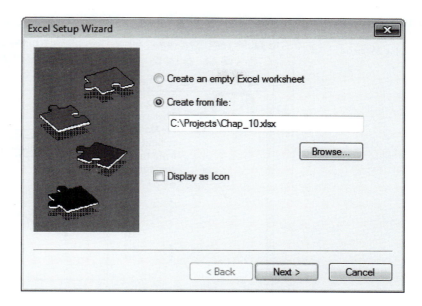

In the Excel Setup Wizard dialog, select **Create from File** and **Browse** to the Excel file containing the data you wish to use in Mathcad. Then click **Next >** to move to the second page of the Excel Setup Wizard, shown in Figure 10.29.

Figure 10.29
The Excel Setup
Wizard (page 2).

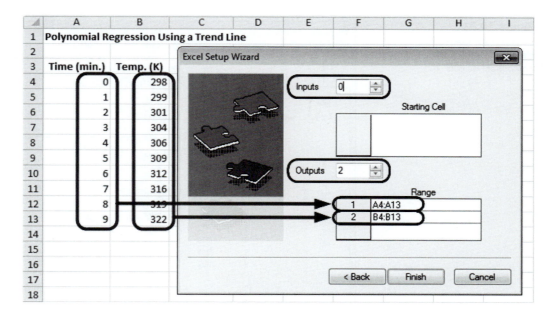

4. In this step, indicate the number of inputs and outputs that the component should manage. The inputs and outputs are from Excel's perspective: to move time and temperature values from Excel into Mathcad will require two (Excel) outputs. Two **Outputs** have been specified in the dialog shown in Figure 10.29. The Excel cell ranges containing the values for each output must be specified. Our time values are in cells **A4:A13**, and the temperature values are in cells **B4:B13**.

Click **Finish** to insert the Excel component on the Mathcad worksheet, as shown in Figure 10.30.

Figure 10.30
The Excel component inserted on the Mathcad worksheet.

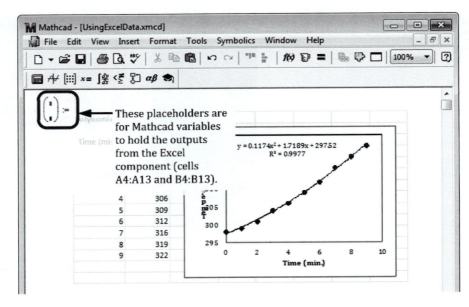

5. Assign the component outputs to Mathcad variables.

The placeholders indicated in Figure 10.30 are for the Mathcad variable names that will hold the outputs from the component; time and temperature. We'll use variable names *time* and *Temp*, as shown in Figure 10.31.

Figure 10.31
Mathcad variables declared to receive the component outputs.

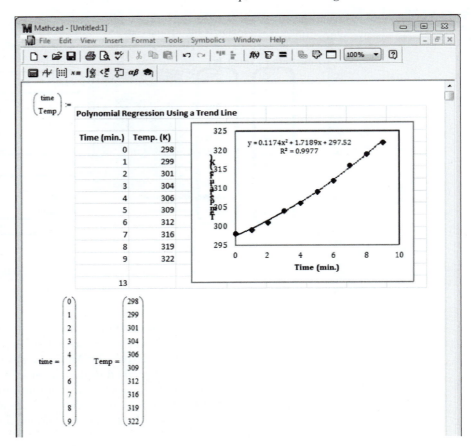

At this point, the Excel data is in Mathcad, so the spline fit problem can be completed in Mathcad, as shown in Figure 10.32.

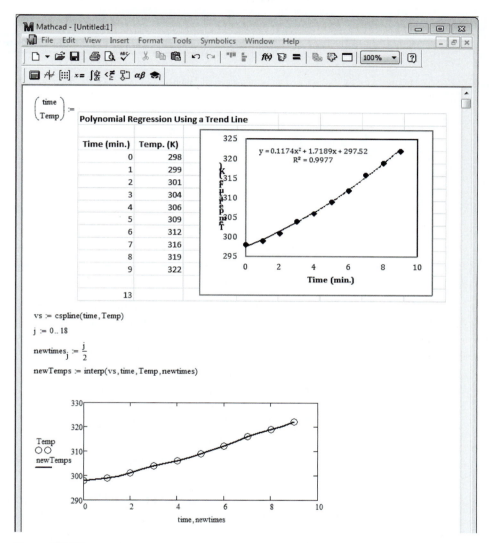

Figure 10.32
The Mathcad spline fit solution using Excel data values.

10.4 EXTERNAL DATA SOURCES

An *external data source* is a data file used as input to an Excel worksheet. The typical use of a data source is to provide automatic updating of inputs to a program. If the Excel worksheet creates a report using an external data source, the report is automatically updated whenever the data change.

It is common for industrial plants to keep a record of all routine plant measurements (e.g., temperatures, pressures, flow rates, and power consumptions) in a data source available for all plant technical staff to refer to and use.

Common data sources are text files and database files. For example, some data from a fictitious industrial plant have been saved as a text file (planData.txt) and a Microsoft Access database table (plantData.mdb). The data in each file are similar.

Title:	Plant Standard Data Sheet
Date:	7/5/2011
Time:	12:02 AM
*T*102:	342.4°C
*T*104:	131.6°C
*T*118:	22.1°C
*T*231:	121.2°C
Q104:	43.3 GPM
Q320:	457.1 GPM
P102:	810.8 PSI
P104:	124.6 PSI
P112:	63.1 PSI
W104:	112.4 KW

The variable names are typical cryptic plant names. $T102$, for example, would be a temperature measurement from unit 102. Q refers to flow rate, P to pressure, and W to power. We will use these files as external data sources for a very simple Excel worksheet that monitors the status of unit 104, a pump. The worksheet checks:

- temperature ($T104 < 150°C$ is OK)
- flow rate ($Q104 > 45$ gpm is OK)
- pressure ($P104 < 180$ psi is OK)
- power consumption ($W104 < 170$ kW is OK)

If all are within normal operating limits, the status reads "OK," but, if not, a warning message is displayed.

Note: The external data source should be saved in a file before the reference to the data source is added to the Excel worksheet.

10.4.1 Text Files as Data Sources for Excel

Text files are very easy-to-use data sources for Excel. To set up a text file as an external data source, use Ribbon options **Data/Get External Data/From Text** as illustrated in Figure 10.33. [Excel 2003: Data/Import External Data/Import Data . . .]

The Import Text File dialog will open. Select the text file that contains the data you want to import. Here, it is in a file called **plantData.txt** in a folder named **DataSource**, as shown in Figure 10.34.

Click the **Import** button to start the Text Import Wizard.

In Step 1, you indicate the file type. This text file is tab delimited (there is a tab character between values on each row) so the **Delimited** option is selected, as shown in Figure 10.35.

Figure 10.33
Ribbon options used to connect to a text file data source.

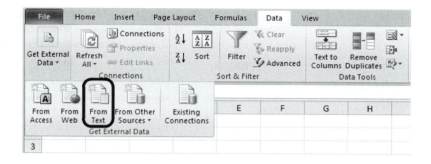

Figure 10.34
Select the data file.

Figure 10.35
The Text Import Wizard, step 1: selecting the file type.

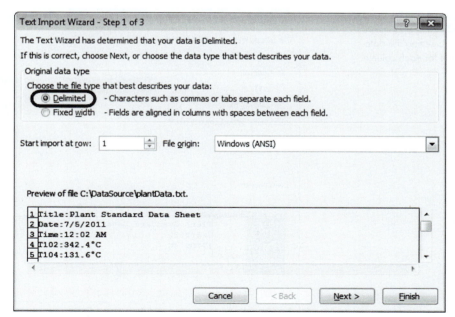

Click **Next** > to move to the next step in the Text Import Wizard.

Step 2 of the wizard shows how the data will be imported (Figure 10.36) and gives you a chance to change the delimiters. For this file, no change was needed.

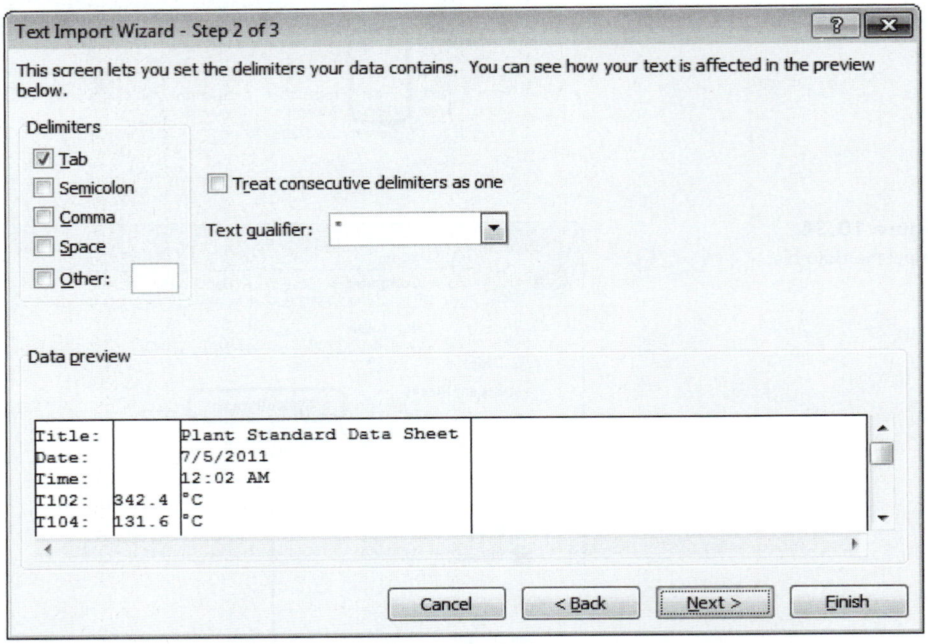

Figure 10.36

The Text Import Wizard, step 2: selecting the delimiters used in the file.

Click **Next** > to move to the last step in the Text Import Wizard.

Step 3 of the wizard allows you to specify the formats used for each column of imported data, as shown in Figure 10.37. **General** format is fine, so no changes are needed.

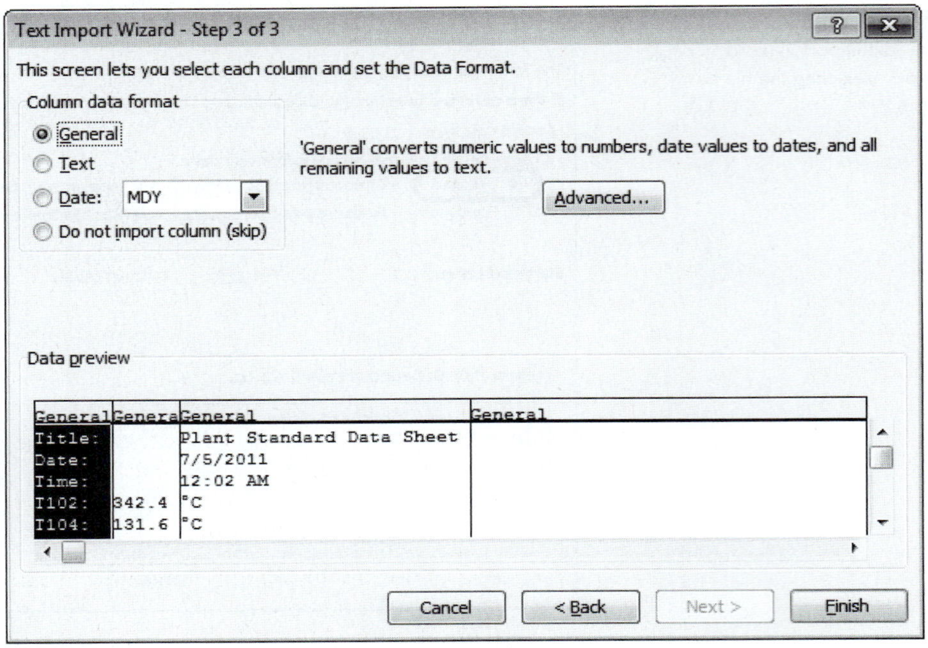

Figure 10.37

The Text Import Wizard, step 3: setting the data format for each column.

Figure 10.38

Indicating where the imported data should be placed.

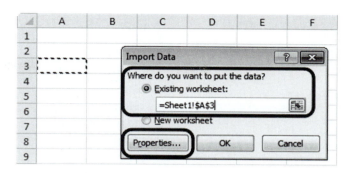

When you click the **Finish** button on the Text Import Wizard, the Import Data dialog is displayed, as shown in Figure 10.38. First indicate where the imported data should go on the worksheet (cell **A3** was used here) and then click the **Properties** button to open the External Data Range Properties dialog, shown in Figure 10.39.

Figure 10.39

The External Data Range Properties dialog.

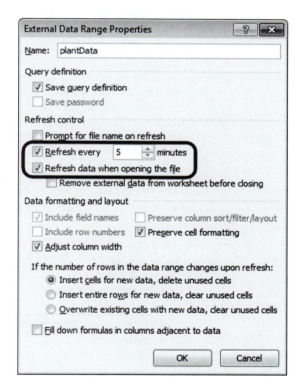

The External Data Range Properties dialog gives you a lot of control over how (and how often) the data should be refreshed. In this example, the data are to be refreshed every 5 minutes and whenever the Excel file is opened. You must select **Save query definition** if you want the data to be updated automatically.

When you click the **OK** button on the External Data Range Properties dialog box, you are returned to the Import Data dialog. Click **OK** again to finish importing the data. The imported data are shown in Figure 10.40.

Note: Automatically importing data every 5 minutes is considered a security risk because someone could replace the usual data file with a virus. Excel will likely show a warning that you should only import files from trusted sources.

Figure 10.40
The data imported from
the text file.

	A	B	C	D
1				
2				
3	Title:		Plant Standard Data Sheet	
4	Date:		7/5/2011	
5	Time:		12:02 AM	
6	T102:	342.4	°C	
7	T104:	131.6	°C	
8	T118:	22.1	°C	
9	T231:	121.2	°C	
10	Q104:	47.2	GPM	
11	Q320:	457.1	GPM	
12	P102:	89.8	PSI	
13	P104:	174.6	PSI	
14	P 12:	63.1	PSI	
15	W104:	112.4	kW	
16				

Finally, the status-checking statements listed below are added, a box is drawn around the imported data (as a reminder not to edit that portion of the worksheet), and the worksheet is complete (Figure 10.41.)

The status-checking statements in column E are as follows:

```
E7:  =IF(B7<150, "OK", "TOO HIGH")
E10: =IF(B10<45, "OK", "TOO LOW")
E13: =IF(B13<180, "OK", "TOO HIGH")
E15: =IF(B15<170, "OK", "TOO HIGH")
```

Figure 10.41
The worksheet with
status-checking statements
in column E.

E10 fx =IF(B10>45,"OK","TOO LOW")

	A	B	C	D	E	F
1	PLANT STATUS SHEET					
2						
3	Title:		Plant Standard Data Sheet		STATUS	
4	Date:		7/5/2011			
5	Time:		12:02 AM			
6	T102:	342.4	°C			
7	T104:	131.6	°C		OK	
8	T118:	22.1	°C			
9	T231:	121.2	°C			
10	Q104:	43.3	GPM		TOO LOW	
11	Q320:	457.1	GPM			
12	P102:	89.8	PSI			
13	P104:	174.6	PSI		OK	
14	P112:	63.1	PSI			
15	W104:	112.4	kW		OK	
16						

Five minutes later, the data are refreshed. The $Q104$ flow rate and $P104$ pressure recorded in the plantData.txt file have gone up, so the status of the pump has changed, as shown in Figure 10.42.

Figure 10.42
The worksheet after the data has automatically refreshed.

	A	B	C	D	E	F
1	PLANT STATUS SHEET					
2						
3	Title:		Plant Standard Data Sheet		STATUS	
4	Date:		7/5/2011			
5	Time:		12:07 AM			
6	T102:	342.5	°C			
7	T104:	131.5	°C		OK	
8	T118:	22.3	°C			
9	T231:	122.1	°C			
10	Q104:	47.1	GPM		OK	
11	Q320:	457.1	GPM			
12	P102:	90.2	PSI			
13	P104:	184.3	PSI		TOO HIGH	
14	P112:	63.4	PSI			
15	W104:	113.3	kW		OK	
16						

10.4.2 Database Files as Data Sources for Excel

Database programs use tables to store information, and these tables have rows and columns, much like an Excel worksheet. It's not surprising, then, that you can use *database tables* as data sources for Excel. The major difference between worksheet files, database tables, and text files is the format used to store the data, but, once the translation algorithms have been written (and many are included with Excel), the various data sources are easy to work with.

The process for using a Microsoft Access database table as an external data source is very similar to that for using a text file. You start by using Ribbon options: **Data/Get External Data/From Access**, as shown in Figure 10.43. This opens the Select Data Source dialog, shown in Figure 10.44.

[Excel 2003: Data/Import External Data/New Database Query]

Figure 10.43
Ribbon options used to connect an Access database data source.

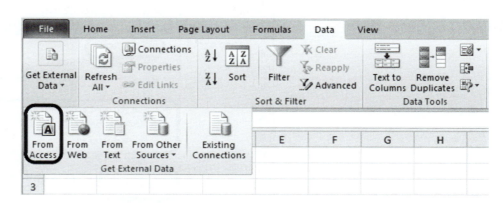

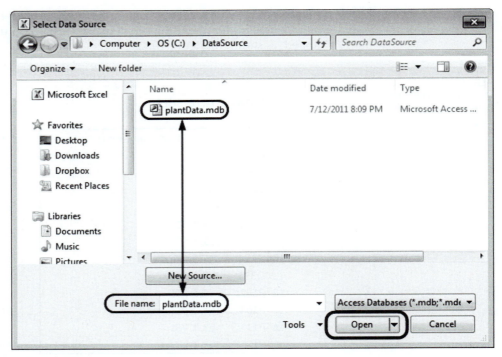

Figure 10.44
The Select Data Source dialog.

Locate the database on your computer's hard drive or a network drive, as shown in Figure 10.44, and then click **Open** to open the Import Data dialog, shown in Figure 10.45.

If the database contains more than one table, Excel will display a Select Table dialog. Since the plant data database contains only one table, Excel used the only available table.

The Import Data dialog is used to tell Excel how to display the imported data. For our purposes, an Excel **Table** will work, and cell **A3** (as before) has been indicated as the top-left cell for the imported data.

Figure 10.45
The Import Data dialog.

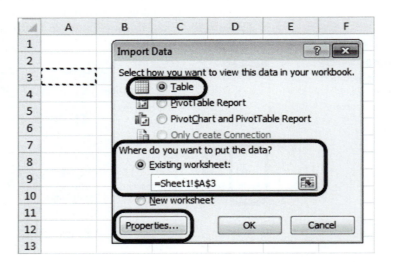

Click the **Properties** button to set the refresh frequency on the Connection Properties dialog, shown in Figure 10.46.

Figure 10.46
The Connection Properties dialog.

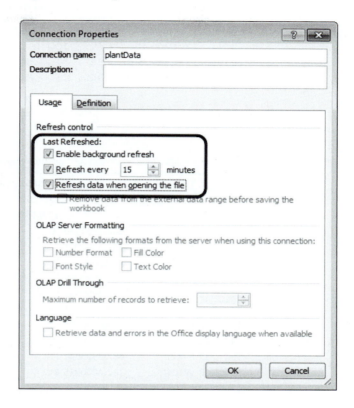

Using the settings shown in Figure 10.46, the database will be accessed every 15 minutes to update the data in the worksheet. Click **OK** to close the Connection Properties dialog and return to the Import Data dialog. Click **OK** (again) to import the data from the database into the worksheet. The result is shown in Figure 10.47.

Figure 10.47
The imported data, as an Excel table.

	A	B	C	D	E
1					
2					
3	ID	VarName	Value	Units	
4	1	Title:		Plant Standard Data Sheet	
5	2	Date:		2/3/2011	
6	3	Time:		12:02 AM	
7	4	T102:	342.3999939	°C	
8	5	T104:	131.6000061	°C	
9	6	T118:	22.10000038	°C	
10	7	T231:	121.1999969	°C	
11	8	Q104:	48.09999847	GPM	
12	9	Q320:	457.1000061	GPM	
13	10	P102:	89.80000305	PSI	
14	11	P104:	124.5999985	PSI	
15	12	P112:	63.09999847	PSI	
16	13	W104:	112.4000015	kW	
17					

After adding headings, and the status-checking formulas in column F, the Pump 104 Status Sheet is complete, as illustrated in Figure 10.48.

The status-checking statements in column F are as follows:

```
F8:  =IF(C8<150, "OK", "TOO HIGH")
F11: =IF(C11>45, "OK", "TOO LOW")
F14: =IF(C14<180, "OK", "TOO HIGH")
F16: =IF(C16<170, "OK", "TOO HIGH")
```

Figure 10.48

The completed worksheet with status-checking statements in column E.

	A	B	C	D	E	F	G
1	PLANT STATUS SHEET						
2						STATUS	
3	ID	VarName	Value	Units			
4	1	Title:		Plant Standard Data Sheet			
5	2	Date:		2/3/2011			
6	3	Time:		12:02 AM			
7	4	T102:	342.3999939	°C			
8	5	T104:	131.6000061	°C		OK	
9	6	T118:	22.10000038	°C			
10	7	T231:	121.1999969	°C			
11	8	Q104:	48.09999847	GPM		OK	
12	9	Q320:	457.1000061	GPM			
13	10	P102:	89.80000305	PSI			
14	11	P104:	124.5999985	PSI		OK	
15	12	P112:	63.09999847	PSI			
16	13	W104:	112.4000015	kW		OK	
17							

KEY TERMS

Clipboard	Excel object	Object Linking and
Copy and paste	External data source	Embedding (OLE)
Data source	Link	Paste special
Database tables	Linked objects	Source
Destination	Mathcad object	Text files
Embedded objects	Object	Windows clipboard
Excel component		

SUMMARY

Embedded and Linked Objects

- *Linked Object:* Stored with the source file; all the destination file knows is how to find the object.

    Note: The object to be linked must be in a saved file before the creation of the link.

- *Embedded Object:*   The destination file receives and stores a copy of the object. No link is maintained to the source file.

    You typically use the *paste special* operation to link objects.

Moving Information between Excel and Word

Copying Cell Contents from Excel to Word

1. Select the cells containing the information to be copied;
2. Copy the cell contents to the Windows clipboard using Excel Ribbon options **Home/Clipboard/Copy** [Excel 2003: Edit/Copy].

To paste the information into Word as a table:

3. Position the cursor in Word at the location where the table should be created.
4. Paste the contents of the Windows clipboard using Word Ribbon options **Home/Clipboard/Paste** [Word 2003: Edit/Paste].

Copying an Excel Graph to Word

1. In Excel, select the graph.
2. Copy it to the Windows clipboard using Excel Ribbon options **Home/Clipboard/Copy** [Excel 2003: Edit/Copy].
3. Position the Word cursor at the location where the table should be pasted.
4. Paste the contents of the Windows clipboard using Word Ribbon options **Home/Clipboard/Paste** [Word 2003: Edit/Paste].

Moving Information between Excel and Mathcad

Copying Excel Values to Mathcad

1. Enter the values in Excel.
2. Select the Excel cell range containing the values.
3. Copy the values to the clipboard from Excel (Ribbon options **Home/Clipboard/Copy**).
4. Enter the variable name that will be used with the matrix in Mathcad and then the define as equal to key [: (colon)].
5. Click on the placeholder that Mathcad creates on the right side of the "define as equal to" symbol.
6. Paste the values into the placeholder, using **Edit/Paste** from Mathcad.

Copying an Array from Mathcad to Excel

1. Select the values in the Mathcad array (not the entire definition, just the values on the right side of the equation).
2. Use **Edit/Copy** from Mathcad.
3. Use Excel Ribbon options **Home/Clipboard/Paste** [Excel 2003: Edit/Paste] to paste the values into Excel cells.

Moving Information between Excel and the World Wide Web

Copying Tabular Data from HTML Pages to Excel

Select the cell values you want to copy (you must copy entire rows in HTML files) and copy them to the clipboard by selecting Edit/Copy from a browser. Then, in Excel, use one of the following approaches:

- Use Excel Ribbon options **Home/Clipboard/Paste** to paste the data into Excel.
- Use Excel Ribbon options **Home/Clipboard/Paste (menu)** to open the Paste menu. Then select Paste Special . . . to open the Paste Special dialog. Then select **Text** or **Unicode Text** from the **As** list.

Creating HTML Pages from Excel

Use the following options to create an HTML version of your Excel worksheet.

- Excel 2010: **File tab/Save As/Save As Type: Web Page**
- Excel 2007: **Office/Save As/Other Formats/ Save As Type: Web Page**
- Excel 2003: **File/Save as Web Page . . .**

External Data Sources

Text Files

1. Use Ribbon options **Data/Get External Data/From Text**. The Import Text File dialog will open.
2. Select the text file that contains the data you want to import. Click the **Import** button to start the Text Import Wizard.
3. Indicate the file type (e.g., **Delimited**).
4. Specify the delimiters used in the date file.
5. Specify the formats used for each column of data (e.g., **General**).
6. Click the **Finish** button on the Text Import Wizard, the Import Data dialog is displayed.
7. Indicate where the imported data should go on the worksheet.
8. Click the **Properties** button to open the External Data Range Properties dialog.
9. Indicate how often the data are to be refreshed. You must select **Save query definition** if you want the data to be updated automatically.
10. Click the **OK** button on the External Data Range Properties dialog box.
11. Click **OK** again to finish importing the data.

Access Database Files

1. Use Ribbon options: **Data/Get External Data/From Access**. This opens the Select Data Source dialog.
2. Locate the database on your computer's hard drive or a network drive and then click **Open** to open the Import Data dialog.
3. If the database contains more than one table, Excel will display a Select Table dialog.
4. Indicate how the imported data should be displayed in Excel (e.g., Excel **Table**).
5. Click the **Properties** button to set the refresh frequency on the Connection Properties dialog.
6. Click **OK** to close the Connection Properties dialog.
7. Click **OK** again to import the data from the database into the worksheet.

PROBLEMS

10.1 Moving Data from Microsoft Word to Excel for Analysis

Requirements: Microsoft Word and Excel

Practicing engineers routinely get requests for assistance from colleagues and clients, often in the form of a memo. In this example, the request to analyze some data came in the form of a Microsoft Word memo, shown in Figure 10.49.

Figure 10.49

Request for data analysis.

> To: Engineering Services
>
> Fr: Sales
>
> Re: We need data for one and two weeks, not every three days!
>
>
> R&D ran an experiment on the cryotron and created a table of performance data every three days for three weeks. We spent $1800 on this experiment and all we really needed was performance data after 1 week and 2 weeks. We got lots of data, but not the two values we really need!
>
> Is it possible to figure out the values at 7 and 14 days from the data R&D provided, or do we have to spend another $1800 to get R&D to give us the values we asked for?
>
> **Cryoton Performance Data**
>
> | Day | Performance |
> |-----|-------------|
> | 0 | 100% |
> | 3 | 75% |
> | 6 | 57% |
> | 9 | 42% |
> | 12 | 32% |
> | 15 | 24% |
> | 18 | 18% |
> | 21 | 14% |

The Word memo is available at www.chbe.montana.edu/excel.

1. Copy and paste the data from Word into Excel.
2. Create an XY scatter graph.
3. Fit a trendline to the data.
4. Use the equation of the trendline to determine the values at 7 and 14 days.
5. Prepare a Word memo back to sales, showing
   a. The two values they requested.
   b. The graph you prepared showing the trendline and regression equation.
   c. The bill for your engineering services.

10.2 Helping a Friend with Loan Calculations

Requirements: Microsoft Word and Excel

A lot of people have no idea how to determine the payment on a loan, but it is an easy calculation with Excel's **PMT** function.

You have a friend who is looking into getting a new vehicle and is wondering what the payments might be for a few scenarios. Help him out by calculating the monthly payments for the following options:

1. The car I really want costs $32,000 and I'd like to keep the down payment to about $2000. I can get a loan for up to 5 years at 6.5% APR. What would the payment be on a 5-year loan?
2. If I cranked the down payment up to $6000, could I afford to make payments for only 4 years?
3. If I go for a 6-year loan, the interest rate jumps to 6.9% APR. What would that do to the payments?

4. There's a 2-year-old car for $24,000 that is OK too, but the interest rates are higher on used cars. What would the payment be on a 4-year loan at 7.5% APR?

5. If I cranked the down payment up to $6000 for the used car, could I afford to make payments for only 3 years?

Help your friend evaluate his options by completing Table 10.1. The first line has been completed as an example. Then, copy the table from Excel to a Word document and present the table in the context of a letter to your friend.

Table 10.1 Monthly payments for various options

Option	Cost	Down	Term (years)	APR	Borrowed	Payment
1	$32,000	$2,000	5	6.50%	$30,000	$586.98
2	$32,000	$6,000	4	6.50%		
3	$32,000	$2,000	6	6.90%		
4	$24,000	$2,000	4	7.50%		
5	$24,000	$6,000	3	7.50%		

10.3 Convincing a Company to Install Energy Efficient Lighting

Requirements: Microsoft Word and Excel

You've been hired as a consultant to provide an outside opinion on replacing the existing incandescent lighting in a company's warehouses. Right now each of the eight warehouses is lit by 88, 100W incandescent light bulbs. Each bulb costs $0.60 (in bulk), has a rated output of 1680 lumens, and has an expected life of 750 hours. The lights are on in the warehouses for about 16 hours per day, five days a week. The owner of the warehouses pays $0.09 per kWh for electricity.

A lighting salesman has suggested switching to long-life incandescent bulbs which cost a little more, but have a 5000 hour life. They also have a light output of 1000 lumens.

You want the company to consider CFL (compact fluorescent light) bulbs. 26W CFL bulbs would be more expensive, at $2.20 each, but they would put out 1700 lumens and have an expected life of 10,000 hours.

Table 10.2 Light bulb comparison table

	Standard Incandescent	Long-Life Incandescent	Compact Fluorescent
Cost per bulb	$0.60	$1.70	$2.20
Power consumption (Watts)	100	100	26
Light output (lumens)	1680	1000	1700
Average life (hours)	750	5000	10,000

1. Do a cost analysis on the three lighting options in Excel. Be sure to consider
   a. Transition cost: the cost to replace all of those light bulbs
   b. Bulb replacement costs
   c. Energy costs
2. Prepare a report (Word document) for the warehouse owner, including
   a. The cost analysis you developed in Excel
   b. Your recommendation

11

Excel Pivot Tables

Objectives

After reading this chapter you will know

- What a pivot table is
- How a pivot table can help you better understand your data
- How to create a pivot table in Excel

- How to sort and filter data in a pivot table
- How to pivot a pivot table
- Some of the capabilities of pivot tables
- How to create an Excel pivot chart

11.1 INTRODUCTION

A *pivot table* is essentially a summary table of a much larger data set. Pivot tables are used to summarize and analyze large data sets. It is very easy to change the configuration of a pivot table to look at the data from a different perspective; it is the ability to easily modify the pivot tables to get different views of the data that is the real power of pivot tables.

Pivot tables are not going to be useful for all data sets. If your data set contains the following features, pivot tables might be very useful for helping you better understand the data:

- You need a large data set; if you have only a few rows of data, a summary table is not going to be helpful.
- There needs to be a lot of repetition in at least some fields (columns).

Pivot tables summarize data using the repeated values in at least some columns; if there are no repeated values, a summary table will not be useful.

If you have a large data set with lots of repeated values, pivot tables can be very useful.

11.2 PREPARING TO CREATE A PIVOT TABLE

Before you create a pivot table, you need:

1. A data set.
2. *Headings* for each column in the data set; these will be used as *field names* when you create the pivot table.
3. Eliminate empty rows and columns; the occasional empty cell is ok, but there should be no empty rows or columns in the data set. (Excel will automatically determine the size of your data set, but an empty row or column will cause Excel to select less than the full data set.)

We will need some data to work within this chapter.

EXAMPLE 11.1

An engineering project management firm has collected a data set containing information on the last 500 completed projects that the firm has supervised. The information includes:

- Project manager
- Estimated time (weeks)
- Actual time (weeks)
- Estimated cost
- Actual cost
- Category (Commercial Construction, Residential Construction, ...)

The Excel worksheet, shown in Figure 11.1, has a few additional columns:

- **OnTime?**—a calculated field that reports "Delayed" if the actual time is more than 5% over the estimated time.
- **OnBudget?**—a calculated field that reports "Over Budget" if the actual cost is more than 2% over the estimated cost.
- **Prj**—there is a column in the worksheet that simply contains a 1 for each project listed. This column will be used to count the number of projects that meet certain criteria. (This is a simple way to count projects that meet certain criteria, but this column is not essential; there are other ways to count projects.)

The worksheet contains 500 rows of data, one per completed project. Only the first 10 rows are shown in Figure 11.1.

	A	B	C	D	E	F	G	H	I
1	Prj.	Project Manager	Est. Time	Act. Time	OnTime?	Est. Cost	Act. Cost	OnBudget?	Category
2	1	Wolff, Carolyn	106.3	106.3	On Time	$2,043,000	$2,043,000	As Budgeted	Municipal Construction
3	1	Turnage, Leslie	100.9	100.9	On Time	$ 506,000	$ 506,000	As Budgeted	Residential Construction
4	1	Wolff, Carolyn	51.9	51.9	On Time	$ 390,000	$ 390,000	As Budgeted	Municipal Construction
5	1	Jaffe, Jerry	60.0	60.0	On Time	$1,917,000	$1,917,000	As Budgeted	Municipal Construction
6	1	Boyce, Bruce	83.7	104.1	Delayed	$1,190,000	$1,880,200	Over Budget	Transportation
7	1	Turnage, Leslie	62.6	62.6	On Time	$2,214,000	$2,214,000	As Budgeted	Transportation
8	1	Jaffe, Jerry	61.3	61.3	On Time	$ 600,000	$ 600,000	As Budgeted	Residential Construction
9	1	Weaver, Diane	65.0	73.3	Delayed	$ 658,000	$ 658,000	As Budgeted	Water/Sewer
10	1	Woods, Susan	27.9	27.9	On Time	$ 352,000	$ 352,000	As Budgeted	Residential Construction

Figure 11.1
Excel worksheet containing the data for the pivot table (10 of 500 rows shown).

This data set is available at the text's website at www.chbe.montana.edu/excel.

We have a data set, there are headings in every column, and there are no empty rows or columns—we are ready to create a pivot table.

11.3 A BASIC PIVOT TABLE

To create a pivot table, click anywhere inside the data set (cell B2, for example), then use the following Ribbon options:

Insert tab/Pivot Table (menu)/Pivot Table (button)

The Create Pivot Table dialog shown in Figure 11.2 will open.

Figure 11.2
The Create Pivot Table dialog.

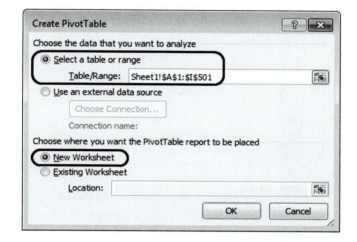

The **Table/Range:** field is automatically filled in by Excel. This is the range of contiguous nonempty cells around the cell you selected (B2, for example) before inserting the pivot table. If your data set contained an empty row or column, Excel would not be able to select the data set correctly.

The default location for the pivot table is a **New Worksheet**. This is recommended because the size of the pivot table will change as you change criteria, and the easiest way to ensure that you do not overwrite existing data is to give the pivot table a worksheet of its own.

Click **OK** to create the new worksheet, so that we can begin creating the pivot table. The new worksheet will be displayed, as shown in Figure 11.3.

There is a lot of information on this worksheet, and we will use all of it eventually, but for now notice the areas marked "1" and "2" in Figure 11.3.

1. The area that will hold the pivot table.
2. The *Pivot Table Field List* shows all of the fields (columns in the data set) that can be used in the pivot table.

11.3.1 Numeric and Nonnumeric Fields

Be default, Excel will treat fields that contain numbers differently than fields that contain labels when creating a pivot table.

- *Nonnumeric Fields:*   Fields that contain labels will, by default, be used to create *row labels* in the pivot table.
- *Numeric Fields:*   Fields that contain numbers will, by default, be used to create *column labels* in the pivot table.

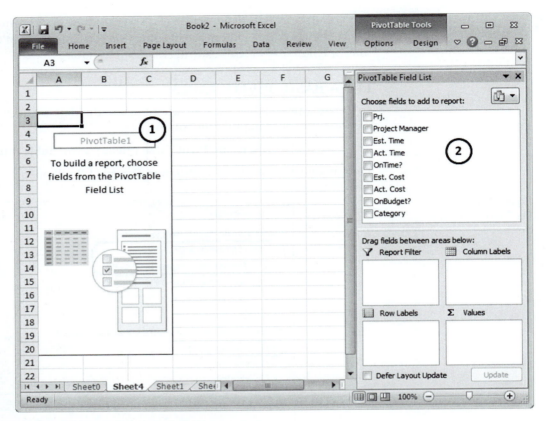

Figure 11.3
The new worksheet, ready to create a pivot table.

In our data set, numeric fields include:

- **Prj** (the "1" values)
- **Est. time**—the estimated project time, in weeks
- **Act. time**—the actual project time, in weeks
- **Est. cost**—the estimated project cost, in dollars
- **Act. time**—the actual project cost, in dollars

The nonnumeric fields include:

- **Project Manager**—the name of the person who managed the project
- **OnTime?**—indicates whether or not the project was completed on time
- **OnBudget?**—indicates whether or not the project was completed within budget
- **Category**—the general type of project

The quickest way to create a pivot table is to choose one numeric field and one non-numeric field. As an example, let's choose the **Prj** and **Project Manager** fields, as shown in Figure 11.4.

When the **Project Manager** (nonnumeric) field was selected, Excel automatically used the nonnumeric values as row labels (in cells A4:A18).

Since the **Prj** field is numeric, Excel used that field as a column heading (cell B3) and displayed the sum of the values in the **Prj** field in column B. Remember,

Figure 11.4
The first pivot table showing how many projects each person has managed.

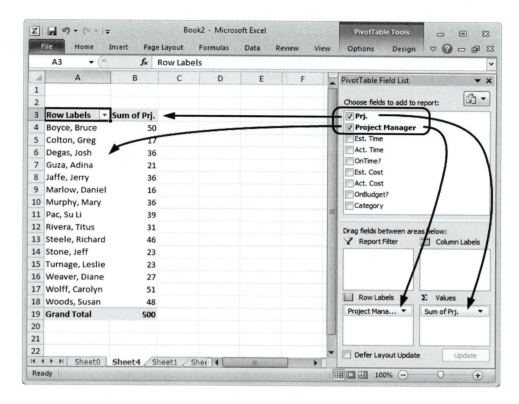

pivot tables are *summary tables*, so summing the values in a category is the default; there are other options, which will be presented later.

11.4 SORTING AND FILTERING A PIVOT TABLE

To help you better understand the data, you can easily sort and filter the information in the pivot table.

11.4.1 Sorting

To see which managers have handled the most projects, we can *sort* the pivot table. To do so:

1. Right-click on the column heading or any of the values in the **Sum of Prj.** column of the pivot table. A pop-up menu will be displayed, as shown in Figure 11.5.
2. Use menu options **Sort/Sort Largest to Smallest**.

The pivot table will be sorted by the number of projects each manager has handled. The result is shown in Figure 11.6.

11.4.2 Filtering

If you want to focus further analysis on only those projects that were completed on time, then you can *filter* the data. Here's how:

1. In the Pivot Table Field List, right-click on the field you want to use in the filter (**OnTime?** In our example). A list of options will be displayed as shown in Figure 11.7.

Figure 11.5
Sorting the pivot table.

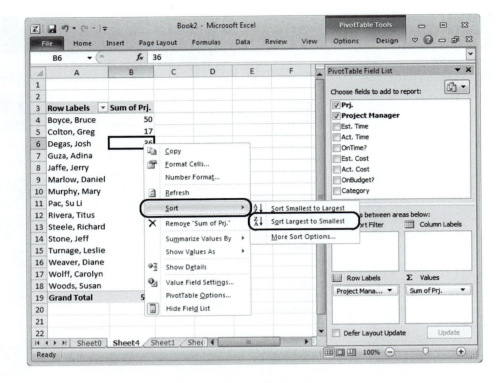

Figure 11.6
The sorted pivot table.

	A	B	C
1			
2			
3	**Row Labels**	**Sum of Prj.**	
4	Wolff, Carolyn	51	
5	Boyce, Bruce	50	
6	Woods, Susan	48	
7	Steele, Richard	46	
8	Pac, Su Li	39	
9	Murphy, Mary	36	
10	Jaffe, Jerry	36	
11	Degas, Josh	36	
12	Rivera, Titus	31	
13	Weaver, Diane	27	
14	Stone, Jeff	23	
15	Turnage, Leslie	23	
16	Guza, Adina	21	
17	Colton, Greg	17	
18	Marlow, Daniel	16	
19	**Grand Total**	500	
20			

2. Select **Add to Report Filter**.

When the **OnTime?** field is added to the Report Filter category, there are also changes in cells A1 and B1 on the worksheet, as shown in Figure 11.8. Cell A1 shows the name of the field that can be used to filter the data shown

Figure 11.7
Asking Excel to filter using the OnTime? field.

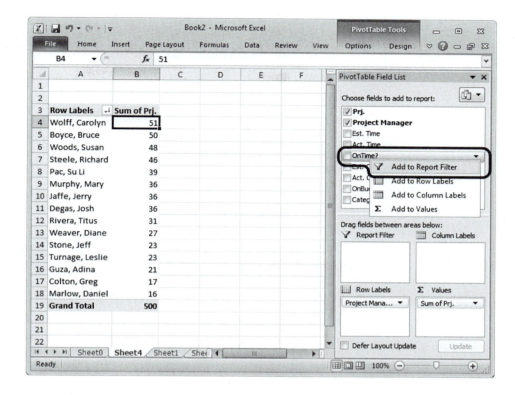

Figure 11.8
The OnTime? field can now be used to filter the data in the pivot table.

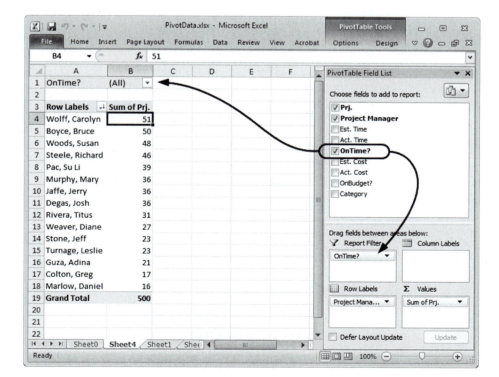

in the pivot table, and the drop-down list in cell B1 will be used to select the filter value.

3. Choose the filter value using the drop-down list tied to cell B1.

To show only projects that were completed on time, select **On Time** from the drop-down list connected to cell B1 (Figure 11.9).

Note: Be careful with nomenclature here:

- **OnTime?** is a column heading in the original data and becomes a field name in the pivot table.
- **On Time** is one of the two labels (text strings) that appear in the OnTime? column (the other is **Delayed**).

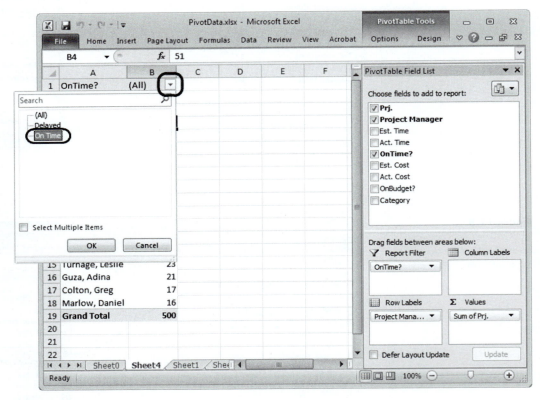

Figure 11.9
Filtering on On Time projects.

The drop-down list of filter options shows **Delayed** and **On Time** because those two labels appear in the **OnTime?** column in the original data (see Figure 11.1).

The filtered pivot table is shown in Figure 11.10. Notice that the icon on the button in cell B1 now looks like a funnel; the *funnel symbol* is used to indicate that a filter is in place in Excel.

Figure 11.10
The filtered pivot table.

11.5 PIVOTING THE PIVOT TABLE

To prepare to *pivot* the pivot table, we will set it up as follows:

- Deselect the **OnTime?** Field—this turns off the filter.
- Select the **OnBudget?** Field—selecting this nonnumeric field, by default, causes Excel to add a second set of row labels. This is shown in Figure 11.11.

Figure 11.11
Preparing to pivot the pivot table.

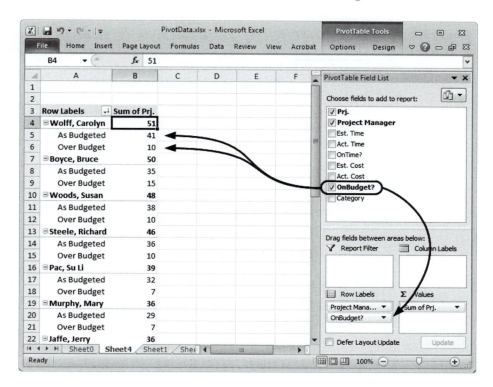

Right now, in Figure 11.11 you can see that Carolyn Wolff has managed 51 projects, and 41 have come in "As Budgeted," but 10 were "Over Budget." The data in the pivot table are easy to understand, but much of it is off screen.

We can pivot the pivot table to present the same information more concisely. To do so, grab the **OnBudget?** marker in the Row Labels category and drag it to the Column Labels category, as illustrated in Figure 11.12.

The result is shown in Figure 11.13.

Figure 11.12
Drag the OnBudget? marker from Row Labels to Column Labels to pivot the pivot table.

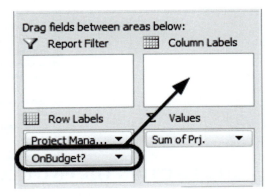

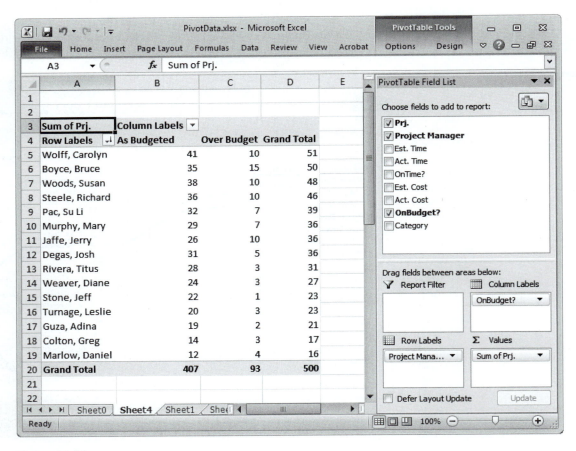

Figure 11.13
The pivoted pivot table.

By pivoting the pivot table, we can now see all project managers listed in column A, and the number of "As Budgeted" and "Over Budget" projects for each manager in columns B and C.

11.6 WHAT CAN BE DONE WITH A PIVOT TABLE?

Pivot tables are typically used to try to understand large data sets and to answer questions. Questions that the pivot table shown in Figure 11.13 can be used to answer include:

- Which project manager has handled the most projects? ANS: Carolyn Wolff
- Which project manager has had the most projects go over budget? ANS: Bruce Boyce

We can start manipulating the fields in the pivot table to answer other questions.

Which project manager has handled the most money (actual costs) with their projects?

To answer this question, we set up the pivot table using the **Act. Cost** field instead of the **OnBudget?** field, and then sort the **Act. Cost** column. The result is shown in Figure 11.14; Bruce Boyce has handled the most money on his projects. (The pivot table values in the **Sum of Act. Cost** column (Column C) were sorted in descending order to make the result more apparent.)

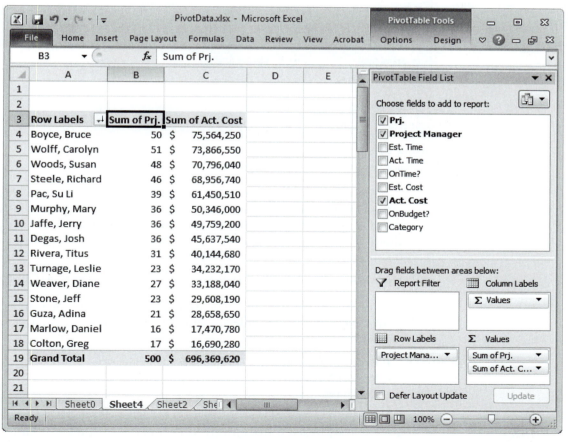

Figure 11.14

Pivot table set to look at total value of projects handled by each manager.

It was mentioned in Section 11.3.1 that, while summing values is the default way to present numeric data in a pivot table, there are other options. If you right-click on a column heading, such as the "Sum of Act. Cost" heading in cell C3, a pop-up menu will be displayed as shown in Figure 11.15.

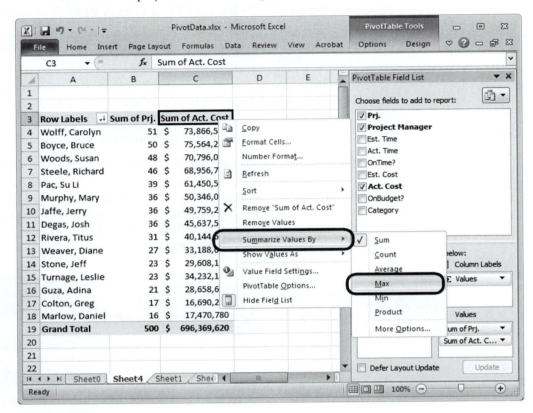

Figure 11.15
Changing the way the data is summarized.

You can select other ways to summarize the data, such as the **Max** option selected in Figure 11.15. When this option is selected, the pivot table will now show the maximum value in the Act. Cost column for each project manager. The result is shown in Figure 11.16; notice that the column heading in cell C3 now indicates that the maximum values are being displayed.

How do the "On Time" and "Delayed" projects vary with the type (Category) of project?

If you start bringing additional fields into the pivot table, you can begin to see how pivot tables can help in understanding the data. For example, if we use the following fields:

- **Prj**—for the number of projects in each category
- **Project Manager**
- **OnTime?**
- **Category**

We can start to see if the various project managers do better at particular types of projects. From the pivot table shown in Figure 11.17, we can see that in the

Figure 11.16
The pivot table displaying
the maximum actual project
cost by manager.

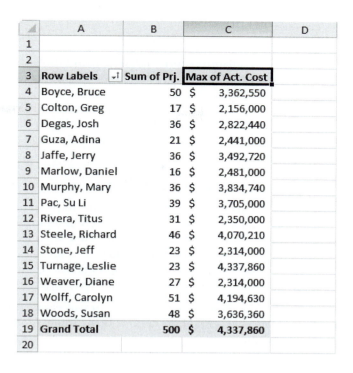

	A	B	C	D
1				
2				
3	**Row Labels** ↓	**Sum of Prj.**	**Max of Act. Cost**	
4	Boyce, Bruce	50	$ 3,362,550	
5	Colton, Greg	17	$ 2,156,000	
6	Degas, Josh	36	$ 2,822,440	
7	Guza, Adina	21	$ 2,441,000	
8	Jaffe, Jerry	36	$ 3,492,720	
9	Marlow, Daniel	16	$ 2,481,000	
10	Murphy, Mary	36	$ 3,834,740	
11	Pac, Su Li	39	$ 3,705,000	
12	Rivera, Titus	31	$ 2,350,000	
13	Steele, Richard	46	$ 4,070,210	
14	Stone, Jeff	23	$ 2,314,000	
15	Turnage, Leslie	23	$ 4,337,860	
16	Weaver, Diane	27	$ 2,314,000	
17	Wolff, Carolyn	51	$ 4,194,630	
18	Woods, Susan	48	$ 3,636,360	
19	**Grand Total**	500	$ 4,337,860	
20				

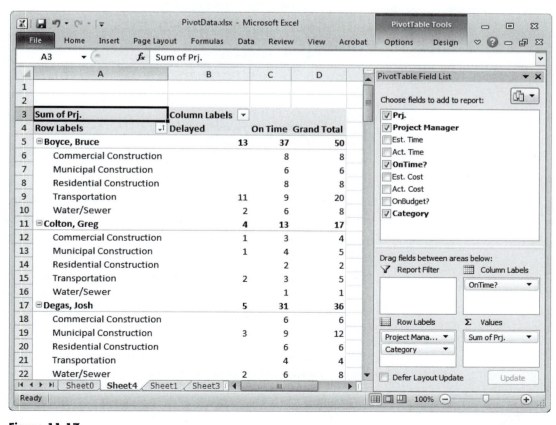

Figure 11.17
Pivot table set to show number of on time and delayed projects by category for each manager.

Transportation category, Bruce Boyce has more "Delayed" projects than "On Time" projects, while Josh Degas has never had a Transportation project come in late.

The possible pivot table variations are too numerous to show, even with the relatively small number of fields in our data set. For very large data sets, pivot tables can be a big help in interpreting your data.

11.7 PIVOT CHARTS

Excel also provides *pivot charts* to provide a graphical representation of the type of data summary that is provided by a pivot table.

To create a pivot chart, click anywhere within the data to be charted (i.e., click in the data on the original worksheet (see Figure 11.1), not the data in the pivot table) and use Ribbon options **Insert/Pivot Table (menu)/Pivot Chart**. This opens the Create Pivot Table with Pivot Chart dialog shown in Figure 11.18.

Note: Alternatively, you can create a pivot chart from an existing pivot table. Use Ribbon options **Pivot Table Tools/Options/Tools/PivotChart**.

Figure 11.18
Create Pivot Table with
Pivot Chart dialog.

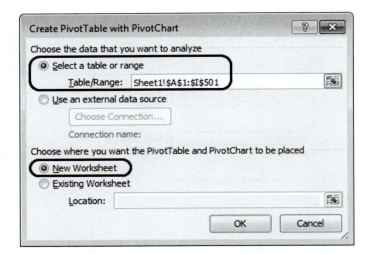

Excel automatically fills the **Table/Range:** field with the range of contiguous cells around the selected cell when the dialog was opened. The default is to create the pivot table and pivot chart on a new worksheet. Since the size of the pivot table changes as various criteria are established, giving the pivot table its own worksheet makes sense.

When you click the **OK** button to create the pivot table and pivot chart, the new worksheet is displayed, as shown in Figure 11.19.

Select fields from the Pivot Table Field List to simultaneously create the pivot table and the pivot chart. Selecting the **Project Manager, OnTime?,** and **Act. Cost** fields, as shown in Figure 11.20, caused the pivot chart shown in Figure 11.21 to be created.

In Excel 2010, the Pivot Chart includes some buttons that allow you to modify the way the data are presented. For example, the **Project Manager** button shown in

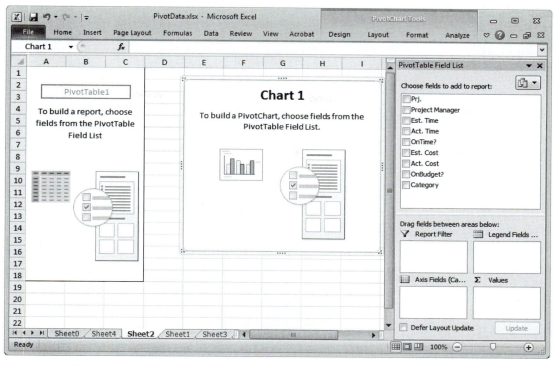

Figure 11.19
The new worksheet for the pivot table and pivot chart.

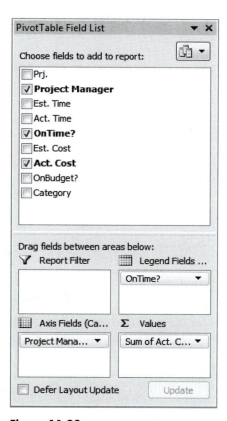

Figure 11.20
Selecting fields for creating the pivot chart.

Figure 11.21
The pivot chart.

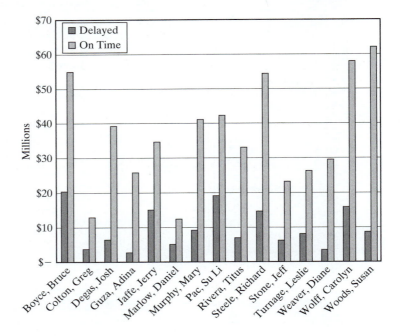

Figure 11.21 allows you to sort and filter the values on the *x* axis (the project manager names). The **OnTime?** button above the legend allows you to filter the displayed data to show only projects marked as Delayed, only projects marked as On Time, or both (as shown in Figure 11.21).

KEY TERMS

Pivot table	Row labels	Filter
Headings	Numeric Fields	Funnel symbol (filter)
Field names	Column labels	Pivot
Pivot Table Field List	Summary tables	Pivot charts
Nonnumeric Fields	Sort	

SUMMARY

Preparing to Create a Pivot Table

1. Select a data set.
2. Provide Headings for each column in the data set.
3. Eliminate empty rows and columns from the data set.

Create a Pivot Table

1. Click anywhere inside the data set.
2. Use Ribbon options **Insert/Pivot Table/Pivot Table** to open the Create Pivot Table dialog.

3. Verify that the data set has been identified correctly.
4. Choose where the pivot table should be placed.
5. Click [OK] on the Create Pivot Table dialog.
6. Select one numeric and one nonnumeric field to create the basic pivot table.

Numeric and Nonnumeric Fields

- Nonnumeric Fields will, by default, be used to create row labels in the pivot table.
- Numeric Fields will, by default, be used to create column labels in the pivot table.

You can move a field marker from the Row Labels category to the Column Labels category (or vice versa) to override the default.

Sorting Data

1. Right-click on any of the values in a column of the pivot table. A pop-up menu will be displayed.
2. Use menu options **Sort/Sort Largest** to **Smallest or Sort/Sort Smallest to Largest**.

Filtering Data

1. In the Pivot Table Field List, right-click on the field you want to use to filter the data.
2. Select Add to Report Filter.
3. Choose the filter value using the drop-down list tied to the field name.

Pivoting the Pivot Table

Pivoting refers to moving a field being used as a row label to act as a column label, or vice versa. It does not change the data that are displayed, but it changes the organization of the displayed data.

To pivot, grab the field marker in the Row Labels category on the Pivot Table Field List and drag it to the Column Labels category (or vice versa).

Pivot Charts

Pivot chart provide a graphical representation of the type of data summary that is provided by a pivot table. A pivot chart is created simultaneously with a pivot table, if the chart is requested when the pivot table is created. To insert a pivot chart with a pivot table, use Ribbon options **Insert/Pivot Table (menu)/Pivot Chart**.

PROBLEMS

Because a fairly large data set is needed to create useful pivot tables, each of the problems listed here use the 500 Completed Engineering Projects data set used throughout this chapter. This data set is available on the text's website at www.chbe .montana.edu/excel.

11.1 Most Transportation Projects

Use the following fields in a pivot table (see Figure 11.22):

- **Prj.**
- **Project Manager**
- **Category**

Figure 11.22
Pivot Table Field List for
parts a and b.

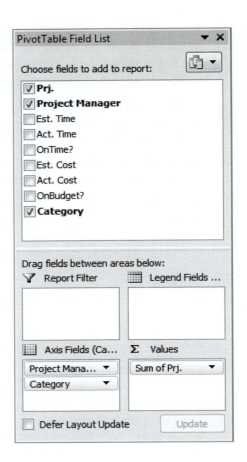

Filter the data so that only Transportation projects are included.

    a. Which project manager has handled the most Transportation projects?

    b. Which project manager has handled the fewest?

Next, bring in the **OnTime?** field as a column label.

    c. Which project manager has had the most "On Time" Transportation projects?

    d. Which project manager has had the fewest "On Time" Transportation projects?

11.2 Most Municipal Construction Projects

Use the following fields in a pivot table (see Figure 11.22):

- **Prj.**
- **Project Manager**
- **Category**

Filter the data so that only Municipal Construction projects are included.

    a. Which project manager has handled the most Municipal Construction projects?

    b. Which project manager has handled the fewest?

Next, bring in the **OnBudget?** field as a column label.

    c. Which project manager has had the most "As Budgeted" Municipal Construction projects?

    d. Which project manager has had the fewest "As Budgeted" Municipal Construction projects?

11.3 Most Severely Over Time Project

Before creating the pivot table, add another column to the data set, one that calculates the difference between the actual project time and the estimated project time. Then create a pivot table and include the new column in the pivot table.

> **Note:** If you have already created a pivot table in the worksheet, without the new column, you may need to refresh the pivot table [**Pivot Table Tools/ Options/Refresh**] to see the new column listed in the Pivot Table Field List.

    a. Which project manager has had the project that went most severely over the estimated project time?

    b. How many weeks over time was the project?

    c. Was the project over budget as well?

11.4 Most Severely Over Budget Project

Before creating the pivot table, add another column to the data set, one that calculates the percent difference between the actual project cost and the estimated project cost.

$$\%\text{Over Budget} = \frac{\text{actual} - \text{estimated}}{\text{actual}} \times 100\%$$

Then create a pivot table and include the new column in the pivot table. If you have already created a pivot table in the worksheet, without the new column, you may need to refresh the pivot table [**Pivot Table Tools/Options/Refresh**] to see the new column listed in the Pivot Table Field List.

    a. Which project manager has had the project that had the highest over budget percentage?

    b. By what percentage was the project over budget?

11.5 Delayed Projects

Use the following fields in a pivot table (see Figure 11.22):

- **Prj.**
- **Category**
- **OnTime?**

Use a pivot table to answer these questions:

    a. Which category has the fewest "Delayed" projects?

    b. Which category has the most "Delayed" projects?

12

Macros and User-Written Functions for Excel

Objectives

After reading this chapter, you will know

- What an Excel macro is and how one can be created and used
- How macros have been used to create computer viruses, and the precautions that Excel users should take to protect their work from macro viruses

- How to create recorded macros
- How to access Excel's built-in Visual Basic programming environment
- How to program simple macros
- How to write your own functions for Excel

12.1 INTRODUCTION

Many years ago, a decision was made at Microsoft to use *Visual Basic for Applications (VBA)* as the *macro language* in the Microsoft Office products. The intent was to give Excel users the ability to create very powerful and useful macros. That power has been misused by some virus programmers, and the usefulness of macros in Excel has suffered.

Macros are usually short programs or sets of recorded keystrokes (stored as a program) that can be reused as needed. There are basically two reasons for writing a macro:

- Excel doesn't automatically do what you need. You can write macros to extend Excel's capabilities.
- You need to perform the same task (usually a multistep task) over and over again. Complex operations can be combined in a macro and invoked with a simple keystroke.

There are two ways to create a macro in Excel:

1. You can record a set of keystrokes.
2. You can write a VBA subprogram.

Programming macros gives you a lot of control over the result, but takes more effort than recording.

12.2 MACROS AND VIRUSES

In the days before toolbars, macros were commonly used to simplify common tasks like formatting cells. Developers would use *auto-start macros* that ran automatically when a workbook was opened to customize the worksheets. But some people have taken advantage of the power of Excel macros and written *macro viruses* that try to run automatically to do their damage when a worksheet is opened. All Excel users who might obtain a workbook from an unknown source, such as the Internet, e-mail, or a disk, should take security against macro viruses very seriously.

With the 2007 version, Excel added a new level of security against macro viruses by using two different *file extensions* for macro-enabled and macro-disabled workbooks:

* .xlsx—the Excel file extension for macro-disabled workbooks (default)
* .xlsm—the Excel file extension for macro-enabled workbooks

Excel will allow you to write and use a macro in a macro-disabled workbook; but Excel won't allow you to save the workbook with the active macro. If you save the workbook with the .xlsx extension, any macros you have written will be disabled. (Excel will warn you if you try to save macros in a macro-disabled workbook.) You must save workbooks containing macros with the .xlsm file extension.

Recent versions of Excel take one of the following four actions when a workbook containing macros is opened:

* Disable all macros without notification.
* Disable all macros with notification (default).
* Disable all macros except digitally signed macros.
* Enable all macros.

You can change the default action taken when macros are encountered using the Excel Options dialog as shown in Figure 12.1.

Use the **Trust Center** panel and Click the **Trust Center Settings...** button (shown in Figure 12.1) to access the Trust Center dialog (Figure 12.2), and choose the **Macro Settings** panel to see the current action taken when macros are encountered.

Basically, unless you choose to enable all macros (not recommended), any workbook containing macros that you open will have the macros disabled. If you accept the default action (shown in Figure 12.2), to notify when macros are disabled, the Excel screen when you first open a macro-enabled workbook will show the Security Warning seen in Figure 12.3.

Clicking the **Enable Content** button indicated in Figure 12.3 will enable all macros in the workbook. If you want a little more information before enabling the macros, you can click on the **Macros have been disabled.** statement, which links to the Excel Options dialog and provides additional information. An excerpt of this information is shown in Figure 12.4. You should only enable macros from trusted sources.

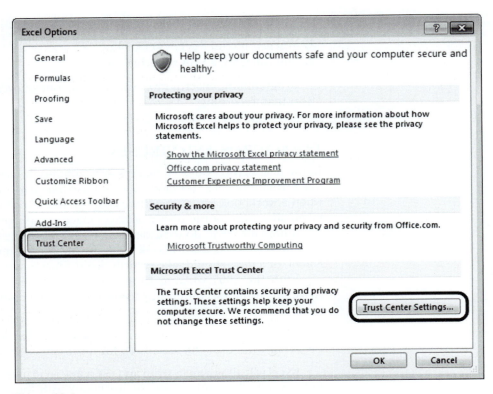

Figure 12.1
Excel Options dialog, Trust Center panel.

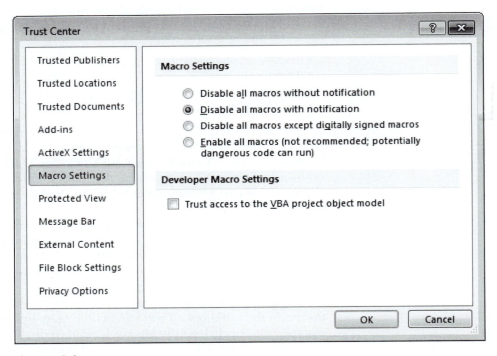

Figure 12.2
Trust Center dialog, Macro Settings panel.

Figure 12.3
Security Warning
displayed when a
workbook containing
macros is opened (macros
disabled).

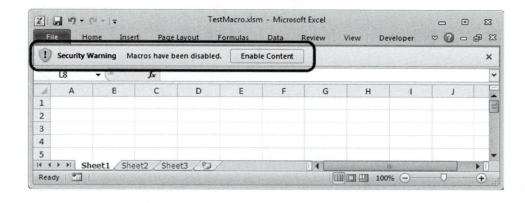

Figure 12.4
Re-enabling disabled
macros.

Information about TestMacro
C:\TestMacro.xlsm

Security Warning
Active content might contain viruses and other security hazards. The following content has been disabled:

- Macros

You should enable content only if you trust the contents of the file.

Trust Center Settings

Learn more about Active Content

Note: Excel 2007 has a slightly different way of notifying you that a workbook has macros. There is still a security warning (as in Figure 12.3), but instead of an **Enable Content** button there is an **Options…** button that opens a dialog. In Excel 2007, you must use the dialog to enable macros.

It is appropriate to be concerned about macros coming from unknown sources, but that shouldn't stop you from creating your own macros when they can help you get your work done faster and easier.

12.3 RECORDED MACROS

Recording of keystrokes is the easiest way to create a macro, so we will consider *recorded macros* first.

When you ask Excel to record a macro, it actually writes a program in VBA, using Visual Basic statements that are equivalent to the commands you enter via the keyboard or mouse. Because the macro is stored as a program, after recording you can edit the program. Recording is a good way to create simple macros, and it is also an effective way to begin writing a VBA programmed macro.

We'll need an example to demonstrate the process of recording a macro. As an easy example, consider a macro that converts a temperature in degrees Fahrenheit to degrees Celsius. The conversion equation is

$$T_C = \frac{T_F - 32}{1.8}.$$

$$(12.1)$$

Some benchmarks will also be helpful in making sure the macro is working correctly. Here are four well-known temperatures in the two temperature systems:

212°F =	100°C	(boiling point of water)
98.6°F =	37°C	(human body temperature)
32°F =	0°C	(freezing point of water)
−40°F =	−40°C	(equivalency temperature)

12.3.1 Recording a Macro

We begin by entering a Fahrenheit temperature into a cell (B3 in this example) and selecting the adjacent cell (C3), as indicated in Figure 12.5. The Macros group on the Ribbon's **View** tab is indicated in Figure 12.5. This is the access point for all recorded macros.

Figure 12.5
Preparing to record a macro.

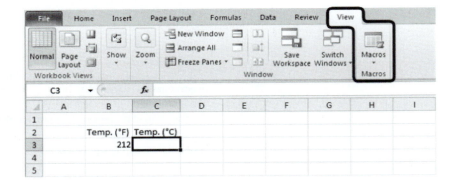

Then we begin recording the macro:

1. Tell Excel you want to record a macro using Ribbon options **View/Macros (group)/Macros (menu)/Record Macro...** as shown in Figure 12.6. [Excel 2003: Tools/Macro/Record New Macro.]

Figure 12.6
Use the **Record Macro...** button in the Macros menu to start recording a macro.

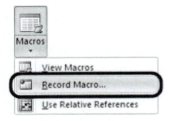

The Record Macro dialog will be displayed, as shown in Figure 12.7.

Use the Record Macro dialog to set the **Macro name, Shortcut key** (optional), storage location, and **Description** (optional). Here the macro has been named **F_to_C**. The shortcut key was set as [Ctrl-f]. (Use a lowercase "f"; shortcut keys are case sensitive.) The macro will be stored in **This Workbook,** which means it is available to any sheet in the current workbook. Other options include **New Workbook** and **Personal Macro Workbook**.

Figure 12.7
Record Macro dialog.

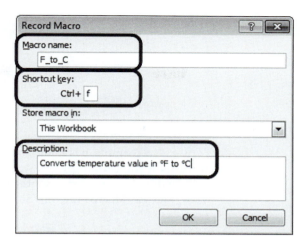

- Macros stored in workbooks are available to any open workbook. One way to collect your commonly used macros and make them available to other workbooks is to create a Personal Macro Workbook.
- If the macro was stored in a new workbook, the macro would be available only if the new workbook is open.

If your macro will be used frequently from more than one workbook, you probably want to save the macro in your Personal Macro Workbook. However, if the macro will be used only with the data in a particular workbook, you probably want to store the macro in the workbook with the data.

Once you click the **OK** button on the Record Macro dialog, every mouse click and key stroke that you make will be recorded.

2. Select absolute or relative cell referencing.

Note: This is important. Many of the frustrations that new macro programmers face are the result of using the wrong type of cell referencing.

If you want the temperature conversion macro to use the value "in the cell to the left of the currently selected cell," then you need to record the macro using *relative referencing*. If you want the temperature conversion macro to use the value "in cell B3," then you want to use *absolute referencing*. Relative referencing is more common.

To indicate that relative cell referencing should be used, be sure that the **Use Relative References** button on the macro menu is selected (Figure 12.8). If it is not selected, the macro will be recorded with absolute cell addresses.

Figure 12.8
Selecting relative cell references for the macro.

3. Enter the conversion equation in cell C3, as shown in Figure 12.9.
   This formula should be entered just like any formula in Excel. The only difference is that the macro recorder is running as the formula as entered. Once the

Figure 12.9
Recording the formula.

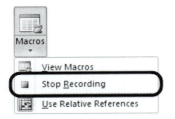

conversion formula has been completed, press the [Enter] key. (The macro recorder is still running.)

Stop the macro recorder by pressing the **Stop Recording** button on the Ribbon: **View/Macros/Macros (menu)/Stop Recording,** as shown in Figure 12.10.

Figure 12.10
Stop recording the macro.

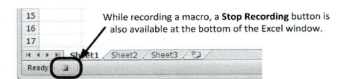

While the Stop Recording button in the Macros group on the Ribbon works, there is an alternative. Any time a macro is being recorded, Excel makes it apparent by displaying a Stop Recording button at the bottom of the Excel window (indicated in Figure 12.11). You can use either Stop Recording button when you have finished recording your macro.

Figure 12.11
Excel also provides a **Stop Recording** button at the bottom of the Excel window.

At this point, the macro has been recorded. The next step is to see whether it works. We'll use our temperature benchmarks to test the macro.

12.3.2 Testing the Recorded Macro

1. First, the other benchmark temperatures are entered in cells B4:B6. Then cell C4 is selected to hold the next calculated value. The list of currently available macros is displayed if using Ribbon options **View/Macro/Macro (menu)/View Macros** (Figure 12.12) [Excel 2003: Tools/Macro/Macros…].

   The **F_to_C** macro is the only macro currently available, as seen in Figure 12.13.

2. Run (replay) the **F_to_C** macro by selecting the macro name and pressing the **Run** button. The value in cell B4 (98.6°F) is converted from Fahrenheit to Celsius, and the result is placed in cell C4, as illustrated in Figure 12.14. (Note: Cell C4 contains the conversion formula =(B4-32)/1.8.)

   The other temperature benchmarks can be converted by using the assigned shortcut key [Ctrl-f] or by again using the Macro dialog. The results are shown in Figure 12.15.

Figure 12.12
Opening the list of available macros.

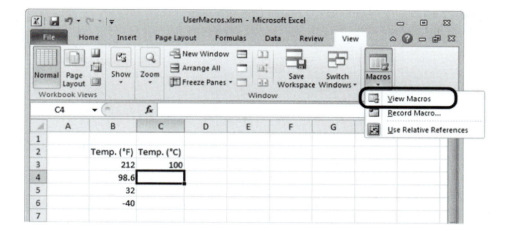

Figure 12.13
Available macros.

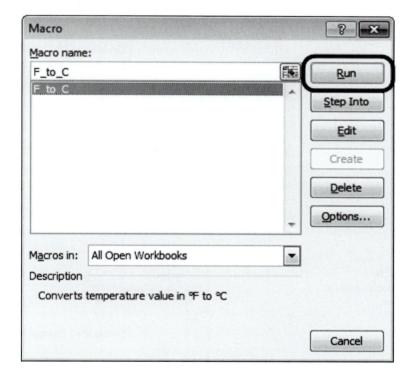

Figure 12.14
The result of running the macro in cell C4.

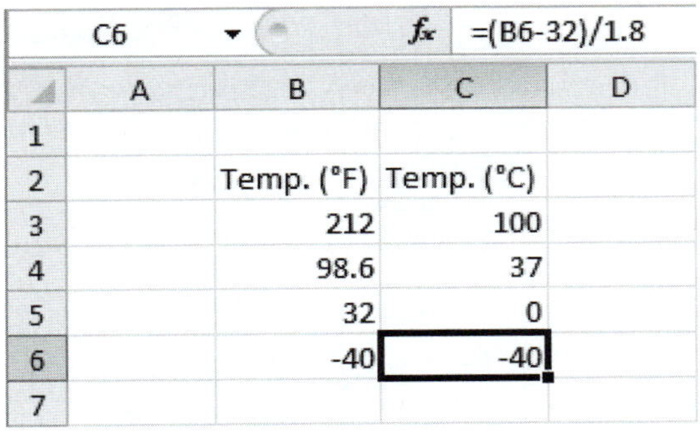

Figure 12.15
Using [Ctrl-f] to run the macro.

Using Absolute Cell References When Recording Macros

If the macro had been recorded with absolute cell references (with the **Use Relative References** button deselected), then the macro would have been recorded using absolute addresses such as "B3" rather than "the cell to the left." No matter which cell you selected before running the macro, the temperature in cell B3 would be converted and the result would always be put in cell C3. Absolute cell references are not commonly used in macros, but if your recorded macro always tries to do the same thing, no matter what cell you select, you might check to see if the **Use Relative References** button is deselected.

Including an Absolute Cell Reference in a Macro Recorded Using Relative Referencing

If you want to include an absolute address on a cell in a macro recorded with relative referencing, just use the dollar signs in the absolute address. For example, if the constant (32) were moved out of the conversion formula and into cell D3, the temperature conversion macro should be recorded with relative referencing selected and absolute address $D$3 used in the recorded formula. Press [F4] immediately after entering D3 to add the dollar signs as the macro is being recorded, as illustrated in Figure 12.16.

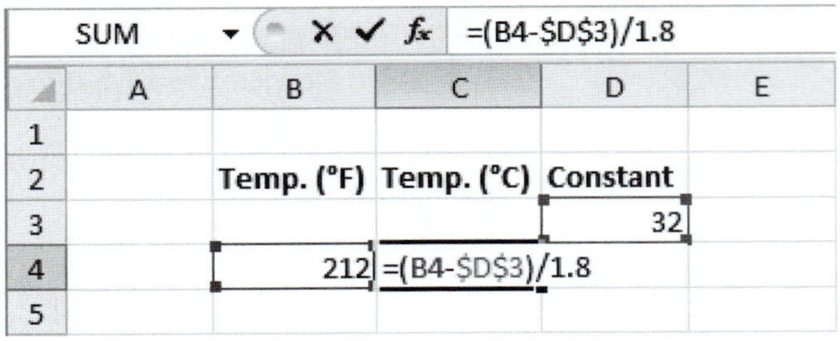

Figure 12.16
Use $ to indicate absolute cell addresses.

12.3.3 Editing a Recorded Macro

Recorded macros are stored as VBA subprograms, or Subs. The **F_to_C** macro was stored like this:

```
Sub F_to_C()
'
' F_to_C Macro
' Converts °F to °C
'
```

```
' Keyboard Shortcut: Ctrl+f
'
  ActiveCell.FormulaR1C1 = "=(RC[-1]-32)/1.8"
  ActiveCell.Offset(1, 0).Range("A1").Select
End Sub
```

The VBA subprogram was given the assigned macro name, **F_to_C**. The parentheses indicate that the macro is stored as a VBA subprogram, or Sub. The single quote ['] at the start of a line indicates a *comment line* in the program. Comment lines are ignored when the macro is run, but are included to provide information about the macro. The first six lines after the title line are comments, so there are only two operational lines in the **F_to_C** Sub.

The two operational lines do the following:

ActiveCell.FormulaR1C1 = "=(RC[−1]−32)/1.8"

- ActiveCell.FormulaR1C1 =

     This portion of the programming statement assigns (=) a formula in R1C1 notation (used for relative cell references) to the active cell.
- "=(RC[−1]−32)/1.8"

     The characters inside the quotes are the formula that is assigned to the active cell.
- RC[−1]

     The RC[-1] tells Excel to use the cell in the current row (R) and one cell to the left (C[-1]) of the active cell.

This programming line puts the formula =(B4-32)/1.8) in cell C4 in Figure 12.14.

ActiveCell.Offset(1, 0).Range("A1").Select

- ActiveCell.

     Starting from the current active cell location…
- Offset(1, 0).

     Move one row down and zero columns right…
- Range("A1").

     On the currently selected worksheet…
- Select

     Make the offset cell the selected (active) cell.

This programming line is VBA's way of moving the Excel cursor down one row after entering the formula.

So, the two programming lines

```
        ActiveCell.FormulaR1C1 = "=(RC[-1]-32)/1.8"
        ActiveCell.Offset(1, 0).Range("A1").Select
```

are used to put the formula in the currently selected cell using relative cell references and to move the cursor down one row.

The macro can be edited as a VBA subprogram by using the VBA editor. To open the VBA editor, use Ribbon options **View/Macro (group)/Macro (menu)/View Macros** [Excel 2003: Tools/Macro/Macros…] to set the list of currently available macros, as shown in Figure 12.17.

Choose the macro you want to edit and then click the **Edit** button to open the *VBA editor* shown in Figure 12.18. The program code for the currently selected macro will be displayed in the editor.

Figure 12.17
Use the [Edit] button to edit the macro as a VBA subprogram.

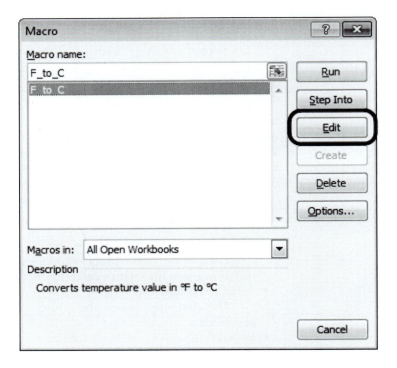

Figure 12.18
The F_to_C Sub in the VBA editor.

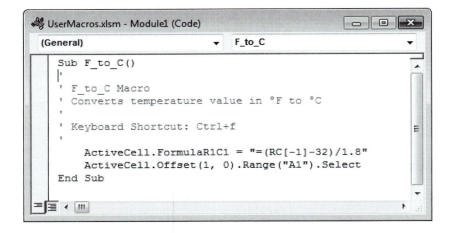

You can modify the existing program (e.g., correct errors) or create entirely new macros. For example, a new macro for converting temperatures in degrees Celsius to degrees Fahrenheit was created by copying Sub **F_to_C,** renaming it Sub **C_to_F,** and modifying it as follows:

```
Sub C_to_F()                        'changed from F_to_C()
'
' C_to_F Macro                      'changed from F_to_C Macro
' Converts °C to °F                 'changed from °F to °C
'
' Keyboard Shortcut: Ctrl+Shift+c   'changed from Ctrl+f
'
```

```
ActiveCell.FormulaR1C1 = "=RC[-1]*1.8+32"   'changed from "
                                              =(RC[-1]-32)/1.8"
ActiveCell.Offset(1, 0).Range("A1").Select
End Sub
```

The result is shown in Figure 12.19.

Figure 12.19
The C_to_F Sub created
in the VBA editor.

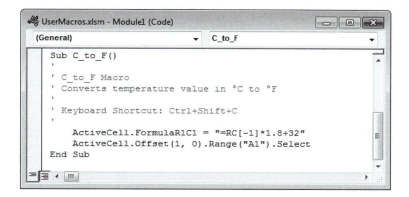

Notice that the keyboard shortcut for the new macro is listed as [Ctrl-Shift-c], not [Ctrl-c]. [Ctrl-c] is a standard shortcut for copying to the clipboard. Excel will allow you to reassign [Ctrl-c] to the new macro, but wiping out a commonly used keyboard shortcut is not a good idea.

Creating a new Sub in VBA makes the macro available to the worksheet. The macro name (**C_to_F**) will be the same as the name of the VBA Sub (with added parentheses). However, the shortcut key and description are not assigned to the macro when the macro is created by using VBA. To assign the shortcut key and description, use Ribbon options **View/Macro/Macro (menu)/View Macros** [Excel 2003: Tools/Macro/Macros...], select the new macro's name from the list of available macros, and press the **Options...** button, as illustrated in Figure 12.20.

Figure 12.20
Setting options on a macro
created in VBA.

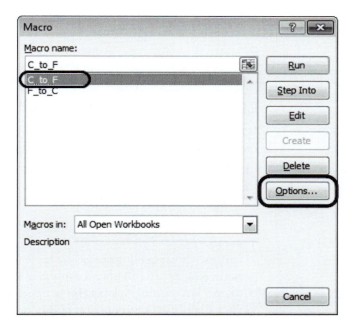

On the Macro Options dialog box, you can set the **Shortcut key** and **Description,** as shown in Figure 12.21.

Figure 12.21
Defining the C_to_F macro.

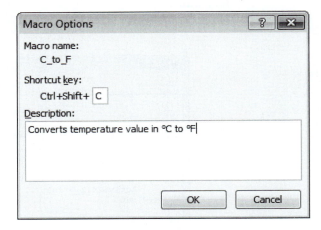

Finally, we test the new macro in the worksheet by entering 100 in cell B3, selecting cell C3, and running the **C_to_F** macro by using the [Ctrl-Shift-c] shortcut. As you can see in Figure 12.22, it worked!

Figure 12.22
The result of running the
C_to_F macro.

	A	B	C	D
1				
2		Temp. (°C)	Temp. (°F)	
3		100	212	
4				
5				

EXAMPLE 12.1

CONVERTING FORMULAS TO VALUES

It is a fairly common to need to convert the formulas in a cell range into values. The usual way to do this in an Excel worksheet is to:

1. Select the cells containing the formulas to be converted
2. Copy the cell contents to the clipboard
3. Paste from the clipboard back into the same cells, as values

It is an easy task to accomplish, but it is also a simple task to automate with a macro.

In the thermocouple response data shown in Figure 12.23, the simulated temperature values are calculated assuming a first-order response with a 3 second time constant. We want to convert the temperature formulas in cells C4:C19 into values. We'll actually use only a portion of the formulas (C4:C8) when recording the macro and save the rest for testing the macro after it has been recorded.

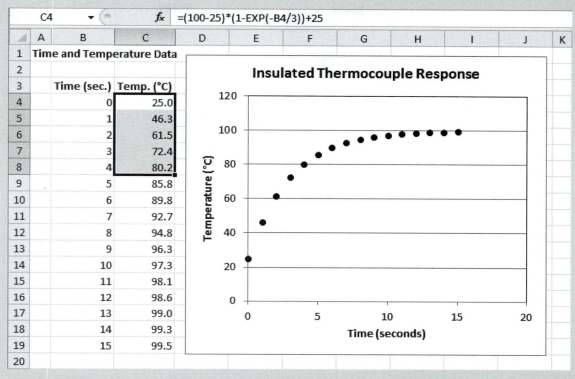

Figure 12.23
Simulated thermocouple response data.

The process for recording the macro is:

1. Select the cells containing the formulas to be converted. We will select only the first 5 temperature values, cells C4:C8, as indicated in Figure 12.23.
2. Start recording the macro with Ribbon options **View/Macros/Macros (menu)/ Record Macro…**; this opens the Record Macro dialog shown in Figure 12.24.

Figure 12.24
Record Macro dialog, completed.

(*continued*)

3. Assign the macro a name, a keyboard shortcut, and a description. This has been done in Figure 12.24.
4. Click **OK** to close the Record Macro dialog and start recording.
5. Press [Ctrl-c] to copy the selected values to the clipboard.
6. Use Ribbon options **Home/Clipboard/Paste (menu)/Paste Values** to paste the values into the selected cells.

Note: In Excel 2010 the Paste menu is now displayed as icons rather than text options. The basic **Paste Values** button is the left button in the Paste Values category (indicated in Figure 12.25).

Figure 12.25
The **Paste Values** button in Excel 2010.

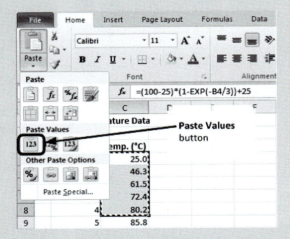

7. The dashed border around the pasted cells indicates that the values are still on the clipboard, ready to be pasted again. Press [Esc] to get rid of the dashed border.
8. Stop recording the macro with Ribbon options **View/Macros/Macros (menu)/Stop Recording**.

At this point, the macro has been recorded and assigned to keyboard shortcut [Ctrl-Shift-B]. The first five formulas have already been converted to values, but the last 11 formulas are still unchanged (indicated in Figure 12.26). To convert the last 11 formulas, select them, then press [Ctrl-Shift-B]. This runs the **Formula_to_Value** macro and saves the temperature values in the selected cells.

Note: There is no reason to write a macro to convert one column of formulas to values; you could use standard Excel commands to convert the formulas in less time than it takes to write the macro. But once the macro has been written, you can use it any time you need to convert a range of formulas to values. Over time, having the macro might prove useful.

The **Formula_to_Value** macro was stored as a VBA subprogram, as listed below:

```
Sub Formula_to_Value()
'
' Formula_to_Value Macro
```

Figure 12.26
Selecting the last 11 formulas before running the macro.

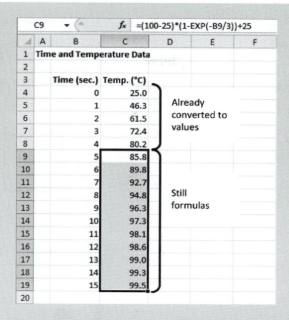

```
' Converts a range of formulas to values (stored in the same cells)
'
' Keyboard Shortcut: Ctrl+Shift+V
'
    Selection.Copy
    Selection.PasteSpecial Paste:=xlPasteValues, Operation:=xlNone, _
    SkipBlanks:=False, Transpose:=False
    Application.CutCopyMode = False
End Sub
```

The three programming lines that do the work of the macro are:

- **Selection.Copy**
  This copies the contents of the currently selected cells to the clipboard.
- **Selection.PasteSpecial Paste:=xlPasteValues, Operation:=xlNone, SkipBlanks:=False, Transpose:=False**
  This command tells Excel to paste values [PasteSpecial Paste:=xlPasteValues] from the clipboard into the currently selected cells [Selection.]
- **Application.CutCopyMode = False**
  This program line was added when the [Esc] key was pressed to get rid of the dashed border around the copied cells; technically we were deactivating Cut Copy Mode.

EXAMPLE 12.2

INSERTING A RANDOM NUMBER

Sometimes you just need a random number. Excel's **RAND** function returns a random number between 0 and 1, and you can enter the formula =RAND() into a cell. But, if you find yourself typing that a lot, it's a quick process to write a macro that will enter =RAND() into all currently selected cells.

(continued)

1. Select the cells that should contain random numbers (Figure 12.27).

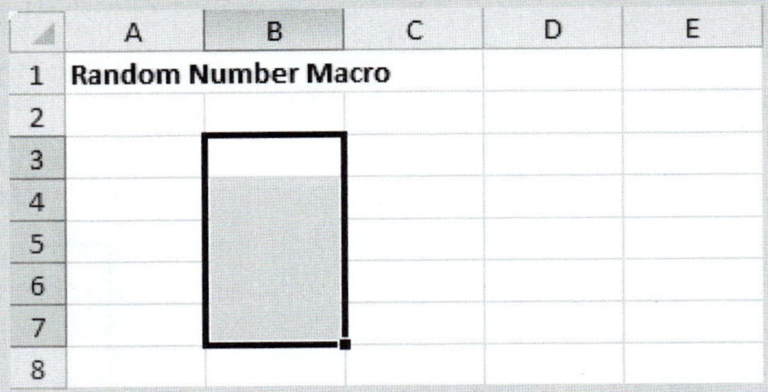

Figure 12.27
Step 1. Select the cells to
receive random number
formulas.

◢	A	B	C	D	E
1	**Random Number Macro**				
2					
3					
4					
5					
6					
7					
8					

2. Start recording the macro. The Record Macro dialog will open.
3. Provide the macro a name, keyboard shortcut, and description. This has been done in Figure 12.28.

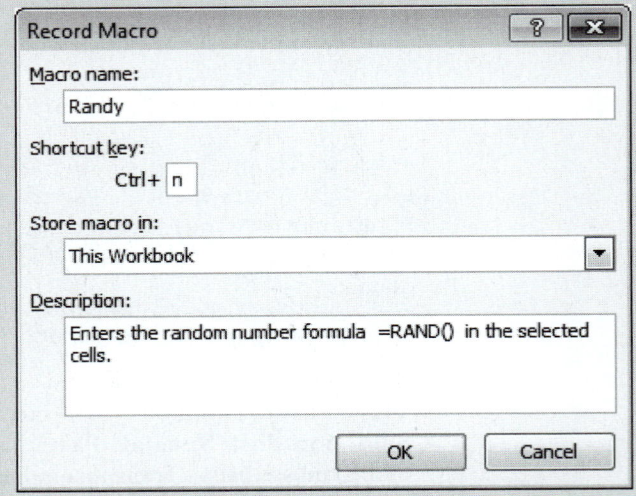

Figure 12.28
Step 3. Complete the
Record Macro dialog.

Record Macro

Macro name:
Randy

Shortcut key:
Ctrl+ n

Store macro in:
This Workbook

Description:
Enters the random number formula =RAND() in the selected cells.

OK   Cancel

4. Enter the formula =RAND() into one cell in the selected range (Figure 12.29) but do not press [Enter]. We are going to fill in the entire selected range at once.

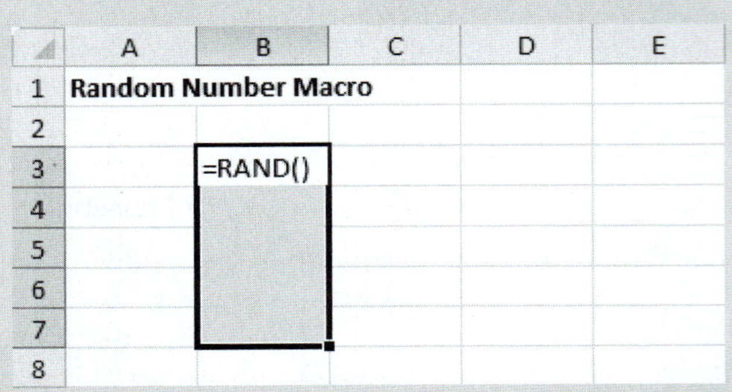

Figure 12.29
Step 4. Enter the formula in
one cell.

◢	A	B	C	D	E
1	**Random Number Macro**				
2					
3		=RAND()			
4					
5					
6					
7					
8					

5. Press [Ctrl-Shift-Enter] to enter the formula into all selected cells (Figure 12.30).

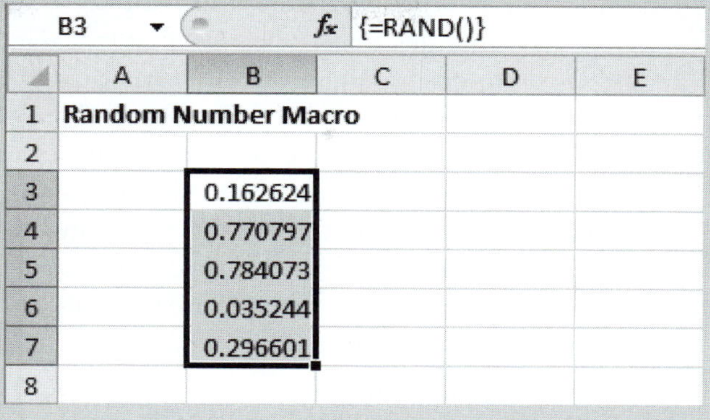

Figure 12.30
Step 5. Use [Ctrl-Shift-Enter] to enter formula in all selected cells.

B3			f_x {=RAND()}		
	A	B	C	D	E
1	Random Number Macro				
2					
3		0.162624			
4		0.770797			
5		0.784073			
6		0.035244			
7		0.296601			
8					

6. Stop recording the macro.

   At this point, we should have a working macro for entering random number formulas in cells wherever we want. So, try entering a single random number formula in cell D4.

   First, we select cell D4 as illustrated in Figure 12.31.

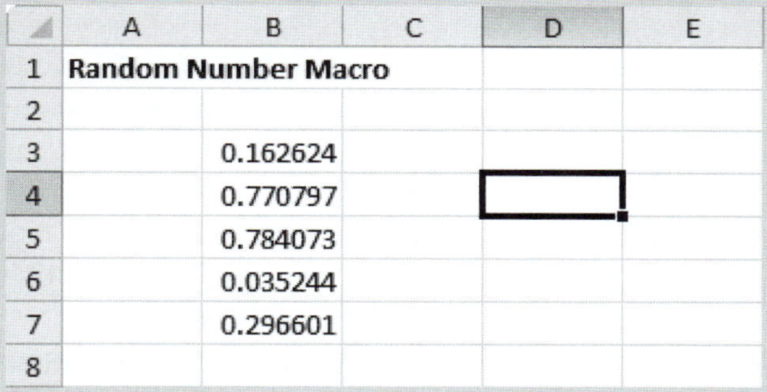

Figure 12.31
Select a cell to try out the random number macro.

	A	B	C	D	E
1	Random Number Macro				
2					
3		0.162624			
4		0.770797			
5		0.784073			
6		0.035244			
7		0.296601			
8					

Then, press [Ctrl-n] to run the **Randy** macro and put the formula =RAND() in cell D4. The result is shown in Figure 12.32.

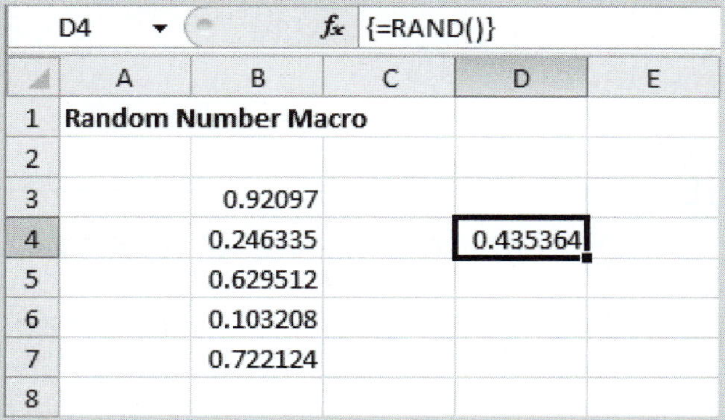

Figure 12.32
Using the Randy macro.

D4			f_x {=RAND()}		
	A	B	C	D	E
1	Random Number Macro				
2					
3		0.92097			
4		0.246335		0.435364	
5		0.629512			
6		0.103208			
7		0.722124			
8					

(continued)

Notice that all of the random values in column B have changed? Every time Excel recalculates the worksheet, which is every time a change is made to any cell, all of the **RAND** functions will be recalculated and send a new random number. Sometimes that is exactly what you need; other times you want random numbers entered into cells, but then you do not want them changing. To generate random values once and then keep them from changing each time the worksheet recalculates, we need to convert the results of the =RAND() formulas into values. Because we already have a macro that turns formulas into values (the **Formula_to_Value** macro), and because Excel stores macros as VBA subprograms, we can just call the Formula_to_Value() subprogram at the end of the **Randy()** subprogram to convert the random number formulas to values. The modified Randy macro is listed below, with the new line indicated.

```
Sub Randy()
'
' Randy Macro
' Enters the random number formula =RAND() in selected cells
'
' Keyboard Shortcut: Ctrl+r
'
    Selection.FormulaArray = "=RAND()"
    Call Formula_to_Value
End Sub
```

Now, when you select a range of cells and press [Ctrl-n] random number values are placed in the selected cells. This is illustrated in Figure 12.33.

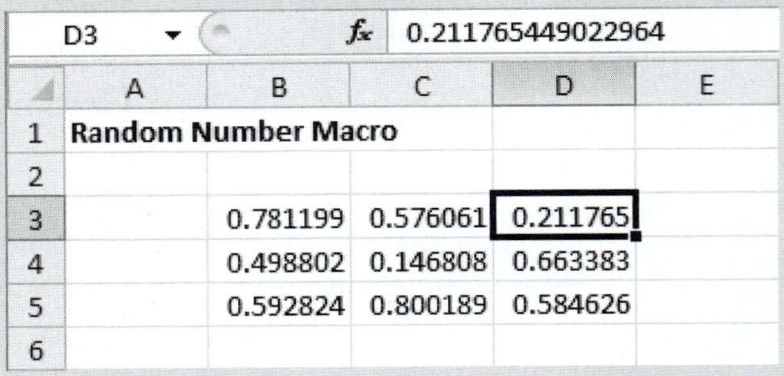

Figure 12.33
Using the modified **Randy** macro to insert random values instead of formulas.

	D3	▼	f_x	0.211765449022964	
	A	B	C	D	E
1	Random Number Macro				
2					
3		0.781199	0.576061	0.211765	
4		0.498802	0.146808	0.663383	
5		0.592824	0.800189	0.584626	
6					

EXAMPLE 12.3

ADDING A TRENDLINE TO A GRAPH

Macros are not restricted to working with numbers in cells. To illustrate this, we will create a macro that adds a linear trendline to a graph, called **Trendy**. The process is:

1. Start with an XY scatter graph, such as that shown in Figure 12.34. Select the chart before beginning to record the macro.

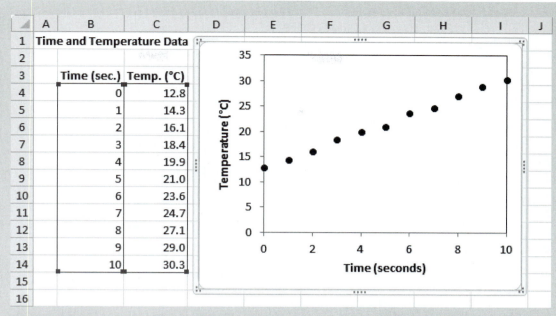

Figure 12.34
Step 1. Select the data series.

2. Start recording the macro. The Record Macro dialog will open.
3. Complete the Record Macro dialog as shown in Figure 12.35, click **OK** to start recording.

Figure 12.35
Step 3. Complete the
Record Macro dialog.

Record Macro	? ✕
Macro name:	
Trendy	
Shortcut key:	
Ctrl+Shift+ T	
Store macro in:	
This Workbook ▾	
Description:	
Adds a linear trendline to the selected data series.	
OK	Cancel

4. Right-click on the data series and select **Add Trendline** from the pop-up menu. The Format Trendline dialog will open (Figure 12.36).
5. Select the desired trendline options; as illustrated in Figure 12.36 we have requested a **Linear** trendline, with the equation of the line and the R^2 value displayed on the graph.
6. Click **Close** to close the Format Trendline dialog.
7. Stop recording.

(continued)

Figure 12.36
Steps 4 and 5. Choose trendline options.

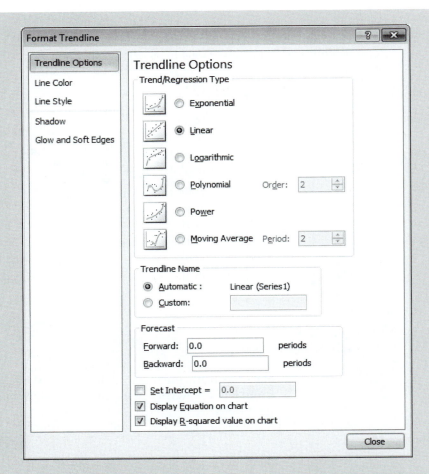

At this point, the XY scatter graph has a trendline, but to test the **Trendy** macro, we delete the existing trendline, and select the data series, as indicated in Figure 12.37.

Figure 12.37
Selecting the graph to test the **Trendy** macro.

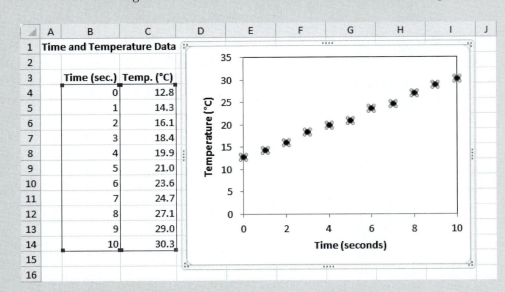

Next, press [Ctrl-Shift-T] to run the **Trendy** macro. The result is shown in Figure 12.38.

Figure 12.38
The trendline added using the **Trendy** macro.

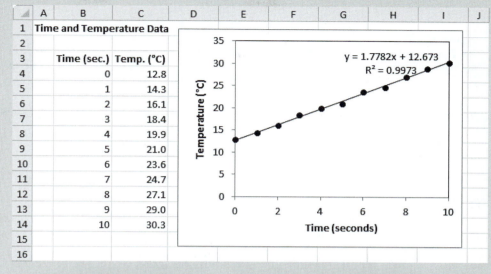

It looks like it worked great…

…but the **Trendy** macro actually has a couple of serious limitations. Two issues and their solutions will be presented.

Problem 1: The Trendy macro always adds the trendline to "Chart 1."

When you have multiple graphs in a workbook, as shown in Figure 12.39, you might like to add a trendline to either graph. When you try to use the **Trendy** macro with Chart 2, there's a problem.

Figure 12.39
Selecting Chart 2 for a trendline.

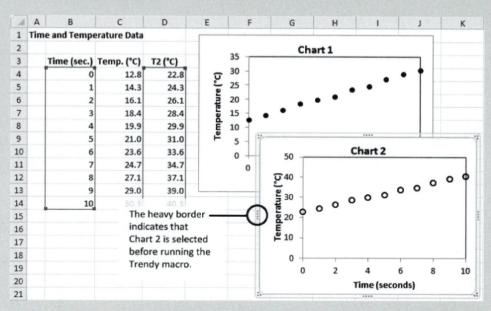

(*continued*)

When we select Chart 2 (as shown in Figure 12.39) and press [Ctrl-T] to run the **Trendy** macro, we do not get the trendline on Chart 2. The (undesired) result is shown in Figure 12.40.

Figure 12.40
The **Trendy** macro applied the trendline to Chart 1.

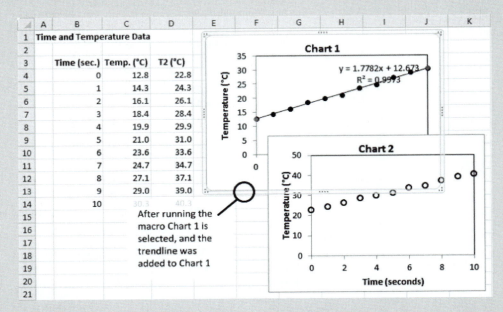

Even though Chart 2 was selected when the **Trendy** macro was run, the macro applied the trendline to Chart 1. A look at the listing of the **Trendy** macro will show why...

```
Sub Trendy()
'
' Trendy Macro
' Adds a linear trendline to a graph - VERSION ONE
'
' Keyboard Shortcut: Ctrl+Shift-T
'
    ActiveSheet.ChartObjects("Chart 1").Activate
    ActiveChart.SeriesCollection(1).Select
    ActiveChart.SeriesCollection(1).Trendlines.Add
    ActiveChart.SeriesCollection(1).Trendlines(1).Select
    Selection.Type = xlLinear
    Selection.DisplayEquation = True
    Selection.DisplayRSquared = True

End Sub
```

The problem line has been boxes in the program listing. The **ActiveSheet. ChartObjects("Chart 1").Activate** statement is exactly like clicking on Chart 1 to select it. No matter which chart we select before running the **Trendy** macro, the macro itself selects Chart 1.

When Excel recorded the macro, it added the **ActiveSheet.ChartObjects("Chart 1"). Activate** statement to ensure that a chart was selected before attempting to select a data series, add a trendline, or select the trendline. That's conservative programming, but it also hardwires the **Trendy** macro to Chart 1.

If we remove the boxed line in the **Trendy** macro listing, the reference to Chart 1 goes away. This means we will need to be sure to select a graph before running the macro, but we can add a trendline to any selected graph.

Figure 12.41 shows the result of running the modified **Trendy** macro after selecting Chart 2.

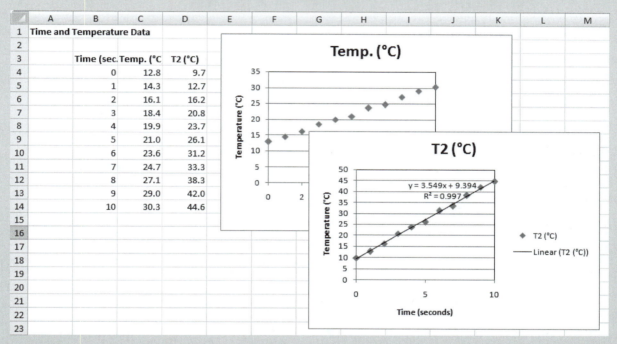

Figure 12.41
After running the modified **Trendy** macro.

Problem 2: The Trendy macro always adds the trendline to the first curve on the graph.

When a graph shows two data sets (data series), as illustrated in Figure 12.42, it would be nice to be able to add a trendline to either or both of the series.

But when the second curve is selected before running the **Trendy** macro, the trendline is still added to the first curve, as shown in Figure 12.43.

The problem is that the **Trendy** macro directly accesses the first data series, no matter which series is selected before running the macro. Each of the boxed lines indicated in the following macro listing points directly to the first data series.

(continued)

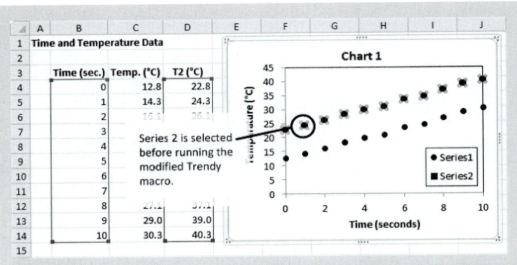

Figure 12.42
After running the modified **Trendy** macro.

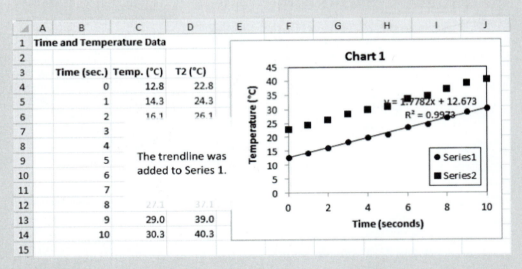

Figure 12.43
After running the modified **Trendy** macro.

```
Sub Trendy()
'
' Trendy Macro
' Adds a linear trendline to a graph - VERSION TWO
'
' Keyboard Shortcut: Ctrl+Shift+T
'
    ActiveChart.SeriesCollection(1).Select
    ActiveChart.SeriesCollection(1).Trendlines.Add
    ActiveChart.SeriesCollection(1).Trendlines(1).Select
```

```
            Selection.Type = xlLinear
            Selection.DisplayEquation = True
            Selection.DisplayRSquared = True

    End Sub
```

The reference to **SeriesCollection(1)** directly references the first curve added to the graph. This version of the **Trendy** macro will always add the trendline to the first data series.

One solution is to add a trendline to every data series on the graph. The following version of the **Trendy** macro is designed to do just that.

```
Sub Trendy()
'
' Trendy Macro
' Adds a linear trendline to each data series - VERSION THREE
'
' Keyboard Shortcut: Ctrl+Shift+T
'
    Dim CurrSeries As Long
    For CurrSeries = 1 To ActiveChart.SeriesCollection.Count
        ActiveChart.SeriesCollection(CurrSeries).Trendlines.Add
        ActiveChart.SeriesCollection(CurrSeries).Trendlines(1).
          Select
        With Selection
            .Type = xlLinear
            .DisplayEquation = True
            .DisplayRSquared = True
        End With
    Next
End Sub
```

This final version of the **Trendy** macro uses a **For/Next** loop to add a trendline to each data series in the graph. The result of using this version of the macro is shown in Figure 12.44.

Figure 12.44

After running version three of the **Trendy** macro.

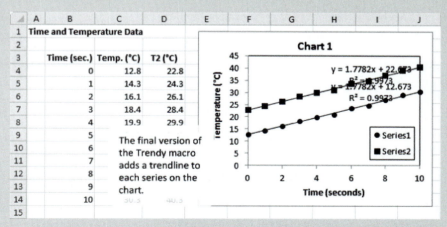

(*continued*)

EXAMPLE 12.4

A MACRO FOR LINEAR INTERPOLATION: **LININTERP**

A linear interpolation macro might be useful for interpolating in data tables. The formula that the macro will create will depend on how the data to be interpolated are arranged. We will assume that the data are arranged in columns, with the known *x* values on the left and *y* values, including a space for the unknown value, on the right. The *x* values will include a low, middle, and high value. The low and high *y* values will be known, and we will solve for the middle *y* value by linear interpolation:

x Values (Knowns)	y Values (Low and High Values are Known)
x_{LOW}	y_{MID}
x_{MID}	$[y_{MID}]$
x_{HIGH}	y_{HIGH}

A quick way to write a linear interpolation equation uses the ratios of differences of *x* values and *y* values:

$$\frac{x_{MID} - x_{LOW}}{x_{HIGH} - x_{LOW}} = \frac{[y_{MID}] - y_{LOW}}{y_{HIGH} - y_{LOW}}. \tag{12.2}$$

Solving for the unknown, y_{MID}, gives

$$y_{MID} = y_{LOW} + (y_{HIGH} - y_{LOW}) \times \left[\frac{x_{MID} - x_{LOW}}{x_{HIGH} - x_{LOW}} \right]. \tag{12.3}$$

We'll develop the macro with a really obvious interpolation (see Figure 12.45) to make sure the macro works correctly.

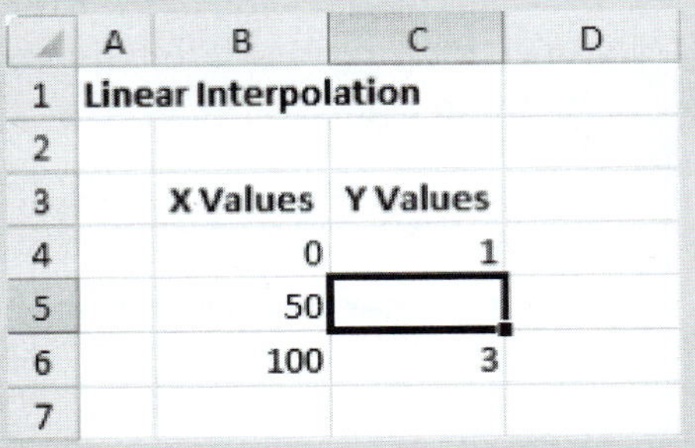

Figure 12.45
Preparing to record a linear interpolation macro.

	A	B	C	D
1	Linear Interpolation			
2				
3		X Values	Y Values	
4		0	1	
5		50		
6		100	3	
7				

The macro should interpolate and get the value 2 for the unknown *y* value in cell C5. To record the macro:

1. Click on cell C5 to select it (as shown in Figure 12.45).
2. Begin recording the macro.
3. In the Record Macro dialog, give the macro a name and assign it to a shortcut key, and enter a description. This is illustrated in Figure 12.46.

Figure 12.46
Define the linear
interpolation macro.

> **Record Macro**
>
> Macro name:
>
> LinInterp
>
> Shortcut key:
>
> Ctrl+Shift+ L
>
> Store macro in:
>
> This Workbook
>
> Description:
>
> Linear Interpolation macro
>
> OK   Cancel

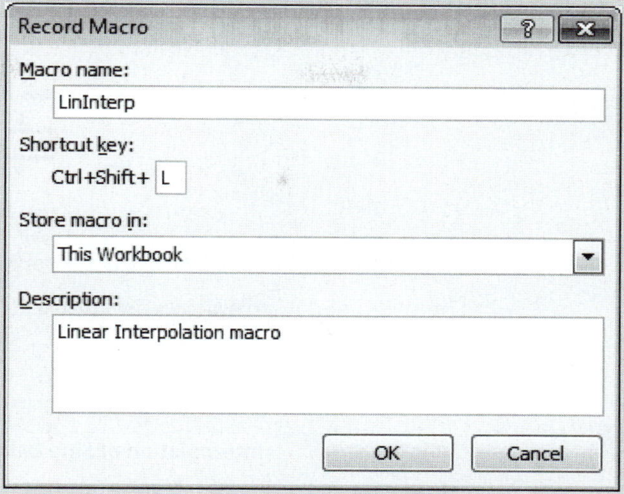

4. Be sure that the **Use Relative References** button in the Macro group on the Ribbon's **View** tab is selected, then
5. Enter the interpolation formula by clicking on the cells containing the following values

$$= 1+(3-1)*((50-0)/(100-0))$$

Excel will take care of the cell addresses, but the formula is actually

$$=C4+(C6-C4)*((B5-B4)/(B6-B4))$$

as shown in Figure 12.47.

Figure 12.47
Entering the interpolation
formula.

SUM ▼	✕ ✓ fx	=C4+(C6-C4)*((B5-B4)/(B6-B4))			
A	B	C	D	E	F
1	**Linear Interpolation**				
2					
3	**X Values**	**Y Values**			
4	0	1			
5	50	=C4+(C6-C4)*((B5-B4)/(B6-B4))			
6	100	3			
7					

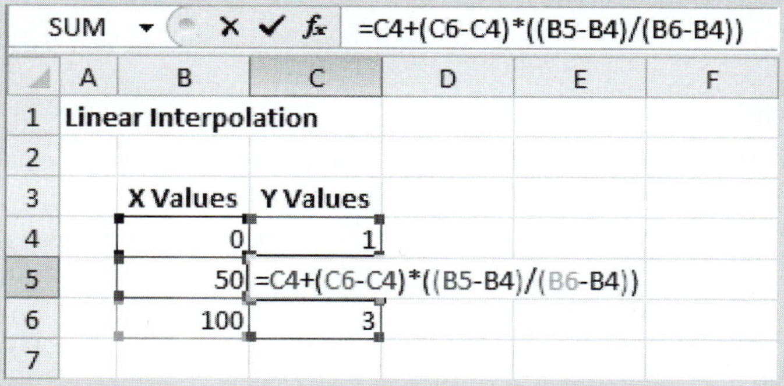

6. Press [Enter] to complete the formula, then stop the macro recorder.

   The interpolation result is shown in Figure 12.48. The interpolation formula got the value 2 correctly, but the real test is how the interpolation will work on other tables.

   Here are a couple of examples of the interpolation macro in use.

   (continued)

Figure 12.48
The interpolation result.

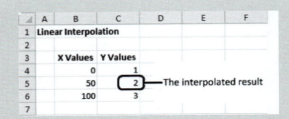

	A	B	C	D	E	F
1	Linear Interpolation					
2						
3		X Values	Y Values			
4		0	1			
5		50	2	—The interpolated result		
6		100	3			
7						

Sine Data between 0° and 90°: Interpolate for Sin(45°)

The original data are shown in Figure 12.49.

Figure 12.49
Original sine data.

C4	▼	f_x	=SIN(RADIANS(B4))

	A	B	C	D	E
1	Interpolation of Sine Data				
2					
3		X Values (°)	Y Values (sin(X))		
4		0	0.000		
5		10	0.174		
6		20	0.342		
7		30	0.500		
8		40	0.643		
9		50	0.766		
10		60	0.866		
11		70	0.940		
12		80	0.985		
13		90	1.000		
14					

Then, we insert a row to add the *x* value 45°, as shown in Figure 12.50.

Figure 12.50
Adding a row for 45°.

	A	B	C	D	E
1	Interpolation of Sine Data				
2					
3		X Values (°)	Y Values (sin(X))		
4		0	0.000		
5		10	0.174		
6		20	0.342		
7		30	0.500		
8		40	0.643		
9		45			
10		50	0.766		
11		60	0.866		
12		70	0.940		
13		80	0.985		
14		90	1.000		
15					

Finally, we select cell C9 and run the **LinInterp** macro to interpolate for $\sin(45°)$. The result is shown in Figure 12.51.

Figure 12.51
The interpolation result.

C9		f_x	=C8+(C10-C8)*((B9-B8)/(B10-B8))		
	A	B	C	D	E
1	Interpolation of Sine Data				
2					
3		X Values (°)	Y Values (sin(X))		
4		0	0.000		
5		10	0.174		
6		20	0.342		
7		30	0.500		
8		40	0.643		
9		45	0.704		
10		50	0.766		
11		60	0.866		
12		70	0.940		
13		80	0.985		
14		90	1.000		
15					

The actual value is 0.707, so linear interpolation on this nonlinear data resulted in a 0.4% error.

Superheated Steam Enthalpy Data: Interpolate for Specific Enthalpy at 510 K

The original data are shown in Figure 12.52.

Figure 12.52
Original steam enthalpy data.

	A	B	C	D	E	F
1	Steam Enthalpy at 5 Bars					
2						
3		Temp. (K)	Specific Enthalpy (kJ/kg)			
4		300	3065			
5		350	3168			
6		400	3272			
7		450	3379			
8		500	3484			
9		550	3592			
10		600	3702			
11						

(*continued*)

To interpolate for the enthalpy at 510 K, we insert a row for 510 K, as shown in Figure 12.53.

Figure 12.53
Inserted row for 510 K.

	A	B	C	D	E	F
1	Steam Enthalpy at 5 Bars					
2						
3		Temp. (K)	Specific Enthalpy (kJ/kg)			
4		300	3065			
5		350	3168			
6		400	3272			
7		450	3379			
8		500	3484			
9		510				
10		550	3592			
11		600	3702			
12						

The result of the interpolation is shown in Figure 12.54.

Figure 12.54
The interpolated result.

C9 f_x =C8+(C10-C8)*((B9-B8)/(B10-B8))

	A	B	C	D	E	F
1	Steam Enthalpy at 5 Bars					
2						
3		Temp. (K)	Specific Enthalpy (kJ/kg)			
4		300	3065			
5		350	3168			
6		400	3272			
7		450	3379			
8		500	3484			
9		510	3506			
10		550	3592			
11		600	3702			
12						

Using the linear interpolation macro, we have determined that the specific enthalpy of steam at 510 K and 5 bars is 3506 kJ/kg.

STATICS

Resolving Forces into Components: Part 1

A standard procedure in determining the net force on an object is breaking down, or resolving, the applied forces into their components in each coordinate direction. In this example (see Figure 12.55), a 400-N force is pulling on a stationary hook at an angle of 20° from the *x* axis. A second force of 200 N acts at 135° from the *x* axis.

Figure 12.55
Forces applied to a stationary hook.

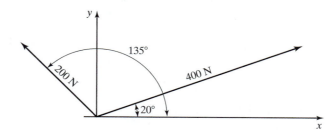

We will record a macro, called **Resolve,** which will resolve forces into their components. Then we will compute the magnitude and direction of the resultant force.

1. We begin setting up the worksheet by entering the first force and its angle from the *x* axis, as shown in Figure 12.56.

Figure 12.56
Preparing to record the macro.

	A	B	C	D	E	F
1	**Resolving Forces**					
2						
3		**Force (N)**	**Angle (°)**	**X-comp (N)**	**Y-comp (N)**	
4		400	20			
5						
6						

2. Then we select cell D4 and start recording the macro.
3. We give the macro a name and shortcut key using the Record Macro dialog (see Figure 12.57). (Remember to use relative cell references.)

Figure 12.57
Naming the macro.

Record Macro

Macro name:
Resolve

Shortcut key:
Ctrl+Shift+ R

Store macro in:
This Workbook

Description:
Resolve a force at an angle into two component forces.

OK   Cancel

4. Enter the following formula in cell D4 to compute the *x* component of the 400-N force:

$$=B4*COS(RADIANS(C4))$$

This is illustrated in Figure 12.58.

D4			f_x	=B4*COS(RADIANS(C4))		
	A	B	C	D	E	F
1	Resolving Forces					
2						
3		Force (N)	Angle (°)	X-comp (N)	Y-comp (N)	
4		400	20	375.9		
5						
6						

Figure 12.58
Computing the x component of force.

5. While still recording the macro, click on cell E4 and enter the formula for computing the *y* component of the 400-N force:

$$=B4*SIN(RADIANS(C4))$$

This is illustrated in Figure 12.59.

E4			f_x	=B4*SIN(RADIANS(C4))		
	A	B	C	D	E	F
1	Resolving Forces					
2						
3		Force (N)	Angle (°)	X-comp (N)	Y-comp (N)	
4		400	20	375.9	136.8	
5						
6						

Figure 12.59
Computing the y component of force.

6. Stop the macro recorder.
The new macro wasn't useful at all for finding the components of the 400-N force, because the required formulas were typed into cells D4 and E4 as the macro was recorded. The value of the macro starts to become apparent during calculation of the components of the 200-N force.

First, the force and angles are entered into the worksheet, and cell C5 is selected, as shown in Figure 12.60.

	A	B	C	D	E	F
1	Resolving Forces					
2						
3		Force (N)	Angle (°)	X-comp (N)	Y-comp (N)	
4		400	20	375.9	136.8	
5		200	135			
6						

Figure 12.60
Using the macro with the 200-N force.

Then the macro is run, either from the macro list or by using the assigned shortcut ([Ctrl-Shift-R] in this example). The x and y components are calculated by the macro; the result is shown in Figure 12.61.

	A	B	C	D	E	F
1	Resolving Forces					
2						
3		Force (N)	Angle (°)	X-comp (N)	Y-comp (N)	
4		400	20	375.9	136.8	
5		200	135	-141.4	141.4	
6						

Figure 12.61
The macro's result.

To find the magnitude and direction of the resultant force, first sum the x components and y components of the forces, as shown in Figure 12.62.

D10			f_x	=DEGREES(ATAN(E7/D7))		
	A	B	C	D	E	F
1	Resolving Forces					
2						
3		Force (N)	Angle (°)	X-comp (N)	Y-comp (N)	
4		400	20	375.9	136.8	
5		200	135	-141.4	141.4	
6						
7			Sums:	234.5	278.2	
8						
9			Resultant Force:	363.8 N		
10			Resultant Angle:	49.9 °		
11						

Figure 12.62
Completing the solution.

Then compute the magnitude of the resultant force as

$$F_{res} = \sqrt{\left(\sum F_x\right)^2 + \left(\sum F_y\right)^2}, \tag{12.4}$$

or, as entered in cell D9, =SQRT (D7^2+E7^2).

Finally, compute the angle of the resultant from the x axis as

$$\theta_{res} = \text{ATAN}\left(\frac{\sum F_y}{\sum F_x}\right), \tag{12.5}$$

which is entered in cell D10 as: =DEGREES (ATAN (E7/D7)). Here, Excel's **DEGREES** function was used to convert the radians returned by the **ATAN** function into degrees. The result is shown in Figure 12.62.

Some of the homework problems at the end of this chapter will provide opportunities to use the **Resolve** macro.

12.3.4 Saving Your VBA Project

Your VBA project (including any modules, Subs, and functions) is stored part of the Excel workbook. To save the project, simply save the Excel workbook. Remember that you must save in a macro-enabled workbook with the .xlsm file extension or your VBA project will be disabled.

12.4 PROGRAMMED MACROS (VBA)

The alternative to a recorded macro is a *programmed macro,* using the VBA programming language that is built into Excel.

In a single chapter, we cannot begin to describe the range of commands available through VBA, but there are a few very common tasks that can be accomplished with just a handful of VBA commands. These include the following:

- changing the values in the active cell
- changing the properties of the active cell
- changing the location of the active cell
- using values from cells near the active cell to calculate a value
- using values from fixed cell locations to calculate a value
- selecting a range of cells
- changing the values in the selected range of cells
- changing the properties of the selected range of cells

As a start, we will develop a macro that changes the properties of the active cell by putting a border around the active cell.

12.4.1 A Simple Border Macro

Our macro will put a heavy, black line around the active cell so that it will be easy to see in a black-and-white text. In VBA, we need to consider *methods, properties,* and *objects.*

- Method—causes an action or change
- Property—the thing that gets changed
- Object—the owner of the property that gets changed

We will use VBA's **BorderAround** method to change the **Border** property of the **ActiveCell** object. The **ActiveCell** is the currently selected cell when the macro is run. The **BorderAround** method needs to know what style of line to use, how thick to draw the line (i.e., the line weight), and what color to make the line.

There are several *predefined constants* in VBA used to specify line style:

- xlContinuous
- xlDash
- xlDashDot
- xlDashDotDot
- xlDot
- xlDouble
- xlSlantDashDot
- xlLineStyleNone

For example, to create a simple solid line, use `LineStyle := xlContinuous`. Similarly, VBA uses predefined constants to specify the line weight:

- xlHairline
- xlThin

- xlMedium
- xlThick

There are two ways to specify the line color:

- Specify the *ColorIndex*
- Set the color value by using the **RGB** function

The **ColorIndex** is a code ranging from 1 to 56 that identifies specific colors. The best way to select colors is to use the VBA help files and look up **ColorIndex**. A few commonly used **ColorIndex** values are listed in Table 12.1.

Table 12.1 ColorIndex and equivalent RGB values

	ColorIndex	RGB Values
Black	1	0, 0, 0
White	2	255, 255, 255
Red	3	255, 0, 0
Green	4	0, 255, 0
Blue	5	0, 0, 255
Yellow	6	255, 255, 0
Magenta	7	255, 0, 255
Cyan	8	0, 255, 255

VBA's **RGB(red,green,blue)** function sets the levels of red, green, and blue individually in the range from 0 (least intense) to 255 (brightest). The RGB values required to generate some common colors are listed in Table 12.1.

Including the statement

```
ActiveCell.BorderAround LineStyle:=xlDouble, ColorIndex:=1
```

in a programmed macro will cause a double-line border in black to be drawn around the cell that is currently active when the macro is run.

Before we can edit a macro to include this statement, we need to create the macro. One way to do this is to use Ribbon options **View/Macros (group)/Macros (menu)/View Macros**. This opens the Macro dialog, shown in Figure 12.63.

Figure 12.63
The **Record** Macro dialog.

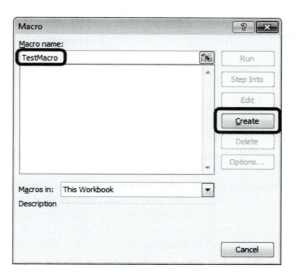

Note: A new Excel workbook has been used to store these programmed macros, so the names of the macros recorded earlier in the chapter do not appear in the Macro dialog.

Enter a name for the new macro, **TestMacro** in this example, and then click the **Create** button to create the VBA Sub that will hold the macro code. The VBA editor will be opened so that the macro code can be entered. In Figure 12.64, the macro code `ActiveCell.BorderAround LineStyle:=xlDouble, ColorIndex:=1` was typed into the VBA editor.

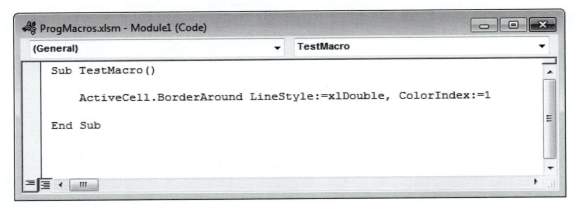

Figure 12.64
Creating the **TestMacro** macro in the VBA editor.

12.4.2 Running the Macro

Excel assumes that any Sub with no parameters (i.e., one with no variables inside the parentheses after the Sub name) is a macro and includes the Sub name in the list of macros that is displayed when you use Ribbon options **View/Macros/Macros (menu)/View Macros** [Excel 2003: Tools/Macro/Macros...].

When you select the **TestMacro** name and click the **Run** button (or use the assigned shortcut key), the **TestMacro()** Sub runs, and a black, double-line border is drawn around the currently active cell (cell B3 in Figure 12.65).

Figure 12.65
After running **TestMacro**.

⊿	A	B	C	D
1	Checking the TestMacro			
2				
3				
4				

12.4.3 Using a Programmed Macro from Another Workbook

Because macros are stored with workbooks, a macro you write (or record) in one workbook will not be immediately available for use in another workbook. However, Excel makes it easy to use macros from other workbooks by showing all of the macros in all open workbooks in the macro list that is displayed on the Macro dialog (Figure 12.66). To use a macro from another workbook, just make sure that the both workbooks are macro enabled, both are open.

For example, if a new macro-enabled workbook is opened and the workbook holding the **TestMacro** macro (called ProgMacros.xlsm here) is still open, the

TestMacro macro is available to the new workbook and displayed in the Macro dialog in the new workbook—if **All Open Workbooks** has been selected as shown in Figure 12.66.

Figure 12.66
The Macro dialog showing the **TestMacro**.

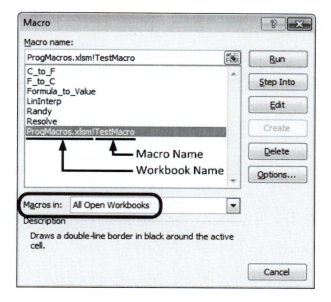

It is possible for the shortcut keys assigned to macros in multiple workbooks to conflict when the workbooks are open together. The shortcut keys for the macros in the current workbook are used if there is a conflict.

The macro names will never conflict, because the macro name is shown with the workbook file name, as file **name!macro name**. In Figure 12.61, the macro named **TestMacro** in file **ProgMacros.xlsm** was listed as **ProgMacros.xlsm!testMacro**.

12.4.4 Common Macro Commands

There are a few very common tasks that can be accomplished with a handful of VBA commands.

Working with the Active Cell
- changing the values in the active cell
- changing the properties of the active cell
- changing the location of the active cell

Using Values from Specified Cells
- using values from cells near the active cell to calculate a value
- using values from fixed cell locations to calculate a value

Working with a Cell Range
- selecting a range of cells
- changing the values in the selected range of cells
- changing the properties of the selected range of cells

Changing the Values in the Active Cell

The *ActiveCell object* tells VBA where the cell currently selected is located and provides access to the contents of the cell (formula and value) and the properties associated with the cell. If you click on a cell in a workbook, the selection box moves to that cell, and the cell becomes the active cell. In Figure 12.67, cell B3 is the active cell.

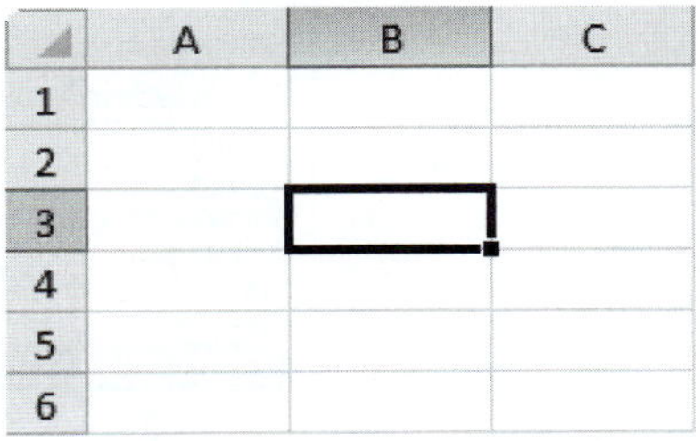

Figure 12.67
Cell B3 is the active cell.

When multiple cells are selected, the active cell is indicated by a background colored differently from the rest of the selected cells. In Figure 12.68, cell B3 is still the active cell; cells B3:B5 comprise the selected range.

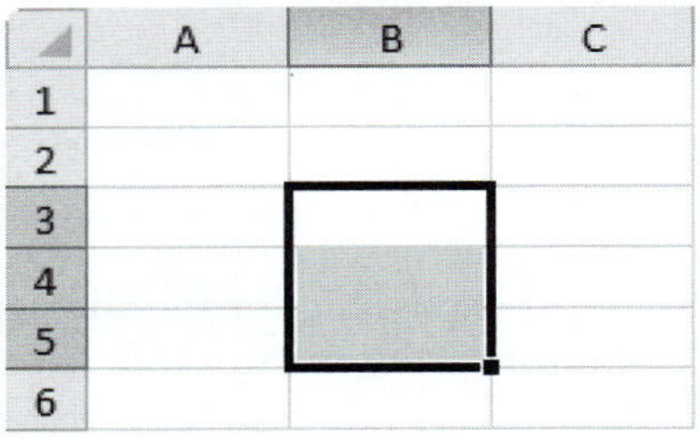

Figure 12.68
Cell B3 is the active cell, cells B3:B5 are the selected range.

If your macro uses the ActiveCell object, only the active cell will be changed by the macro, not the selected range.

To access or change the value of the active cell, use ActiveCell.value. The following macro, called **setValue,** checks the current value of the selected cell and then changes it. If the current value is a number greater than or equal to 5, then the macro sets the value to 100. If the current value is less than 5, then the macro puts a text string, "less than five," in the cell. It's not obvious why you would want to do this, but the macro does illustrate how your Sub can use the value of the active cell (in the If statement), as well as change the value, either to a numeric or to a text value:

```
Public Sub setValue()
        If ActiveCell.Value >= 5 Then
                ActiveCell.Value = 100
        Else
                ActiveCell.Value = "Less than five"
        End If
End Sub
```

In Figure 12.69, the cells in columns B and C initially held the same values. The **setValue** macro was run in each cell in column C.

Figure 12.69
Before and after running the **setValue** macro in column C.

	A	B	C	D
1				
2		Before	After	
3		3	Less than five	
4		4	Less than five	
5		5	100	
6		6	100	
7				

EXAMPLE 12.5

MODIFIED LINEAR INTERPOLATION

Sometimes, instead of leaving a formula in a cell, you might want just the interpolated value. The **LinInterp** macro developed in Example 12.4 can be modified so that it leaves the interpolated value in the active cell.

As a reminder, the following is the original **LinInterp** macro:

```
Sub LinInterp()
'
' LinInterp Macro
' linear interpolation
'
' Keyboard Shortcut: Ctrl+Shift+L
'
    ActiveCell.FormulaR1C1 = "=R[-1]C+(R[1]C-R[-1]
    C)*((RC[-1]-R[-1]C[-1])/(R[1]C[-1]-R[-1]C[-1]))"
'   <insert new line here>

End Sub
```

The formula with all of the R's and C's is fairly complicated, but we're not going to modify that. Instead, we will simply allow Excel to put the formula into the cell (as it already does). Then we will grab the value from the cell and assign it back to the cell—effectively overwriting the formula with the calculated value.

To accomplish this, simply replace

```
'   <insert new line here>
```

with

```
ActiveCell.Value = ActiveCell.Value
```

The ActiveCell.Value on the right reads the current value in the cell (after the formula was entered). The ActiveCell.Value= on the left sets the cell's value to whatever is on the right side of the equal sign (in this case the cell's own value, thus replacing the formula as the cell's contents).

After this modification, the final result in the Sine Data example (previously in Figure 12.51) is shown in Figure 12.70.

The worksheet looks about the same, but cell C9 now contains a value rather than the interpolation formula.

Figure 12.70
Sine data interpolation leaving a value, not a formula.

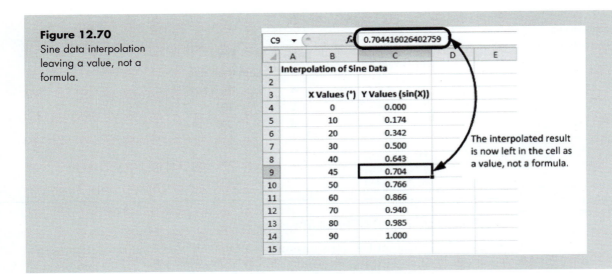

Changing the Properties of the Active Cell

The macro developed earlier, called **TestMacro**, was an example of changing the properties of the active cell. In that macro, the border of the cell was changed by using the BorderAround method. Common cell properties that can be changed include those in Table 12.2.

Table 12.2 Cell properties

ActiveCell.Borders	.Color or .ColorIndex	Color := RGB(red, blue, green), where red, blue, and green are values between 0 and 255
		ColorIndex := val where val is an integer between 1 and 56
	.LineStyle	xlContinuous, xlDash, xlDashDot, xlDashDotDot, xlDot, xlDouble, xlSlantDashDot, xlLineStyleNone
	Weight	xlHairline, xlThin, xlMedium, xlThick
ActiveCell.font	.Bold	True or False
	.Color or .ColorIndex	(Same as described above for ActiveCell.Borders)
	.Italic	True or False
	.Name	A font name in quotes, such as "Courier"
	.Size	Numeric values (8–12 are common)
	.Subscript	True or False
	.Superscript	True or False
ActiveCell.Interior	.Color or .ColorIndex	(Same as described above for ActiveCell.Borders)
	.Pattern	xlPatternAutomatic, xlPatternChecker, xlPatternCrissCross, xlPatternDown, xlPatternGray16, xlPatternGray25, xlPatternGray50, xlPatternGray75, xlPatternGray8, xlPatternGrid, xlPatternHorizontal, xlPatternLightDown, xlPatternLightHorizontal, xlPatternLightUp, xlPatternLightVertical, xlPatternNone, xlPatternSemiGray75, xlPatternSolid, xlPatternUp, xlPatternVertical
	.PatternColor or .PatternColorIndex	(Same as described above for ActiveCell.Borders)

The **setProperty** macro listed below changes the active cell's font color to blue [RGB(0,0,255)], its font name to "Courier," its Font size to 24, and its interior color to yellow (ColorIndex = 6) with a green (PatternColorIndex = 4) diagonal stripe (Pattern = xlPatternLightUp):

```
Public Sub setProperty()
        ActiveCell.Font.Color = RGB(0, 0, 255)
        ActiveCell.Font.Name = "Courier"
        ActiveCell.Font.Size = 24
        ActiveCell.Interior.ColorIndex = 6
        ActiveCell.Interior.Pattern = xlPatternLightUp
        ActiveCell.Interior.PatternColorIndex = 4
End Sub
```

Making a Different Cell the Active Cell

The *Offset property* of the ActiveCell object allows you to select cells a specified number of rows and columns away from the active cell. The *Activate method* causes the selected cell to become the active cell.

The following macro finds a cell two rows below and one row to the right of the currently active cell and makes the new cell the active cell and puts the word "New" in the cell:

```
Public Sub MoveActive()
        Dim rowOffset As Integer
        Dim colOffset As Integer

        rowOffset = 2
        colOffset = 1

        ActiveCell.Offset(rowOffset, colOffset).Activate
        ActiveCell.Value = "New"
End Sub
```

The following macro is functionally equivalent, but uses numbers in the ActiveCell.Offset statement instead of variables:

```
Public Sub MoveActive()
        ActiveCell.Offset(2, 1).Activate
        ActiveCell.Value = "New"
End Sub
```

In the worksheet shown in Figure 12.71, the cell labeled "Original" is selected before running the **MoveActive** macro. After we run the macro, the cell labeled "New" is active.

Figure 12.71
After running the **MoveActive** macro.

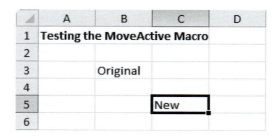

The most common use of this technique is to move the selection box to the next cell after completing a task—to prepare for the next task. The **convertUnits** macro, listed below, takes a length in inches in the active cell, converts it to centimeters, marks the new units in the adjacent cell, and moves the active cell down to the next row to prepare for the next value to be entered or converted:

```
Public Sub convertUnits()

        ' convert inches in active cell to cm, then store new
        value back in active cell

        ActiveCell.Value = ActiveCell.Value * 2.54
        ' mark the units as "cm" in adjacent column
        ActiveCell.Offset(0, 1).Value = "cm"

        ' move selection box down to prepare for next conversion
        ActiveCell.Offset(1, 0).Activate

End Sub
```

In the worksheet shown in Figure 12.72, the first three values have already been converted.

Figure 12.72
Using the **convertUnits** macro to update a list of values.

	A	B	C	D
1	Length Values			
2				
3		2.54	cm	
4		5.08	cm	
5		7.62	cm	
6		4	inches	
7		5	inches	
8		6	inches	
9		7	inches	
10				

Using Values from Cells near the Active Cell to Calculate a Value

The Offset property is also used to grab values from nearby cells to perform calculations. For example, the volumes of ideal gas at various temperatures can be calculated by running the **calcVolume** macro:

```
Public Sub calcVolume()

        Dim P As Single
        Dim V As Single
        Dim N As Single
        Dim R As Single
        Dim T As Single

        R = 0.08206    'liter atm/gmol K
        P = 1#         'atm
        N = 1#         'gmol

        '  get temperature from cell to left of active cell
        T = ActiveCell.Offset(0, -1).Value

        '  calculate volume using ideal gas law
        V = N * R * T / P
```

```
'   assign volume to active cell
ActiveCell.Value = V

'   move active cell down one row to prepare for next
calculation
ActiveCell.Offset(1, 0).Activate
```

    End Sub

In the worksheet shown in Figure 12.73, the **calcVolume** macro has been run in cells B6 through B8 (admittedly, it is easier to calculate these volumes without using a macro).

Figure 12.73
Calculating ideal gas volumes with the **calcVolume** macro.

	A	B	C	D	E
1	Ideal Gas Volume of 1 mole at 1 atm abs.				
2					
3		Temp. (K)	Volume (L)		
4		200	16.41		
5		250	20.52		
6		300	24.62		
7		350			
8					

Using Values from Fixed Cell Locations

In the **calcVolume** macro, the pressure and number of moles were specified in the macro code. The macro would be more useful if the user could specify the pressure and number of moles of gas on the worksheet. To fix this, we modify the worksheet to include the pressure and number of moles, as shown in Figure 12.74.

Figure 12.74
Modified worksheet allowing P and N to be specified.

	A	B	C	D	E
1	Ideal Gas Volumes				
2					
3		Pressure:	2	atm abs	
4		Moles:	4	moles	
5					
6		Temp. (K)	Volume (L)		
7		200			
8		250			
9		300			
10		350			
11					

Then, in the macro code, we need to get the values for pressure and number of moles from cells C3 and C4, respectively. This is done by using the Value property of the Cells object:

```
P = Cells(3, 3).Value   ' pressure in cell C3 (row 3, column 3)
N = Cells(4, 3).Value   ' moles in cell C4 (row 4, column 3)
```

Note that Cells(3,3).Value refers to the value in the cell in row 3, column 3—or cell C3 using the standard cell-referencing nomenclature.

The modified macro, called **calcVolume2,** is listed here:

```
Public Sub calcVolume2()
        Dim P As Single
        Dim V As Single
        Dim N As Single
        Dim R As Single
        Dim T As Single

        R = 0.08206             ' liter atm/gmol K
        P = Cells(3, 3).Value ' atm
        N = Cells(4, 3).Value ' gmol

        ' get temperature from cell to left of active cell
        T = ActiveCell.Offset(0, -1).Value
        ' calculate volume using ideal gas law
        V=N*R*T/P

        ' assign volume to active cell
        ActiveCell.Value = V

        ' move active cell down one row to prepare for next
        calculation
        ActiveCell.Offset(1, 0).Activate
End Sub
```

The result of using the **calcVolume2** macro is shown in Figure 12.75.

Figure 12.75
The result of applying macro **calcVolume2**.

	A	B	C	D	E
1	Ideal Gas Volumes				
2					
3		Pressure:	2	atm abs	
4		Moles:	4	moles	
5					
6		Temp. (K)	Volume (L)		
7		200	32.82		
8		250	41.03		
9		300	49.24		
10		350	28.72		
11					

Selecting a Range of Cells

The **setRangeProp** macro uses the *Range object* to set the background of cells B3 through C5 to a light-gray pattern and uses the Select property to select the entire range:

```
Public Sub setRangeProp()
        Range("B3:C5").Interior.Pattern = xlPatternGray16
End Sub
```

The result is shown in Figure 12.76.

Figure 12.76
After running the
setRangeProp macro.

	A	B	C	D	E
1	Setting a Property Value of a Range of Cells				
2					
3					
4					
5					
6					

This approach isn't very useful because the cell addresses are coded directly into the macro. In the next macro, we will make changes to the cell range selected by the user before running the macro.

Changing the Values in a Selected Range of Cells

If you have selected a range of cells before running a macro, you can use the RangeSelection.Address property of the *ActiveWindow object* to make changes to all of the selected cells. Here, the **setRangeValues** macro sets the value of each cell in the selected range to 10:

```
Public Sub setRangeValues()
        Range(ActiveWindow.RangeSelection.Address).Value = 10
End Sub
```

Before running the **setRangeValues** macro, a cell range is selected, as shown in Figure 12.77.

Figure 12.77
Cells B3:D5 selected
before running the
setRangeValues macro.

	A	B	C	D	E
1	Changing the Currently Selected Range of Cells				
2					
3					
4					
5					
6					

After running the macro, the selected cells have new values, as shown in Figure 12.78.

Figure 12.78
After running the
setRangeValues macro.

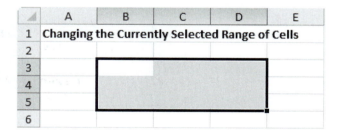

	A	B	C	D	E
1	Changing the Currently Selected Range of Cells				
2					
3		10	10	10	
4		10	10	10	
5		10	10	10	
6					

Changing the Properties of a Selected Range of Cells

If a range of cells is selected before running the macro, use Range(ActiveWindow. RangeSelection.Address) to access the range of cells and then append the desired property change. The following example changes the background pattern of the selected range to diagonal stripes:

```
Public Sub setSelRangeProp()

        Range(ActiveWindow.RangeSelection.Address).Interior.
        Pattern = xlPatternUp

End Sub
```

Before running the **setSelRangeProp** macro, we select as cell range (Figure 12.79).

Figure 12.79
Cells C2:D8 selected before running the **setSelRangeProp** macro.

	A	B	C	D	E	F
1	Changing the Currently Selected Range of Cells					
2						
3			10	10	10	
4			10	10	10	
5			10	10	10	
6						
7						
8						

The worksheet after running the macro and clicking outside of the selected region is shown in Figure 12.80.

Figure 12.80
After running the **setSelRangeProp** macro.

	A	B	C	D	E
1	Changing the Currently Selected Range of Cells				
2					
3			10	10	10
4			10	10	10
5			10	10	10
6					
7					
8					

APPLICATION

STATICS

Calculating Resultant Forces and Angles: Part 2
We will use VBA to write a macro to take the x and y components of force and determine the resultant force and the direction of the resultant force vector.

The resultant force is computed as

$$F_{\text{res}} = \sqrt{\left(\sum F_x\right)^2 + \left(\sum F_y\right)^2}, \tag{12.6}$$

and the angle of the resultant from the x axis is

$$\theta_{\text{res}} = \text{Atan}\left(\frac{\sum F_y}{\sum F_x}\right). \tag{12.7}$$

For this macro, we will assume that the x and y components of force will be available in adjacent cells, and we will place the calculated results in cells to the right of the force components. The worksheet shown in Figure 12.81 shows the summed force components in cells D7 and E7. The active cell is cell F7, which will contain the computed resultant force. The computed angle will be put in cell G7.

	A	B	C	D	E	F	G	H
1	Resolving Forces							
2								
3		Force (N)	Angle (°)	X-comp (N)	Y-comp (N)			
4		400	20	375.9	136.8			
5		200	135	-141.4	141.4			
6						Resultant (N)	Res. Angle (°)	
7			Sums:	234.5	278.2			
8								

Figure 12.81
Preparing to calculate the resultant force and angle.

The following VBA Sub (or macro) called **Resultant** computes the resultant force and angle:

```
Public Sub Resultant()
        Dim Fx As Single
        Dim Fy As Single
        Dim Fres As Single
        Dim Theta As Single
        Dim Pi As Single

        Pi = 3.1416

        ' the cell containing Fx is two columns to the left of
        the active cell
        Fx = ActiveCell.Offset(0, -2).Value

        ' the cell containing Fy is one column to the left of the
        active cell
        Fy = ActiveCell.Offset(0, -1).Value

        ' calculate the resultant force, store result in active
        cell
        Fres = Sqr(Fx ^ 2 + Fy ^ 2)        ' Note: Sqr() is a VBA
        function, not an Excel function
        ActiveCell.Value = Fres

        ' calculate the angle in radians, convert to degrees,
        then store in cell
        ' one column to the right of the active cell
        Theta = (Atn(Fy / Fx)) * (180 / Pi)    ' Note: Atn is a
        VBA function
        ActiveCell.Offset(0, 1).Value = Theta
End Sub
```

Notice that this macro uses VBA's **Sqr** and **Atn** functions rather than Excel's **SQRT** and **ATAN** functions. Also, VBA does not have a **DEGREES** function, so the angle in radians was converted to degrees by using (180/Pi).

The resulting worksheet is shown in Figure 12.82.

Note that the results in cells F7 and G7 are values, not formulas. VBA did the calculation, not Excel.

	A	B	C	D	E	F	G	H
1	Resolving Forces							
2								
3		Force (N)	Angle (°)	X-comp (N)	Y-comp (N)			
4		400	20	375.9	136.8			
5		200	135	-141.4	141.4			
6						Resultant (N)	Res. Angle (°)	
7			Sums:	234.5	278.2	363.8	49.9	
8								

Figure 12.82
The result of running the **Resultant** macro.

KEY TERMS

Absolute cell referencing
Activate method (VBA)
ActiveCell object (VBA)
ActiveWindow object
    (VBA)
Auto-start macros
ColorIndex (VBA)
Comment line (VBA)
File extensions
    (.xlsx, .xlsm)

Macro
Macro language
Macro viruses
Methods
Objects
Offset property (VBA)
Predefined constants
    (VBA)
Programmed macro
Properties

Range object (VBA)
Recorded macro
Relative cell referencing
Shortcut key
Subprogram (Sub)
    (VBA)
Virus
VBA

SUMMARY

Recorded Macros

1. Tell Excel you want to record a macro using Ribbon options **View/Macros/ Macros (menu)/Record Macro...** [Excel 2003: Tools/Macro/Record New Macro].

2. Establish the macro name, shortcut key (optional), storage location, and description (optional).

3. Select absolute or relative cell referencing.

4. Record the macro.

5. Stop the recorder.

You can edit a recorded macro, using the Visual Basic editor from Excel. Use:

1. Use Ribbon options **View/Macro/Macro (menu)/View Macros** [Excel 2003: Tools/Macro/Macros...] to set the list of currently available macros.

2. Choose the macro you want to edit and then click the [Edit] button to open the VBA editor.

Programmed Macros

Macros are Visual Basic Subs (subprograms) that receive no arguments. They are stored in a module that is saved as part of a Visual Basic project with the Excel workbook.

To create an empty Sub that can be edited to create a programmed macro:

1. Tell Excel you want to create a blank macro using Ribbon options **View/Macros/ Macros (menu)/View Macros** [Excel 2003: Tools/Macro/Macros…].

2. Enter the macro name, then click **Create**. The VBA editor will open.

Common Macro Commands

In a single chapter, we cannot begin to describe the range of commands available through VBA, but there are a few very common tasks that can be accomplished through VBA commands. These include the following:

- To change the values in the active cell, use ActiveCell.Value.
- To change the properties, use the following:

ActiveCell.Borders	.Color or .ColorIndex	Color: = RGB (red, blue, green), where red, blue, and green are values between 0 and 255 ColorIndex: = val, where val is an integer between 0 and 55
	.LineStyle	xlContinuous, xlDash, xlDashDot, xlDashDotDot, xlDot, xlDouble, xlSlantDashDot, xlLineStyleNone
	.Weight	xlHairline, xlThin, xlMedium, xlThick
ActiveCell.Font	.Bold	True or False
	.Color or .ColorIndex	(Same as described above for ActiveCell.Borders)
	.Italic	True or False
	.Name	A font name in quotes, such as "Courier"
	.Size	Numeric values (8–12 are common)
	.Subscript	True or False
	.Superscript	True or False
ActiveCell.Interior	.Color or .ColorIndex	(Same as described above for ActiveCell.Borders)
	.Pattern	xlPatternAutomatic, xlPatternChecker, xlPatternCrissCross, xlPatternDown, xlPatternGray16, xlPatternGray25, xlPatternGray50, xlPatternGray75, xlPatternGray8, xlPatternGrid, xlPatternHorizontal, xlPatternLightDown, xlPatternLightHorizontal, xlPatternLightUp, xlPatternLightVertical, xlPatternNone, xlPatternSemiGray75, xlPatternSolid, xlPatternUp, xlPatternVertical
	.PatternColor or .PatternColorIndex	(Same as described above for ActiveCell.Borders)

- To move the active cell, use ActiveCell.Offset(rowOffset, colOffset).Activate.
- To use values from cells near the active cell to calculate a value, use ActiveCell. Offset(rowOffset, colOffset).Value.
- To use values from fixed cell locations, use Cells(row, col).Value.
- To select a range of cells (e.g., B2:C5), use Range("B2:C5").Select.
- To change the values in the selected range of cells, use Range(ActiveWindow. RangeSelection.Address).Value = newValue.

PROBLEMS

Note: The first few problems are attempts to create some potentially useful macros for students. These need to be stored in an accessible location if they are to be useful. One solution is to store commonly used macros in a worksheet file designated for that purpose, such as Macros.xls. All macros in currently open worksheets are available to all open worksheets, so you could simply open Macros.xls to have all of your macros available for use.

12.1 Recorded Macro: Name and Date

Record a macro that will insert your name and the current date in the active cell. The current date is available via the **TODAY** function.

12.2 Recorded Macro: Highlight Your Results

Record a macro that will add a colored background to the selected range and add a border.

12.3 Recorded Macro: Insert the Gravitational Constant

Record a macro that will insert the gravitational constant g_c in the active cell. The commonly used value of g_c is

$$g_c = 32.174 \frac{\text{ft lb}_m}{\text{lb}_f \, s^2}. \tag{12.8}$$

12.4 Recorded Macro: Insert the Ideal Gas Constant

Record a macro that will insert the ideal gas constant R in the active cell. The value of R depends on the units. Select the value you use most frequently:

Ideal Gas Constants	
8.314	$\dfrac{m^3 \, Pa}{gmol \, K}$
0.08314	$\dfrac{L \, bar}{gmol \, K}$
0.08206	$\dfrac{L \, atm}{gmol \, K}$
0.7302	$\dfrac{ft^3 \, atm}{lbmol \, °R}$
10.73	$\dfrac{ft^3 \, psia}{lbmol \, °R}$
8.314	$\dfrac{J}{gmol \, K}$
1.987	$\dfrac{BTU}{lbmol \, °R}$

12.5 Computing Ideal Gas Volume

Create a macro (either by recording or by programming) that will compute the volume of an ideal gas when given the absolute temperature, the absolute pressure, the number of moles, and an ideal gas constant. Use your macro to complete the following table.

Ideal gas volume

Temperature	Pressure	Moles	Gas Constant	Volume
273 K	1 atm	1 gmol	$0.08206 \dfrac{\text{L atm}}{\text{gmol K}}$	
500 K	1 atm	1 gmol	$0.08206 \dfrac{\text{L atm}}{\text{gmol K}}$	
273 K	10 atm	1 gmol	$0.08206 \dfrac{\text{L atm}}{\text{gmol K}}$	
273 K	1 atm	10 gmol	$0.08206 \dfrac{\text{L atm}}{\text{gmol K}}$	
500 K	1 atm	10 gmol	$0.08206 \dfrac{\text{L atm}}{\text{gmol K}}$	
500 °R	1 atm	1 lbmol	$0.7302 \dfrac{\text{ft}^3 \text{ atm}}{\text{lbmol °R}}$	

What are the units of volume in each row?

12.6 Computing Gas Heat Capacity

There are standard equations for calculating the heat capacity (specific heat) of a gas at a specified temperature. One common form of a heat capacity equation is a simple third-order polynomial in T:[1]

$$C_p = a + bT + cT^2 + dT^2. \tag{12.9}$$

If the coefficients a, b, c, and d are known for a particular gas, you can calculate the heat capacity of the gas at any T (within an allowable range). The coefficients for a few common gases are listed in the following table:

Heat Capacity Coefficients

Gas	a	b	c	d	Units on T	Valid Rwange
Air	28.94×10^{-3}	0.4147×10^{-5}	0.3191×10^{-8}	-1.965×10^{-12}	°C	0–1500°C
CO_2	36.11×10^{-3}	4.233×10^{-5}	-2.887×10^{-8}	7.464×10^{-12}	°C	0–1500°C
CH_4	34.31×10^{-3}	5.469×10^{-5}	0.3661×10^{-8}	-11.00×10^{-12}	°C	0–1200°C
H_2O	33.46×10^{-3}	0.6880×10^{-5}	0.7607×10^{-8}	-3.593×10^{-12}	°C	0–1500°C

The units on the heat capacity values computed from this equation are kJ/gmol°C.

Create a macro (either by recording or by programming) that will compute the heat capacity of a gas, using coefficients and temperature values in nearby cells, such as in the following table:

Temperature	a	b	c	d	C_p
300	33.46×10^{-3}	0.6880×10^{-5}	0.7604×10^{-8}	-3.593×10^{-12}	

Use your macro to find the heat capacity of water vapor at 300°C (shown above), 500°C, and 750°C. Does the heat capacity of steam change significantly with temperature?

---

[1]From Elementary Principles of Chemical Processes, 3d ed., Felder, R. M. and R. W. Rousseau, New York: Wiley, 2000.

12.7 Resolving Forces into Components

Create a macro to compute the horizontal and vertical components for the following forces:

a. A 400-N force acting at 40° from horizontal.

Figure 12.83a

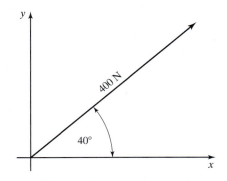

b. A 400-N force acting at 75° from horizontal.

Figure 12.83b

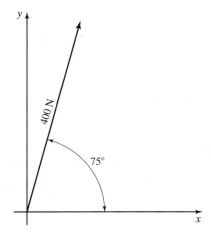

c. A 2000-N force acting at 210° from horizontal.

Figure 12.83c

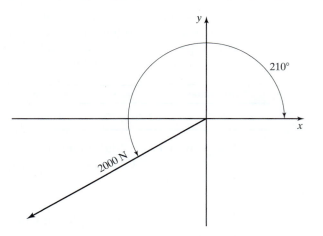

12.8 Calculating Resultant Force and Angle

Create one or two macros to calculate the horizontal and vertical components and the resultant force and angle for the following forces:

Figure 12.84a

a.

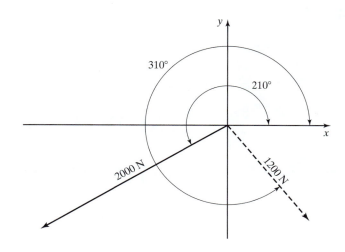

Figure 12.84b

b.

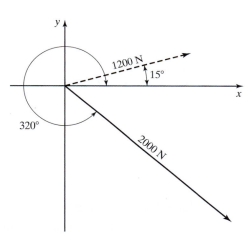

Figure 12.84c

c.

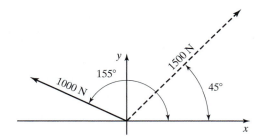

13 User-Written Functions for Excel

Objectives

After reading this chapter, you will know

- How to access Excel's built-in Visual Basic programming environment

- How to write your own functions for Excel

13.1 INTRODUCTION

Many years ago, a decision was made at Microsoft to use *VBA* as the *macro language* in the Microsoft Office products. This decision had significant impacts:

1. Excel comes with a built-in, full-featured programming language. Many engineers use the programming capabilities of Excel because it is so readily available.
2. It is easy to write your own functions in VBA and use them in your worksheets.
3. You can use Excel's built-in functions in your own macros and functions.
4. VBA can be used to write virus programs that can be transmitted with Microsoft Office documents, including Excel workbooks.

The original intent was to give Excel users the ability to write their own functions and to create very powerful and useful macros. That power has been misused by some *virus* programmers.

13.1.1 Formulas, Functions, Macros, VBA Programs: Which One and When?

Excel provides a lot of ways to get work done:

- Formulas
- Functions
- Macros
- VBA programs

The last three are closely related. Excel stores macros as VBA subprograms (called Subs). Both subprograms and functions are part of a VBA program.

- A *function* is a self-contained piece of a program that receives input through arguments (or parameters), performs calculations, and returns a result.
- A *subprogram* is just like a function, except it does not return a result.
- A *macro* is a subprogram that does not receive any inputs; that is, there are no arguments with macro subprograms.

There are situations in which each of these approaches is preferred over the others, but there is a lot of overlap—situations where more than one approach could be used. In some cases, any of these approaches could be used. The following discussion is intended to provide some very general guidelines for selecting the most appropriate approach for some common types of problems.

Formulas

Formulas work directly with the data on the worksheet, using basic math operations. They tend to be short, simple calculations. If they get very complex they become hard to enter, edit, and debug. They can be copied from one cell to another cell or cell range.

Conclusion: Formulas are preferred when the needed operation is a simple math operation using nearby cells as operands or when the same operation will be performed on many cells.

Functions

Functions accept arguments (or parameters), can perform complex sequences of math or other operations on the arguments, and return a result. The arguments provide a lot of flexibility: they can contain a value, a cell reference, or another function. Because functions are used in formulas, they can also be copied from one cell to another or many others.

Built-In functions are included with Excel to handle many common tasks. They are available to any workbook and worksheet. They can be used in combination (one function can call another). There is little on the down side, except that there might be no built-in function that meets your need.

User-written functions can be written whenever needed. They can be tailored to meet very specific needs. But the writing takes time, and the functions typically are available only in the workbook in which they are stored.

Conclusion: Built-in functions should be used if they are available. User-written functions typically are used for very specific, complex calculations.

Macros

Macros can use information from nearby cells (e.g., add three to the value in the cell to the left), but they do not accept arguments. They can output results to the active cell or a related cell (e.g., place the result in the cell three positions to the right), but do not actually return a value. Macros are often used to store sequences of nonmathematical operations.

Many of the operations that used to be coded into macros have been replaced by toolbars and pop-up menus.

Macros can be tied to a keystroke, so they are easy to call when needed, but typically they are available only in the workbook in which they are stored. They cannot be copied from one cell to another.

Most Excel workbooks have macros disabled.

Excel stores macros as VBA programs (actually, subprograms), and once the macro has been saved you can edit it as a program.

Conclusion: Macros are not as commonly used as they once were. They are primarily used for repeated sequences of commands, often nonmathematical commands.

VBA Programs

VBA programs offer the most flexibility and power. You can create *graphical user interfaces (GUIs)* to get information from the user. You can use standard programming methods to make decisions, control operations, and store values. The down side is that programming requires much more effort than the other approaches listed here.

VBA is designed to complement the application (Excel) in which it resides. You can access cells and cell ranges using VBA, and you can use Excel's built-in functions from within VBA.

Conclusion: Usually, VBA programs are written for special applications that require either speed to handle lots of calculations, complex structures to decide how to handle various situations, or a more elaborate user interface than is possible with the simpler methods.

In this chapter, we will write some simple functions. We will use the VBA programming environment to write the functions, but we will barely touch the surface of programming. We will use basic arithmetic, not complex control algorithms—we'll save that for the next chapter.

13.2 MACRO-ENABLED WORKSHEETS

With the 2007 version, Excel added a new level of security against macro viruses by using two different file extensions for macro-enabled and macro-disabled workbooks:

- .xlsx—the Excel file extension for macro-disabled workbooks (default)
- .xlsm—the Excel file extension for macro-enabled workbooks

While the term is "macro-enabled," Excel actually controls access to the VBA *project*. A project is a collection of forms, functions, and subroutines needed to make a program work. Projects are also used to store macros and user-written functions in Excel. Basically, you have to use a macro-enabled workbook to store your user-written functions.

Excel will allow you to write and use your own functions in a macro-disabled workbook; but Excel won't allow you to save the workbook. You must save workbooks containing user-written functions with the .xlsm file extension.

Since 2007, Excel takes one of the following four actions when a workbook containing macros or user-written functions is opened. The options describe macros, but Excel takes the same action in response to user-written functions stored in a workbook.

- Disable all macros without notification.
- Disable all macros with notification (default).
- Disable all macros except digitally signed macros.
- Enable all macros.

You can change the default action taken when macros are encountered using the Excel Options dialog, as shown in Figure 13.1.

- Excel 2010: **File tab/Options**
- Excel 2007: **Office/Excel Options**
- Excel 2003: **File/Options**

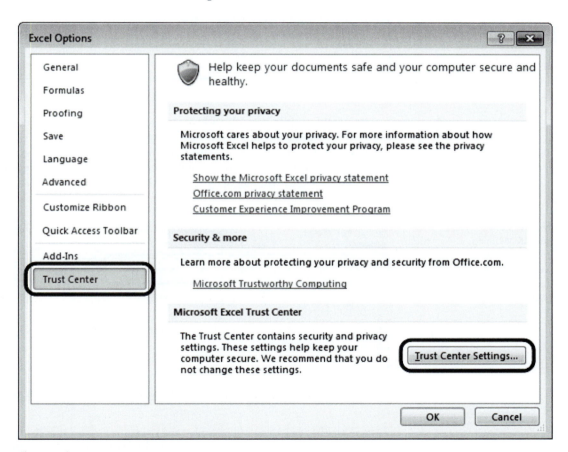

Figure 13.1
Excel Options dialog, Trust Center panel.

Use the **Trust Center** panel and Click the **Trust Center Settings...** button (shown in Figure 13.1) to access the Trust Center dialog (Figure 13.2) and choose the **Macro Settings** panel to see the current action taken when macros are encountered.

Figure 13.2
Trust Center dialog,
Macro Settings panel.

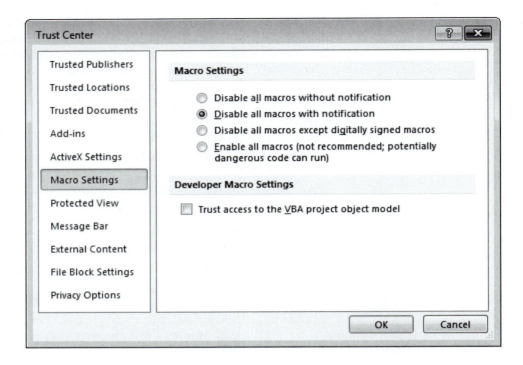

Basically, unless you choose to enable all macros (not recommended), any workbook containing macros or user-written functions that you open will have the macros disabled. If you accept the default action (shown in Figure 13.2), to notify when macros are disabled, the Excel screen when you first open a macro-enabled workbook will show the Security Warning seen in Figure 13.3.

Figure 13.3
Security Warning
displayed when a
workbook containing
macros is opened
(macros disabled).

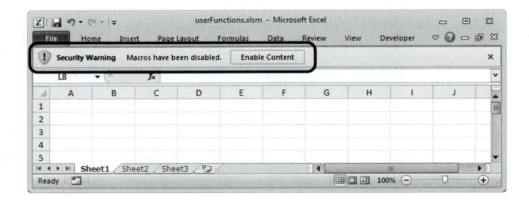

Again, the warning mentions macros, but you will get the same warning if your workbook contains user-written functions.

Clicking the **Enable Content** button indicated in Figure 13.3 will enable all macros in the workbook. If you want a little more information before enabling the macros, you can click on the **Macros have been disabled.** statement, which links to the Excel Options dialog and provides additional information.

Note: Excel 2007 has a slightly different way of notifying you that a workbook has macros. There is still a security warning (as in Figure 13.3), but instead of an **Enable Content** button there is an **Options...** button that opens a dialog. In Excel 2007, you must use the dialog to enable macros.

It is appropriate to be concerned about macros coming from unknown sources, but that shouldn't stop you from writing your own functions when they can help you get your work done faster and easier.

13.3 INTRODUCTION TO VISUAL BASIC FOR APPLICATIONS

Visual Basic is a complete programming language and programming environment built into the Microsoft Office programs as VBA, or *Visual Basic for Applications.* The "for Applications" part of the name implies that VBA in Excel (for example) has added features that allow it to work with the Excel worksheets. VBA can access information in a cell, for example.

Excel stores user-written function in a VBA project, and we will use the VB editor to write and edit our functions.

13.3.1 Starting Visual Basic for Applications

Lots of Excel users never use VBA, so, by default, the Ribbon tab that provides access to VBA isn't shown. Before you can access VBA, you first have to show the *Developer tab.* This is done from the Excel Options dialog, but it is handled differently in Excel 2010 and Excel 2007.

Excel 2010: To show the **Developer** tab, use Ribbon options **File tab/Options** to display the Excel Options dialog, then select the **Customize Ribbon** panel and check the box labeled **Developer** in the **Main Tabs** list. This is illustrated in Figure 13.4.

In Excel 2007: To show the Developer tab, use the Microsoft Office button with options **Office/Excel Options** to display the Excel Options dialog, then select the **Popular** panel and check the box labeled **Show Developer tab in the Ribbon**.

The Ribbon's **Developer** tab is shown in Figure 13.5. The Code group provides access to VBA.

The **Developer** tab's Code group provides a **Visual Basic** button to access to the Visual Basic editor. To start the *VBA editor,* click the **Visual Basic** button.

Developer/Code/Visual Basic   [Excel 2003: Tools/Macro/Visual Basic Editor]

The Visual Basic editor opens in a new window, as shown in Figure 13.6.

The *VBA editor* is a multipanel window, and you can control the layout of the various panels. The three standard panels are shown in Figure 13.6. The main area (empty in Figure 13.6) is the *development area.* When you are adding buttons and text fields to a form, the form is displayed in the development area. This panel is also used for writing program code. If you have already recorded one or more macros, the VBA code for the macros will be displayed in the development area.

The *project panel* lists all of the items in the *project.* A project contains all of the functions, forms, and subprograms need to make a program work. In Excel VBA, a project also contains the worksheets in the workbook. By default, a workbook contains three sheets (worksheets). By selecting a project item, you can connect program code with that item (only). For example, you could have Sheet1 and Sheet2 respond differently to mouse clicks.

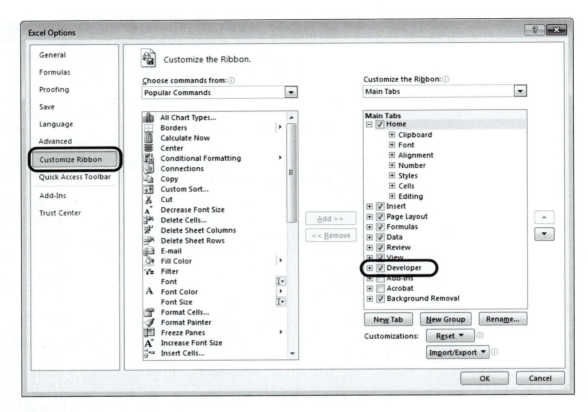

Figure 13.4
Activating the Developer tab in Excel 2010.

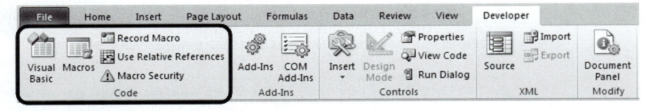

Figure 13.5
The Developer tab.

The *properties panel* is used to access and modify the various properties of the currently selected object. When working with a button on a form, for example, you would use the properties panel to change the text displayed on the button (the button's *caption property*).

13.3.2 Inserting a Module

If you want to create a user-written function, you should write the code in a *module*. A module is simply a project item that stores program code, such as Subs and functions. If you have already recorded a macro using the current workbook, then Excel has already added a module and named it Module1. If there is no Module1 listed in the project list (as in Figure 13.6), then you should insert a module before writing your function.

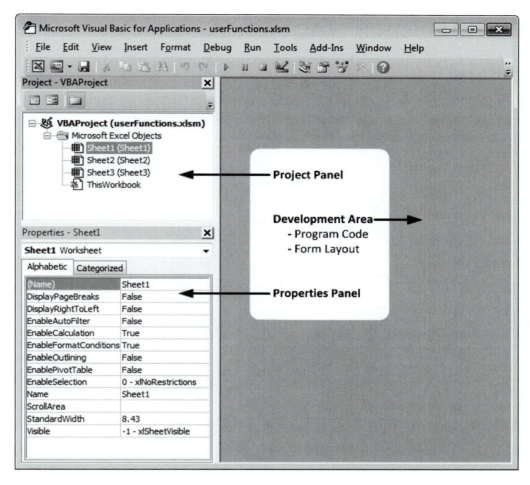

Figure 13.6
Visual Basic Editor.

To insert a module, select **Insert/Module** from the VBA menu (Figure 13.7).

Figure 13.7
Inserting a new module.

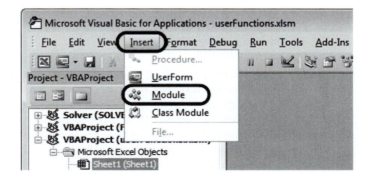

A module named Module1 will then be added to the project list. (You can click twice on the name in the project list to rename the module, if desired.) Once the module has been added to the project, the development area shows the code stored in the module (which is empty for now), as illustrated in Figure 13.8.

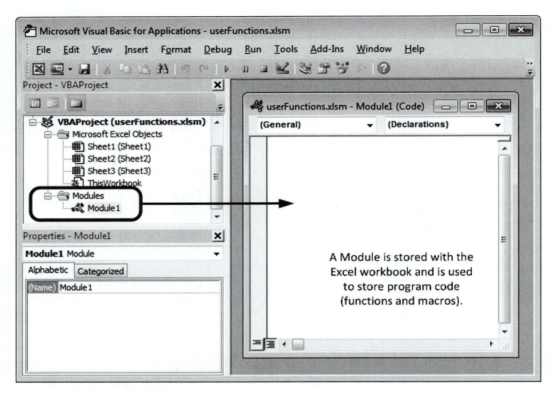

Figure 13.8
Module1 displayed in the Development area.

Once you have Module1 in your project, you have a place to write your own functions. Functions need some keywords on the first and last line so that VBA can recognize the beginning and end of each function. VBA will help make sure the function syntax is correct if you **Insert** the function into the module using the **Insert** menu.

13.3.3 Inserting the Function Procedure into the Module

To have VBA create the first and last lines of your function, select **Insert/Procedure...** from the VBA menu, as illustrated in Figure 13.9.

The Add Procedure dialog box is displayed (Figure 13.10).

- Every procedure (e.g., a function or a subprogram (Sub)) must have a unique *name*. Here, the name **MyFunction** has been entered into the dialog box.
- For a user-written function, the **Function** *type* is used.
- Set the *scope* to **Public** (both Private and Public functions can be used in the worksheet in which they are stored).
- Finally, you must decide whether you need *static variables*. By default, VBA resets all variables declared inside a function to zero each time the function is used. If

you want the variables used in the function to retain their values (not be reinitialized), then check the **All Local variables as Statics** box. For example, if you wanted to count the number of times a function was used, the variable inside the function that holds the count would have to be declared static or the counter would be reset each time the function was used. Static local variables are rarely needed.

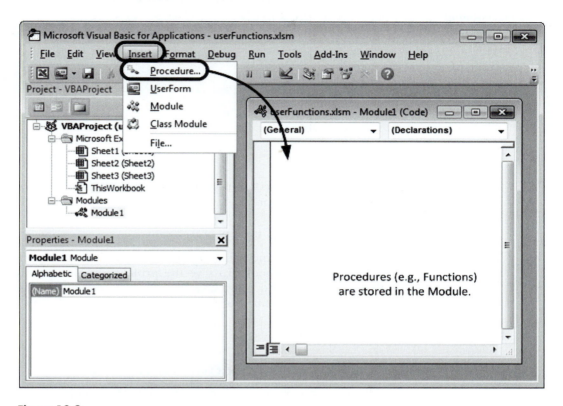

Figure 13.9
Inserting the Function procedure into Module1.

Figure 13.10
The Add Procedure dialog.

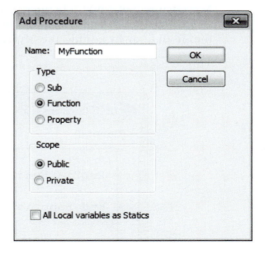

When you click the **OK** button, VBA creates the first and last lines of the new function in Module1. The program code you write goes between the **Public Function MyFunction()** line and the **End Function** line, as shown in Figure 13.11.

Figure 13.11
The first and last lines of the **MyFunction()** user-written function.

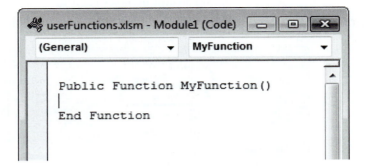

Before we go too far into writing functions, now might be a good time to think about saving your work.

13.3.4 Saving Your VBA Project

Your VBA project (including any modules, Subs, and functions) is stored as part of the Excel workbook. To save the project, simply save the Excel workbook. Remember that you must save in a macro-enabled workbook with the .xlsm file extension. In this example, we will save our work with the file name user Functions.xlsm.

13.4 WRITING YOUR OWN FUNCTION

Before we can go much further, we need to decide what the function called **MyFunction** is supposed to do. In this chapter, we use some of the same examples used in the chapter on macros so that the two approaches can be more easily compared. Let's begin with a function to convert temperatures in °F to °C. The **MyFunction** name doesn't convey much information about what the function does, so let's rename it, **degC**. This can be done either by inserting a new function into the module or by editing the **MyFunction** program listing. The result is shown in Figure 13.12.

Figure 13.12
Starting the **degC()** function.

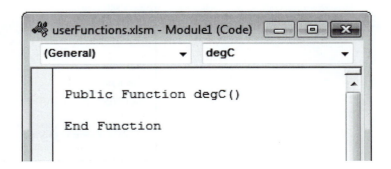

Functions receive input information through *arguments* (also called *parameters*). This function needs to know the temperature in °F, so we add the *degF* variable as an argument inside the function's parentheses, as degC(*degF*). This is illustrated in Figure 13.13.

Figure 13.13
Adding a parameter (*degF*) to the function declaration.

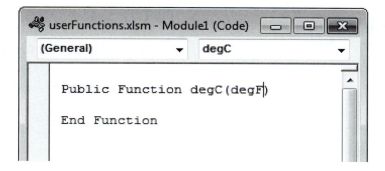

Finally, we add the math that converts a temperature in °F to °C. This is shown in Figure 13.14.

Figure 13.14
Completing the function.

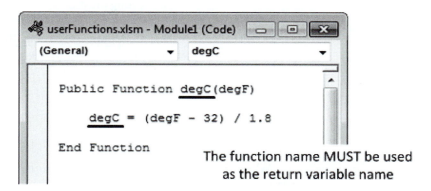

IMPORTANT! Notice that the *variable name* (*degC*) used to hold the calculated result is the same as the *function name*. Remember that functions return values; in VBA the value that is returned is the value assigned to the variable with the same name as the function. Here, the function is named **degC**, so we used a variable named *degC* to hold the calculated temperature value. VBA will return the value of variable *degC* when function **degC** is used.

13.4.1 Testing Your Function

To make sure the function is working correctly, we will test it against some temperature benchmarks:

212°F =	100°C	(boiling point of water)
98.6°F =	37°C	(human body temperature)
32°F =	0°C	(freezing point of water)
−40°F =	−40°C	(equivalency temperature)

The Fahrenheit temperatures have been entered into the worksheet shown in Figure 13.15.

Next, build the **degC** function into a formula in cell C4, grabbing the temperature to be converted from cell B4. This is illustrated in Figure 13.16.

$$\text{C4:} \quad \text{=degC(B4)}$$

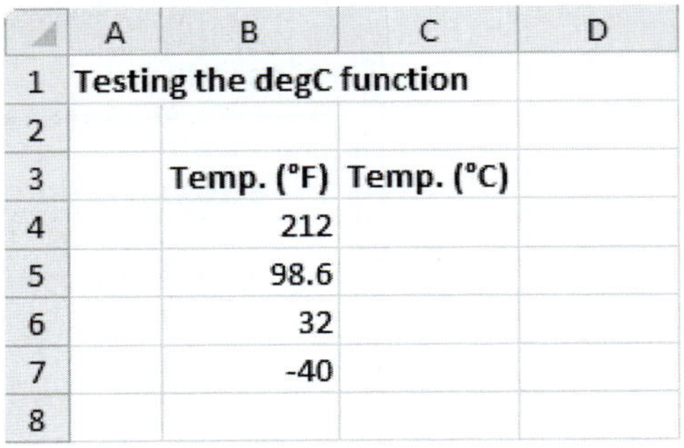

Figure 13.15
Preparing to test the **degC** function.

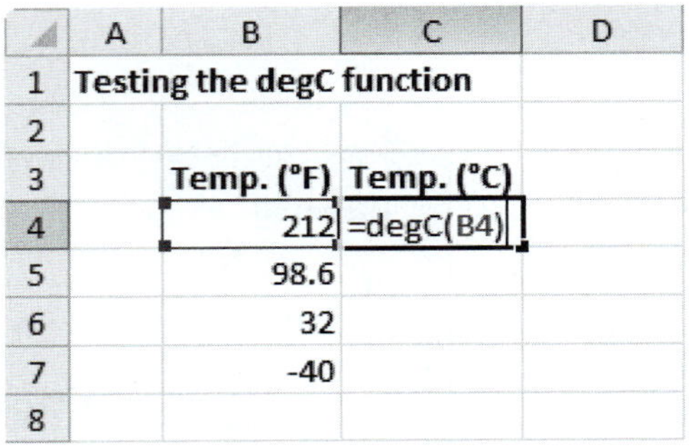

◢	A	B	C	D
1	Testing the degC function			
2				
3		Temp. (°F)	Temp. (°C)	
4		212		
5		98.6		
6		32		
7		-40		
8				

Figure 13.16
Using the **degC** function in a formula.

◢	A	B	C	D
1	Testing the degC function			
2				
3		Temp. (°F)	Temp. (°C)	
4		212	=degC(B4)	
5		98.6		
6		32		
7		-40		
8				

When the [Enter] key is pressed to complete the formula, Excel uses the user-written function to convert the temperature in B4, showing the result in cell C4 (Figure 13.17).

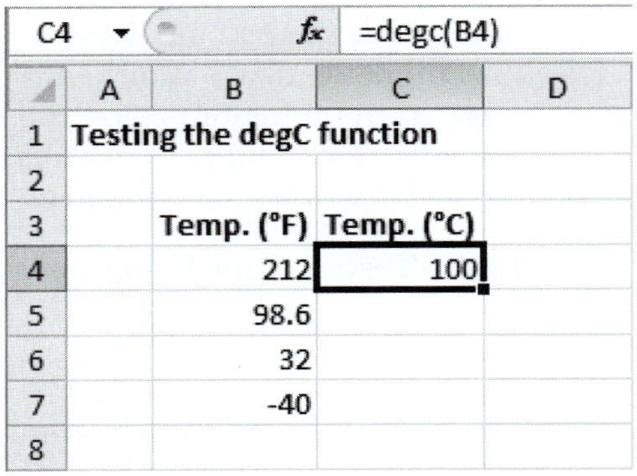

Figure 13.17
The completed formula in cell C4.

C4	▼		f_x	=degc(B4)	

◢	A	B	C	D
1	Testing the degC function			
2				
3		Temp. (°F)	Temp. (°C)	
4		212	100	
5		98.6		
6		32		
7		-40		
8				

Excel treats user-written functions just like its built-in functions; to convert the other temperatures in column B, simply copy the formula in cell C4 to cells C5:C7, as illustrated in Figure 13.18.

Figure 13.18 illustrates one of the features that make functions more convenient than macros in many instances: a function can be copied from one cell to many others in a single step. The macro must be repeatedly executed in each cell.

Figure 13.18
Copying the temperature conversion formula.

C7		f_x	=degc(B7)

	A	B	C	D
1	**Testing the degC function**			
2				
3		**Temp. (°F)**	**Temp. (°C)**	
4		212	100	
5		98.6	37	
6		32	0	
7		-40	-40	
8				

13.4.2 Better Programming Style

VBA is pretty flexible when it comes to declaring variables, but good programming style requires that variables be declared to be of particular *data types*. (Data types are covered in more detail in the next chapter.) The variables in this function are *degF* and *degC*, and both variables are expected to contain numeric values. We can tell Excel this by declaring each variable to be of type *Single* (single precision, numeric data). It is also considered a good programming practice to add a few comment lines to indicate who wrote the function and what it is used for. This has been done in Figure 13.19.

Figure 13.19
Declaring the data types of *degC* and *degF* variables.

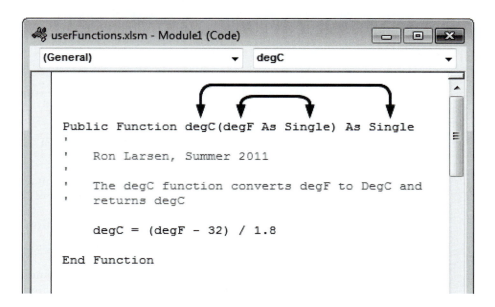

```
userFunctions.xlsm - Module1 (Code)

(General)                          degC


   Public Function degC(degF As Single) As Single
   '
   '   Ron Larsen, Summer 2011
   '
   '   The degC function converts degF to DegC and
   '   returns degC

       degC = (degF - 32) / 1.8

   End Function
```

13.5 USING YOUR FUNCTIONS IN OTHER WORKBOOKS

Since Excel stores user-written functions in a workbook, you may be concerned that your functions will have to be rewritten every time you use a new Excel workbook—they don't.

Prior to Excel 2007, this was a true statement:

Excel gives any open workbook access to all of the functions in every open workbook.

With Excel 2007 and 2010, it must be modified slightly:

Excel gives any open macro-enabled workbook access to all of the functions in every open macro-enabled workbook.

If you have some useful user-written functions stored in a macro-enabled workbook, just make sure that the workbook is open and you can use those functions in other macro-enabled workbooks, you just have to tell Excel where to find the function by pre-pending the workbook name to the function name, and connecting them with an exclamation point. In our example, this would be **userFunctions.xlsm!degC()** since the **degC** function has been stored in the userFunctions.xlsm workbook.

To test this, we will open another macro-enabled workbook, called newWB.xlsm, and use the **degC** function from the original file with the formula `=userFunctions.xlsm!degC(B6)`. The result is shown in Figure 13.20.

Figure 13.20
Using a user-written function from another workbook.

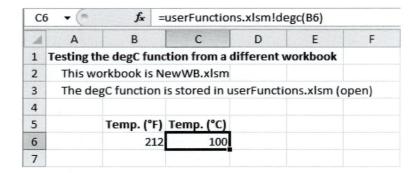

If you do this very often, it would pay to store your functions in a workbook with a short file name.

13.6 EXAMPLES OF USER-WRITTEN FUNCTIONS

Writing your own functions is pretty straightforward in Excel, and with VBA you can write some very powerful functions. In this section, we will develop a few additional user-written functions to illustrate some features of Excel and VBA.

13.6.1 An Enhanced Random Number Function

Excel provides two random number generating functions:

- **RAND**—returns a random value between 0 and 1
- **RANDBETWEEN(Low, High)**—returns a random integer between Low and High

You either get noninteger values between 0 and 1 or get integers over a specified range. You can't get noninteger values between 7 and 13, for example. We'll fix that with our own **randVal(Low, High)** function.

This example also illustrates how to call an Excel function from VBA, or how to use an Excel function inside your own function.

To do this, we will use the **RANDBETWEEN** function, but we will modify the limits. If we want noninteger values between 7 and 13, we will ask **RANDBETWEEN** to provide a random integer between 7000 and 13,000. Then we'll divide the result by 1000 and return the value.

The new **randVal(Low, High)** function looks like this:

```
Public Function randVal(Low As Single, High As Single) As Single
    randVal = Application.WorksheetFunction.RandBetween
    (Low * 1000, High * 1000) / 1000
End Function
```

Notice that we included the statement.

```
Application.WorksheetFunction.RandBetween(Low * 1000, High *
1000)
```

to use Excel's **RANDBETWEEN** function in our VBA function. The "Application" is the open Excel workbook, and we have asked VBA to look in the Excel workbook for a "WorksheetFunction" called "RandBetween."

In Figure 13.21, the **randVal** function has been used to create five random numbers between 7 and 13.

Figure 13.21
Using the **randVal(Low, High)** function.

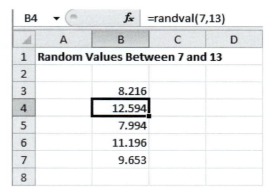

13.6.2 A Linear Interpolation Function

A linear interpolation function is useful for interpolating in data tables. The known x values will include a low, middle, and high value. The low and high y values will be known, and we will solve for the middle y value by linear interpolation:

x Values (Knowns)	y Values (Low and High Values are Known)
x_{LOW}	y_{LOW}
x_{MID}	$[y_{MID}]$ unknown
x_{HIGH}	y_{HIGH}

A quick way to write a linear interpolation equation uses the ratios of differences of x values and y values:

$$\frac{x_{MID} - x_{LOW}}{x_{HIGH} - x_{LOW}} = \frac{[y_{MID}] - y_{LOW}}{y_{HIGH} - y_{LOW}}. \qquad (13.1)$$

Solving for the unknown, y_{MID}, gives

$$y_{MID} = y_{LOW} + (y_{HIGH} - y_{LOW}) \times \left[\frac{x_{MID} - x_{LOW}}{x_{HIGH} - x_{LOW}}\right]. \qquad (13.2)$$

Next, we'll write a linear interpolation function that will receive the five known values as arguments and return the calculated y_{MID} value.

Function: **linearInterp(xLow, xMid, xHigh, yLow, yHigh).**

Purpose: Performs linear interpolation.

Note: The "As Single" type declarations on all five arguments, and variable linearInterp have been omitted to save space.

Definition:

```
Public Function linearInterp(xLow, xMid, xHigh, yLow, yHigh)
    linearInterp = yLow + (yHigh - yLow) * (xMid - xLow)/
    (xHigh - xLow)
End Function
```

Usage:

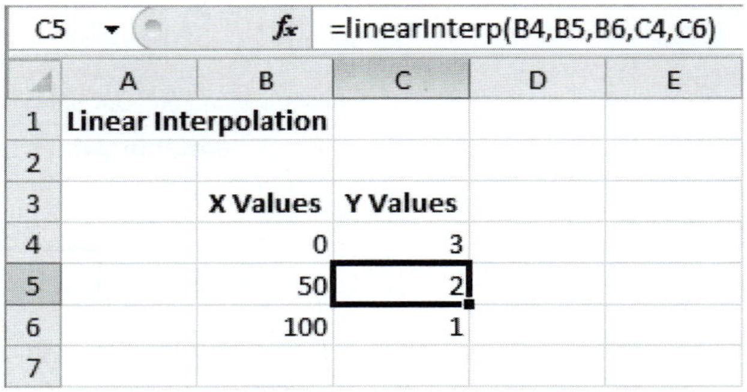

Figure 13.22
Using the **linearInterp** function.

13.6.3 User-Written Functions to Calculate Resultant Force and Angle

In the last chapter, we wrote a macro to compute the resultant force and angle, given the summed x and y components of force. Here we'll write two functions to accomplish the same tasks.

A worksheet where the force components have been calculated is shown in Figure 13.23.

In order to compute a resultant force (equation 13.3), the sums of the x and y components of force must be known (cells D7 and E7 in Figure 13.23).

$$F_{res} = \sqrt{\left(\sum F_x\right)^2 + \left(\sum F_y\right)^2}, \qquad (13.3)$$

To pass these into a function named **Resultant,** two variables would be included as arguments in the parentheses following the function name, as **Resultant**(*SumFx, SumFy*).

F7	▾	f_x	=Resultant(D7,E7)				

⊿	A	B	C	D	E	F	G	H
1	Resolving Forces							
2								
3		Force (N)	Angle (°)	X-comp (N)	Y-comp (N)			
4		400	20	375.9	136.8			
5		200	135	-141.4	141.4			
6						Resultant (N)	Res. Angle (°)	
7			Sums:	234.5	278.2			
8								

Figure 13.23
Worksheet with x and y components of force calculated.

We will need VBA's **Sqr** function to take the necessary square root. The **Resultant** function listing looks like this:

```
Public Function Resultant(SumFx As Single, SumFy As Single) As
Single
    Resultant = Sqr(SumFx^2 + SumFy^2)
End Function
```

In Figure 13.24, the **Resultant** function has been entered into cell F7, and the values in cells D7 and E7 were passed in as the **Resultant** function's arguments. The calculated resultant force is 363.8N.

F7	▾	f_x	=Resultant(D7,E7)				

⊿	A	B	C	D	E	F	G	H
1	Resolving Forces							
2								
3		Force (N)	Angle (°)	X-comp (N)	Y-comp (N)			
4		400	20	375.9	136.8			
5		200	135	-141.4	141.4			
6						Resultant (N)	Res. Angle (°)	
7			Sums:	234.5	278.2	363.8		
8								

Figure 13.24
The resultant force calculated with the **Resultant** function.

The angle of the resultant force, measured from the x axis, is determined as

$$\theta_{res} = \text{ATAN}\left(\frac{\sum F_y}{\sum F_x}\right), \tag{13.4}$$

A user-written function to determine the angle of the resultant force, called **ResAngle(SumFx, SumFy),** might look like this:

```
Public Function ResAngle(SumFx As Single, SumFy As Single) As Single
    ResAngle = Atn(SumFy / SumFx)
End Function
```

The **ResAngle** function returns the resultant angle in radians. So, we'll use Excel's **DEGREES** function when the angle is calculated. The result is shown in Figure 13.25.

G7	▾	f_x	=DEGREES(ResAngle(D7,E7))				
	A	B	C	D	E	F	G	H
1	**Resolving Forces**						
2							
3		**Force (N)**	**Angle (°)**	**X-comp (N)**	**Y-comp (N)**		
4		400	20	375.9	136.8		
5		200	135	-141.4	141.4		
6						**Resultant (N)**	**Res. Angle (°)**
7			**Sums:**	234.5	278.2	363.8	49.9
8							

Figure 13.25
The **ResAngle** function in use.

KEY TERMS

Argument (VBA)
Built-in function
Caption property
Data types (VBA)
Developer tab (Ribbon)
Development area (VBA)
File extensions (.xlsx, .xlsm)
Formula
Function
Function name
Graphical user interface (GUI)
Macro
Macro language
Macro-enabled workbook
Module (VBA)
Parameter (VBA)
Project (VBA)
Project panel (VBA)
Properties panel (VBA)
Scope (VBA)
Single (data type) (VBA)
Static variables (VBA)
Subprogram
User-written function
Variable name (VBA)
VBA editor
Virus
Visual Basic
Visual Basic for Applications (VBA)

SUMMARY

Macro-Enabled Worksheets

- .xlsx—the Excel file extension for macro-disabled workbooks (default)
- .xlsm—the Excel file extension for macro-enabled workbooks

Showing the Ribbon's Developer Tab

Excel 2010

1. Use Ribbon options **File tab/Options** to display the Excel Options dialog, then select the **Customize Ribbon** panel.
2. Check the box labeled **Developer** in the **Main Tabs** list.

Excel 2007

1. Use the Microsoft Office button with options **Office/Excel Options** to display the Excel Options dialog, then select the **Popular** panel.
2. Check the box labeled **Show Developer tab in the Ribbon**.

Starting the Visual Basic Editor

Use Ribbon options **Developer/Code/Visual Basic**

[Excel 2003: Tools/Macro/Visual Basic Editor]

Inserting a Module into a Project

A module is a VBA project item that holds program code. Your functions are written and stored in a module. A module can store any number of functions.

1. Click on "VBA Project" in the Project panel
2. Select **Insert/Module** from the VBA menu

Inserting a Function into a Module

1. Click in the Module to be sure it is selected
2. Select **Insert/Procedure…** from the VBA menu
3. Specify:
   a. function **Name**
   b. Type **Function**
   c. Scope (typically) **Public**
   d. **All Local Variables as Statics** (generally not checked)

Data Types (Common)

Data Type	Description	Use
Single	Single precision real numbers (4 bytes) $-3.402823E^{38}$ to $3.402823E^{38}$	General, low-precision math
Double	Double-precision real numbers (8 bytes) $-1.79769313486231E^{308}$ to $1.79769313486231E^{308}$	General, high-precision math
Integer	Small integer values (2 bytes) $-32,768$ to $32,767$	Counters, index variables
Long	Long integer values (4 bytes) $-2,417,483,648$ to $2,147,483,647$	Used whenever an integer value could exceed 32,767
Boolean	Logical values (2 bytes) True or False	Status variables
String	Text strings	Words and phrases, file names
Date	Date values (8 bytes) Floating point number representing days, hour, minutes, and seconds since 1/1/100	Dates and times

Using Your Functions in Other Workbooks

To use your functions in other workbooks:

1. Both workbooks must be macro enabled (look for the .xlsm file extension).

2. You must include the workbook name with the function name.

Example: To use a function named **MyFunction** that is stored in a workbook named MyWorkbook.xlsm, refer to the function (in workbooks other than MyWorkbook.xlsm) as **MyWorkbook.xlsm!MyFunction()**.

PROBLEMS

13.1 The Gravitational Constant

Write a function (without arguments) that returns the gravitational constant, g_c. A commonly used value of g_c is

$$g_c = 32.174 \frac{\text{ft lb}_m}{\text{lb}_f \text{s}^2}. \tag{13.5}$$

13.2 Ideal Gas Constant

Write a function (without arguments) that returns the ideal gas constant, R. The value of R depends on the units. Select the value you use most frequently:

Ideal Gas Constants	
8.314	$\dfrac{m^3\,\text{Pa}}{\text{gmol K}}$
0.08314	$\dfrac{\text{L bar}}{\text{gmol K}}$
0.08206	$\dfrac{\text{L atm}}{\text{gmol K}}$
0.7302	$\dfrac{\text{ft}^3\,\text{atm}}{\text{lbmol °R}}$
10.73	$\dfrac{\text{ft}^3\,\text{psia}}{\text{lbmol °R}}$
8.314	$\dfrac{\text{J}}{\text{gmol K}}$
1.987	$\dfrac{\text{BTU}}{\text{lbmol °R}}$

13.3 Computing Ideal Gas Volume

Create a user-written function that will compute the volume of an ideal gas when the absolute temperature, the absolute pressure, the number of moles, and an ideal gas constant are passed into the function as arguments. Use your function to complete Table 13.3P.

Table 13.3P Ideal Gas Volume

Temperature	Pressure	Moles	Gas Constant	Volume
273 K	1 atm	1 gmol	$0.08206 \dfrac{\text{L atm}}{\text{gmol K}}$	
500 K	1 atm	1 gmol	$0.08206 \dfrac{\text{L atm}}{\text{gmol K}}$	
273 K	10 atm	1 gmol	$0.08206 \dfrac{\text{L atm}}{\text{gmol K}}$	
273 K	1 atm	10 gmol	$0.08206 \dfrac{\text{L atm}}{\text{gmol K}}$	
500 K	1 atm	10 gmol	$0.08206 \dfrac{\text{L atm}}{\text{gmol K}}$	
500 °R	1 atm	1 lbmol	$0.7302 \dfrac{\text{ft}^3 \text{ atm}}{\text{lbmol °R}}$	

Create a macro (either by recording or by programming) that will compute the volume of an ideal gas when given the absolute temperature, the absolute pressure, the number of moles, and an ideal gas constant. Use your macro to complete Table 13.3P.

What are the units of volume in each row?

13.4 Computing Gas Heat Capacity

There are standard equations for calculating the heat capacity (specific heat) of a gas at a specified temperature. One common form of a heat capacity equation is a simple third-order polynomial in T:[1]

$$C_p = a + bT + cT^2 + dT^2. \tag{13.6}$$

If the coefficients a, b, c, and d are known for a particular gas, you can calculate the heat capacity of the gas at any T (within an allowable range). The coefficients for a few common gases are listed in the following table:

Heat Capacity Coefficients						
Gas	a	b	c	d	Units on T	Valid Range
Air	28.94×10^{-3}	0.4147×10^{-5}	0.3191×10^{-8}	-19.65×10^{-12}	°C	0–1500°C
CO_2	36.11×10^{-3}	4.233×10^{-5}	-2.887×10^{-8}	7.464×10^{-12}	°C	0–1500°C
CH_4	34.31×10^{-3}	5.469×10^{-5}	0.3661×10^{-8}	-11.00×10^{-12}	°C	0–1200°C
H_2O	33.46×10^{-3}	0.6880×10^{-5}	0.7607×10^{-8}	-3.593×10^{-12}	°C	0–1500°C

The units on the heat capacity values computed with these coefficients are kJ/gmol °C.

Write a function that computes the heat capacity of a gas. Your function will need to receive five parameters (four coefficients and one temperature).

---

[1] *From Elementary Principles of Chemical Processes*, 3d ed., Felder, R. M. and R. W. Rousseau, New York: Wiley, 2000.

Use your function to find the heat capacity of water vapor at 300°C, 500°C, and 750°C. Does the heat capacity of steam change significantly with temperature?

13.5 Resolving Forces into Components I

Write a function to compute the horizontal and vertical components for the following forces:

    a. A 400-N force acting at 40° from horizontal.
    b. A 400-N force acting at 75° from horizontal.
    c. A 2000-N force acting at 210° from horizontal.

13.6 Converting Temperatures

The **degC()** function to convert temperature in °F to °C is presented in Section 13.4. Write the companion function **degF()** that will convert temperature in °C to °F. Use your function to complete Table 13.6P.

Table 13.6P Converting Temperature Values

Temperature (°C)	Temperature (°F)
100	
37	
0	
−40	

14

Programming in Excel with VBA

Objectives

After reading this chapter, you will know

- How to obtain access to and use the VBA programming environment
- The fundamental elements of programming
- How to read and create programming flowcharts

- How to create and edit functions and subprograms
- How to create and use forms for data entry and for presenting results
- How to use an Excel macro to open a form from the Excel workbook

14.1 INTRODUCTION

Many years ago, Microsoft made the decision to use a version of Visual Basic, called *Visual Basic for Applications,* or *VBA,* as the macro language in Microsoft Office products. That decision gave *macro programming* a lot of power in these applications. It also created the potential for *macro viruses.*

VBA has developed over time into a very powerful programming language. The "A" in VBA is important; VBA is a version of Visual Basic that is highly integrated with the Application—that is, Excel VBA is a version of Visual Basic that has been designed to work well with Excel. The presence of VBA imbedded in Excel puts a programming environment on the engineer's desktop. The last two chapters have touched on using VBA's editor, but this chapter is about programming.

14.2 VIRUS PROTECTION AND MACRO-ENABLED WORKBOOKS

Before you can do any programming in Excel you have to be able to get past the macro protection. Excel 2007 has added a new level of security against macro viruses by using two different *file extensions* for macro-enabled and macro-disabled workbooks:

- .xlsx—the Excel file extension for *macro-disabled workbooks* (default)
- .xlsm—the Excel file extension for *macro-enabled workbooks*

While the term is "macro-enabled," Excel actually controls access to the VBA *project*. A project is a collection of forms, functions, and subroutines needed to make a program work. Projects are also used to store macros and user-written functions in Excel. Basically, you have to use a macro-enabled workbook to store your programs.

Unless you choose to enable all macros (not recommended), any workbook containing a VBA project (i.e., containing program code) that you open will have the project disabled. By default, Excel will notify you when a VBA project is disabled. The Excel screen when you first open a macro-enabled workbook with a VBA project will show the Security Warning seen in Figure 14.1.

Figure 14.1
Security warning when a workbook includes a VBA project.

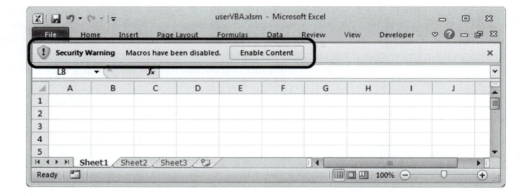

The warning mentions macros, but the same warning applies to VBA programs. To verify or change how Excel responds to workbooks containing VBA projects, use the Excel Options dialog:

- Excel 2010: **File tab/Options**
- Excel 2007: **Office/Excel Options**
- Excel 2003: **File/Options**

Select the **Trust Center** panel and click the **Trust Center Settings...** button. This opens the Trust center dialog which is used to adjust macro settings.

The security warning indicated in Figure 14.1 says that macros have been disabled, but it is actually the VBA project that is disabled. This means that you must enable the project to use or edit your VBA programs.

It is appropriate to be concerned about program code coming from unknown sources, but that shouldn't stop you from writing your own programs when they can help you get your work done faster and easier.

14.3 VISUAL BASIC FOR APPLICATIONS (VBA)

VBA is a complete programming language residing within the Microsoft Office applications, including Excel. Running within Excel, VBA can take input from a worksheet, calculate results, and send the results back to a worksheet. Or, you can create forms to request input from the user and display the results. You can use VBA behind the scenes to add power to your worksheets, or you can use VBA as a programming environment and ignore the rest of Excel.

14.3.1 Showing the Ribbon's Developer Tab

Lots of people use the Excel worksheets and ignore VBA. Because of this, the Ribbon tab that provides access to the VBA editor is, by default, not shown. Before you can start the VBA editor, you have to get Excel to display the *Developer tab*. This is done from the Excel Options dialog, but it is handled differently in Excel 2010 and Excel 2007.

In Excel 2010: To show the **Developer** tab, use Ribbon options **File tab/Options** to display the Excel Options dialog, then select the **Customize Ribbon** panel, and check the box labeled **Developer** in the **Main Tabs** list. This is illustrated in Figure 14.2.

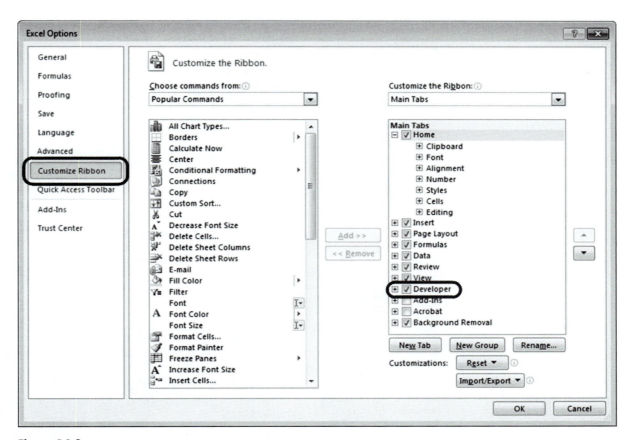

Figure 14.2
Activating the **Developer** tab in Excel 2010.

In Excel 2007: To show the Developer tab, use the Microsoft Office button with options **Office/Excel Options** to display the Excel Options dialog, then select the **Popular** panel, and check the box labeled **Show Developer tab in the Ribbon**.

The Ribbon's **Developer** tab is shown in Figure 14.3. The Code group provides access to VBA.

Figure 14.3
The Developer tab.

14.3.2 Starting the VBA Editor

The **Developer** tab's Code group provides a **Visual Basic** button to access to the Visual Basic editor. To start the *VBA editor,* click the **Visual Basic** button.

Developer/Code/Visual Basic   [Excel 2003: Tools/Macro/Visual Basic Editor]

The editor opens in a new window, as shown in Figure 14.4.

Figure 14.4
Visual Basic Editor.

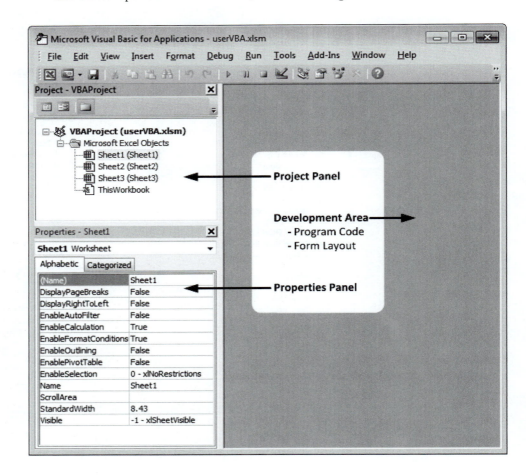

The VBA editor is a multipanel window, and you can control the layout of the various panels. The three standard panels are shown in Figure 14.4. The main area (empty in Figure 14.4) is the *development area*. When you are adding buttons and text fields to a form, the form is displayed in the development area. This panel is also used for writing program code. If you have already recorded one or more macros, the VBA code for the macros will be displayed in the development area.

The *project panel* lists all of the items in the *project*. A project contains all of the functions, forms, and subprograms needed to make a program work. In Excel VBA, a project also contains the worksheets in the workbook. By default, a workbook contains three sheets (worksheets). By selecting a project item, you can connect program code with that item (only). For example, you could have Sheet1 and Sheet2 respond differently to mouse clicks.

The *properties panel* is used to access and modify the various properties of the currently selected object. When working with a button on a form, for example, you would use the properties panel to change the text displayed on the button (the button's *caption property*).

14.4 PROJECTS, FORMS, AND MODULES

A VB project is a collection of programming pieces that are used together to create a complete program. A typical VB project includes

- one or more *forms* to collect and present information to the user; and
- one or more *modules* to hold variable definitions and program code.

An Excel VBA project also includes

- the workbook from which the VBA program is created—this is listed as This Workbook in the project list (see Figure 14.4); and
- the individual sheets that comprise the workbook.

By including a module, the workbook, and each individual sheet in the project, you have a great deal of control over access to your programs:

- Program elements *stored in a module* generally are accessible from any project source—that is, a function stored in a module (for example) can be accessed from a form stored in the project or from any sheet of the workbook.
- Program elements *stored in the workbook* are available to any of the sheets, but generally are not available to the rest of the project.
- Program elements *stored with a specific worksheet* generally are available only to that sheet. This allows the programmer to write multiple versions of a function (same function name, stored with different sheets) to create specialized responses for each sheet. For example, you might create two different **CreateHeader** functions. One might create a simple title, date, and author header on worksheets designed to be used within the company, whereas another version would create a more elaborate header, including a logo and company contact information, on worksheets intended for customers and clients.

You can add additional objects to the project. Program code that is not specifically tied to an object (such as a sheet, form, or form object) normally is stored in a *module*. A module is simply an object that stores program code, such as variable definitions, Subs (subprograms), and functions. You can also insert form objects to create graphical user interfaces for your programs.

14.4.1 Inserting a Module

VBA program code is housed in a *module*. A module is simply a project item that stores program code, such as Subs and functions. If there is no Module1 listed in the project list (as in Figure 14.4), then you should insert a module before writing your function.

To insert a module, select **Insert/Module** from the VBA menu (Figure 14.5).

Figure 14.5
Inserting a new module.

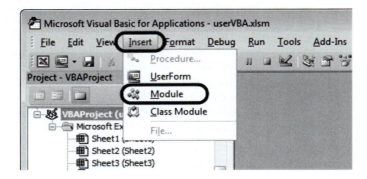

A module named Module1 will then be added to the project list. (You can click twice on the name in the project list to rename the module, if desired.) Once the module has been added to the project, the development area shows the code stored in the module (which is empty for now), as illustrated in Figure 14.6.

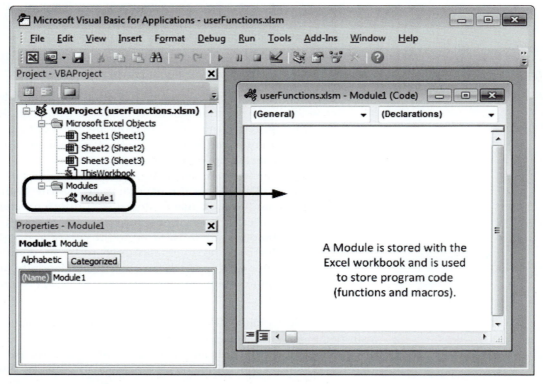

Figure 14.6
Module1 displayed in the Development area.

Once you have Module1 in your project, you have a place to write your own functions. Functions need some keywords on the first and last line so that VBA can recognize the beginning and end of each function. VBA will help make sure the function syntax is correct if you insert the function into the module using the **Insert** menu.

14.4.2 Creating a Function

Later in the chapter, we will present the fundamental elements of all programming languages; first, however, we will create a very simple function just to illustrate the process used to create VBA *functions.*

If you store the code for a function in a VBA module, your function will be available to any of the sheets in the project—that is, you can use your function in any cell on any of the sheets in the workbook. The process for creating a new function is the following:

1. Use VBA menu options **Insert/Procedure...** to
   a. declare the function,
   b. create the first and last lines of the function, and
   c. place the function code in a module.
2. Enter the required lines of program code to accomplish the desired task.
3. Return to the worksheet and use the function in the same way that Excel's built-in functions are used.
4. Return to the VBA editor as needed to modify or debug the program code in the function.

As an example, we will write a **calcVolume** function that is intended to calculate the volume occupied by an ideal gas.

$$PV = NRT$$

so,

$$V = \frac{NRT}{P}$$

The absolute pressure (P), absolute temperature (T), and number of moles of gas (N) are values that must be passed into the function. The value of the gas law constant (R) will be set within the function.

Step 1. Use Insert/Procedure... to Create the Function. To insert the starting and ending lines of a function into a module, first select the module in the project panel, then click in the development area for the module (to let the editor know where to insert the function). Then insert the function, using Insert/ **Procedure...** from the VBA main menu. This is illustrated in Figure 14.7.

Figure 14.7

Inserting a procedure into the project.

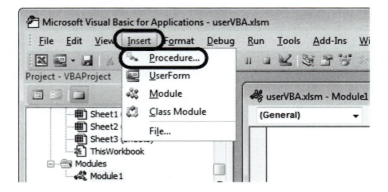

The Add Procedure dialog will be displayed, as shown in Figure 14.8.

Figure 14.8
The Add Procedure dialog.

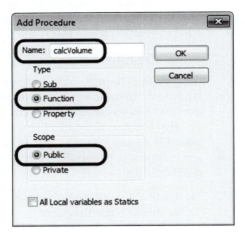

Enter a name for the new function and set the Type to **Function**. Accept the default scope, **Public,** so that the function will be available outside the module (to your worksheets, for example) and leave the **All Local variables as Statics** box unchecked—the variables in this function do not need to retain their value between calls.

When you click the **OK** button, VBA will insert the first and last lines of the function into Module1, as illustrated in Figure 14.9.

Figure 14.9
The first and last lines of
the **calcVolume** function
in Module1.

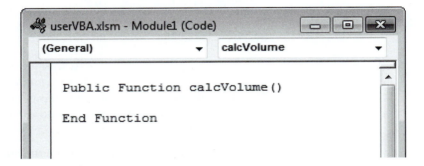

Step 2. Add the Required Program Code.   Right now, the function is not receiving any values; we want it to receive the pressure (P), temperature (T), and number of moles of gas (N). These variable names must be added inside the parentheses after the function name (i.e., inside the argument list), as shown in Figure 14.10.

Figure 14.10
Adding arguments:
P, T, and *N.*

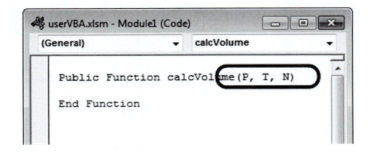

The ideal gas constant is needed for the calculation, so a variable, R, is assigned the value 0.08206 (Figure 14.11). This is the gas constant with units of liter atm/mole Kelvin.

Figure 14.11
Adding the gas constant definition.

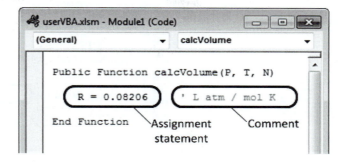

Finally, we calculate the volume and assign the result to the variable *calcVolume*, as shown in Figure 14.12.

Figure 14.12
Calculating the ideal gas volume.

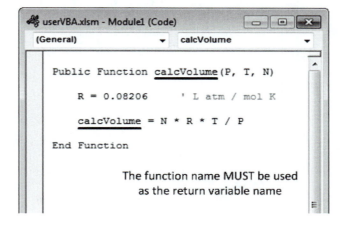

IMPORTANT! In VB (and VBA), the value of the variable with the same name as the function name is the value that is returned when the End Function statement is reached.

Step 3. Return to the Worksheet and Use the Function. The **calcVolume** function is now complete; it can be called from any worksheet in the workbook.

Notice that the **calcVolume** function requires specific units on the P, T, and N values and returns a volume in liters. Those units are shown in column D of the worksheet shown in Figure 14.13.

Better Programming Style

VBA is pretty flexible when it comes to declaring variables, but good programming style requires that variables be declared to be of particular *data types*. (Data types are covered in more detail later in this chapter.) The variables in this function P, T, N, R, and calcVolume are all expected to contain numeric values. We can tell Excel this by declaring each variable to be of type *Single* (single precision, numeric data). This has been done in Figure 14.14. It is also considered a good practice to include a few comment lines indicating who wrote the function, when, and what the function is intended to do.

Figure 14.13
Using the calcVolume()
function.

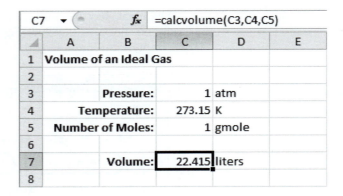

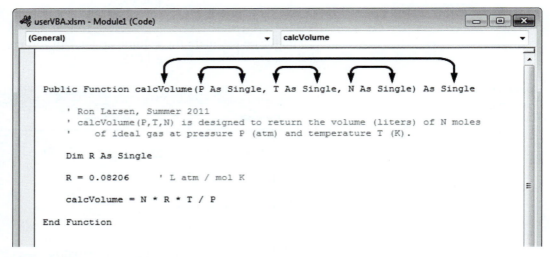

Figure 14.14
Program listing with declared data types.

The more complete version declares variable R through the Dim R statement. Since R is declared inside the **calcVolume** function, it is a local variable; it can be used inside the **calcVolume** function, but no other function or worksheet will know that it exists.

Before we go too far into programming, now might be a good time to think about saving your work.

14.4.3 Saving Your VBA Project

Your VBA project (including any modules, Subs, and functions) is stored as part of the Excel workbook. To save the project, simply save the Excel workbook. Remember that you must save in a macro-enabled workbook with the .xlsm file extension. In this example, we will save our work with the file name userFunctions.xlsm.

14.5 FLOWCHARTS

A *flowchart* is a visual depiction of a program's operation. It is designed to show, step by step, what a program does. Typically, it is created before the writing of the program; it is used by the programmer to assist in the development of the program and by others to help them understand how the program works.

There are standard symbols used in computer flowcharts. These include the following:

Table 14.1 Standard flowchart symbols

Symbol	Name	Usage
(rounded rectangle)	Terminator	Indicates the start or end of a program
(rectangle)	Operation	Indicates a computation step
(parallelogram)	Data	Indicates an input or output step
(diamond)	Decision	Indicates a decision point in a program
(circle)	Connector	Indicates that the flowchart continues in another location

These symbols are connected by arrows to indicate how the steps are connected and the order in which the steps occurs.

Flowcharting Example

As an example of a simple flowchart, consider a **Thermostat** function that is designed to return a code value that is used to turn a heater on or off. Specifically,

- If the temperature is below 23°C, return the value 1 to indicate that the heater should be activated.
- If the temperature is above 25°C, return the value −1 to indicate that the heater should be shut off.
- If the temperature is between 23°C and 25°C, return the value 0 to indicate that there should be no change in the heater's status.

The flowchart for this decision process is shown in Figure 14.15.
The complete function will look like this:

```
Public Function Thermostat(T As Single) as Integer

Dim RV As Integer
RV = 0            ' assign a default value to RV
If T < 23 Then RV = 1
If T > 25 Then RV = -1
Thermostat = RV ' assign the return value to the return variable

End Function
```

The flowchart in Figure 14.15 indicates that the function starts, and the value of *T* is passed in from the parameter list. Then, the first line of the function is an *operation* step, namely, assign variable *RV* the value 0. The program code for this step is:

```
RV = 0            ' assign a default value to RV
```

Figure 14.15
Flowchart for the
Thermostat function.

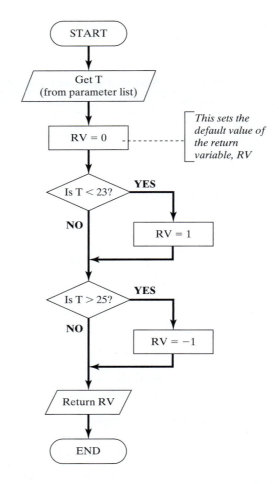

Note: RV stands for *return value*, the value that will be returned when the function terminates. VBA doesn't care what variable you use for the return value, but *RV* makes sense to a lot of programmers.

This step ensures that the return variable, *RV*, has a value no matter what *T* value is received. This step is called *assigning a default value* to *RV*.

The next line is a *decision* step, indicated by the diamond symbol on the flowchart in Figure 14.15.

```
If T < 23 Then RV = 1
```

The "If *T* < 23" portion of the second line is the *condition* that is checked, and the "Then *RV* = 1" is the operation that is performed if the condition is found to be true.

The next line of the program is another decision step:

```
If T > 25 Then RV = -1
```

The condition "If *T* > 25" is tested, and, if the condition is found to be true, the operation *RV* = −1 is performed. Just before the program ends, the value of *RV* is returned through variable *Thermostat* so that it is available to the worksheet:

```
Thermostat = RV        ' assign the return value to the return
                         variable
```

Note: The *Thermostat* variable could have been used throughout the function in place of RV.

You might have noticed that the **Thermostat** function tests *T* twice. Even if it has already been found that *T* < 23, it still checks to see whether *T* > 25. This is not a particularly efficient way to write this program. The last steps of a better version might be flowcharted as shown in Figure 14.16.

Figure 14.16
Improved flowchart for the
Thermostat function.

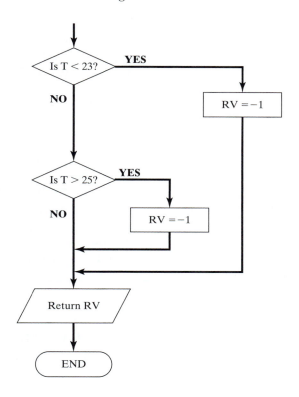

This illustrates two of the reasons to flowchart:

1. to help identify inefficient programming; and
2. to indicate how the program should be written.

PRACTICE!

Flowcharts may be used to depict virtually any multistep process and are frequently used to illustrate a *decision tree* (process leading to a particular decision). This Practice! exercise is about creating a flowchart to show a common decision process: discovering whether someone has a fever and should take some medicine and whether that medicine should be aspirin.

Disclaimer: There is no universally accepted criterion for deciding whether someone's body temperature is high enough to require medication. The values given here are sometimes used, but certainly are no replacement for good medical advice.

- For babies less than a year old, a temperature over 38.3°C (101°F) should receive medication.
- For children less than 12 years old who are not babies a temperature over 38.9°C (102°F) should receive medication.
- For people over 12 years old, a temperature over 38.3°C (101°F) should receive medication.

Oral or ear temperatures are assumed in all cases. The threshold is slightly higher for children because their body temperatures fluctuate more than baby or adult temperatures.

Aspirin is a good fever reducer, but you should not give it to children less than 12 years old (some say less than 19 years) because of the risk of a rare but serious illness known as Reye's syndrome. Acetaminophen and ibuprofen are alternatives for young people.

Create a flowchart that illustrates the input values and the decision steps necessary to learn.

1. whether the person's fever is high enough to warrant medication, and
2. whether the medication should be aspirin.

How would your flowchart need to be modified to include a check for whether the person's temperature is above 41.1°C (106°F)—a level that requires immediate medical assistance?

ANSWER: At the end of chapter.

14.6 FUNDAMENTAL ELEMENTS OF PROGRAMMING

There are seven elements that are common to all programming languages. These include:

- **Data** — Single-valued variables and array variables are used to hold data.
- **Input** — Getting information into the program is an essential first step in most cases.
- **Operations** — These may be as simple as addition and subtraction, but operations are essential elements of programming.
- **Output** — Once you have a result, you need to do something with it. This usually means assigning the result to a variable, saving it to a file, or displaying the result on the screen. Only the first option is available in Mathcad programs.
- **Conditional Execution** — The ability to have a program decide how to respond to a situation is a very important and powerful aspect of programming.
- **Loops** — Loop structures make repetitive calculations easy to perform in a program.
- **Functions** — The ability to create reusable code elements that (typically) perform a single task is considered an integral part of a modern programming language.

14.6.1 Data

What a program does is manipulate data, which is stored in variables. The data values can

- come from cells on the worksheet,
- come from parameter values passed into the function through the parameter list (argument list),
- be read from an external file,
- be computed by calculations within the program.

Variables can hold a single value or an array of values.

Using Worksheet Cell Values in Functions

When you are working in Excel, your data normally are stored in the cells of the worksheets. Being able to use this data in your VBA functions is essential—it is also pretty easy. There are two ways to get worksheet values into a VBA program:

1. Pass the data into a function through a parameter list.
2. Access a cell's contents directly with a VBA function.

Both methods are useful, but using a parameter list is more common and is presented first.

Passing Values through a Parameter List

The items in the parentheses after the function name are the function's *arguments* (also called *parameters*). It is good programming practice to pass all of the information required by a function into the function through the parameter list. This allows the function to be self-contained and ready to be used anywhere. For example, when the **calcVolume** function was used from the worksheet in Figure 14.13, the function received the pressure, P, temperature, T, and number of moles, N, from worksheet cells C3, C4, and C5. Those three values are all the function needs to complete the calculation.

In VBA, argument values are, by default, passed to a function *by reference*, not *by value*. *Passing by reference* means that the memory address of the variable is sent to the function, and the function can look up the value when it needs it. It can also change the value at that memory address. So, when values are passed by reference, the function can change the value. You can explicitly request that variables be passed by reference by including ByRef before the variable name in the argument list, but if ByRef is omitted, the default will be used, so the variable will be passed by reference anyway.

There is one exception to this that is very important for Excel programmers:

Values stored in worksheet cell locations and sent to a function through the argument list can be changed within the function, but the value stored in the worksheet cell will not be changed. The return value from the function can (and usually does) change the value in a worksheet cell, but the cell values sent to the function through the argument list will not be changed by the function call. This means that values stored in cells are passed to functions by value.

Passing by value means the actual value is sent to the function, not the memory address. The function can change the value it receives, but the value stored in memory is not changed, because the function does not know where it is stored. Passing by value is not the default in VBA, except for values stored in cell addresses. You can request that a value be passed by value by including ByVal before the variable name in the argument list.

While you are writing a function, the variables you include in a parameter list and then use in the body of the function are there only to indicate how the various parameter values should be manipulated—that is, while you are defining the function, the variables in the parameter list are *placeholders,* or *dummy variables.* Also, if the variables used in the new function definition already have their own definitions in another function, it doesn't matter. You can use any variable names you want, but well-named variables will make your program easier to understand.

The following two program definitions (flowcharted in Figure 14.17) are functionally equivalent, but the first one is preferred because the variable names have more meaning:

```
Public Function CylinderArea_1(D As Single, L As Single) As
    Single

    Dim AreaEnd As Single
    Dim AreaSide As Single
    Dim AreaTotal As Single
    Dim Pi As Single
    Pi = 3.1416
    AreaEnd = Pi * (D / 2) ^ 2
    AreaSide = Pi * D * L
    AreaTotal = 2 * AreaEnd + AreaSide
    CylinderArea_1 = AreaTotal    ' set return variable value

End Function
```

Figure 14.17
Flowchart of
CylinderArea function.

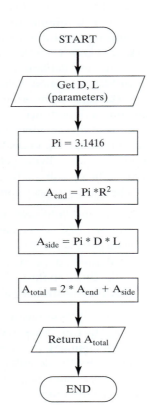

START

Get D, L
(parameters)

$Pi = 3.1416$

$A_{end} = Pi * R^2$

$A_{side} = Pi * D * L$

$A_{total} = 2 * A_{end} + A_{side}$

Return A_{total}

END

```
Public Function CylinderArea_2(v1 As Single, v2 As Single) As
    Single

    Dim v3 As Single
    Dim v4 As Single
    Dim v5 As Single
    Dim v6 As Single
    v6 = 3.1416
    v3 = v6 * (v1 / 2) ^ 2
    v4 = v6 * v1 * v2
    v5 = 2 * v3 + v4
    CylinderArea_2 = v5              ' set return variable value

End Function
```

Here is a quick summary of passing data into a function by means of a parameter list:

- By default, passing information into a function through a parameter causes the memory address of the variables in the parameter list, not their values, to be passed into the function. This allows the function to access the value when needed and to change the value stored in memory. To have a variable's value passed into a function instead of its address, use the ByVal keyword in front of the variable name in the parameter list.
- Values sent into VBA functions from worksheet cells are passed by value—a function can use cell values passed into a function, but cannot change the value in the cells referenced as parameters. (There are VBA program statements that will allow a function to change a cell's contents.)
- Using a parameter list is the preferred way to pass data from a worksheet into a function.
- Parameter lists help make functions self-contained so that they can more easily be reused in other workbooks.

CALCULATING A RESULTANT FORCE AND ANGLE

Force-balance problems often require resolving multiple force vectors into horizontal and vertical force components. The solution may then be obtained by summing the force components in each direction and solving for the resultant force and angle. This multistep solution process (illustrated in Figure 14.18) can be written as an Excel function.

The **Resultant(vh,vv)** function receives two cell ranges: *vh*, a vector of horizontal force components; and *vv*, a vector of corresponding vertical force components. The solution process requires that the components in each direction be summed, the magnitude of the resultant force be computed by using the Pythagorean theorem, and the angle of the resultant force be calculated by using the VBA **Atn** function. Here is the code:

```
Public Function Resultant(vh As Range, vv As Range) As Variant

    Dim i As Integer
    Dim Nrows As Integer
    Dim SumVh As Single
    Dim sumVv As Single
    Dim RF As Single
    Dim Rtheta As Single
```

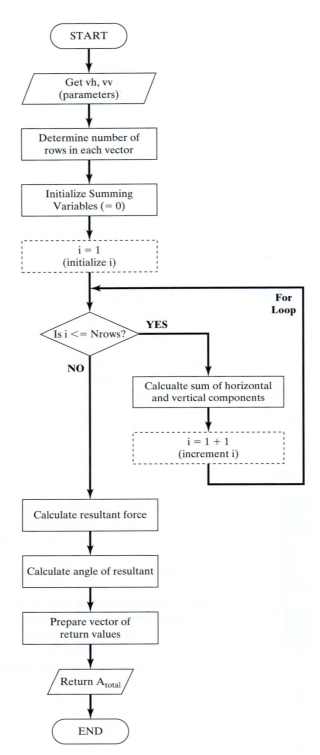

Figure 14.18
Steps required to calculate resultant force and angle.

```
    Dim Result(2) As Single

    Nrows = vh.Cells.Count

    SumVh = 0
    sumVv = 0

    For i = 1 To Nrows
       SumVh = SumVh + vh(i)
       sumVv = sumVv + vv(i)
    Next i

    RF = Sqr(SumVh ^ 2 + sumVv ^ 2)

     ' Rtheta = Atn(sumVv / SumVh)                    ' radians
    Rtheta = Atn(sumVv / SumVh) * 180 / 3.1416        ' degrees

    Result(0) = RF
    Result(1) = Rtheta

    Resultant = Result

End Function
```

Notes:

1. Using the number of cells in *vh* (rather than the number of rows) to establish the number of rows means that the *vh* and *vv* vectors can be either row or column vectors.
2. There are two lines of code for calculating *Rtheta*. The first leaves the angle in radians but has a single quote at the beginning of the line to turn the line into a comment (noncalculated line). The second *Rtheta* calculates the resultant angle in degrees. The intent is to allow the user to easily modify the function if he or she would prefer to have the angle returned in radians.
3. This is an array function because it returns two values. As with any array function, in order to use it, the size of the result must be selected before the entering of the formula containing the function, and the formula is entered by using [Ctrl-Shift-Enter].
4. Excel's default is to place the returned array values side by side (a row vector). You can use the Excel **Transpose** function from VBA to create a column vector if desired. If this is done, the resultant variable is assigned values as follows:

    Resultant = Application.WorksheetFunction.Transpose(Result)

5. Getting the resultant force and angle linked together as an array may not be the handiest way to use these results in other calculations. An alternative to the **Resultant** function is to create two functions, one to calculate the resultant force, and one to calculate the resultant angle. This was done in the previous chapter.
6. The flowchart boxes with dashed-line borders represent parts of the **For... Loop** that are automatically handled by VBA.

An example of the use of the **Resultant** function in a worksheet is shown in Figure 14.19.

E4	▾	*fx*	{=resultant(B4:B8,C4:C8)}			
◢	A	B	C	D	E	F	G
1	Resultant Force and Angle						
2					Resultant		
3		V_H (N)	V_V (N)		Force (N)	Angle (°)	
4		-50	50		686.4	75.7	
5		35	125				
6		75	270				
7		85	100				
8		25	120				
9							

Figure 14.19
Using the **Resultant** function.

Accessing a Cell's Contents Directly

VBA also provides ways of accessing cells directly, either for reading values from cells or for placing formulas or values into cells. The following is a short list of available methods.

Obtaining a Value from a Specific Cell (two options)

- CellVal = Workbooks("Book1").Sheets("Sheet1").Range("B5").Value
- CellVal = Worksheets("Sheet1").Cells(5,2).Value

There are two referencing methods available. One uses the "A1" style used in Excel worksheets; the other uses row and column numbers. Range("B5") and Cells(5,2) refer to the same cell.

Assigning a Value to a Specific Cell (two options)

- Workbooks("Book1").Sheets("Sheet1").Range("B5").Value = CellVal
- Worksheets("Sheet1").Cells(5,2).Value = CellVal

As used here, *CellVal* is a VBA variable that has already been assigned a value. These statements assign the value of *CellVal* to cell B5.

Making a Specific Sheet the Active Sheet

- Worksheets("Sheet1").Activate

Selecting a Range of Cells

1. Worksheets("Sheet1").Activate   First be sure that the correct sheet is active.
2. Range("A1:D4").Select   Then select the desired cell range.

Activating a Particular Cell

1. Worksheets("Sheet1").Activate   First be sure that the correct sheet is active.
2. Range("A1:D4").Select   Then select the desired cell range.
3. Range("B2").Activate   Then activate the desired cell.

Multiple cells can be selected (a selected cell range), but only one cell can be the active cell.

Reading a Value from the Active Cell

- CellVal = ActiveCell.Value

Assigning a Value to the Active Cell

- ActiveCell.Value = CellVal

Note: As used here, *CellVal* is a VBA variable that has already been assigned a value.

14.6.2 Input

There are a variety of *input sources* available on computers: keyboard, disk drives, tape drive, a mouse, microphone, and more. Because Excel functions are housed within a workbook, there are basically two available input sources: data available in the worksheet itself and data in files. The use of worksheet data was presented in the previous section, so this section deals only with reading data files.

One of the simplest ways of dealing with data files is opening them in Excel. Then you can use the values in the cells. But, there are times when you want your program to read a data file. All of the standard Visual Basic file access statements are available in VBA. The process is as follows:

1. Open a data file for input and assign a unit ID.
2. Read data from the file by using the unit ID.
3. Close the file.

In the following example, data are read from file C:\MyData.txt and stored in two array variables, *A* and *B*:

```
Public Function getData()

    Dim A, B
    Dim i as Integer

    Open "C:\MyData.txt" For Input As #1    ' Open file for input as Unit 1
    i = 1                                   ' initialize i
    Do While Not EOF(1)                     ' Loop until end of file
            Input #1, A(i), B(i)            ' Read data
    Loop
    Close #1                                ' Close the file

End Function
```

14.6.3 Operations

VBA uses the typical *mathematical operators* as listed in Table 14.2.

Table 14.2 Standard math operators

Symbol	Name	Shortcut Key
+	Addition	+
−	Subtraction	−
*	Multiplication	[Shift-8]
/	Division	/
^	Exponentiation	[Shift-6]

Operator Precedence Rules

VBA evaluates expressions from left to right (starting at the assignment operator, =), following standard *operator precedence* rules:

Operator Precedence		
Precedence	**Operator**	**Name**
First	^	Exponentiation
Second	*, /	Multiplication, Division
Third	+ , −	Addition, Subtraction

For example, you might see the following equation in a function:

$$C = A \cdot B + E \cdot F.$$

You would need to know that VBA multiples before it adds (operator precedence) in order to understand that the equation would be evaluated as

$$C = (A \cdot B) + (E \cdot F).$$

It is a good idea to include the parentheses to make the order of evaluation obvious.

Assignment Operator, =

The equal sign used in VBA program statements is called the *assignment operator,* and its action is to assign the calculated quantity on the right to the variable on the left (variable *C* in the equation below).

$$C = (A \cdot B) + (E \cdot F).$$

14.6.4 Output

Output from a VBA function can be handled in several ways:

- Information can be returned to the cell(s) in which the function was called.
- The VBA function can place values or formulas directly into cells.
- The VBA function can send data to a file.

Using the Function's Return Value

VBA uses the function name as the variable name that contains the function's *return value*. For example, in the **calcVolume** function listed below, "calcVolume" is both the function name and the name of the variable that holds the function's return value (the calculated ideal gas volume):

```
Public Function calcVolume(P As Single, T As Single, N As
   Single) As Single

     Dim R As Single
     R = 0.08206     ' liter atm / mol Kelvin
     calcVolume = N * R * T / P   ' assign the result to the
                                     return variable

End Function
```

You can return multiple values from a function by returning them as an array. This was done in the **Resultant** function in Section 14.5.1.

Writing Information Directly into a Cell

It is possible to write a value either to a specific cell or to the active cell. VBA also allows a formula to be written to either a specific cell or the active cell.

Assigning a Value to a Specific Cell (two options)

- Workbooks("Book1").Sheets("Sheet1").Range("B5").Value = CellVal
- Worksheets("Sheet1").Cells(5, 2).Value = CellVal

As used here, *CellVal* is a VBA variable that has already been assigned a value. These statements assign the values of *CellVal* to cell B5.

Assigning a Value to the Active Cell

1. Worksheets("Sheet1").Activate
2. ActiveCell.Value = CellVal

As used here, *CellVal* is a VBA variable that has already been assigned a value.

Writing a Formula to a Cell

- Worksheets("Sheet1").Range("C5").Formula = "=B5^2"

This places the formula "=B5^2" in cell C5 on Sheet 1.

- Worksheets("Sheet1").ActiveCell.Formula = "=B5^2"

This places the formula "=B5^2" in the currently active cell on Sheet 1.

Sending Data to a File

The process used to write data to a file is very similar to that used to read a file:

1. Open a data file for output and assign a unit ID.
2. Write data to the file by using the unit ID.
3. Close the file.

In the following example, data are written to file C:\MyData.txt from two array variables, *A* and *B*. The values in arrays *A* and *B*, and the number of rows *(Nrows)* must be set within the program before calling function **saveData**.

```
Public Function saveData()

    Dim i as Integer
    ' A and B each contain Nrows of Data
    Nrows = 10                      ' 10 is arbitrary, but Nrows must
                                    '   have a value
    Open "C:\MyData.txt"            ' Open file for output as Unit 1
      For Output As #1
    For I = 1 to Nrows              ' Loop through all of the data
       Output #1, A(I), B(I)        ' Write data
    Next I
    Close #1                        ' Close the file

End Function
```

EXAMPLE 14.1

LINEAR REGRESSION OF A DATA SET

Here is an example of using an Excel VBA function to perform a series of operations on a data set and then return the results as a vector (a one-row array). This program will perform a linear regression (using Excel's built-in functions) on x and y values stored in two vectors and return the slope, intercept, and R^2 values. First, the data vectors must be entered into the worksheet:

We could calculate the slope, intercept, and value by using three of Excel's built-in functions: **Slope, Intercept,** and **RSq.**

Or, we can combine these three steps into a single function that calculates all three results at the same time. The function is listed below and flowcharted in Figure 14.20.

Figure 14.20
Flowchart of the **Regress** function.

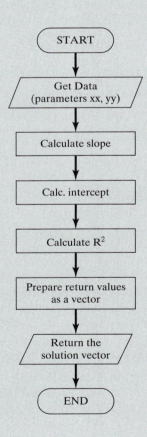

```vba
Public Function Regress(vX As Range, vY As Range) As Variant

    Dim Slope As Single
    Dim Intercept As Single
    Dim R2 As Single
    Dim Result(3)
```

```
Slope = Application.WorksheetFunction.Slope(vY, vX)
Intercept = Application.WorksheetFunction.Intercept(vY, vX)
R2 = Application.WorksheetFunction.RSq(vY, vX)

Result(0) = Slope
Result(1) = Intercept
Result(2) = R2

Regress = Result

End Function
```

The three computed results are collected in the Result array, then returned as a vector through the Regress variable. An example of the use of the **Regress** function is shown in Figure 14.21. The data have been graphed with a linear trendline for comparison.

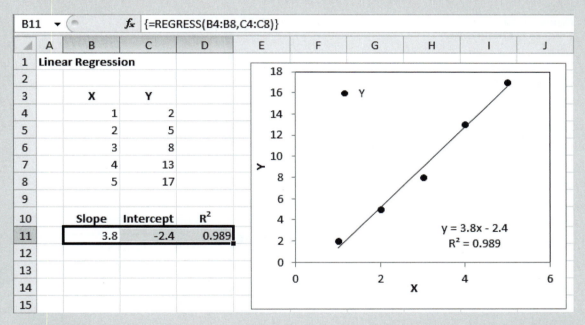

Figure 14.21
Using the **Regress** function.

There is little need to build the **Slope, Intercept,** and **RSq** functions into a function to handle a single data set, but it might be convenient to have a function like **Regress** available if you regularly need to perform a linear regression on many sets of data.

14.6.5 Conditional Execution

It is extremely important for a program to be able to perform certain calculations under specific conditions. For example, in order to determine the density of water at a specific temperature and pressure, you first have to find out whether water is a solid, liquid, or gas at those conditions. A program would use conditional execution statements to select the appropriate equation for density.

If Statement

The classic conditional execution statement is the *If statement*. An If statement is used to select from two options by means of the result of a calculated (logical) condition. In the following example (flowcharted in Figure 14.22), the **CheckForIce** function checks the temperature to see whether freezing is a concern.

```
Public Function CheckForIce(Temp As Single) As String

    Dim RV As String
    RV = "No Problem"
    If Temp < 273.25 Then RV = "Look Out For Ice!"
    CheckForIce = RV

End Function
```

Figure 14.22
Flowchart of **CheckForIce**
function

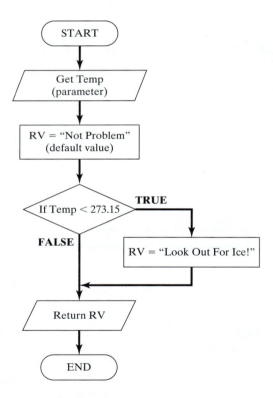

The use of the **CheckForIce** function is illustrated in Figure 14.23.

Block Form of the If Statement

The If statement used in **CheckForIce** was a one-line version of the If statement. An alternative form is the *If block*. In an If block, the check for ice would look like this:

```
If Temp < 273.25 Then
    RV = "Look Out For Ice!"
End If
```

Figure 14.23
Using the **CheckForIce** function.

	A	B	C	D	E	F
	C4		f_x =checkforice(B4)			
1	Checking for Ice					
2						
3		Temp. (K)	Function Response		Formula in Column C	
4		100	Look Out For Ice!		=checkforice(B4)	
5		150	Look Out For Ice!		=checkforice(B5)	
6		200	Look Out For Ice!		=checkforice(B6)	
7		250	Look Out For Ice!		=checkforice(B7)	
8		300	No Problem		=checkforice(B8)	
9		350	No Problem		=checkforice(B9)	
10						

An If block allows multiple lines of code to be executed when the If condition evaluates to True.

COMPUTING THE CORRECT KINETIC ENERGY CORRECTION FACTOR, α FOR A PARTICULAR FLOW

The mechanical energy balance is an equation that is commonly used by engineers for working out the pump power required to move a fluid through a piping system. One of the terms in the equation accounts for the change in kinetic energy of the fluid and includes a kinetic energy correction factor, α. The value of α is 2 for fully developed laminar flow, but approximately 1.05 for fully developed turbulent flow. In order to learn whether the flow is laminar or turbulent, we must calculate the Reynolds number, defined as

$$\text{Re} = \frac{DV_{avg}\rho}{\mu};$$

where

D is the inside diameter of the pipe,
V_{avg} is the average fluid velocity,
ρ is the density of the fluid, and
μ is the absolute viscosity of the fluid at the system temperature.

If the value of the Reynolds number is 2100 or less, then we will have laminar flow. If it is at least 6000, we will have turbulent flow. If the Reynolds number is between 2100 and 6000, the flow is in a transition region and the value of cannot be forecast precisely. We can write a short VBA function to first calculate the Reynolds number and then use two if statements to set the value of α according to the value of the Reynolds number. The **SetAlpha** function is charted in Figure 14.24 and listed below.

```
Public Function SetAlpha(D As Single, Vavg As Single,
Rho As Single, Mu As Single) As Variant

    Dim Re As Single
    SetAlpha = "Cannot Determine"        ' set default response
    Re = (D * Vavg * Rho) / Mu
    If Re > 6000 Then setAlpha = 1.05
    If Re <= 2100 Then setAlpha = 2

End Function
```

Figure 14.24
Flowchart of **SetAlpha** function.

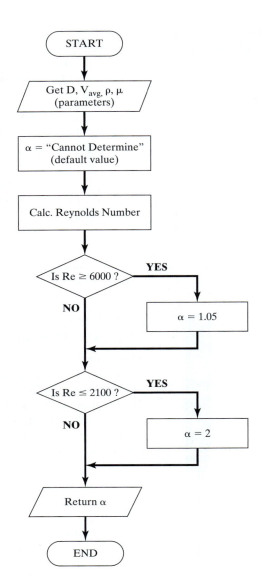

Notice in the function listing and flowchart (Figure 14.24) that the **SetAlpha** return value was initially assigned the text string "Cannot Determine." This is the default case; if both of the If statements evaluate to false, the returned α value will be the warning text string.

Note: The return variable SetAlpha was declared to be of type Variant so that the return "value" could be either the text phrase "Cannot Determine" or a numerical value. The variant data type is an "anything goes" type and will accept either text strings or numerical values.

Next, the Reynolds number is calculated. Then, an If statement is used to see whether the Reynolds number is less than or equal to 2100. If it is, then the flow is laminar and SetAlpha is given the value 2. The next If statement checks to see whether the flow is turbulent (Re > 6000) and, if so, sets $\alpha = 1.05$.

In the following example, the flow of a fluid having a density equal to 950 kg/m^3 and viscosity equal to 0.012 poise, in a 2-inch pipe at an average velocity of 3 m/s, was found to be turbulent, so $\alpha = 1.05$ (the Reynolds number is 120,600, as shown in Figure 14.25).

E10	▾		f_x	=setalpha(E4,E5,E6,E7)		
	A	B	C	D	E	F	G
1	Selecting Kinetic Energy Correction Factor, α						
2							
3			Original Units		Consistent Units		
4	D:		2 inches		0.0508 m		
5	Vavg:		3 m/s		3 m/s		
6	ρ:		950 kg/m^3		950 kg/m3		
7	μ:		0.012 poise		0.0012 Pa sec		
8							
9			Reynolds Number:		120650		
10			α:		1.05		
11							

Figure 14.25
Using the **SetAlpha** function.

If the velocity is lowered to 0.1 m/s, then the Reynolds number falls below 6000, and the program indicates this by sending back the warning string, as shown in Figure 14.26.

	A	B	C	D	E	F	G
1	Selecting Kinetic Energy Correction Factor, α						
2							
3			Original Units		Consistent Units		
4	D:		2 inches		0.0508 m		
5	Vavg:		0.1 m/s		0.1 m/s		
6	ρ:		950 kg/m^3		950 kg/m3		
7	μ:		0.012 poise		0.0012 Pa sec		
8							
9			Reynolds Number:		4022		
10			α:		Cannot Determine		
11							

Figure 14.26
The **SetAlpha** response when the Reynolds number is indeterminant.

Else Statement

The *Else statement* is used in conjunction with an If statement when you want the program to do something else when the condition in the If statement evaluates to false. For example, we might rewrite the **CheckForIce** function to test for temperatures below freezing, but provide an Else to set the text string to "No Problem" when freezing is not a concern.

```
Public Function CheckForIce(Temp As Single) As String

    If Temp < 273.25 Then
        CheckForIce = "Look Out For Ice!"
    Else
        CheckForIce = "No Problem"
    End If

End Function
```

This version is functionally equivalent to the earlier version, but the code and flowchart (Figure 14.27) might be more easily read by someone unfamiliar with programming. In the earlier version, the return value was set to "No Problem" and then overwritten by "Look Out For Ice!" if the temperature was below freezing. In this version of the program, there is no overwriting; the return value is set to one text string or the other based on the result of the If test.

Figure 14.27
Flowchart of modified **CheckForIce** function.

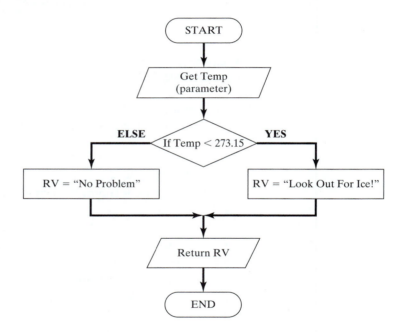

On Error GoTo Statement for Error Trapping

The *On Error GoTo statement* is used for *error trapping* and provides an alternative program-flow path when certain conditions will cause errors in a program. Typical error situations are divisions when the denominator is zero and attempting to open a file that does not exist on the specified drive.

Error trapping involves turning on error handling just before the error-prone step, providing an alternative program-flow path in case an error is found, and turning off the error handling if the error is not detected. The statements used to accomplish these steps are the following:

On Error GoTo MyErrorHandler	*activates error handling*
On Error GoTo 0	*deactivates error handling*

The MyErrorHandler is actually an argument to the On Error GoTo statement; it is a line label indicating where the program flow should go if an error is detected. For example, the **GetData** function could be modified as illustrated in Figure 14.28 to provide error trapping in case the file cannot be opened. Also, return variable GetData now returns a code value: 1 means that the file was read successfully, −1 that the file could not be read.

Figure 14.28
Flowchart for modified **GetData** function.

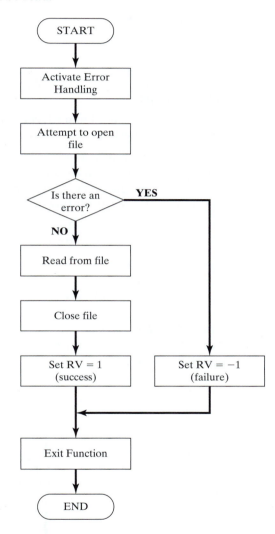

```
Public Function GetData() As Integer

    Dim A, B
    Dim i As Integer

    On Error GoTo MyErrorHandler      ' turn on error handling
    Open "C:\MyData.txt"              ' Open file for input as
      For Input As #1                     Unit 1
    On Error GoTo 0                   ' turn off error handling
    i = 1                            ' initialize i
    Do While Not EOF(1)              ' Loop until end of file
```

```
       Input #1, A(i), B(i)    ' Read data
    Loop

    Close #1                    ' Close the file

    GetData = 1                 ' 1 = success, file was read

    Exit Function               ' if all goes well, exit here
    MyErrorHandler:             ' if you get here, the file could
                                    not be opened
    GetData = -1                ' -1 = failure, file could not be read

End Function
```

14.6.6 Loops

Loop structures are used to perform calculations over and over again. Loops are very commonly used when:

- You want to repeat a series of calculations for each value in a data set or matrix.
- You want to perform an iterative (guess and check) calculation until the difference between the guessed value and the calculated value is within a preset tolerance.
- You want to move through the rows of data in an array until you find a value that meets a particular criterion.

There are three loop structures supported in VBA:

- For... Next loops
- Do... Loop loops
- For Each... Next loops.

For... Next Loops

For... Next loops (flowcharted in Figure 14.29) use a counter to loop a specified number of times:

```
For i = 1 To 10
    x(i) = i + 1
Next i
```

Figure 14.29
Flowchart of a For...
Next loop.

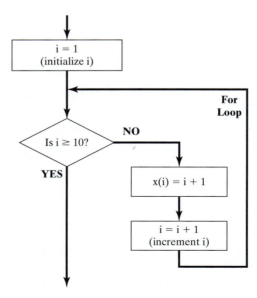

Loops can run backward, or use nonunity steps if the step is specified. The loop listed below decrements i by 2 each time through the loop and stops when i is less than or equal to zero:

```
For i = 10 To 0 Step -2
    x(i) = i + 1
Next i
```

There is an Exit For statement that can be used to send the program flow out of a For... Next loop:

```
For i = 1 To 10
    x(i) = i + 1
    If x(i) = 4 then Exit For
Next i
```

Do... Loop Loops

Do... (or Do While...) loops are used when you want to loop until some condition is met. A flowchart for a Do... While loop is shown in Figure 14.30.

Figure 14.30
Flowchart of a Do... While loop.

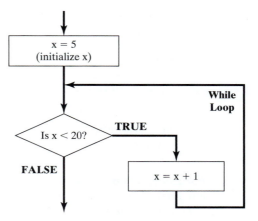

```
x = 5
Do While x < 20
    x = x + 1
Loop
```

The Do While... Loop shown above evaluates the condition (is $x < 20$?) before entering the loop. Because of this, the following loop would never execute, and x would never be incremented:

```
x = 50
Do While x < 20        ' since x > 20 before the loop, the loop
                         never executes at all
    x = x + 1
Loop
```

Alternatively, you can use a Do... Loop While structure:

```
x = 50
Do
    x = x + 1
Loop While x < 20   ' the loop always executes at least one
```

With the evaluation (While $x < 20$) placed after the Loop, the preceding loop always executes at least once.

There are also two Do… Loop loops that use an Until instead of a While. Again, the two versions vary the position of the condition evaluation:

```
x = 5
Do Until x < 20      ' since x < 20 before the loop, the loop
                       never executes at all

    x = x − 1

Loop
```

This loop will never quit (watch out for this type of error):

```
x = 50
Do Until x < 20      ' since x > 20 the loop executes, but
                       since x is increased inside the loop,
                       the loop never terminates

    x = x + 1
Loop


x = 50
Do
    x = x − 1
Loop Until x < 20    ' since the condition follows the loop,
                       this loop always executes at least once
```

There is an Exit Do statement that can be used to send the program flow out of a Do… Loop.

For Each… Next Loops

For Each… Next loops are specialized loop structures designed for use with *collections*. A collection is a set of related objects. A For Each… Next loop is designed to step through each object in the collection. In VBA, a common use of this loop structure is for stepping through each value in a cell range, because a Range is a type of collection in VBA. For example, the following loop counts the number of cells in the range A1:D10 that have values greater than 100:

```
Counter = 0
For Each cellObject In Worksheets("Sheet1").Range("A1:D10").Cells
    If cellObject.Value > 100 Then counter = counter + 1
Next
```

> **Note:** *cellObject* is a variable name in this function, and you can use any name you want. It is simply a way of identifying each element in the collection. In the For Each statement, *cellObject* is assigned a value corresponding to a specific element (i.e., cell) in the cell range. Then, in the If cellObject.value statement, *cellObject* is used to identify the particular cell being tested.

14.6.7 Functions

Functions are an indispensable part of modern programming because they allow a program to be broken down into pieces, each of which ideally handles a single task (i.e., performs a single function). The programmer can then call upon the functions as needed to complete a more complex calculation.

Functions are an indispensable part of Excel as well. Excel provides built-in functions that can be used as needed to complete a lot of computational tasks. If Excel's built-in functions cannot perform a calculation, you can write your own function.

The Excel worksheet provides such a convenient place in which to call functions that, most of the time, you will never need to write a complete program. The most common type of programming in Excel is simply writing additional functions for use in the worksheets.

The Difference between Subs (Subprograms) and Functions

Subprograms or Subs are basically functions that don't return values (except through the argument list). Both Subs and functions basically the same structure, with a start line and an end line that are created by the VBA editor when the Sub or function is inserted. Both can have argument lists. But a function is designed to return a value, and a Sub isn't.

In Excel, a macro is simply a Sub that does not accept any arguments. This means you can write a Sub with no arguments in VBA, and Excel will recognize the no-argument Sub and automatically display your Sub in the list of available macros.

14.6.8 Declaring Data Types for Variables

VBA provides a lot of *data types* to describe the type of value being stored and tell VBA how much memory to allocate for each variable. If you do not explicitly declare variables to be of a particular type, VBA declares them to be *Variant* by default, which is an "anything goes" data type that can hold either numbers or text strings. Variants are handy, but they use a lot of memory and can slow down your program.

Some commonly used data types include those in the following table:

Data Type	Description	Examples	Use
Single	Single precision real numbers (4 bytes) $-3.402823E^{38}$ to $3.402823E$	12.6, 1213.432, -836.5	General, low-precision math
Double	Double-precision real numbers (8 bytes) $-1.79769313486231E^{308}$ to $1.79769313486231E$	12.631245, -836.50001246	General, high-precision math
Integer	Small integer values (2 bytes) $-32,768$ to 32,767	1, 2, 2148, 16324	Counters, index variables
Long	Long integer values (4 bytes) $-2,147,483,648$ to $2,147,483,647$	32768, 64000, 1280000	Used whenever an integer value could exceed 32,767
Boolean	Logical values (2 bytes) True or False	True, False	Status variables
String	Text strings	"yes," "the result is: ", "C:\My Documents"	Words and phrases, file names
Date	Date values (8 bytes). Floating point number representing days, hour, minutes, and seconds since 1/1/100	Values are stored and used as floating point numbers, but displayed as dates and times	Dates and times

14.6.9 The Scope of Variables and Functions

The *scope* of a variable describes its availability to other objects. A variable declared within a function (or Sub) by using a Dim statement is available only within that function (or Sub). An example of this is the variable *R* in the **calcVolume** function:

```
Public Function calcVolume(P As Single, T As Single, N As
Single) As Single

   Dim R As Single
   R = 0.08206     ' liter atm / mol Kelvin
   calcVolume = N * R * T / P

End Function
```

Variable *R* was defined within the **calcVolume** function by using a Dim statement and so can be used anywhere within the **calcVolume** function, but no other function or object would know that variable *R* even exists.

The statement

```
             Public Function calcVolume()
```

declares the scope of the **calcVolume** function to be Public. This means that other functions or objects can call the **calcVolume** function. (Technically, even objects in other workbooks can call the **calcVolume** function, but they must specifically refer to the function within its defining workbook.)

As a general rule, variables that will be used only within a function should have *local scope* (defined with a Dim statement within the function and available only within the function). Variables that must be known to multiple functions (or Subs) must have *public scope*, and these are generally defined in the "Declarations" area of the "General" section of the project's module.

Declaring a Variable That Will Be Available to All Functions in a Module

When you insert a module into your VBA project, the module contains a Declarations section. All Public variables are declared in that section.

Example: Declare a variable named *convFactor* to be of type Double and Public in scope (i.e., available to any function or Sub in the module). This is illustrated in Figure 14.31.

Figure 14.31
Declaring convFactor to be a double-precision variable with Public scope.

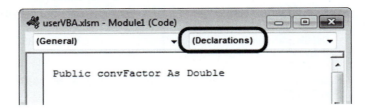

In this example, the Declarations section contains only a single line. It will expand as additional variables are declared. The editor shows all of the module's program code together, to make editing easier, but, if you click on the **calcVolume** function code, the editor will show you that you have left the Declarations section and are now editing the CalcVolume section.

Declaring a Variable within a Function with Local Scope

To declare a *local variable* (a variable that cannot be used except within the function in which it is declared), you use a *Dim statement* inside the function. Declaration statements, such as Dim statements, are the first statements in a function.

Example:

```
Public Function calcValue(Dx as Single) as Single

    Dim ABC as Integer
    Public XYZ as Long        —this statement doesn't work here!
    Dim DateVar as Date

End Function
```

Because the variables *ABC* and *DateVar* were declared inside **calcValue** with Dim statements, they cannot be used outside the **calcValue** function.

You cannot make a variable declared within a function available to other functions by using the Public statement. The statement

```
                    Public XYZ as Long
```

is a valid statement, but not inside of a function. Variables declared with the Public statement, to be available to any function (or Sub) in a module, must be declared in the "Declarations" section of a module (see Figure 14.31).

14.7 WORKING WITH FORMS

VBA also allows you to create *forms* (e.g., dialog boxes). Forms can contain standard programming objects, such as the following:

- Labels (display text, not for data entry)
- Text Fields (for display and data entry)
- List Boxes
- Check Boxes
- Option Buttons (a.k.a. radio buttons)
- Frames (for collecting objects, especially option buttons)
- Button (a.k.a. command buttons)
- RefEdit Boxes (allow you to easily get a cell address from an Excel worksheet)

You can tie program code to the objects to make the form do what you want. VBA gives you a lot of programming power, and using forms can be very useful for some tasks. Forms can be used with data on an Excel worksheet (like Excel's dialog boxes), or by themselves.

14.7.1 Creating a Form for a Simple Calculation

Once again, let's calculate the volume of an ideal gas, only this time we will use a VBA form that is totally disconnected from the Excel worksheets. As a preview, here's the form we are going to create (Figure 14.32).

The form includes four labels, four text boxes, and one button.

The process we will use to create the form is as follows:

1. Insert a UserForm into the VBA project.
   a. Change the caption from "Form1" to "Ideal Gas Volume."
2. Add a label to the form and set the caption to "Pressure (atm.)"
   b. Change the font size and set the bold attribute of the label.

Figure 14.32
The Ideal Gas Volume
form.

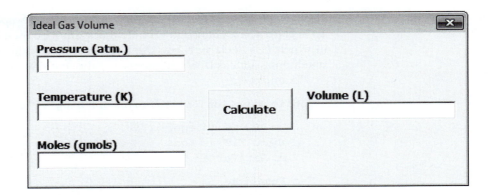

3. Make three copies of the label:
   a. Set one caption to "Temperature (K)"
   b. Set one caption to "Moles (gmols)"
   c. Set one caption to "Volume (L)"
4. Add a text box to the form.
5. Make three copies of the text box.
6. Change the names of each text box, to:
   a. txtPressure
   b. txtTemp
   c. txtMoles
   d. txtVolume
7. Add a button to the form
   a. Change the button name to btnCalc
   b. Change the caption to Calculate
   c. Change the font size and set the bold attribute
8. Add code to the button to calculate the volume.

Don't get intimidated by eight steps; the process of creating the form is easier to do than to describe.

Step 1. Insert a UserForm into Module1. Use VBA menu options **Insert/Userform** (Figure 14.33) to add a form (called Form1) to your project. The empty form is shown in Figure 14.34.

Figure 14.33
Use **Insert/UserForm** to
add a form to your project.

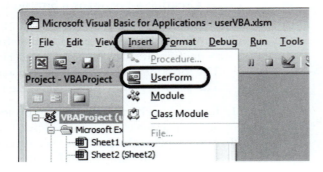

Figure 14.34
The empty form.

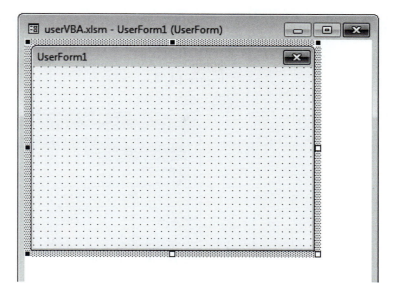

You can resize the form with the handles on the right and bottom edges. The dots on the form are there to help align objects on the form; they do not show when the form is used.

The title displayed in the form ("UserForm1" by default) is the form's *caption property*. You change properties using the *Properties box* in the VBA editor. In Figure 14.35, the caption property has been indicated, and the values of the caption property has been changed to "Ideal Gas Volume."

Figure 14.35
Properties box in the VBA editor.

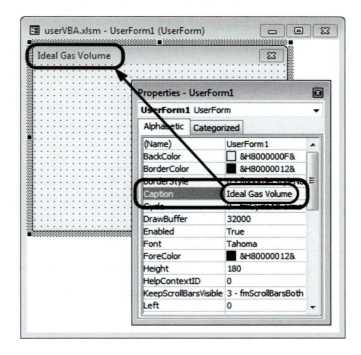

Step 2. Add a Label to the Form and Set the Label's Caption to "Pressure (atm.)." When a form is selected (indicated by the gray border and resize handles, as in Figure 13.34), the VBA editor displays a Controls Toolbox, as shown in Figure 14.36. The label tool has been indicated in Figure 14.36.

Figure 14.36
Controls Toolbox.

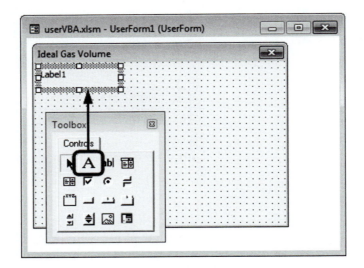

To add a label to the form, click on the label tool (the capital A), then draw a box the approximate size of the desired label on the form. When you release the mouse, the label, with a default caption of "Label1" will be left on the form. This is illustrated in Figure 14.36.

The heavy gray border and grab handles indicate that the label is selected. (Click on the label to select it if the border is not visible.) When the label is selected, the Property Box displays the properties of the label (Figure 14.37). We are going to change two properties: the *caption* and the *font*.

Figure 14.37
Properties of the pressure label.

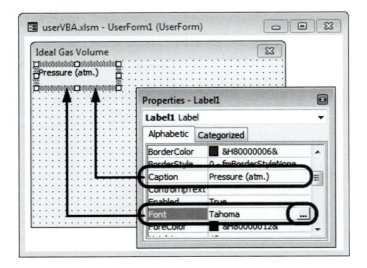

In Figure 14.37, the label's caption has already been changed to "Pressure (atm.)," and the Font property has been selected. When the Font property is selected, a small button with three dots (an ellipsis) appears at the right edge of the Property Box (indicated in Figure 14.37); click this button to open the Font dialog to change the font, as illustrated in Figure 14.38. Here, we have changed the font style to Calibri, increased the font size to 11 points, and set the Bold attribute.

Figure 14.38
The Font dialog.

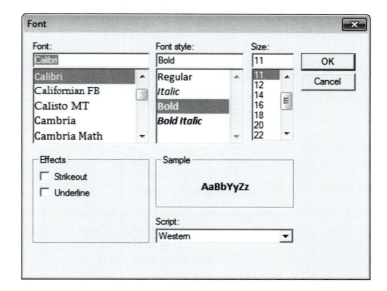

Step 3. Make Three Copies of the Label. With the pressure label selected, copy [Ctrl-c] and paste [Ctrl-v] the label to create three copies, or a total of four labels, as shown in Figure 14.39.

Figure 14.39
The form with the four needed labels.

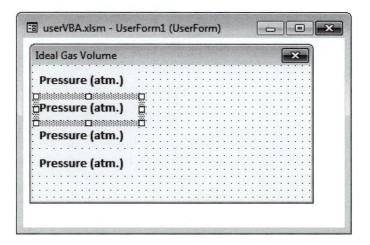

Select each of the new labels and set the caption property to "Temperature (K)," "Moles (gmols)," and "Volume (L)," as illustrated in Figure 14.40. The labels have been located in positions that approximate where they will end up on the form.

Figure 14.40
The labels with updated captions.

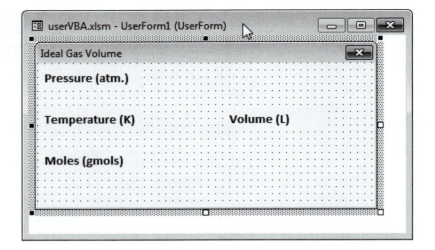

Note: By copying the pressure label rather than adding additional labels from the Controls Toolbox we avoided having to change the font on the three additional labels.

Step 4. Add a Text Box to the Form. A *text box* is used to collect information when the form is used. Click on the text box tool (indicated in Figure 14.41) and draw a box underneath the pressure label on the form. When you release the mouse, VBA will place a text box on the form, as shown in Figure 14.41.

Figure 14.41
Controls Toolbox with text box tool indicated.

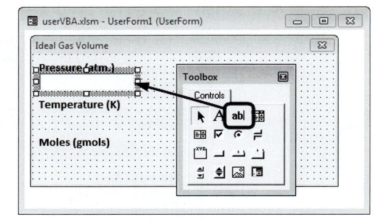

Step 5. Make Three Copies of the Text Box. With the pressure text box selected (as in Figure 14.41), copy [Ctrl-c] and paste [Ctrl-v] the text box to create three copies, or a total of four text boxes, as shown in Figure 14.42.

Step 6. Change the Names of Each Text Box. The text boxes have names like Textbox1, Textbox2, etc. We will use the textbox names when we calculate the volume, so let's change the Name property of each textbox to something more meaningful:

- txtPressure
- txtTemp
- txtMoles
- txtVolume

Figure 14.42

The form after adding the four text boxes.

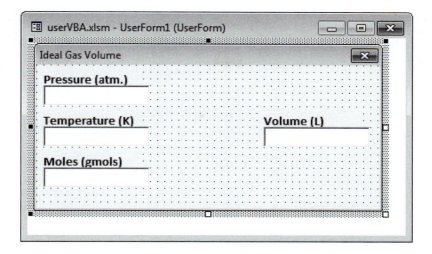

The Property box for the Temperature text box is shown as an illustration in Figure 14.43.

Figure 14.43

Changing the name property for the volume text box.

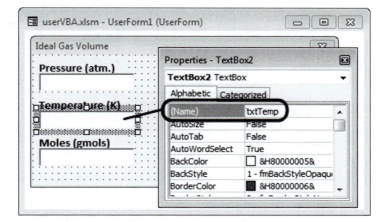

Starting the text box names with txt is traditional in VB programming; using object names that include an indication of the type of object (like txt for text box) can help you keep track of your objects in large projects.

Step 7. Add a Button to the Form. Use the **command button** tool on the Controls Toolbox (indicated in Figure 14.44) and draw a box on the form. When you release the mouse, VBA will place a button on the form as shown in Figure 14.44.

Next, change the name, caption, and font properties of the command button.

- Change the button name to "btnCalc" (many programmers would use "cmdCalc" since it is a command button)
- Change the caption to "Calculate"
- Change the font style and size, and set the bold attribute

These changes are indicated on the Property Box shown in Figure 14.45.

Figure 14.44
Controls Toolbox,
command button tool
indicated.

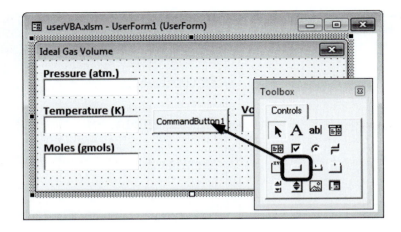

Figure 14.45
Property changes for the
Calculate button.

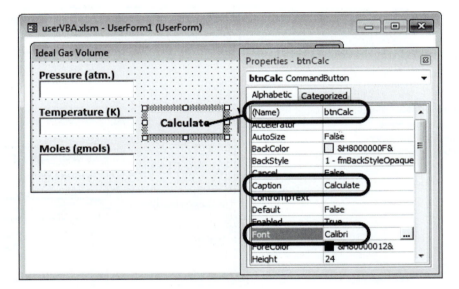

At this point, the form is complete, as shown in Figure 14.46. If you run the VBA program, the form will open and you can enter values in the text boxes. You can even click the **Calculate** button; it just doesn't calculate anything. We need to tie some code to the **Calculate** button so that it calculates something.

Figure 14.46
The form with all objects
in place.

Ideal Gas Volume

Pressure (atm.)

Temperature (K)

Calculate

Volume (L)

Moles (gmols)

Step 8. Add Code to the Button to Calculate the Volume.   If you double-click on the **Calculate** button, a code page for the form will open and VBA will add the first and last lines of a new Sub; the Sub that will run when someone clicks the **Calculate** button. The new Sub, called **btnCalc_Click()**, is shown in Figure 14.47.

Figure 14.47
The **btnCalc_Click()** function will run when someone clicks the Calculate button.

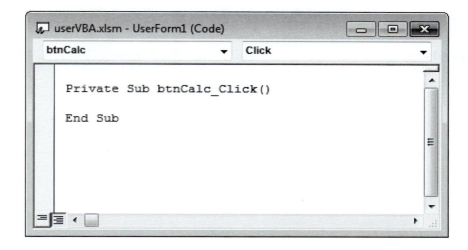

We need to add code to this Sub to calculate the ideal gas volume using the pressure, temperature, and moles data in the text boxes. Then, the calculated volume needs to be displayed in the volume text box.

The code we need looks like this:

```
Private Sub btnCalc_Click()

    Dim P As Single
    Dim V As Single
    Dim N As Single
    Dim R As Single
    Dim T As Single

    R = 0.08206                 ' L atm / mol K

    P = txtPressure.Text
    T = txtTemp.Text
    N = txtMoles.Text

    V = (N * R * T) / P

    txtVolume.Text = V

End Sub
```

In this sub, we:

1. Declared five variables (P, V, N, R, T) to be of type Single (single-precision, numeric), and local scope.
2. Assigned the R variable a value (the ideal gas constant).
3. Read the .Text property of the pressure, temperature, and moles text boxes and assigned the values to variables P, T, and N.

4. Calculated the ideal gas volume and assigned it to variable *V*.
5. Assigned the value of variable *V* to the .Text property of the volume text box. This causes the calculated volume to be displayed on the form.

Running the Program and Using the Form

Use VBA menu options **Run/Run Sub/UserForm** (or press [F5]) to run the program and display the form. Enter values in the input text boxes, as shown in Figure 14.48.

Figure 14.48
The form in use, input values have been entered.

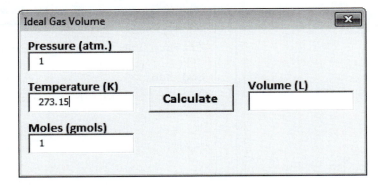

The values used here ($P = 1$ atm, $T = 273.15$K, $N = 1$ mol) should generate the standard molar volume, or 22.414 L. We can use this known value to verify that our program is working correctly.

Press the **Calculate** button to solve for the volume. The result is shown in Figure 14.49.

Figure 14.49
The calculated volume.

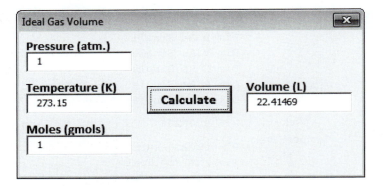

14.7.2 Creating a Form that Uses Worksheet Values

Usually, if you are working in Excel, you want to access the data on the worksheets. VBA provides a variation on a text box that allows you to jump to a worksheet to get a cell address. (This is commonly done on Excel's dialog boxes.) The tool is called a *RefEdit tool,* and it is indicated on the Controls Toolbox shown in Figure 14.50.

You add it to a form just like a text box, but it has a button on the right side of the field. In Figure 14.51, the Ideal Gas Volume form has been updated to get the inputs from the worksheet using RefEdit fields. The fields have been rearranged to make the form a bit more compact.

Figure 14.50
Controls Toolbox, RefEdit
tool indicated.

Figure 14.51
Updated form, using
RefEdit fields for the inputs.

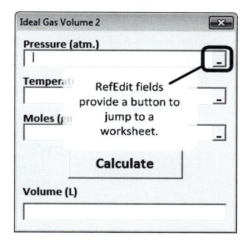

When the program is run, the form appears over the Excel worksheet (like all VBA forms), as shown in Figure 14.52.

	A	B	C	D
1	Volume of an Ideal Gas			
2				
3		Pressure:	1	atm.
4		Temperature:	273.15	K
5		Moles:	1	gmoles
6				
7				
8				
9				
10				
11				
12				
13				
14				

Ideal Gas Volume 2

Pressure (atm.)

Temperature (K)

Moles (gmols)

Calculate

Volume (L)

Figure 14.52
Running the new form.

When you click one of the RefEdit field buttons, the form will collapse to show only that field, and you can use the mouse to select a cell (or range of cells) on the worksheet. In Figure 14.53, the cell containing the pressure value is being selected.

Figure 14.53
Selecting the pressure value using the RefEdit field.

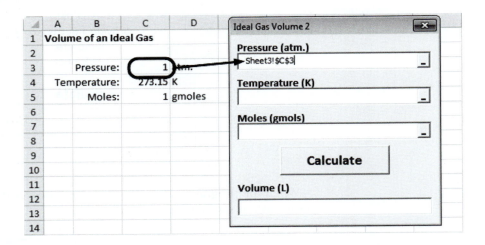

When you click the button at the right side of the collapsed form, the original form is shown (Figure 14.54), with the cell address containing the pressure value indicated in the pressure RefEdit field.

Figure 14.54
The form showing the cell address containing the pressure value.

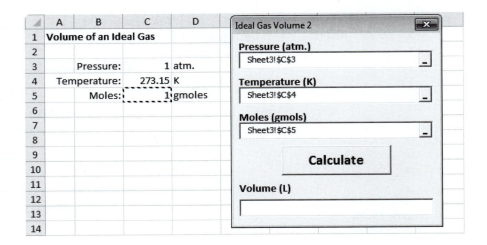

The same process is used to identify the cells containing the temperature and number of moles. The form, after setting all of the input values, is shown in Figure 14.55.

Figure 14.55
The form after indicating where all of the input values can be found.

Then, click the **Calculate** button to solve for the ideal gas volume using the worksheet values. The result is shown in Figure 14.56.

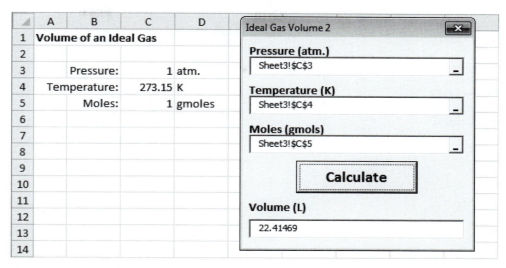

Figure 14.56
The calculated volume.

The code behind the **Calculate** button is a little different than in the first form, because the .Value property of the RefEdit field holds the cell address (exactly as shown in Figure 14.56), not the cell contents. For example, the pressure RefEdit field is named refPressure, and refPressure.Value = "Sheet1!$C$3." In the code behind the **Calculate** button, we need to use this cell address to go find the pressure value on the worksheet. This is accomplished with the command

```
Range(refPressure.Value).Value
```

This is equivalent to

```
Range("Sheet1!$C$3").Value
```

And the .Value in the cell (Range) at "Sheet1!$C$3" is the pressure value we need. The code behind the **Calculate** button is listed below.

```
Private Sub btnCalc_Click()

    Dim P As Single
    Dim V As Single
    Dim N As Single
    Dim R As Single
    Dim T As Single

    R = 0.08206        ' L atm / mol K

    P = Range(refPressure.Value).Value
    T = Range(refTemp.Value).Value
    N = Range(refMoles.Value).Value

    V = (N * R * T) / P

    txtVolume.Text = V

End Sub
```

This is very similar to the code used with the text boxes, except for getting the cell values from the cell addresses in the refEdit fields for *P, T,* and *N.*

We've obviously only scratched the surface of what can be accomplished using VBA in Excel. Hopefully, we have at least whetted your appetite to see if VBA can be useful in your own work.

Answer to Practice Problem

Figure 14.57
Flowchart to decide if OK to give aspirin.

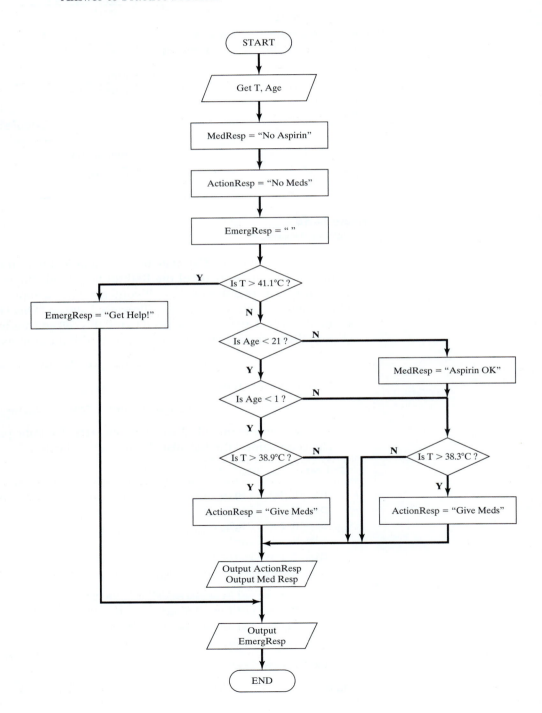

KEY TERMS

Application.
WorksheetFunction.
Argument
Assignment operator
Button (command
button)
By reference (passing
variables)
By value (passing
variables)
Caption property
Caption property (VBA)
Check box
Condition
Conditional execution
Data
Data type (VBA)
Decision (flowchart)
Decision tree (flowchart)
Default value
Developer tab (Ribbon)
Development area
(VBA)
Dim statement
Do… Loop
Double (data type) (VBA)

Dummy variable
Else statement
Error trapping
File extensions (.xlsx,
.xlsm)
Flowchart
Font property
For Each… Next
loop
For… Next loop
Form (VBA)
Frame
Function (VBA)
If block (block for of
If statement)
If statement
Input
Input source
Label
List box
Local variable
Loop structure
Macro programming
Macro virus
Macro-disabled
workbook

Macro-enabled
workbook
Mathematical operator
Module (VBA)
Name property
On Error GoTo
statement
Operation (flowchart)
Operator precedence
Option button
Output
Parameter
Placeholder
Project (VBA)
Project panel (VBA)
Properties panel
(VBA)
RefEdit Box
Return value
Scope
Single (data type)
(VBA)
Subprogram (Sub)
Text field (text box)
VBA editor
VBA

SUMMARY

Macro-Enabled Worksheets
- .xlsx—the Excel file extension for macro-disabled workbooks (default)
- .xlsm—the Excel file extension for macro-enabled workbooks

Showing the Ribbon's Developer Tab

Excel 2010

1. Use Ribbon options **File tab/Options** to display the Excel Options dialog, then select the **Customize Ribbon** panel.
2. Check the box labeled **Developer** in the **Main Tabs** list.

Excel 2007

1. Use the Microsoft Office button with options **Office/Excel Options** to display the Excel Options dialog, then select the **Popular** panel.
2. Check the box labeled **Show Developer tab in the Ribbon**.

Starting the Visual Basic Editor

Use Ribbon options **Developer/Code/Visual Basic**

[Excel 2003: Tools/Macro/Visual Basic Editor]

Inserting a Module into a Project

A module is a VBA project item that holds program code. Your functions are written and stored in a module. A module can store any number of functions.

1. Click on "VBA Project" in the Project panel.
2. Select **Insert/Module** from the VBA menu.

Inserting a Function into a Module

1. Click in the Module to be sure it is selected.
2. Select **Insert/Procedure...** from the VBA menu.
3. Specify:
   a. function **Name**
   b. Type **Function**
   c. Scope (typically) **Public**
   d. **All Local Variables as Statics** (generally not checked)

Data Types (Common)

Data Type	Description	Use
Single	Single-precision real numbers (4 bytes) $-3.402823E^{38}$ to $3.402823E^{38}$	General, low-precision math
Double	Double-precision real numbers (8 bytes) $-1.79769313486231E^{308}$ to $1.79769313486231E^{308}$	General, high-precision math
Integer	Small integer values (2 bytes) $-32,768$ to $32,767$	Counters, index variables
Long	Long integer values (4 bytes) $-2,147,483,648$ to $2,147,483,647$	Used whenever an integer value could exceed 32,767
Boolean	Logical values (2 bytes) True or False	Status variables
String	Text strings	Words and phrases, file names
Date	Date values (8 bytes) Floating point number representing days, hour, minutes, and seconds since 1/1/100	Dates and times

Function Summary

Code for functions is stored in a module.

Function Scope:

- Public—visible to entire workbook (and other workbooks)
- Private—visible to module only

Variable Scope:

- Public—visible to entire workbook (and other workbooks)
- Private—visible to function only

Store public variables in a module (Declarations section)

- Passing values—through an argument list
- Data types for variables—commonly: single, double, string, Boolean, variant (default)
- Returning a value—Use the function name as a variable holding the function's return value.

Form Summary

Standard Form Objects

- Labels (display text, not for data entry)
- Text Fields (for display and data entry)
- List Boxes
- Check Boxes
- Option Buttons (a.k.a. radio buttons)
- Frames (for collecting objects, especially option buttons)
- Buttons (a.k.a. command buttons)
- RefEdit Boxes

General Development Process

1. Decide what you want the form to do.
2. Develop a general layout.
3. Insert a user form and a module into the current project.
4. Build the form by dragging and dropping control objects from the controls toolbox onto the form.
5. Declare public variables in the Declarations section of a module.
6. Set initial values by using UserForm_Initialize().
7. Add code to the controls on the form.
8. Set any return value(s); usually through code attached to an OK button.

   If the form is to be opened from an Excel worksheet...
9. Create a Sub that shows the form.
10. Tie the Sub that shows the form to a shortcut key, so that the form can be displayed from an Excel worksheet.

PROBLEMS

14.1 Ideal Gas Volume

Write a VBA function that receives the absolute temperature, absolute pressure, number of moles, and ideal gas constant (optional) and then returns the calculated ideal gas volume. Use the equation

$$PV = NRT$$

Use an IF() statement to see whether the function received a zero for the gas constant:

```
IF (R = 0) THEN
R = 0.08206            ' use your default R value here
END IF
```

If R is equal to zero, use your favorite default value from this list:

Ideal Gas Constants	
8.314	$\dfrac{m^3 Pa}{gmol\ K}$
0.08314	$\dfrac{L\ bar}{gmol\ K}$
0.08206	$\dfrac{L\ atm}{gmol\ K}$
0.7302	$\dfrac{ft^3\ atm}{lbmol\ °R}$
10.73	$\dfrac{ft^3\ psia}{lbmol\ °R}$
8.314	$\dfrac{J}{gmol\ K}$
1.987	$\dfrac{BTU}{lbmol\ °R}$

14.2 Absolute Pressure from Gauge Pressure

Write a VBA function that receives a gauge pressure and the barometric pressure and returns the absolute pressure. Have the function check whether the barometric pressure is zero, indicating that the program should use a default value of 1 atmosphere as the barometric pressure. If the barometric pressure is zero, add 1 atmosphere (or 14.696 psia) to the gauge pressure to calculate the absolute pressure.

14.3 Ideal Gas Solver

Write a VBA function that receives a variable name plus four known values and then solves the ideal gas equation for a specified variable. Use a SELECT CASE statement to select the correct equation to solve. The first two options (solve for P and V) are listed next. You will need to add additional cases (for N, R, and T) ahead of the Case Else statement to complete the function.

>   **Note: PVNRT** is declared as a Variant type in the next line so that it can return either a value (the solution) or a text string (the error message from the Case Else statement).

```
Public Function PVNRT(SolveFor As String, V1 As Single, V2 As
Single, V3 As Single, V4 As Single) As Variant

    Dim P As Single
    Dim V As Single
    Dim N As Single
    Dim R As Single
    Dim T As Single
```

```
Select Case SolveFor
        Case "P"
                V = V1
                N = V2
                R = V3
                T = V4
                PVNRT = N * R * T / V
        Case "V"
                P = V1
                N = V2
                R = V3
                T = V4
                PVNRT = N * R * T / P
        Case Else
                PVNRT = "Ouch! Could not recognize the variable
                to solve for."
    End Select
End Function
```

14.4 Computing Gas Heat Capacity II

There are standard equations for calculating the heat capacity (specific heat) of a gas at a specified temperature. One common form of a heat capacity equation is a simple third-order polynomial in T:[1]

$$C_p = a + bT + cT^2 + dT^3$$

If the coefficients a through d are known for a particular gas, you can calculate the heat capacity of the gas at any T (within an allowable range).
The coefficients for a few common gases are listed here:

Heat Capacity Coefficients

Gas	a	b	c	d	Units on T	Valid Range
Air	28.94×10^{-3}	0.4147×10^{-5}	0.3191×10^{-8}	-1.965×10^{-12}	°C	0–1500°C
CO_2	36.11×10^{-3}	4.233×10^{-5}	-2.887×10^{-8}	7.464×10^{-12}	°C	0–1500°C
CH_4	34.31×10^{-3}	5.469×10^{-5}	0.3661×10^{-8}	-11.00×10^{-12}	°C	0–1200°C
H_2O	33.46×10^{-3}	0.6880×10^{-5}	0.7607×10^{-8}	-3.593×10^{-12}	°C	0–1500°C

The units on the heat capacity values computed by using this equation are kJ/gmol °C.

   a. Write a VBA function that receives a temperature and four coefficients and then returns the computed heat capacity at the temperature.
   b. Use your macro to find the heat capacity of water vapor at 300°, 500°, and 750°C.
   c. Does the heat capacity of steam change significantly with temperature?

_____

[1] *From Elementary Principles of Chemical Processes,* 3rd ed., Felder, R. M. and R. W. Rousseau, New York: Wiley, 2000.

Note: Because of the very small values of coefficients c and $d,$ you should use double-precision variables in this problem.

14.5 Resolving Forces into Components

Write two functions, **xComp** and **yComp**, which will receive a force and an angle in degrees from horizontal and return the horizontal or vertical component, respectively. Test your functions with the following data:

A force $F = 1000$ N, acting at 30° (0.5235 rad) from the horizontal, resolves into the horizontal component 866 N, and the vertical component 500 N:

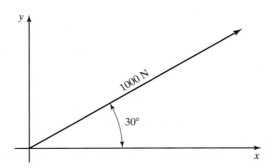

Use the functions you have written to find the horizontal and vertical components of the following forces:

a.

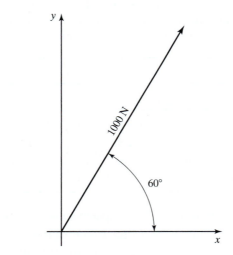

b.

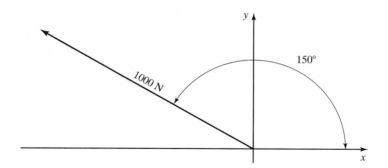

c.

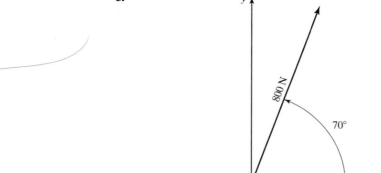

d.

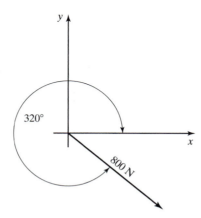

14.6 Calculating Magnitude and Angle of a Resultant Force

Write two functions, **resForce** and **resAngle**; each is to receive horizontal and vertical force components. One returns the magnitude, the other the angle, of the resultant force (in degrees from horizontal), respectively. Test your functions on this example: A horizontal component, 866 N, and a vertical component, 500 N, produce as a resultant force 1000 N acting at 30° from horizontal.

Solve for the resultant force and angle for the following:

  a. $f_H = 500$ N, $f_V = 500$ N
  b. $f_H = 200$ N, $f_V = 600$ N
  c. $f_H = -200$ N, $f_V = 300$ N
  d. $f_H = 750$ N, $f_V = -150$ N

Combine the following horizontal and vertical components and then solve for the magnitude and angle of the resultant force:

$$f_{H1} = 200 \text{ N}, f_{H2} = -120 \text{ N}, f_{H3} = 700 \text{ N}$$
$$f_{V1} = 600 \text{ N}, f_{V2} = 400 \text{ N}, f_{V3} = -200 \text{ N}$$

15

Numerical Differentiation Using Excel

Objectives

After reading this chapter, you will know
- How to calculate derivative values from a data set by using finite differences

- How to use Excel's Data Analysis package to filter noisy data
- How to calculate derivative values from a data set by using curve fitting

15.1 INTRODUCTION

One of the basic data manipulations is *differentiation*—it can also be one of the more problematic if there is noise in the data. This chapter presents two methods for numerical differentiation: using finite differences on the data values themselves and using *curve fitting* to obtain a best-fit equation through the data. It is then possible to take the derivative of the equation. Curve fitting is one way to handle differentiation of noisy data; another is *filtering*. Data-filtering methods are also presented in this chapter.

15.2 FINITE DIFFERENCES

Finite differences are simply algebraic approximations of derivatives and are used to calculate approximate derivative values from data sets. These may be used directly or as a first step in integrating a differential equation. However, some care should be taken when calculating derivatives from data, because noise in the data can have a significant impact on the derivative values, even to the extent of making them meaningless.

15.2.1 Derivatives as Slopes

In Figure 15.1, a tangent line has been drawn to the curve at $x = 0.8$:

Figure 15.1

Evaluating the slope at $x = 0.8$.

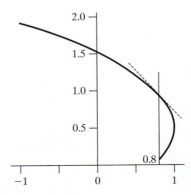

The slope of the tangent line is the value of the *derivative* of the function (a parabola) evaluated at $x = 0.8$. When you have a continuous curve, the tangent line is fairly easy to draw, and the slope can be calculated.

Because the plotted function is a parabola, described by the equation

$$y = 0.5 + \sqrt{(1-x)}, \tag{15.1}$$

it is possible to take the derivative of the function,

$$\frac{dy}{dx} = -\frac{1}{2}(1-x)^{-1/2}, \tag{15.2}$$

and evaluate the derivative at $x = 0.8$:

$$\left.\frac{dy}{dx}\right|_{x=0.8} = -\frac{1}{2}(1-0.8)^{-1/2} = -1.118. \tag{15.3}$$

But when the function is represented by a series of data points (Figure 15.2) instead of a continuous curve, how do you calculate the slope at $x = 0.8$?

Figure 15.2

How do you obtain the slope at $x = 0.8$ with discrete data?

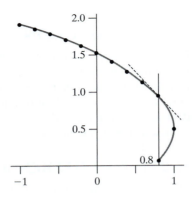

The usual method for estimating the slope at $x = 0.8$ is to use the x and y values at adjacent points and calculate the slope, using

$$\text{slope} \approx \frac{\Delta y}{\Delta x}. \tag{15.4}$$

The "approximately equal to" symbol is a reminder that the deltas in the equation make this a *finite-difference* approximation of the slope and will give the true value of the slope only in the limit as Δx goes to zero. There are several options for using adjacent points to estimate the slope at $x = 0.8$.

- You could use the points at $x = 0.6$ and $x = 0.8$. This is called a *backward difference* (technically, a "backward finite difference approximation of a first derivative"; see Figure 15.3).

Figure 15.3
Estimating slope with a backward finite difference.

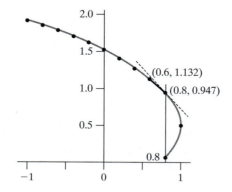

$$\text{slope}_B \approx \frac{\Delta y}{\Delta x} = \frac{0.947 - 1.132}{0.8 - 0.6} = -0.925. \qquad (15.5)$$

- You could use the points at $x = 0.8$ and $x = 1.0$ to estimate the slope at $x = 0.8$. This is called a *forward-difference* approximation (Figure 15.4).

Figure 15.4
Estimating slope with a forward finite difference.

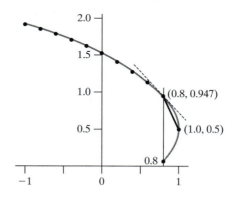

$$\text{slope}_F \approx \frac{\Delta y}{\Delta x} = \frac{0.5 - 0.947}{1.0 - 0.8} = -2.235. \qquad (15.6)$$

- You could use the points around $x = 0.8$: $x = 0.6$ and $x = 1.0$. This is called a *central difference* approximation (Figure 15.5).

Figure 15.5

Estimating slope with a central finite difference.

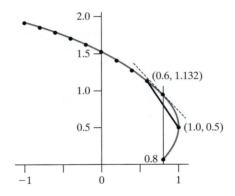

$$\text{slope}_C \approx \frac{\Delta y}{\Delta x} = \frac{0.5 - 1.132}{1.0 - 0.6} = -1.58. \tag{15.7}$$

Which is the correct way to calculate the slope at $x = 0.8$? Well, all three finite differences are approximations, so none is actually "correct." But all three methods can be and are used. Central difference approximations are the most commonly used.

In this case, none of the methods gives a particularly good approximation, because the points are widely spaced and the curvature (slope) changes dramatically between the points. These finite-difference approximations will give better results if the points are closer together (and approximate the true slope as Δx goes to zero).

15.2.2 Caution on Using Finite Differences with Noisy Data

Finite-difference calculations can give very poor results with *noisy data*. For example, if a thermocouple is left in a pot of boiling water, you would expect the temperature to remain constant over time, as shown in Figure 15.6. If the thermocouple measures precisely, the temperature vs. time plot would be a horizontal line, and the slope calculated using any finite-difference approximation would be zero:

Figure 15.6

Ideal temperature vs. time graph for a constant temperature source.

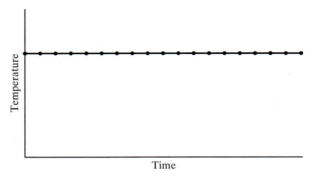

But if the thermocouple is returning a noisy signal because of corrosion, or loose connections, or because the thermocouple is bouncing in and out of the water, the data might look more like the points plotted in Figure 15.7.

Figure 15.7
Constant temperature
source, but noisy data.

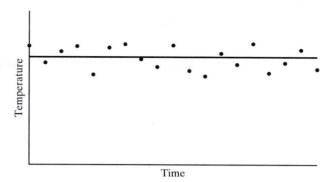

If you try to use finite-difference approximations on adjacent points to estimate the derivative with data like these, you will get all sorts of values, but none is likely to be correct. If you have noisy data, your options are as follows:

- Try to clean up the data as they are being taken (fix the equipment or filter the signal electronically)
- Try to clean up the data after they have been taken (numerical filtering of the data)
- Fit a curve to the data and then differentiate the equation of the curve

Some of these options will be discussed here.

15.2.3 Common Finite-Difference Forms

We've already seen three equations for common finite-difference forms for first derivatives. Now let's write them in standard mathematical form. If you want to evaluate the derivative at point i, the adjacent points can be identified as points $i-1$ (to the left) and $i+1$ (to the right), as illustrated in Figure 15.8.

Figure 15.8
Nomenclature using in
defining finite differences.

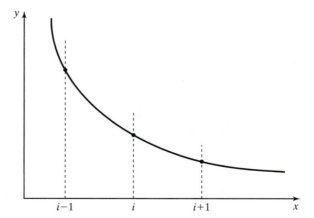

The finite-difference equations can then be written in terms of these generic points and applied wherever needed. The most commonly used finite-difference approximations for first derivatives are listed in Table 15.1.

Table 15.1 First-derivative finite-difference approximations

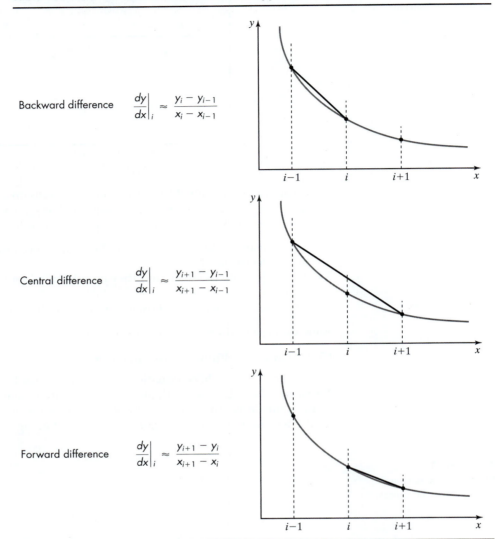

Backward difference $\dfrac{dy}{dx}\bigg|_i \approx \dfrac{y_i - y_{i-1}}{x_i - x_{i-1}}$

Central difference $\dfrac{dy}{dx}\bigg|_i \approx \dfrac{y_{i+1} - y_{i-1}}{x_{i+1} - x_{i-1}}$

Forward difference $\dfrac{dy}{dx}\bigg|_i \approx \dfrac{y_{i+1} - y_i}{x_{i+1} - x_i}$

Finite-difference approximations involving more points are sometimes used to obtain better approximations of derivatives. These multipoint approximations are called *higher order approximations*. A few higher order approximations for first derivatives are listed in Table 15.2.

Table 15.2 Higher order finite-difference approximations for first derivatives

Asymmetric	$\dfrac{dy}{dx}\bigg|_i \approx \dfrac{-y_{i-2} + 4y_{i-1} - 3y_i}{2\Delta x}$
Asymmetric	$\dfrac{dy}{dx}\bigg|_i \approx \dfrac{-3y_i + 4y_{i+1} - y_{i+2}}{2\Delta x}$
Central	$\dfrac{dy}{dx}\bigg|_i \approx \dfrac{-y_{i-2} + 8y_{i-1} - 8y_{i+1} + y_{i+2}}{12\,\Delta x}$

Second-Derivative Finite-Difference Approximations

The commonly used *finite-difference approximation for a second derivative* is the central difference approximation:

$$\left.\frac{d^2y}{dx^2}\right|_i \approx \frac{y_{i+1} - 2y_i + y_{i+1}}{(\Delta x)^2}. \tag{15.8}$$

This equation is derived by assuming uniform spacing between x values (Δx = constant) and so is written with $(\Delta x)^2$ in the denominator. The parentheses in the denominator are a reminder that it is Δx that is squared, not the difference of squared x values.

This equation can be derived fairly easily, and the process will be helpful if other finite-difference approximations are required (e.g., higher order derivatives or nonuniformly spaced x values).

We begin the derivation by adding two temporary points, call them A and B, halfway between the existing points, as shown in Figure 15.9.

Figure 15.9

Adding two intermediate points: A and B.

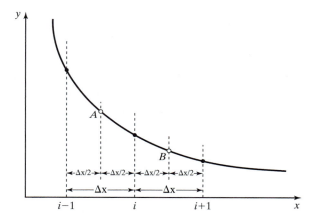

A central difference first-derivative approximation at point A uses points at i and $i-1$ (Figure 15.10):

$$\left.\frac{dy}{dx}\right|_A \approx \frac{y_i - y_{i-1}}{2\frac{\Delta x}{2}}. \tag{15.9}$$

Figure 15.10

Central difference approximation of first derivative at A.

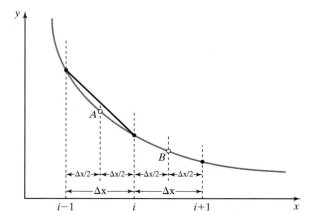

Similarly, a central difference first-derivative approximation at point B uses points at $i+1$ and i (Figure 15.11):

$$\frac{dy}{dx}\bigg|_B \approx \frac{y_{i+1} - y_i}{2\frac{\Delta x}{2}}. \tag{15.10}$$

Figure 15.11
Central difference approximation of first derivative at B.

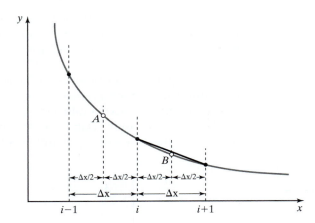

But a second derivative is the derivative of a first derivative, so a second-derivative central finite-difference approximation can be created from the first-derivative equations:

$$\frac{d^2y}{dx^2}\bigg|_i = \frac{d}{dx}\frac{dy}{dx}\bigg|_i \approx \frac{\frac{dy}{dx}\big|_B - \frac{dy}{dx}\big|_A}{2\frac{\Delta x}{2}} = \frac{\frac{y_{i+1} - y_i}{2\frac{\Delta x}{2}} - \frac{y_i - y_{i-1}}{2\frac{\Delta x}{2}}}{2\frac{\Delta x}{2}} = \frac{y_{i+1} - 2y_i + y_{i-1}}{(\Delta x)^2}. \tag{15.11}$$

Central difference approximations for second derivatives are the most commonly used, but forward and backward approximations are also available. These are listed in Table 15.3.

Table 15.3 Second-derivative finite-difference approximations

Backward difference	$\dfrac{d^2y}{dx^2}\bigg|_i \approx \dfrac{y_i - 2y_{i-1} + y_{i-2}}{(\Delta x)^2}$
Central difference	$\dfrac{d^2y}{dx^2}\bigg|_i \approx \dfrac{y_{i+1} - 2y_i + y_{i-1}}{(\Delta x)^2}$
Forward difference	$\dfrac{d^2y}{dx^2}\bigg|_i \approx \dfrac{y_{i+2} - 2y_{i+1} + y_i}{(\Delta x)^2}$

Higher order approximations involving additional points are available for second derivatives also. A few of these are listed in Table 15.4.

Many other finite-difference approximations of various orders of derivatives are available. Only a few of the more common forms have been shown here.

15.2.4 Accuracy and Error Management

Finite differences are approximations of derivatives, so some error associated with the use of these equations is to be expected. The finite differences turn into derivatives in

Table 15.4 Higher order finite-difference approximations for second derivatives

Asymmetric	$\dfrac{d^2y}{dx^2}\bigg|_i \approx \dfrac{-y_{i-3} + 4y_{i-2} - 5y_{i-1} + 2y_i}{(\Delta x)^2}$
Asymmetric	$\dfrac{d^2y}{dx^2}\bigg|_i \approx \dfrac{2y_i - 5y_{i+1} + 4y_{i+2} - y_{i+3}}{(\Delta x)^2}$
Central	$\dfrac{d^2y}{dx^2}\bigg|_i \approx \dfrac{-y_{i-2} + 16y_{i-1} - 30y_i + 16y_{i+1} - y_{i+2}}{12(\Delta x)^2}$

the limit as Δx goes to zero. So the first rule of error management is to *keep Δx small.* If you are taking the data that will have to be differentiated, take lots of data points.

Central difference approximations are more accurate than forward and backward differences, so rule two is to *use central difference approximations for general-purpose calculations.* There are some situations for which forward or backward approximations are preferable. For example, if point i is at a right-side boundary, then point $i + 1$ doesn't exist. Also, backward differences are often used for wave-propagation problems, to keep from performing calculations using information from regions the wave has not yet reached (assuming the wave is moving in the $+x$ direction).

Finally, *you need clean data* if you are going to use finite-difference approximations to calculate derivatives. Filtering noisy data is an option, but filtering can change the calculated derivative values. (Technically, it *does* change the calculated derivative values, but a little bit of filtering is possible in many cases. Heavy-handed filtering will always change the calculated derivative values.)

15.3 FILTERING DATA

Excel actually provides a very simple filter called a *moving average,* available as a trendline and as part of the Data Analysis package (Analysis ToolPak). *Exponential smoothing* is also available through the Data Analysis package.

15.3.1 Moving Average

Because the moving average is available as a trendline on a graph, you can observe the effect of the moving average on your data, but you can't use the calculated point values, because they are not recorded in the worksheet. Fortunately, Excel's Data Analysis package also provides a moving-average filter, and it does output the results to the worksheet.

Excel's moving-average trendline displays the averaged value at the x value at the end of the N-point region, where N is the number of values used in the moving average. Because of this, there is an N-point gap at the left side of the graph when a moving average is used, and the trendline appears to be N points to the right of the data points. If N is small, this offset might not be visible.

To show how a moving average can be used to clean up a noisy data set and what happens when N is large, consider a cosine curve ($0 \leq x \leq 2\pi$) with some random noise added, shown in Figure 15.12.

First, a moving-average trendline is added to the graph by right-clicking on any data point and selecting Add Trendline from the pop-up menu. On the Add Trendline dialog box (Figure 15.13), Moving Average is selected, and the period (the number of averaged values) is set to $N = 5$, a reasonable value for a moving-average filter.

Figure 15.12
Cosine data, with added
noise.

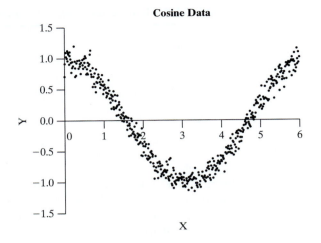

Figure 15.13
The Format Trendline
dialog.

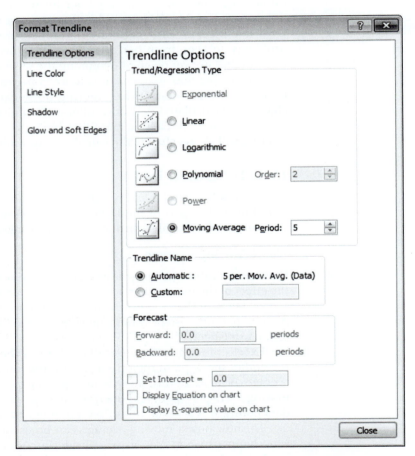

In Figure 15.14, the original data values are presented in gray so that the moving-average results can be more easily observed.

Now, the number of values used in the moving average is changed to a very large value, $N = 60$, by right-clicking on the trendline and selecting **Format Trendline** from the pop-up menu.

Figure 15.14
Cosine data in gray,
moving-average trendline
in black.

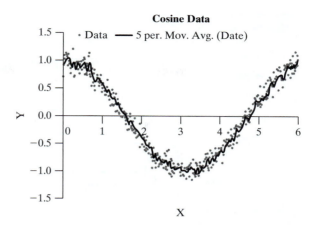

Figure 15.15
The moving-average
trendline with 60 points
averaged.

The trendline is moving to the right of the data (Figure 15.15). If you used the moving average to calculate derivatives, you would find that the derivative would not be zero at $x = \pi$ (as expected with cosine data), but closer to $x = 3.5$. This is an artifact of the moving-average filter; it becomes most apparent when a lot of points are averaged.

Moving-Average Filtering with the Data Analysis Package

Excel's Data Analysis Package also provides a moving-average filter, but you may need to activate the Data Analysis Package before it can be used.

Activating the Data Analysis Package By default, the Data Analysis Package is installed, but is not an activated part in Excel. To see if the package has already been activated, look in the Ribbon. If the **Data Analysis** button does not appear in the **Data** tab's Analysis group, it has not been activated.

To activate the Data Analysis Package follow these steps:

1. Open the Excel options dialog and select the **Add-Ins** panel (shown in Figure 15.16). The method for accessing the Excel Options dialog varies depending on the version of Excel that you are using:
    *   Excel 2010: **File tab/Options**
    *   Excel 2007: **Office/Excel Options**
    *   Excel 2003: **File/Options**

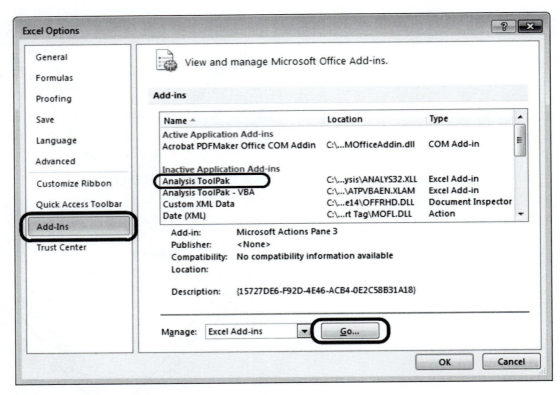

Figure 15.16
Excel Options dialog, **Add-Ins** panel.

2. Select the **Analysis Toolpak** and click the **Go...** button. This opens the Add-Ins dialog shown in Figure 15.17.

Figure 15.17
The Add-Ins dialog.

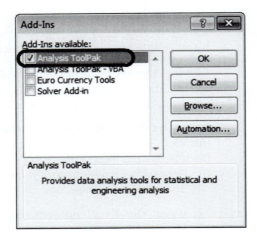

3. Check the box next to the **Analysis ToolPak** and click the **OK** button.

Once the Analysis ToolPak has been activated, a **Data Analysis** button will be displayed on the Ribbon, in the **Data** tab's Analysis group, as shown in Figure 15.18.

Figure 15.18
The Ribbon's Data tab.

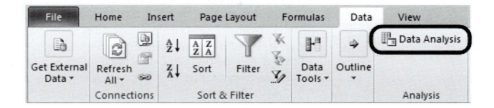

[Excel 2003: TOOLS/ADD-INS …, and activate the Analysis option.]

Click the **Data Analysis** button to open the Data Analysis dialog, shown in Figure 15.19. The Data Analysis dialog provides access to a lot of analysis tools, including a moving-average filter. Select **Moving Average** from the list of analysis options.

Figure 15.19
Data Analysis dialog.

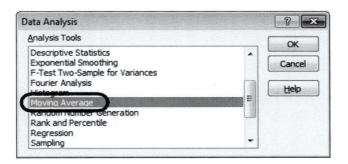

Selecting **Moving Average** on the Data Analysis dialog and clicking **OK** opens the Moving Average dialog shown in Figure 15.20.

Figure 15.20
Moving Average dialog.

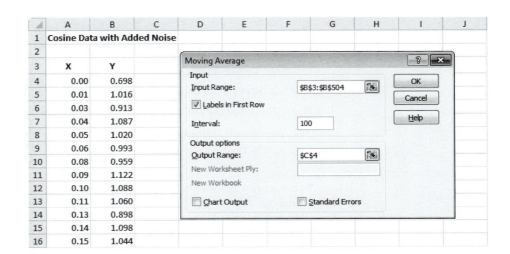

The Moving Average dialog asks for the **Input Range** (the cell range containing the values to be averaged) with (optional) a label at the top of the column and averaging **Interval** (the number of values to include in the average.) In Figure 15.20, the Y values (noisy cosine values) in column B are to be averaged, with a very large interval, $N = 100$.

When you use the Data Analysis package moving-average filter, the filtered Y values are stored in the worksheet. This means you can edit the worksheet to move the filtered Y values by $N/2$ cells to realign the original and filtered data. The result has been graphed in Figure 15.21.

Figure 15.21

Moving-average filtering, realigned with data.

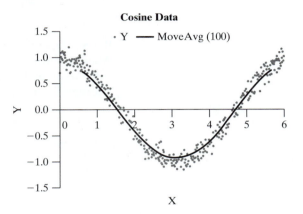

Notice that the filtered curve has been realigned with the original data, but the minimum value is not $Y = -1$. Heavy filtering flattens curves.

The moving-average filter weights each value in the average equally. An alternative is to weight nearby points more heavily than more distant points. One implementation of this approach is *exponential smoothing*, which is also available as part of the Data Analysis package.

15.3.2 Exponential Smoothing

Excel's version of the exponential smoothing equation is

$$y_{F_i} = (1-\beta)y_i + \beta y_{F_{i-1}}, \tag{15.12}$$

where

- y is an original data value,
- y_F is a filtered data value,
- β is the *damping factor* ($0 < \beta < 1$; small β means little smoothing)

The exponential-smoothing equation is often written in terms of a *smoothing constant*, α, where

$$\alpha = 1-\beta. \tag{15.13}$$

Exponential smoothing is also available using Ribbon options **Data/Analysis/ Data Analysis** [Excel 2003: Tools/Data Analysis] and then choosing **Exponential Smoothing** from the Data Analysis list of options.

The Exponential Smoothing dialog (Figure 15.22) asks for the **Input Range** (the column of values to be smoothed; B3:B632 has been used here), the desired **Damping factor** ($\beta = 0.75$ is used here, fairly heavy filtering), and the **Output Range** (where the filtered values should be saved). In this example, cell C4 is

Figure 15.22
Exponential Smoothing dialog.

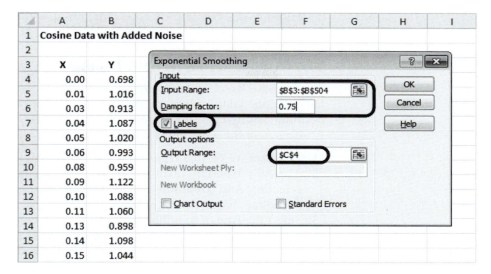

specified as the top-most cell in the output range. The result of applying heavy exponential smoothing to the cosine data is shown in Figure 15.23.

Figure 15.23
Exponential smoothing ($\beta = 0.75$) of the noisy cosine data.

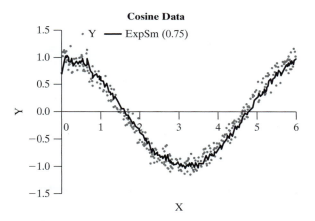

Notice that the filtered data curve, even with $\beta = 0.75$, is not particularly smooth. If we increase β to try to smooth out the curve, we can get a smooth curve, but it may not represent the original data well. A graph with $\beta = 0.95$ is shown in Figure 15.24.

Figure 15.24
Exponential smoothing ($\beta = 0.95$) of the noisy cosine data.

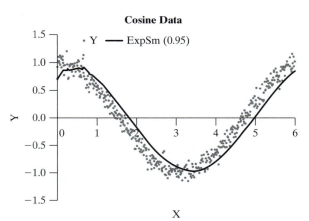

Filtering is an option for noisy data, but it needs to be done carefully. Heavy-handed filtering will skew the data.

15.4 CURVE FITTING AND DIFFERENTIATION

You can have Excel perform a *linear regression* on a data set either by adding a *trendline* to graphed data or by using the Data Analysis package (Regression option). The trendline approach is simpler and will be used here.

To obtain the equation of a regression line via a trendline:

1. The data are plotted (the cosine data will be used again).
2. The Add Trendline dialog is called up by right-clicking on a data point and selecting **Add Trendline...** from the pop-up menu. This opens the Format Trendline dialog, shown in Figure 15.25.
3. Select the type of trendline desired.
   - The **Logarithmic, Power,** and **Exponential** fits, are not available for the cosine data, because they require operations that are invalid for these data.
   - The **Moving Average** trendline is available, but it does not produce an equation.
   - The **Linear** fit will put a straight line through the data set, so that's out.
   - That leaves **Polynomial** as the only viable option for this data set.
4. Indicate that Excel should show the equation and R^2 value on the graph.

Figure 15.25
Format Trendline dialog.

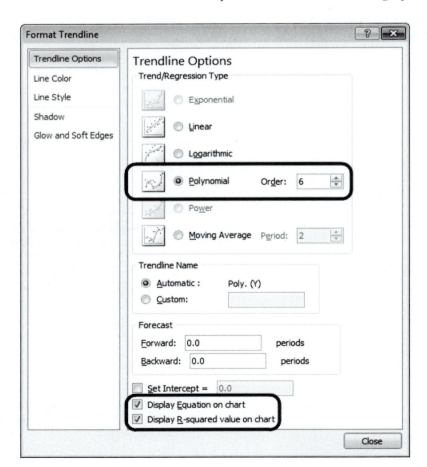

With the polynomial fit, you need to try various orders of polynomial to see which fits the data best. Excel's polynomial trendline allows orders between 2 and 6. Order 6 will be used here, but the fit is only minimally better than a fourth-order fit.

Figure 15.26

Cosine data with polynomial regression line.

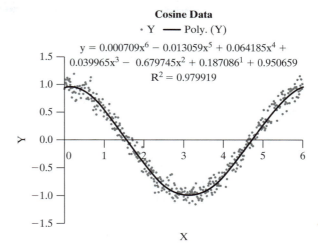

Cosine Data

· Y —— Poly. (Y)

$$y = 0.000709x^6 - 0.013059x^5 + 0.064185x^4 + 0.039965x^3 - 0.679745x^2 + 0.187086^1 + 0.950659$$

$$R^2 = 0.979919$$

The regression equation shown in Figure 15.26 was modified to increase the number of displayed decimal places. To increase the number of displayed digits in the regression equation:

1. Right-click on the displayed equation.
2. Select **Format Trendline Label** from the pop-up menu. This opens the Format Trendline Label dialog shown in Figure 15.27.
3. Select the **Number** panel.
4. Select the **Number** category.
5. Set the Decimal places to a larger value; **6** was used in Figure 15.27.

Figure 15.27

The Format Trendline Label dialog.

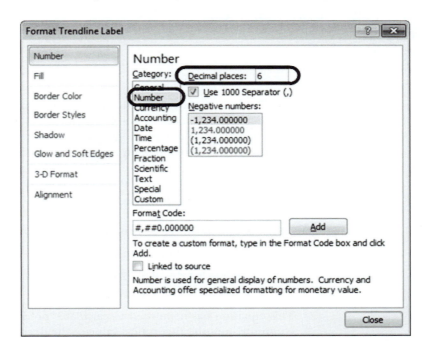

The fit in Figure 15.26 doesn't look too bad. A residual plot can be used as a check on the fit. To calculate residuals, we first have to calculate predicted y values at each x using the regression equation.

$$y_p = 0.000709x^6 - 0.013059x^5 + 0.064185x^4 + 0.039965x^3$$
$$- 0.679745x^2 + 0.187186x + 0.950659$$

Then, we can calculate residuals as $y - y_p$ for every x. The results of these calculations are shown in columns C and D of Figure 15.28.

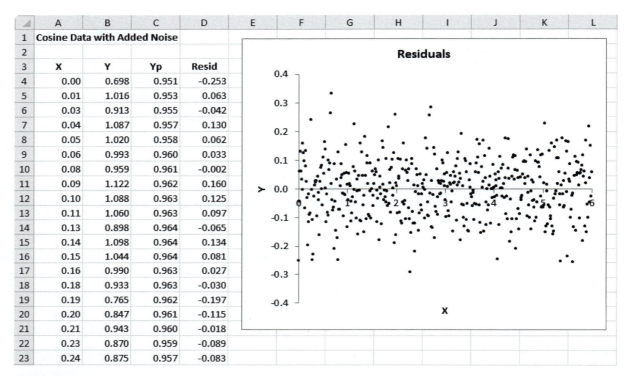

	A	B	C	D
1	Cosine Data with Added Noise			
2				
3	X	Y	Yp	Resid
4	0.00	0.698	0.951	-0.253
5	0.01	1.016	0.953	0.063
6	0.03	0.913	0.955	-0.042
7	0.04	1.087	0.957	0.130
8	0.05	1.020	0.958	0.062
9	0.06	0.993	0.960	0.033
10	0.08	0.959	0.961	-0.002
11	0.09	1.122	0.962	0.160
12	0.10	1.088	0.963	0.125
13	0.11	1.060	0.963	0.097
14	0.13	0.898	0.964	-0.065
15	0.14	1.098	0.964	0.134
16	0.15	1.044	0.964	0.081
17	0.16	0.990	0.963	0.027
18	0.18	0.933	0.963	-0.030
19	0.19	0.765	0.962	-0.197
20	0.20	0.847	0.961	-0.115
21	0.21	0.943	0.960	-0.018
22	0.23	0.870	0.959	-0.089
23	0.24	0.875	0.957	-0.083

Figure 15.28
A residual plot shows no apparent patterns.

The lack of apparent patterns in the residual plot suggests that this polynomial is fitting the data about as well as possible.

The polynomial regression equation (shown on the graph in Figure 15.26) can be differentiated to obtain an equation for calculating derivative values at any x:

$$y = 0.000709x^6 - 0.013059x^5 + 0.064185x^4 + 0.039965x^3$$
$$- 0.679745x^2 + 0.187186x + 0.950659$$

$$\frac{dy}{dx} = 6(0.000709)x^5 - 5(0.013059)x^4 + 4(0.064185)x^3 + 3(0.039965)x^2$$
$$- 2(0.679745)x + 0.187186$$

$$= 0.004254x^5 - 0.065295x^4 + 0.256740x^3 + 0.119895x^2$$
$$- 1.359490x + 0.187186. \tag{15.14}$$

This equation has been used to calculate "predicted" derivatives for each x value in column E, and the theoretical value of the derivative $[-\sin(x)]$ has been calculated in column F. In Figure 15.29, the actual and predicted derivative values are compared.

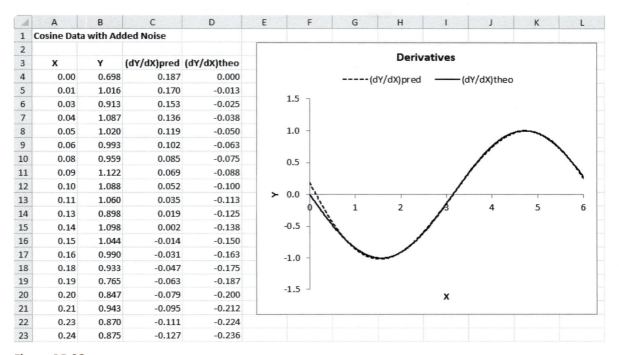

	A	B	C	D	E	F	G	H	I	J	K	L
1	Cosine Data with Added Noise											
2												
3	X	Y	(dY/dX)pred	(dY/dX)theo								
4	0.00	0.698	0.187	0.000								
5	0.01	1.016	0.170	-0.013								
6	0.03	0.913	0.153	-0.025								
7	0.04	1.087	0.136	-0.038								
8	0.05	1.020	0.119	-0.050								
9	0.06	0.993	0.102	-0.063								
10	0.08	0.959	0.085	-0.075								
11	0.09	1.122	0.069	-0.088								
12	0.10	1.088	0.052	-0.100								
13	0.11	1.060	0.035	-0.113								
14	0.13	0.898	0.019	-0.125								
15	0.14	1.098	0.002	-0.138								
16	0.15	1.044	-0.014	-0.150								
17	0.16	0.990	-0.031	-0.163								
18	0.18	0.933	-0.047	-0.175								
19	0.19	0.765	-0.063	-0.187								
20	0.20	0.847	-0.079	-0.200								
21	0.21	0.943	-0.095	-0.212								
22	0.23	0.870	-0.111	-0.224								
23	0.24	0.875	-0.127	-0.236								

Figure 15.29
Comparing the predicted and theoretical derivative values.

APPLICATION

TRANSPORT PHENOMENA

Conduction Heat Transfer

Figure 15.30 depicts a device that could be used to measure the thermal conductivity of a material. A rod of the material to be tested is placed between a resistance heater (on the right) and a block containing a cooling coil (on the left). Five thermocouples are inserted into the rod at evenly spaced intervals. The entire apparatus is placed under a bell jar, and the space around the rod is evacuated to reduce energy losses.

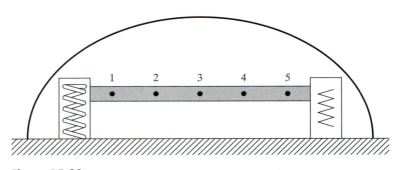

Figure 15.30
Experimental apparatus for determining thermal conductivity.

To run the experiment, a known amount of power is sent to the heater, and the system is allowed to reach steady state. Once the temperatures are steady, the power level and the temperatures are recorded. A data sheet from an experiment with the device is reproduced here:

Thermal Conductivity Experiment

Rod Diameter:	2 cm
Thermocouple Spacing:	5 cm
Power:	100 Watts

Thermocouple	Temperature (K)
1	348
2	387
3	425
4	464
5	503

The thermal conductivity can be determined via Fourier's law,

$$\frac{q}{A} = -k\frac{dT}{dx},$$ (15.15)

where

q is the power applied to the heater and

A is the cross-sectional area of the rod.

The q/A is the *energy flux* and is a vector quantity, having both magnitude and direction. With the energy source on the right, the energy moves to the left (in the negative x direction), so the flux in this problem is negative.

Calculate the Cross-Sectional Area of the Rod and the Energy Flux

	A	B	C	D	E	F	G
1	Conduction Heat Transfer						
2							
3		Rod Diameter:	2	cm			
4		Thermocouple Spacing:	5	cm			
5		Power:	100	watts			
6		Cross-Section:	3.142	cm^2			
7		Energy Flux:	31.8	W/cm^2			
8							

Figure 15.31
Calculating the energy flux down the rod.

C6: =PI()/4*C3^2

Estimate dT/dx

Central difference approximations have been used on the interior points (thermocouple locations) in Figure 15.32.

E11	▼	f_x	=(D12-D10)/(C12-C10)				
	A	B	C	D	E	F	G
1	Conduction Heat Transfer						
2							
3		Rod Diameter:	2	cm			
4	Thermocouple Spacing:		5	cm			
5		Power:	100	watts			
6		Cross-Section:	3.142	cm^2			
7		Energy Flux:	31.8	W/cm^2			
8							
9		Thermocouple	Position (cm)	Temp. (K)	dT/dx (K/cm)		
10		1	0	348			
11		2	5	387	7.7		
12		3	10	425	7.7		
13		4	15	464	7.8		
14		5	20	503			
15							

Figure 15.32
Calculating temperature derivatives.

Calculate the Thermal Conductivity of the Material

F11	▼	f_x	=-$C$7/E11*100				
	A	B	C	D	E	F	G
1	Conduction Heat Transfer						
2							
3		Rod Diameter:	2	cm			
4	Thermocouple Spacing:		5	cm			
5		Power:	100	watts			
6		Cross-Section:	3.142	cm^2			
7		Energy Flux:	31.8	W/cm^2			
8							
9		Thermocouple	Position (cm)	Temp. (K)	dT/dx (K/cm)	k (W/m K)	
10		1	0	348			
11		2	5	387	7.7	-413.4	
12		3	10	425	7.7	-413.4	
13		4	15	464	7.8	-408.1	
14		5	20	503			
15							

Figure 15.33
Calculating the thermal conductivity of the rod material.

The "100" in the formula in cell F11 is used to convert centimeters to meters to allow the thermal conductivity to be reported in watts/m K (Figure 15.33).

$$F11: = (\$C\$7/E11)*100$$

KEY TERMS

Backward difference
Central difference
Curve fitting
Damping factor
Derivative
Differentiation

Exponential smoothing
(filter)
Filter
Finite difference
Forward difference
Linear regression

Moving average (filter)
Noisy data
Slope
Smoothing constant
Trendline

SUMMARY

Common Finite-Difference Approximations

First-Derivative Finite-Difference Approximations

Backward difference $\quad \dfrac{dy}{dx}\bigg|_i \approx \dfrac{y_i - y_{i-1}}{x_i - x_{i-1}}$

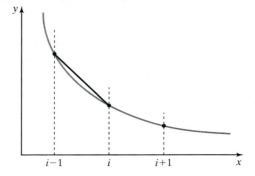

Central difference $\quad \dfrac{dy}{dx}\bigg|_i \approx \dfrac{y_{i+1} - y_{i-1}}{x_{i+1} - x_{i-1}}$

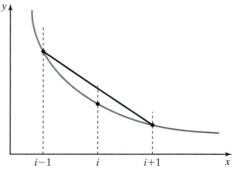

Forward difference $\quad \dfrac{dy}{dx}\bigg|_i \approx \dfrac{y_{i+1} - y_i}{x_{i+1} - x_i}$

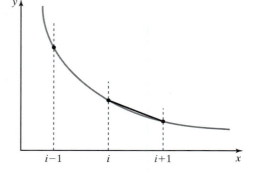

Second-Derivative Finite-Difference Approximations		
Backward difference	$\left.\dfrac{d^2y}{dx^2}\right|_i \approx$	$\dfrac{y_i - 2y_{i-1} + y_{i-2}}{(\Delta x)^2}$
Central difference	$\left.\dfrac{d^2y}{dx^2}\right|_i \approx$	$\dfrac{y_{i+1} - 2y_i + y_{i-1}}{(\Delta x)^2}$
Forward difference	$\left.\dfrac{d^2y}{dx^2}\right|_i \approx$	$\dfrac{y_{i+2} - 2y_{i+1} + y_i}{(\Delta x)^2}$

Filtering Data

- Excel provides **moving-average filters** as trendlines and as part of the Data Analysis package.
- **Exponential smoothing** is available in the Data Analysis package.

Curve Fitting and Differentiation

- Use a trendline or regression analysis to fit a curve to the data, then differentiate the regression equation.
- Particularly useful (necessary) for noisy data.

PROBLEMS

15.1 Finite Differences or Regression and Differentiation

Given the following data sets, which are the candidates for finite-difference calculations and which should be fit with a regression function to obtain derivative values?

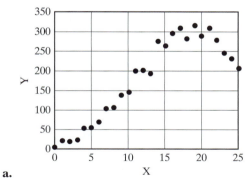

a.

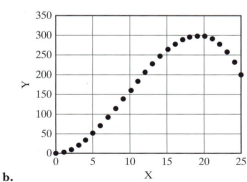

b.

c.

15.2 Conduction Heat Transfer II

The application example in this chapter described a device that could be used to measure the thermal conductivity of a material:

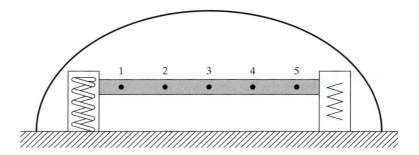

A data sheet from an experiment with the device is reproduced here:

Thermal Conductivity Experiment	
Rod Diameter:	2 cm
Thermocouple Spacing:	5 cm
Power:	100 Watts

Thermocouple	Temperature (K)
1	287
2	334
3	380
4	427
5	474

a. Calculate the cross-sectional area of the rod and the energy flux.
b. Use central finite-difference approximations to estimate the dT/dx at points 2 through 4.
c. Use Fourier's law and the results of part (b) to calculate the thermal conductivity of the material at points 2 through 4.

15.3 Calculating Heat Capacity I

The heat capacity at constant pressure is defined as

$$C_P = \left(\frac{\partial \hat{H}}{\partial T}\right)_P,$$

(15.16)

where

$\hat{H}$ is specific enthalpy.
T is absolute temperature.

Data for carbon dioxide at a pressure of 1 atmosphere are tabulated as follows:[1]

Temperature (°C)	$\hat{H}$ (KJ/MOL)
100	2.90
200	7.08
300	11.58
400	16.35
500	21.34
600	26.53
700	31.88
800	37.36
900	42.94
1000	48.60
1100	54.33
1200	60.14
1300	65.98
1400	71.89
1500	77.84

a. Use finite-difference approximations to calculate the molar heat capacity of CO_2 at each temperature.
b. Plot the specific enthalpy on the y axis against absolute temperature on the x axis and add a linear trendline with the data. Compare the slope of the trendline with the heat capacity values you calculated in part (a). Are they equal? Should they be equal?
c. Does it appear that the heat capacity of CO_2 is constant over this temperature range? Why, or why not?

15.4 Calculating Heat Capacity II

The heat capacity at constant pressure is defined as

$$C_p = \left(\frac{\partial \hat{H}}{\partial T}\right)_p,$$

(15.17)

where

$\hat{H}$ is specific enthalpy and
T is absolute temperature.

---

[1]This data is from a steam table in *Elementary Principles of Chemical Engineering*, 3rd ed., Felder, R. M. and R. W. Rousseau, New York: Wiley, 2000.

If enthalpy data are available as a function of temperature at constant pressure, the heat capacity can be computed. For common combustion gases, the data are available. Enthalpy data for carbon dioxide at a pressure of 1 atmosphere are tabulated in the previous problem.

a. Use finite-difference approximations to calculate the molar heat capacity of CO_2 at each temperature.

b. Plot the specific enthalpy on the y axis against absolute temperature on the x axis and add a polynomial trendline to the data. Ask Excel to display the equation of the line and the R^2 value.

c. Record the equations for second- and third-order polynomial fits to the data.

Note: You might need to have Excel display a lot of decimal places to see anything other than zero for the higher order coefficients:

1. Right-click the regression equation.

2. Select **Format Trendline Label...** from the pop-up menu.

3. Use the **Number** panel on the dialog.

4. Choose the **Number** format.

5. Enter the number of desired decimal places.

d. Differentiate the equations from part (c) to obtain two equations for the heat capacity as a function of temperature.

e. Calculate heat capacity values with each equation from part (d) at each temperature.

f. The heat capacity of carbon dioxide at temperatures between 0 and 1500°C can be calculated from the equation[2]

$$C_p = 36.11 + 0.04233\ T - 2.887 \times 10^{-5}\ T^2 + 7.464\ \times 10^{-9}\ T^3.$$
$$+ 7.464\ \times 10^{-9}\ T^3. (15.18)$$

Calculate heat capacity values at each temperature, using this equation. Be aware that the units on the heat capacity from this equation are J/kg K rather than kJ/kg K.

g. Do your calculated heat capacity values from parts (a), (e), and (f) agree? If not, which of the numerical methods gives the best agreement with the result from part (f)?

---

[2]*Equation from Basic Principles and Calculations in Chemical Engineering,* 6th ed., Himmelblau D. M., Upper Saddle River, NJ: Prentice Hall PTR, 1996.

16

Numerical Integration Using Excel

Objectives

After reading this chapter, you will know

- How integrating a function is equivalent to calculating the area between a plot of the function and the x axis

- How to use geometric regions to approximate the area between a curve and the x axis
- How to fit an equation to a curve via regression and then integrate that equation to determine the area

16.1 INTRODUCTION

There are two types of *integration* that engineers routinely perform: integration for the *area under a curve* and integration of a *differential equation*. The two are related mathematically, but procedurally they are handled quite differently. This chapter presents two *numerical methods* to find the area beneath a curve:

- Using small regions (e.g., trapezoids) to compute the total area beneath the curve
- Using regression to fit the curve. The regression equation can then be integrated to determine the area beneath the curve.

Integrating differential equations is the subject of the next chapter.

16.2 INTEGRATING FOR AREA UNDER A CURVE

The function $y = 3 + 1.5x - 0.25x^2$ is plotted as the heavy curve in Figure 16.1.

Figure 16.1

Integrating for area beneath a curve.

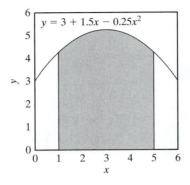

The integral

$$\int_1^5 (3 + 1.5x - 0.25x^2)\, dx \qquad (16.1)$$

is physically represented by the area between the x axis and the curve representing the function—the shaded area in Figure 16.1. Any method we can find that calculates that area can be used to integrate the function. The geometric integration methods we will use in this chapter are all methods for estimating the area under the function.

16.3 INTEGRATING FOR AREA BETWEEN TWO CURVES

There are times when the area you need to calculate does not go all the way to the x axis. In the example shown in Figure 16.2, the desired area is the area between two functions: $y = 3 + 1.5x - 0.25x^2$ and $y = 0 + 1.5x - 0.25x^2$. (The zero is obviously not needed in the last equation; it is there as a reminder that the two functions have the same form, just a different offset from the x axis.)

Figure 16.2

Finding the area between two curves.

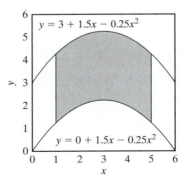

We can calculate the hatched area in this latter plot by first calculating the area between $y = 3 + 1.5x - 0.25x^2$ and the x axis, then subtracting out the area between $y = 0 + 1.5x - 0.25x^2$ and the x axis, as illustrated in Figure 16.3.

Mathematically, this can be written as

$$\text{Area} = \int_1^5 (3 + 1.5x - 0.25x^2)\, dx - \int_1^5 (0 + 1.5x - 0.25x^2)\, dx. \qquad (16.2)$$

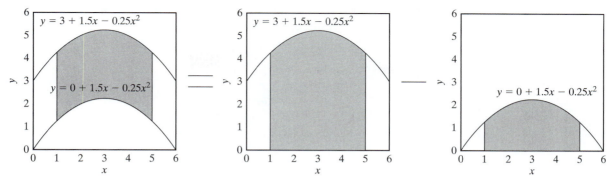

Figure 16.3
The area between two curves can be calculated in two steps.

In this case, most of the terms cancel out, leaving a very simple result:

$$\text{Area} = \int_1^5 3\, dx$$
$$= 12.$$ (16.3)

In general, however, the two functions would be integrated separately, then subtracted to find the area between the curves.

16.4 NUMERICAL INTEGRATION METHODS

Implementation of the following *numerical integration techniques,* or "rules," will be discussed here:

- Approximating areas by using rectangles (*rectangular integration method*)
- Approximating areas by using trapezoids (*trapezoidal integration method*)
- *Simpson's rule*

Each of these methods will be illustrated on the set of cosine data listed in Table 16.1.

Table 16.1 Cosine data set

X	Y = cos(X)
0.0000	1.0000
0.1571	0.9877
0.3142	0.9511
0.4712	0.8910
0.6283	0.8090
0.7854	0.7071
0.9425	0.5878
1.0996	0.4540
1.2566	0.3090
1.4137	0.1564
1.5708	0.0000

Using cosine data will allow comparison of the numerical integration results with the analytical result of integrating

$$y = \cos(x) \tag{16.4}$$

between 0 and $\pi/2$:

$$\int_0^{\pi/2} y\,dx = \int_0^{\pi/2} \cos(x)\,dx$$

$$= \sin(\pi/2) - \sin(0)$$

$$= 1. \tag{16.5}$$

Approximating the area under a curve with a series of rectangular or trapezoidal shapes defined by a series of data pairs is very straightforward in a worksheet. The process requires the following steps:

1. Enter or import the data.
2. Enter the formula for the area of one rectangle or trapezoid.
3. Copy the formula over all *intervals* (not all data points).
4. Sum the areas.

Implementation of Simpson's rule is slightly more involved because two data intervals are used for each integration step; thus, a distinction must be made between an *interval* (between data points) and an *integration step* (two intervals). Simpson's rule is presented in detail in Section 16.4.3.

16.4.1 Integration by Using Rectangles

Step 1. Enter or Import the Raw Data. The eleven data pairs of cosine data have been entered in the worksheet shown in Figure 16.4.

Figure 16.4
The data set for integration.

	A	B	C	D	E	F
1	Integration Using Rectangles					
2	Height set by using Y value as left side of rectangle.					
3						
4		X	Y			
5		0.0000	1.0000			
6		0.1571	0.9877			
7		0.3142	0.9511			
8		0.4712	0.8910			
9		0.6283	0.8090			
10		0.7854	0.7071			
11		0.9425	0.5878			
12		1.0996	0.4540			
13		1.2566	0.3090			
14		1.4137	0.1564			
15		1.5708	0.0000			
16						

Step 2. Enter the Formula for One Rectangle. The area below the cosine data can be approximated with rectangles as shown in Figure 16.5.

Figure 16.5

Approximating the area below the curve with rectangles.

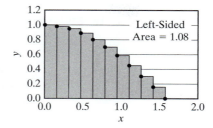

The height of the first rectangle is 1.0 (using the y value on the left side of the first interval, which is in cell C5). The integration step is the width of the rectangle $x_2 - x_1$ (cell B6 minus cell B5). The formula for the area of the first rectangle is =C5*(B6-B5). This formula is entered into cell D5, as shown in Figure 16.6.

Figure 16.6

Entering the formula for the first rectangle.

D5	▾	f_x	=C5*(B6-B5)		

◢	A	B	C	D	E	F
1	**Integration Using Rectangles**					
2	Height set by using Y value as left side of rectangle.					
3						
4		**X**	**Y**	**Area**		
5		0.0000	1.0000	0.1571		
6		0.1571	0.9877			
7		0.3142	0.9511			
8		0.4712	0.8910			
9		0.6283	0.8090			
10		0.7854	0.7071			
11		0.9425	0.5878			
12		1.0996	0.4540			
13		1.2566	0.3090			
14		1.4137	0.1564			
15		1.5708	0.0000			
16						

Step 3. Copy the Area Formula to All Intervals. *Remember that the number of intervals is one less than the number of data pairs.*

There are 11 data pairs, so there are 10 intervals. Copy the formula in cell D5 over the range D5:D14. This causes the areas of all 10 rectangles to be computed, as shown in Figure 16.7.

Step 4. Sum the Individual Areas. Summing the rectangle areas in cells D5 through D14 yields the estimated area under the entire curve. For this data set, using rectangles with the left side aligned to the data points overestimates the true area (obtained from analytical integration) by 7.65%, giving the area 1.0765 instead of 1, as shown in Figure 16.8.

Figure 16.7
Determining the area of each of the rectangles.

D14			f_x	=C14*(B15-B14)		
	A	B	C	D	E	F
1	**Integration Using Rectangles**					
2	Height set by using Y value as left side of rectangle.					
3						
4		**X**	**Y**	**Area**		
5		0.0000	1.0000	0.1571		
6		0.1571	0.9877	0.1551		
7		0.3142	0.9511	0.1494		
8		0.4712	0.8910	0.1400		
9		0.6283	0.8090	0.1271		
10		0.7854	0.7071	0.1111		
11		0.9425	0.5878	0.0923		
12		1.0996	0.4540	0.0713		
13		1.2566	0.3090	0.0485		
14		1.4137	0.1564	0.0246		
15		1.5708	0.0000			
16						

Figure 16.8
Summing the areas of the rectangles to approximate the area beneath the cosine curve.

D17			f_x	=SUM(D5:D14)		
	A	B	C	D	E	F
1	**Integration Using Rectangles**					
2	Height set by using Y value as left side of rectangle.					
3						
4		**X**	**Y**	**Area**		
5		0.0000	1.0000	0.1571		
6		0.1571	0.9877	0.1551		
7		0.3142	0.9511	0.1494		
8		0.4712	0.8910	0.1400		
9		0.6283	0.8090	0.1271		
10		0.7854	0.7071	0.1111		
11		0.9425	0.5878	0.0923		
12		1.0996	0.4540	0.0713		
13		1.2566	0.3090	0.0485		
14		1.4137	0.1564	0.0246		
15		1.5708	0.0000			
16						
17			**Total Area:**	1.0765		
18						

Aligning the Right Side of the Rectangles with the Data

Using the y values on the left side of the rectangles to determine the rectangle area was an arbitrary choice. Next, we'll see how the result changes if we use the y value on the right side of the interval to determine the height of the rectangle. As evident in the graph in Figure 16.9, for this cosine data, using y values on the right should cause the rectangles to underestimate the area under the cosine curve.

Figure 16.9

Approximating the area with rectangles, using Y on the right side to determine the height.

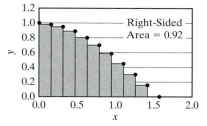

The y value on the right side of the first rectangle is in cell C6. The width of the rectangle is still B6 − B5. The formula for the area of the first rectangle is then =C6*(B6-B5), which is placed in cell D6, as shown in Figure 16.10.

Figure 16.10

Calculating the area of the first rectangle.

D6			f_x	=C6*(B6-B5)		
	A	B	C	D	E	F
1	**Integration Using Rectangles**					
2	Height set by using Y value as right side of rectangle.					
3						
4		**X**	**Y**	**Area**		
5		0.0000	1.0000			
6		0.1571	0.9877	0.1551		
7		0.3142	0.9511			
8		0.4712	0.8910			
9		0.6283	0.8090			
10		0.7854	0.7071			
11		0.9425	0.5878			
12		1.0996	0.4540			
13		1.2566	0.3090			
14		1.4137	0.1564			
15		1.5708	0.0000			
16						

The equation in cell D6 should be copied over the range D6:D15 to find the area of each of the 10 rectangles. The estimate of the total area under the curve is obtained by summing the areas for each interval. In this example, the formula =SUM(D6:D15) was placed in cell D17 to perform this task. The result, Area = 0.9194, is lower than the true value by about 8%, as shown in Figure 16.11.

Figure 16.11
The result of right-sided rectangular integration.

D17		f_x =SUM(D6:D15)				
	A	**B**	**C**	**D**	**E**	**F**
1	**Integration Using Rectangles**					
2	Height set by using Y value as right side of rectangle.					
3						
4		**X**	**Y**	**Area**		
5		0.0000	1.0000			
6		0.1571	0.9877	0.1551		
7		0.3142	0.9511	0.1494		
8		0.4712	0.8910	0.1400		
9		0.6283	0.8090	0.1271		
10		0.7854	0.7071	0.1111		
11		0.9425	0.5878	0.0923		
12		1.0996	0.4540	0.0713		
13		1.2566	0.3090	0.0485		
14		1.4137	0.1564	0.0246		
15		1.5708	0.0000	0.0000		
16						
17			**Total Area:**	0.9194		
18						

At this point, many people want to try to improve the results by averaging the left-aligned and right-aligned results. It is certainly possible to do that—it is also equivalent to using trapezoids to approximate the area below the curve.

16.4.2 Integration by Using Trapezoids

Step 1. Enter or Import the Data. Same as with rectangular integration (Figure 16.4).

Step 2. Enter the Formula for the Area of One Trapezoid. The area of a trapezoid is governed by the data pairs on the left $(x, y)_L$ and on the right $(x, y)_R$. The equation for the area of a trapezoid is

$$A_{\text{trap}} = \frac{y_L + y_R}{2}(x_R - x_L). \tag{16.6}$$

The formula entered into cell D5 is =0.5*(C4+C5)*(B5−B4). This is shown in Figure 16.12.

Step 3. Copy the Formula Over All the Intervals. The formula in cell C5 is copied over all 10 intervals in the range D4:D13, as shown in Figure 16.13.

Figure 16.12
Determining the area
of the first trapezoid.

D4	▾	f_x	=0.5*(C4+C5)*(B5-B4)	

◢	A	B	C	D	E
1	Integration Using Trapezoids				
2					
3		X	Y	Area	
4		0.0000	1.0000	0.1561	
5		0.1571	0.9877		
6		0.3142	0.9511		
7		0.4712	0.8910		
8		0.6283	0.8090		
9		0.7854	0.7071		
10		0.9425	0.5878		
11		1.0996	0.4540		
12		1.2566	0.3090		
13		1.4137	0.1564		
14		1.5708	0.0000		
15					
16			Total Area:		
17					

Figure 16.13
Calculating the area
of each trapezoid.

D13	▾	f_x	=0.5*(C13+C14)*(B14-B13)	

◢	A	B	C	D	E
1	Integration Using Trapezoids				
2					
3		X	Y	Area	
4		0.0000	1.0000	0.1561	
5		0.1571	0.9877	0.1523	
6		0.3142	0.9511	0.1447	
7		0.4712	0.8910	0.1335	
8		0.6283	0.8090	0.1191	
9		0.7854	0.7071	0.1017	
10		0.9425	0.5878	0.0818	
11		1.0996	0.4540	0.0599	
12		1.2566	0.3090	0.0366	
13		1.4137	0.1564	0.0123	
14		1.5708	0.0000		
15					

Step 4. Sum the Areas of All Trapezoids.   The formula =SUM(D4:D13) is entered
into cell D16 to compute the estimate of the area under the cosine curve
as shown in Figure 16.14.

The result, Area = 0.9979, is lower than the true value by 0.2%, which
is a considerable improvement over either of the results using rectangles.

Figure 16.14
Summing the trapezoid
areas to estimate the area
below the curve.

D16 ▾		f_x	=SUM(D4:D13)	
	A	**B**	**C**	**D**	**E**
1	**Integration Using Trapezoids**				
2					
3		**X**	**Y**	**Area**	
4		0.0000	1.0000	0.1561	
5		0.1571	0.9877	0.1523	
6		0.3142	0.9511	0.1447	
7		0.4712	0.8910	0.1335	
8		0.6283	0.8090	0.1191	
9		0.7854	0.7071	0.1017	
10		0.9425	0.5878	0.0818	
11		1.0996	0.4540	0.0599	
12		1.2566	0.3090	0.0366	
13		1.4137	0.1564	0.0123	
14		1.5708	0.0000		
15					
16			**Total Area:**	0.9979	
17					

CALCULATING REQUIRED VOLUME

Concrete Retaining Wall

Forms for a 9-inch-thick retaining wall with a complex shape (illustrated in Figure 16.15) have been constructed, and the contractor is ready to order the concrete. Given the dimensions shown on the following drawing (units are in feet), how much concrete should be ordered?

The solution to this problem will involve a few steps:

1. Enter the dimension data into a worksheet.
2. Calculate the surface area of the upper portion of the form.
3. Calculate the surface area of the lower portion of the form.
4. Calculate the total surface area of the form.
5. Multiply by the depth (9 inches) to find the required volume.

Step 1. Enter the Dimension Data into a Worksheet. The dimensions are entered into the worksheet shown in Figure 16.16 as x values, y_U-(upper) values, and y_L-(lower) values.

Step 2. Calculate the Surface Area of the Upper Portion of the Form. Using trapezoids, the area of the upper portion of the form is calculated as shown in Figure 16.17.

The total area of the upper portion of the form is determined by summing the trapezoid areas. In Figure 16.17, this was done in cell E11 with the formula =SUM(E4:E8).

Step 3. Calculate the Surface Area of the Lower Portion of the Form. Using trapezoids, the area of the lower portion of the form is calculated as shown in Figure 16.18.

APPLICATION

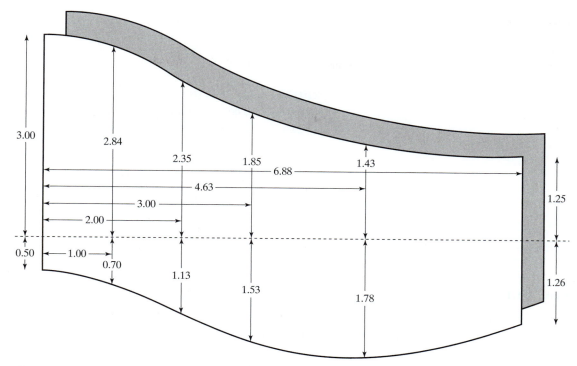

Figure 16.15
Schematic of a complex retaining wall.

▲	A	B	C	D	E	F	G	H
1	**Concrete Retaining Wall**							
2								
3		X (ft)	Y$_U$ (ft)	Y$_L$ (ft)				
4		0.00	3.00	0.50				
5		1.00	2.84	0.70				
6		2.00	2.35	1.13				
7		3.00	1.85	1.53				
8		4.63	1.40	1.78				
9		6.88	1.25	1.26				
10								

Figure 16.16
The dimension data, entered into an Excel worksheet.

Step 4. Calculate the Total Surface Area of the Form.   The total surface area of the form is simply the sum of the upper and lower portions. This is illustrated in Figure 16.19.

Step 5. Multiply by the Depth to Find the Required Volume of Concrete.   The distance between the front and back of the form (the form depth) is 9 inches, or 9/12 feet. Multiplying the area of the form by the form depth gives the required volume of concrete: 16.7 ft^3, as shown in Figure 16.20.

Figure 16.17
Using the trapezoidal method to find the area of the upper portion.

	E4	▼	●	f_x	=0.5*(C5+C4)*(B5-B4)		

	A	B	C	D	E	F	G	H
1	Concrete Retaining Wall							
2								
3		X (ft)	Y_U (ft)	Y_L (ft)	A_U (ft^2)			
4		0.00	3.00	0.50	2.92			
5		1.00	2.84	0.70	2.60			
6		2.00	2.35	1.13	2.10			
7		3.00	1.85	1.53	2.65			
8		4.63	1.40	1.78	2.98			
9		6.88	1.25	1.26				
10					A_U (ft^2)			
11				Totals:	13.25			
12								

Figure 16.18
Calculating the area of the lower portion of the wall.

	F4	▼	●	f_x	=0.5*(D5+D4)*(B5-B4)		

	A	B	C	D	E	F	G	H
1	Concrete Retaining Wall							
2								
3		X (ft)	Y_U (ft)	Y_L (ft)	A_U (ft^2)	A_L (ft^2)		
4		0.00	3.00	0.50	2.92	0.60		
5		1.00	2.84	0.70	2.60	0.92		
6		2.00	2.35	1.13	2.10	1.33		
7		3.00	1.85	1.53	2.65	2.70		
8		4.63	1.40	1.78	2.98	3.42		
9		6.88	1.25	1.26				
10					A_U (ft^2)	A_L (ft^2)		
11				Totals:	13.25	8.96		
12								

Figure 16.19
Calculating the total surface area of the form.

	G11	▼	●	f_x	=E11+F11		

	A	B	C	D	E	F	G	H
1	Concrete Retaining Wall							
2								
3		X (ft)	Y_U (ft)	Y_L (ft)	A_U (ft^2)	A_L (ft^2)		
4		0.00	3.00	0.50	2.92	0.60		
5		1.00	2.84	0.70	2.60	0.92		
6		2.00	2.35	1.13	2.10	1.33		
7		3.00	1.85	1.53	2.65	2.70		
8		4.63	1.40	1.78	2.98	3.42		
9		6.88	1.25	1.26				
10					A_U (ft^2)	A_L (ft^2)	A_{FORM} (ft^2)	
11				Totals:	13.25	8.96	22.21	
12								

G13	▾	f_x	=G11*(9/12)					
	A	B	C	D	E	F	G	H
1	Concrete Retaining Wall							
2								
3		X (ft)	Y_U (ft)	Y_L (ft)	A_U (ft^2)	A_L (ft^2)		
4		0.00	3.00	0.50	2.92	0.60		
5		1.00	2.84	0.70	2.60	0.92		
6		2.00	2.35	1.13	2.10	1.33		
7		3.00	1.85	1.53	2.65	2.70		
8		4.63	1.40	1.78	2.98	3.42		
9		6.88	1.25	1.26				
10					A_U (ft^2)	A_L (ft^2)	A_{FORM} (ft^2)	
11				Totals:	13.25	8.96	22.21	
12								
13						Volume (ft^3):	16.7	
14								

Figure 16.20
Determining the required volume of concrete.

16.4.3 Integration by Using Smooth Curves (Simpson's Rule)

Simpson's rule requires

1. An odd number of data points
2. Evenly spaced data (i.e., uniform Δx)

The procedure for integration using Simpson's rule is quite similar to that used in the preceding cases, but you must be careful to distinguish between the *data interval* between adjacent x values and the *integration step* which, for Simpson's rule, incorporates two data intervals. It is because of this that Simpson's rule is restricted to an even number of intervals (an odd number of data points).

Procedure for Simpson's Rule

1. Enter or import the data.
2. Compute h, the distance between any two adjacent x values (because the x values are uniformly spaced).
3. Enter the area formula for one integration step (not data interval).
4. Copy the formula over all integration regions.
5. Sum the areas of all integration regions.

Step 1. Enter or Import the Data.   No change from previous cases.

Step 2. Compute h, the Distance between Adjacent x Values.   The value of h can be computed by using any two adjacent x values, because Simpson's rule requires uniformly spaced x values. The first two x values in cells A5 and A6 have been used in Figure 16.21, and h is computed with the formula =B7−B6 and stored in cell C3.

Figure 16.21

Calculating the *h* value for Simpson's Rule integration.

	C3	▼	f_x =B7-B6		
	A	B	C	D	E	F

	A	B	C	D	E	F
1	**Simpson's Rule Integration**					
2						
3			h:	0.1571		
4						
5		**X**	**Y**	**Int. Step**		
6		0.0000	1.0000	1		
7		0.1571	0.9877	1		
8		0.3142	0.9511	2		
9		0.4712	0.8910	2		
10		0.6283	0.8090	3		
11		0.7854	0.7071	3		
12		0.9425	0.5878	4		
13		1.0996	0.4540	4		
14		1.2566	0.3090	5		
15		1.4137	0.1564	5		
16		1.5708	0.0000			
17						

Note: *h* is the width of the interval between any two data points, not the entire integration step.

$$h = \Delta x. \qquad (16.7)$$

Step 3. Enter the Formula for the Area of One Integration Step. The formula for Simpson's rule is

$$A_s = \frac{h}{3}(y_L + 4y_M + y_R), \qquad (16.8)$$

where A_s is the area of an integration step for Simpson's rule, which covers two data intervals. The subscripts *L, M,* and *R* stand for left, middle, and right sides of the integration step, respectively. (The integration steps are numbered on the worksheet in Figure 16.21 as a reminder.)

The first area formula is entered into cell E6 as = ($C$3/3)*(C6+4*C7+ C8), as shown in Figure 16.22.

Note: The dollar signs on the reference to cell $C$3 indicate that the reference to *h* should not be changed when the formula in cell E6 is copied.

Step 4. Copy the Area Formula to Each Integration Step. The formula in cell E6 must be copied to each integration step (i.e., every other row). A quick way to do this is to:

1. Copy the formula to the Windows clipboard [Ctrl-c].
2. Select the destination cells by first clicking on cell E8 and holding down the [Ctrl] key while clicking on cells E10, E12, and E14. (Excel uses the [Ctrl] key to select multiple, noncontiguous regions.)
3. Paste the formula into the cells from the clipboard [Ctrl-v].

The result is shown in Figure 16.23.

Figure 16.22
Calculating the area of the first integration region.

	fx =($C$3/3)*(C6+4*C7+C8)					
E6						

	A	B	C	D	E	F
1	Simpson's Rule Integration					
2						
3		h:	0.1571			
4						
5		X	Y	Int. Step	Area	
6		0.0000	1.0000	1	0.3090	
7		0.1571	0.9877	1		
8		0.3142	0.9511	2		
9		0.4712	0.8910	2		
10		0.6283	0.8090	3		
11		0.7854	0.7071	3		
12		0.9425	0.5878	4		
13		1.0996	0.4540	4		
14		1.2566	0.3090	5		
15		1.4137	0.1564	5		
16		1.5708	0.0000			
17						

Figure 16.23
Calculating the area of each integration region.

	fx =($C$3/3)*(C14+4*C15+C16)					
E14						

	A	B	C	D	E	F
1	Simpson's Rule Integration					
2						
3		h:	0.1571			
4						
5		X	Y	Int. Step	Area	
6		0.0000	1.0000	1	0.3090	
7		0.1571	0.9877	1		
8		0.3142	0.9511	2	0.2788	
9		0.4712	0.8910	2		
10		0.6283	0.8090	3	0.2212	
11		0.7854	0.7071	3		
12		0.9425	0.5878	4	0.1420	
13		1.0996	0.4540	4		
14		1.2566	0.3090	5	0.0489	
15		1.4137	0.1564	5		
16		1.5708	0.0000			
17						

Step 5. Sum the Areas for Each Region. You can use =SUM(E6:E14) to compute this sum because empty cells are ignored during the summing process. This is illustrated in Figure 16.24.

The resulting estimate for the area under the curve is 1.0 (actually 1.000003), which is virtually equivalent to the true value, 1:

Figure 16.24
The area under the cosine curve, determined with Simpson's Rule integration.

	fx =SUM(E6:E14)					
E18	▼					
	A	B	C	D	E	F
1	Simpson's Rule Integration					
2						
3		h:	0.1571			
4						
5		X	Y	Int. Step	Area	
6		0.0000	1.0000	1	0.3090	
7		0.1571	0.9877	1		
8		0.3142	0.9511	2	0.2788	
9		0.4712	0.8910	2		
10		0.6283	0.8090	3	0.2212	
11		0.7854	0.7071	3		
12		0.9425	0.5878	4	0.1420	
13		1.0996	0.4540	4		
14		1.2566	0.3090	5	0.0489	
15		1.4137	0.1564	5		
16		1.5708	0.0000			
17						
18				Total Area:	1.0000	
19						

Simpson's Rule: Summary
Simpson's rule is pretty straightforward to apply and does a good job of fitting most functions, but there are two important limitations on Simpson's rule. You must have the following:

1. An odd number of data points (even number of intervals)
2. Uniformly spaced *x* values

The trapezoidal rule is slightly less accurate, but it is very simple to use and does not have the restrictions just listed. Sometimes Simpson's method is used with an even number of data points (with uniformly spaced *x* values) by using the Simpson's rule formula for all intervals except the last one. A trapezoid is used to compute the area of the final interval (half-Simpson integration step).

16.5 USING REGRESSION EQUATIONS FOR INTEGRATION

Another general approach to integrating a data set is to:

1. Use *regression* to fit a curve to the data, then
2. Analytically integrate the *regression equation.*

Excel provides a couple of alternatives for linear regression: *trendlines* with limited regression forms and regression by using the Data Analysis package.

For the cosine data used here ($y = \cos(x)$, $0 \leq x \leq \pi/2$), the obvious fitting function is a cosine. To give the regression analysis something to work with, significant random noise has been added to the data values shown in Figure 16.25.

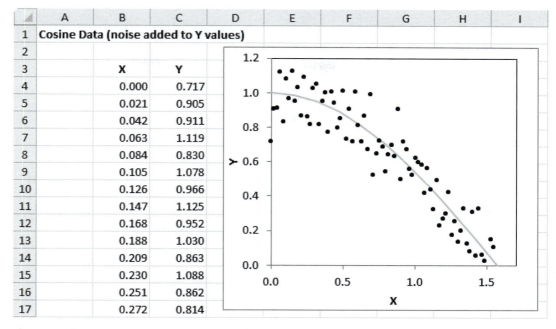

	A	B	C	D	E	F	G	H	I
1	Cosine Data (noise added to Y values)								
2									
3		X	Y						
4		0.000	0.717						
5		0.021	0.905						
6		0.042	0.911						
7		0.063	1.119						
8		0.084	0.830						
9		0.105	1.078						
10		0.126	0.966						
11		0.147	1.125						
12		0.168	0.952						
13		0.188	1.030						
14		0.209	0.863						
15		0.230	1.088						
16		0.251	0.862						
17		0.272	0.814						

Figure 16.25
A new set of cosine data with 76 data points and noise added to the y values.

A cosine fit is not an option with Excel's trendlines, so we will use the regression option in the Data Analysis package.

Activating the Data Analysis Package

By default, the Data Analysis Package is installed, but is not an activated part in Excel. To see if the package has already been activated, look in the Ribbon. If the **Data Analysis** button does not appear in the **Data** tab's Analysis group, it has not been activated.

To activate the Data Analysis Package follow these steps:

1. Open the Excel options dialog as follows:
   - Excel 2010: **File tab/Options**
   - Excel 2007: **Office/Excel Options**
   - Excel 2003: **File/Options**
2. Select the **Add-Ins** panel as illustrated in Figure 16.26.
3. Select **Analysis Toolpak** and click the **Go...** button. This opens the Add-Ins dialog shown in Figure 16.27.
4. Check the box next to the **Analysis ToolPak** and click the **OK** button.

Once the Analysis ToolPak is active, a **Data Analysis** button will be displayed on the Ribbon, in the **Data** tab's Analysis group, as shown in Figure 16.28. [Excel 2003: Tools/Data Analysis.]

Using the Regression Tool

The regression model for our cosine data is

$$y_p = b_0 + b_1 \cos(x).$$ (16.9)

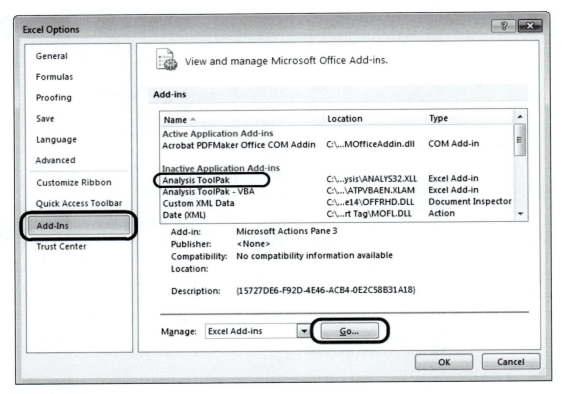

Figure 16.26
Excel Options dialog, Add-Ins panel.

Figure 16.27
The Add-Ins dialog.

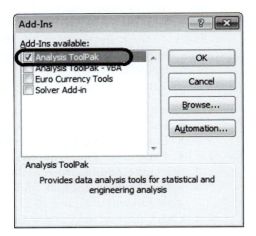

Figure 16.28
The Ribbon's **Data** tab.

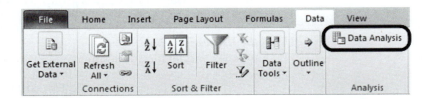

We are fitting cosine data, so the intercept b_0 should be unnecessary and will be set equal to zero during the regression process. To prepare for the regression, we first take the cosine of each x value in the data set, as shown in Figure 16.29.

Figure 16.29
Taking cos(X) in preparation for regression analysis (only 6 of 76 data points shown).

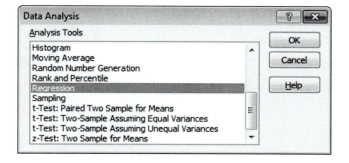

	A	B	C	D	E
1	Cosine Data (noise added to Y values)				
2					
3		X	Y	cos(X)	
4		0.000	0.717	1.000	
5		0.021	0.905	1.000	
6		0.042	0.911	0.999	
7		0.063	1.119	0.998	
8		0.084	0.830	0.996	
9		0.105	1.078	0.994	

D4 — fx =COS(B4)

Use Ribbon options **Data/Analysis/Data Analysis** [Excel 2003: Tools/Data Analysis] to open the Data Analysis dialog (shown in Figure 16.30). Select **Regression** from the list of analysis options.

Figure 16.30
Data Analysis dialog.

This opens the Regression dialog box shown in Figure 16.31. The required input is as follows:

- *Input Y Range*: The cell range containing the y values (C3:C79, including the column heading).
- *Input X Range*: The cell range containing the x values for the regression model. In this case, the regression model uses $\cos(x)$ as the independent variable, so the *X* **Input Range** is D3:D79 (with column heading).
- *Labels*: The **Labels** checkbox is checked because column headings were included with the input ranges.
- *Output Options*: Results will be sent to a **New Worksheet Ply**. (The output can be extensive.)
- *Constant is Zero*: The **Constant is Zero** checkbox is checked to force the intercept (b_0) to zero.
- *Plots*: **Residual** and **Line Fit** plots are requested.

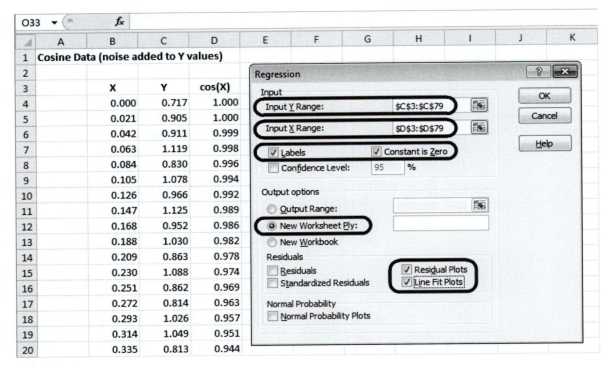

Figure 16.31
The Regression dialog.

When the **OK** button is clicked, the regression is performed, and the output is sent to a new worksheet. A portion of the output is shown in Figure 16.32.

Figure 16.32
Regression results.

	A	B	C
1	SUMMARY OUTPUT		
2			
3	*Regression Statistics*		
4	Multiple R	0.9876	
5	R Square	0.9753	
6	Adjusted R Square	0.9619	
7	Standard Error	0.1112	
8	Observations	76	
9			
10	ANOVA		
11		*df*	*SS*
12	Regression	1	36.5397
13	Residual	75	SSE 0.9270
14	Total	76	SSTo 37.4667
15			
16		*Coefficients*	*Standard Error*
17	Intercept	0	#N/A
18	cos(X)	0.9806	0.0180
19			

The intercept is zero (as requested), and $b_1 = 0.9806$. The R^2 value is reported as 0.9753.

Unfortunately, in Excel 2003, the R^2 value calculated using the Regression tool in the Data Analysis package on the same data was 0.8842. R^2 can be calculated directly using the formula

$$R^2 = 1 - \frac{\text{SSE}}{\text{SSTo}}.$$

The values for SSE (sum of squared error, also known as the residual sum of squares) and SSTo (total sum of squares) are listed in the Summary Output from the regression (see Figure 16.31).

As reported by Excel

SSE = 0.9270
SSTo = 37.4667

Using these values yields the R^2 value reported by Excel, $R^2 = 0.9753$. But SSE and SSTo can also be calculated from the data values:

As calculated from the Data

SSE = 0.9270
SSTo = 8.005

Using these values yields an R^2 value of 0.8842, as reported by Excel 2003.

I'm afraid there is a bug in Excel 2007 and Excel 2010 in the calculation of R^2 in the Data Analysis package. Hopefully this will be remedied soon.

Interestingly, when Excel is asked to calculate the intercept, instead of forcing it to zero, the R^2 value was found to be 0.8842, as shown in Figure 16.33.

Figure 16.33
Regression results when Excel calculated the intercept.

	A	B	C
1	SUMMARY OUTPUT		
2			
3	*Regression Statistics*		
4	Multiple R	0.9403	
5	R Square	0.8842	
6	Adjusted R Square	0.8826	
7	Standard Error	0.1119	
8	Observations	76	
9			
10	ANOVA		
11		*df*	*SS*
12	Regression	1	7.0780
13	Residual	74	SSE 0.9270
14	Total	75	SSTo 8.0050
15			
16		*Coefficients*	*Standard Error*
17	Intercept	0.00069	0.0291
18	cos(X)	0.97971	0.0412
19			

The requested line-fit plot is shown in Figure 16.34. (Excel creates a line graph by default; the chart type was changed to XY scatter and the plot was reformatted for clarity.)

Figure 16.34

Comparing the regression line with the data.

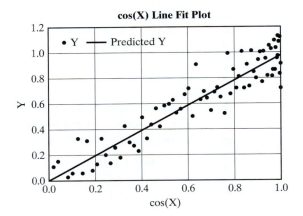

The requested residual plot (Figure 16.35) shows no apparent patterns, suggesting the regression model is adequate.

Figure 16.35

Residual plot of the regression result.

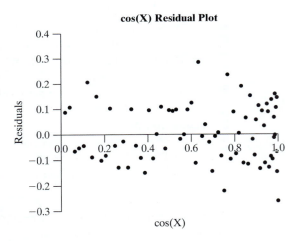

The $\cos(x)$ label on the x axis in Figure 16.34 is as a reminder that the regression analysis used $\cos(x)$ as the independent variables (the x values in the regression analysis). The regressed slope ($b_1 = 0.9806$) can be used to calculate predicted y values with the regression equation

$$y_p = 0 + 0.9806 \cdot \cos(x)$$

The curve of predicted values has been superimposed on the original data in Figure 16.36.

The regression equation used in cell E4 was =0+0.9806*COS(B4). (The zero intercept serves no purpose except to remind us that the intercept was forced to zero as part of the regression analysis.)

Once the regression equation is obtained, the integration for the area under the curve is straightforward:

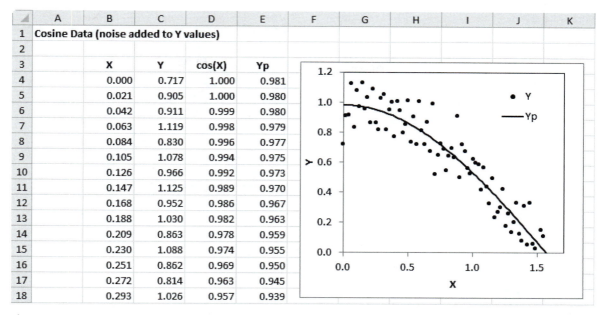

	A	B	C	D	E
1	Cosine Data (noise added to Y values)				
2					
3		X	Y	cos(X)	Yp
4		0.000	0.717	1.000	0.981
5		0.021	0.905	1.000	0.980
6		0.042	0.911	0.999	0.980
7		0.063	1.119	0.998	0.979
8		0.084	0.830	0.996	0.977
9		0.105	1.078	0.994	0.975
10		0.126	0.966	0.992	0.973
11		0.147	1.125	0.989	0.970
12		0.168	0.952	0.986	0.967
13		0.188	1.030	0.982	0.963
14		0.209	0.863	0.978	0.959
15		0.230	1.088	0.974	0.955
16		0.251	0.862	0.969	0.950
17		0.272	0.814	0.963	0.945
18		0.293	1.026	0.957	0.939

Figure 16.36
The predicted values (curve) superimposed on the original data (points).

$$A = \int_0^{\pi/2} [0.9806 \cos(x)] \, dx$$
$$= 0.9806 \int_0^{\pi/2} \cos(x) \, dx$$
$$= 0.9806 \left[\sin\left(\tfrac{\pi}{2}\right) - \sin(0) \right]$$
$$= 0.9806 \, [1 - 0]$$
$$= 0.9806. \tag{16.10}$$

The result is low by about 2%, which is not too bad considering the noise in the data.

16.5.1 Trying an Excel Trendline for the Fitting Equation

Integration for the area under a curve is one situation where having a fitting equation (a.k.a., regression model) with a strong theoretical underpinning is not necessary; we just need an equation of a curve that fits the data well. If one of Excel's trendlines will fit the data, that is a much quicker way to get a regression equation.

In Figure 16.37, a second-order polynomial trendline has been fit to the data. Let's integrate the regression equation and see how the area comes out.

$$A = \int_0^{\pi/2} [0.972 - 0.040x - 0.384x^2] dx$$
$$= 0.972 \int_0^{\pi/2} dx - 0.040 \int_0^{\pi/2} x \, dx - 0.384 \int_0^{\pi/2} x^2 \, dx$$
$$= 0.972 \left(\frac{\pi}{2}\right) - \frac{0.040}{2}\left(\frac{\pi}{2}\right)^2 - \frac{0.384}{3}\left(\frac{\pi}{2}\right)^3$$
$$= 1.527 - 0.0493 - 0.496$$
$$= 0.981.$$

The result is not bad, and it took less work than using the Data Analysis package.

Figure 16.37
Fitting the data with an
Excel polynomial trendline.

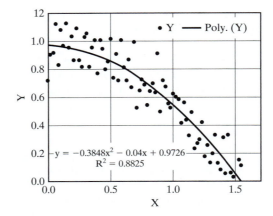

SUMMARY

Numerical Integration by Using Geometric Regions

Rectangles

$$\text{Area} = \sum_{i=1}^{N-1} \left[y_i(x_{i+1} - x_i) \right], \text{ or}$$

$$\text{Area} = \sum_{i=1}^{N-1} \left[y_{i+1}(x_{i+1} - x_i) \right]. \tag{16.11}$$

Trapezoids

$$\text{Area} = \sum_{i=1}^{N-1} \left[\left(\frac{y_i + y_{i+1}}{2} \right)(x_{i+1} - x_i) \right]. \tag{16.12}$$

Simpson's Rule

Requirements

1. An odd number of data points (even number of intervals)
2. Uniformly spaced x values.

Area Equation for One Integration Region (**2 Δx**)

$$A_s = \frac{h}{3}(y_L + 4y_M + y_R), \tag{16.13}$$

where $h = \Delta x$ (must be a constant).

Total Area Equation

$$\text{Area} = \sum_{j=1}^{M} [A_{S_j}], \tag{16.14}$$

where

M is the number of integration regions $M = \frac{N-1}{2}$ and
N is the number of data points.

Integration by Using Regression Equations

1. Perform a regression analysis on the data to fit a curve to the data. Options:
   • Add a trendline to an XY scatter graph of your data and request the equation of the trendline.
   • Use the Regression tool in the Data Analysis package.
2. Integrate the regression equation analytically.

PROBLEMS

16.1 Area Beneath a Curve

Approximate the area under curve 1 in Figure 16.38, using trapezoids:

Figure 16.38
Area beneath curves.

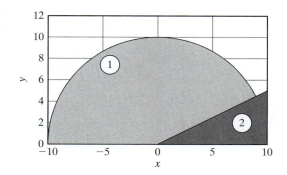

x	y₁	y₂
−10.0	0.00	0.0
−9.8	1.99	0.0
−9.4	3.41	0.0
−9.0	4.36	0.0
−8.0	6.00	0.0
−7.0	7.14	0.0
−6.0	8.00	0.0
−5.0	8.66	0.0
−4.0	9.17	0.0
−3.0	9.54	0.0
−2.0	9.80	0.0
−1.0	9.95	0.0
0.0	10.00	0.0
1.0	9.95	0.5
2.0	9.80	1.0
3.0	9.54	1.5
4.0	9.17	2.0
5.0	8.66	2.5
6.0	8.00	3.0
7.0	7.14	3.5
8.0	6.00	4.0
9.0	4.36	4.5
9.4	3.41	4.7
9.8	1.99	4.9
10.0	0.00	5.0

16.2 Area Between Curves

Using the data in the previous problem, approximate the area between curves 1 and 2, using numerical integration.

16.3 Work Required to Stretch a Spring

The device shown in Figure 16.39 can be used to estimate the work required to extend a spring. This device consists of a spring, a spring balance, and a ruler. Before the stretching of the spring, its length is measured and found to be 1.3 cm. The spring is then stretched 0.4 cm at a time, and the force indicated on the spring balance is recorded. The resulting data set is shown in the following table:

Spring data			
Measurement (cm)	Unextended Length (cm)	Extension (cm)	Force (N)
1.3	1.3	0.0	0.00
1.7	1.3	0.4	0.88
2.1	1.3	0.8	1.76

Measurement (cm)	Unextended Length (cm)	Extension (cm)	Force (N)
2.5	1.3	1.2	2.64
2.9	1.3	1.6	3.52
3.3	1.3	2.0	4.40
3.7	1.3	2.4	5.28
4.1	1.3	2.8	6.16
4.5	1.3	3.2	7.04
4.9	1.3	3.6	7.92

Figure 16.39
Work required to stretch a spring.

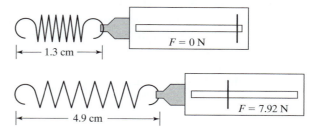

The work required to extend the spring can be computed as

$$W = \int F\, dx, \tag{16.15}$$

where x is the extension (distance stretched) of the spring.
Calculate the work required to extend the spring from 0 to 3.6 cm of extension.

16.4 Enthalpy Required to Warm a Gas

The enthalpy required to warm n moles of a gas from T_1 to T_2 can be calculated from the heat capacity of the gas:

$$\Delta H = n \int_{T_i}^{T_2} C_P(T)\, dT. \tag{16.16}$$

But the heat capacity is a function of temperature. An equation commonly used to describe the change in heat capacity with temperature is a simple third-order polynomial in T:[1]

$$C_p = a + bT + cT^2 + dT^3. \tag{16.17}$$

If the coefficients a through d are known for a particular gas, you can calculate the heat capacity of the gas at any T (within an allowable range). The coefficients for a few common gases are as follows:

Heat capacity coefficients

Gas	a	b	c	d	Units on T	Valid Range
Air	28.94×10^{-3}	0.4147×10^{-5}	0.3191×10^{-8}	-1.965×10^{-12}	°C	0–1500°C
CO_2	36.11×10^{-3}	4.233×10^{-5}	-2.887×10^{-8}	7.464×10^{-12}	°C	0–1500°C
CH_4	34.31×10^{-3}	5.469×10^{-5}	0.3661×10^{-8}	-11.00×10^{-12}	°C	0–1200°C
H_2O	33.46×10^{-3}	0.6880×10^{-5}	0.7604×10^{-8}	-3.593×10^{-12}	°C	0–1500°C

[1] *From Elementary Principles of Chemical Processes*, 3rd ed., Felder, R. M. and R. W. Rousseau, New York: Wiley, 2000.

The units on the heat capacity values computed from Equation (16.17) are kJ/gmol °C.

    a. Calculate the heat capacity of methane (CH_4) between 200 and 800°C, using 20° intervals.

    b. Integrate the heat capacity data, using a numerical integration technique, to find the energy (i.e., enthalpy change) required to warm 100 gmol of methane from 200 to 800°C.

    c. Substitute the polynomial for C_P into the enthalpy integral and integrate analytically, to check your numerical integration result.

16.5 Stress–Strain Curve

Strength testing of materials often involves a tensile test in which a sample of the material is held between two mandrels and increasing force (actually, *stress* = force per unit area) is applied. A stress vs. strain curve for a typical ductile material is shown in Figure 16.40; the sample first stretches reversibly (A to B), then irreversibly (B to D), before it finally breaks (point D):

Figure 16.40

Stress–strain curve.

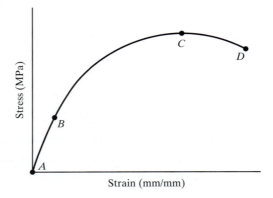

The strain is the amount of elongation of the sample (mm) divided by the original sample length (mm). A little reworking of the data transforms the stress–strain curve to a force–displacement curve, so

$$F = \text{stress} \times A_{\text{cross section}},$$

and

$$D = \text{strain} \times L_{\text{original}}, \tag{16.18}$$

where

F is the force applied to the sample,
$A_{\text{cross section}}$ is the cross-sectional area of the sample (100 mm^2),
D is the displacement, and
L_{original} is the original sample length (40 mm).

For a sample 10 mm by 10 mm by 40 mm, the force and displacement data can be calculated and a force vs. displacement graph can be produced, as illustrated in Figure 16.41.

Figure 16.41

Force–displacement curve.

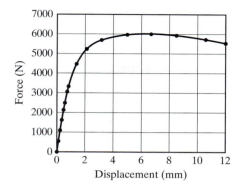

The area under this curve represents the work done on the sample by the testing equipment. Use a numerical integration technique to estimate the work done on the sample. Use the following data:

Strain (mm/mm)	Stress (MPa)	Force (N)	Displacement (mm)
0.0000	0.00	0	0.00
0.0030	5.38	538	0.12
0.0060	10.76	1076	0.24
0.0090	16.14	1614	0.36
0.0120	21.52	2152	0.48
0.0140	25.11	2511	0.56
0.0170	30.49	3049	0.68
0.0200	33.34	3334	0.80
0.0350	44.79	4479	1.40
0.0520	52.29	5229	2.08
0.0790	57.08	5708	3.16
0.1240	59.79	5979	4.96
0.1670	60.10	6010	6.68
0.2120	59.58	5958	8.48
0.2640	57.50	5750	10.56
0.3000	55.42	5542	12.00

These data are available at www.chbe.montana.edu/excel.

16.6 Fluid Velocity from Pitot Tube Data

A pitot tube (see Figure 16.42) is a device that can be used to determine a *local velocity,* that is, a velocity in the immediate vicinity of the measuring device. The theory behind the operation of the pitot tube comes from Bernoulli's equation (without the potential energy terms) and relates the change in kinetic energy to the change in fluid pressure:

$$\frac{p_a}{\rho} + \frac{u_a^2}{2} = \frac{p_b}{\rho} + \frac{u_b^2}{2}.$$

(16.19)

Figure 16.42

Pitot tube.

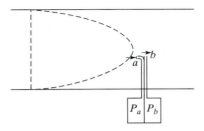

The flow that strikes the tube at point a hits a dead end, and its velocity goes to zero (stagnation). At point b, the flow slides past the pitot tube and does not slow down at all (free-stream velocity). The pressure at point b is the free-stream pressure. At point a, the energy carried by the fluid has to be conserved, so the kinetic energy of the moving fluid is transformed to pressure energy as the velocity goes to zero. The pitot tube measures a higher pressure at point a than at point b, and the pressure difference can be used to calculate the free-stream velocity at point b.

When the velocity at point a has been set to zero and Bernoulli's equation is rearranged to solve for the local velocity at point b, we get

$$u_b = \sqrt{\frac{2}{\rho}(p_a - p_b)}. \tag{16.20}$$

By moving the pitot tube across the diameter of a pipe, we find that a pressure-difference profile can be measured across the pipe. From the pressure differences, local velocities can be calculated. In this way, a pitot tube can be used to measure the velocity profile across a pipe. If the local velocities are integrated, the average velocity of the fluid in the pipe can be approximated:

$$\bar{V} = \frac{\int_{r=0}^{R} \int_{\theta=0}^{2\pi} ur \, dr \, d\theta}{\int_{r=0}^{R} \int_{\theta=0}^{2\pi} r \, dr \, d\theta}$$

$$= \frac{2\pi \int_{r=0}^{R} ur \, dr}{\pi R^2}. \tag{16.21}$$

Given the pitot-tube pressure differences shown in the accompanying table,

   a. calculate and graph the local velocities at each point if the fluid has a density of 810 kg/m³; and

   b. use trapezoids to numerically integrate the local velocities to estimate the average velocity for the flow.

Pitot-Tube Data	
r (mm)	**($p_a - p_b$) (kPa)**
125	0.00
100	1.18
75	3.77
50	6.51
25	8.49
0	9.22
−25	8.50
−50	6.49
−75	3.76
−100	1.19
−125	0.01

17

Numerical Integration Techniques for Differential Equations Using Excel

Objectives

After reading this chapter, you will know

- How to integrate a single, first-order ordinary differential equation (ODE) by using Euler's method
- How to use a fourth-order Runge–Kutta method to integrate a single, first-order ODE
- How to rewrite a second-order ODE as two, first-order ODEs

- How to use a fourth-order Runge–Kutta method to integrate simultaneous ODEs
- How to use Excel's matrix math functions to integrate a partial differential equation (PDE), using an implicit method

17.1 INTRODUCTION

There are two types of *integration* that engineers routinely perform: integration for the *area under a curve* and integration of a *differential equation*. The two are related mathematically, but procedurally they are handled quite differently. This chapter presents numerical methods for integrating differential equations.

Three methods are presented:

1. Euler's method
2. Runge–Kutta methods
3. Implicit method

The process for transforming a second-order differential equation into two first-order equations is also presented, because this technique makes the first-order solution methodologies applicable to second- (and higher) order differential equations.

17.2 EULER'S METHOD

Mathematical models of physical processes are often described by differential equations. Although complex models are better handled outside of Excel, many simple differential equations can be integrated in a worksheet. The simplest numerical integration method is *Euler's method*.

17.2.1 Choosing an Integration Step Size

Numerical integration methods typically work forward step by step through either space or time (the *independent variable*). To demonstrate Euler's method, we will consider a specific example (Example 17.1), shown next.

EXAMPLE 17.1

SINGLE ODE

The physical system being considered in this example is the washout of an inert material from a well-stirred tank. Initially, the tank contains a known concentration of the inert material, called component A. Then, at time $t = 0$, a flow is started both into and out of the tank. This flow contains no component A, so the A in the tank begins to wash out with the outlet flow. This process is described mathematically by the following differential equation:

$$\frac{dC_A}{dt} = \frac{1}{\tau}\left[C_{A_{\text{inf}}} - C_A\right]. \tag{17.1}$$

Here,

C_A is the concentration of A in the tank and in the tank effluent (mg/mL),
$C_{A_{\text{inf}}}$ is the concentration of A in the influent to the tank (mg/mL),
τ is the residence time of the system ($\tau = V/\dot{V}$),
V is the tank volume (constant) (mL), and
$\dot{V}$ is the influent and effluent volumetric flow rate (mL/s).

To integrate the equation, we must know the initial concentration in the tank and some parameter values, such as the concentration of A in the influent, the tank volume, and the flow rate:

Required Information	
Concentration of A in tank initially:	100 mg/mL
Concentration of A in influent:	0 mg/mL
Tank volume:	10 L
Flow rate:	100 mL/s

(continued)

The differential equation can be integrated analytically to yield

$$C_A = C_{A_{\text{inf}}} + (C_{A_{\text{inf}}} - C_{A_{\text{inf}}}) e^{-t/\tau}, \tag{17.2}$$

where

$C_{A_{\text{init}}}$ is the concentration of A initially in the tank (mg/mL).

We will use the analytical solution to check the accuracy of our numerical integration results.

In this example, we have a time derivative of concentration, so we will integrate forward in time. Euler's method will be used to work forward from a known concentration at a point in time to compute new concentrations (*dependent variable*) at new times; this is the solution to our problem. We must choose an *integration step size,* or *time step,* Δt. The choice needs to be made with some care. Euler's method uses current information and an algebraic approximation of the differential equation to predict what will happen in the future. The method will be much more successful at predicting into the near future than into the distant future.

Problems generally have some kind of information that gives you a clue about the *time scale* of the problem. In this example, it is in the vessel *residence time,* τ. The residence time can be viewed as the length of time required to fill the tank (10 L, or 10,000 mL) at the stated flow rate (100 mL/s). Euler's method will succeed at predicting the changing concentration in the tank only if the integration time step is small compared with τ. The residence time for Example 17.1 is $\tau = 100$ seconds. Our choice of Δt should be small compared with this.

For this problem, a reasonable time step size might be 1 second; that is only 1% of τ. Instead, we will use a 20-second time step to demonstrate the negative impact of a relatively large time step on the accuracy of the solution. The time step will be built into a cell of the worksheet, so that it can easily be changed later to obtain a more accurate solution, which will be demonstrated later in this chapter.

Solution Procedure

The following steps are required to use Euler's method to integrate the differential equation:

1. Enter problem parameters (tank volume, flow rate, influent concentration) into the worksheet.
2. Enter the integration parameter (time step, Δt).
3. Compute a working variable (residence time, τ)
4. Enter the initial condition (concentration in the tank at time zero).
5. Compute the new time and concentration in the tank (and tank effluent) after one time step.
6. Copy the formulas used in Step 5 to compute the effluent concentrations at later times.

    The numerical solution is complete at this point; however, the following steps will be performed to check the accuracy of the computed solution:
7. Compute the concentration at each time, using the analytical solution.
8. Calculate the percentage error at each time.

Step 1. Entering Problem Parameters into the Worksheet. It is common to place *parameter* values (values common to the entire problem) near the top of the worksheet. This is simply a matter of style, not a requirement, but it does keep the parameters easy to reach when entering formulas.

The parameters for our problem include the 10 L (10,000 mL) volume of the tank, the influent and effluent flow rates of 100 mL/s (they are equal in this problem, so the liquid volume stays constant), and the concentration of component A in the influent stream, which is zero in this problem. These are entered into the worksheet as shown in Figure 17.1.

Figure 17.1
Setting parameter values in the worksheet.

	A	B	C	D	E	F
1	Single ODE: Euler's Method					
2						
3		Parameter	Value	Units		
4		V	10000	mL		
5		V_{dot}	100	mL/sec		
6		C_{A_inf}	0	mg/mL		
7						

Step 2. Enter the Integration Parameter, Δt. The size of the integration step (a time step in this problem) was chosen to be 20 seconds for this example, which, as mentioned before, is a fairly large time step relative to the residence time. This large step size was chosen primarily to demonstrate a noticeable error in the numerical solution results when compared with the analytical results. The integration step size can easily be reduced later to produce a more accurate solution. The chosen time step is entered in cell C7, as shown in Figure 17.2.

Figure 17.2
Setting the integration step size.

	A	B	C	D	E	F
1	Single ODE: Euler's Method					
2						
3		Parameter	Value	Units		
4		V	10000	mL		
5		V_{dot}	100	mL/sec		
6		C_{A_inf}	0	mg/mL		
7		Δt	20	sec		
8						

Step 3. Compute Working Variables. The residence time τ appears in the differential equation and will be used for each calculation, so it is convenient to calculate it from the problem parameter values. This is illustrated in Figure 17.3.

Step 4. Enter the Initial Conditions. The initial conditions begin the actual presentation of the solution. To integrate the differential equation, a starting value of the dependent variable (concentration in this example) must be available. The problem statement specifies that the concentration of A in

Figure 17.3
Calculating the tank's
residence time.

	A	B	C	D	E	F
1	**Single ODE: Euler's Method**					
2						
3		**Parameter**	**Value**	**Units**		
4		V	10000	mL		
5		V_{dot}	100	mL/sec		
6		C_{A_inf}	0	mg/mL		
7		Δt	20	sec		
8		τ	100	sec		
9						

the tank initially is 100 mg/mL. This value is entered in cell C12 as the first concentration value in the solution, and a zero is entered as the initial time in cell B13, as shown in Figure 17.4.

Figure 17.4
Establishing the initial
condition.

	A	B	C	D	E	F
1	**Single ODE: Euler's Method**					
2						
3		**Parameter**	**Value**	**Units**		
4		V	10000	mL		
5		V_{dot}	100	mL/sec		
6		C_{A_inf}	0	mg/mL		
7		Δt	20	sec		
8		τ	100	sec		
9						
10	**Solution**					
11		**Time (sec)**	**C_A (mg/mL)**			
12		0	100.00			
13						

Step 5. Compute Time and Effluent Concentrations after One-Integration Step.
The time at the end of the integration step is simply the previous time plus the time step

$$t_{new} = t_{previous} + \Delta t, \tag{17.3}$$

which can be entered into cell B13 as

B13:   =B12+$C$7

The dollar signs in this formula will allow it to be copied down the worksheet to compute future times; the copied formulas will always reference the specified time step in cell C7.

To compute the new effluent concentration at the end of the integration step, we use Euler's method. Euler's method simply involves replacing the differential equation by an algebraic approximation. Using a forward finite-difference approximation for the time derivative, our differential equation becomes

$$\frac{C_{A_{new}} - C_{A_{old}}}{\Delta t} = \frac{1}{\tau}(C_{A_{inf}} - C_{A_{old}}). \qquad (17.4)$$

Either the new or the old concentration value could be used on the right side of this equation. When the old (i.e., currently known) concentration value is used, the integration technique is called an *explicit technique* because the new (unknown) concentration can be solved for directly. When the new (currently unknown) concentration is used, the technique is called *implicit* because the new concentration value appears on both sides of the equation and (typically) cannot be solved for directly.

For now, we will use an explicit technique, with the known value of concentration, $C_{A_{old}}$, on the right side of the finite-difference equation.

Solving the finite-difference equation for the new concentration at the end of the time step gives

$$C_{A_{new}} = C_{A_{old}} + \frac{\Delta t}{\tau}(C_{A_{inf}} - C_{A_{old}}). \qquad (17.5)$$

This equation is built into cell C13 as the formula

C13:   =C12+($C$7/$C$8)*($C$6-C12)

as shown in Figure 17.5.

Figure 17.5
Determining time and concentration after one time step.

C13 ▼		f_x	=C12+($C$7/$C$8)*($C$6-C12)			
	A	B	C	D	E	F
1	Single ODE: Euler's Method					
2						
3		Parameter	Value	Units		
4		V	10000	mL		
5		V_{dot}	100	mL/sec		
6		C_{A_inf}	0	mg/mL		
7		Δt	20	sec		
8		τ	100	sec		
9						
10	Solution					
11		Time (sec)	C_A (mg/mL)			
12		0	100.00			
13		20	80.00			
14						

When the formula in cell C13 is copied down the worksheet to new cells, it will always reference the cell just above it as the "old" C_A, but the absolute referencing (dollar signs) on $\Delta t/\tau$ and $C_{A_{inf}}$ ensures that the parameter values will always be used in the copied formulas.

Step 6. Copy the Formulas Down the Worksheet. At this point, the rest of the integration can be accomplished simply by copying the formulas entered in Step 5 down to rows 14 and below; each new row represents a new integration step. Copying the formulas in cells B13 and C13 down through row

37 results in computing the results for 25 integration steps (500 s). The first 18 time steps of the complete solution are shown in Figure 17.6, and the entire solution is shown graphically.

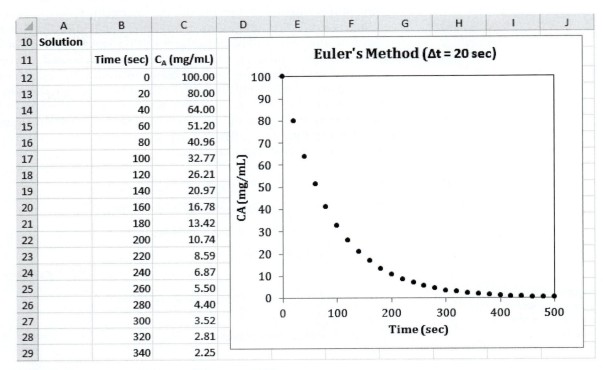

	A	B	C
10	Solution		
11		Time (sec)	C_A (mg/mL)
12		0	100.00
13		20	80.00
14		40	64.00
15		60	51.20
16		80	40.96
17		100	32.77
18		120	26.21
19		140	20.97
20		160	16.78
21		180	13.42
22		200	10.74
23		220	8.59
24		240	6.87
25		260	5.50
26		280	4.40
27		300	3.52
28		320	2.81
29		340	2.25

Figure 17.6
The numerical solution.

Five hundred seconds was chosen as the span of the integration, because it represents five residence times (5τ) for this system. For a wash-out from a perfectly mixed tank, the concentration should fall off by 95% after three residence times. (That is easily seen from the analytical solution.) Therefore, most of the "action" would be expected to occur within the first three residence times, and the effluent concentration should be nearly zero by 5τ. This appears to be the case in Figure 17.6.

Step 7. Calculate the Analytic Result. It is not common to have an *analytical solution* available. (If an analytical solution is available, there is little point in integrating the differential equation numerically.) It is common, however, to develop solution procedures using differential equations for which analytical solutions are available to provide a true value against which to judge the accuracy of the solution method. This is the case here.

The analytical solution, shown here, will be incorporated into the worksheet:

$$C_A = C_{A_{inf}} + (C_{A_{inf}} - C_{A_{inf}})e^{-t/\tau}. \tag{17.6}$$

This equation is entered into cell D13 as the formula

D13:   =$C$6+($D$12-$C$6)*EXP(-B13/$C$8)

and then copied to cells D13 through D37. The result (numerical and graphical) is shown in Figure 17.7.

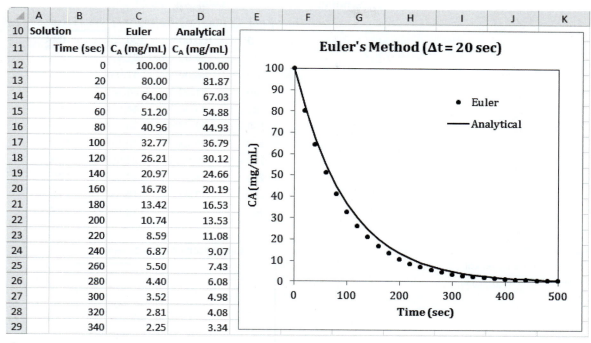

	A	B	C	D
10	Solution		Euler	Analytical
11		Time (sec)	C_A (mg/mL)	C_A (mg/mL)
12		0	100.00	100.00
13		20	80.00	81.87
14		40	64.00	67.03
15		60	51.20	54.88
16		80	40.96	44.93
17		100	32.77	36.79
18		120	26.21	30.12
19		140	20.97	24.66
20		160	16.78	20.19
21		180	13.42	16.53
22		200	10.74	13.53
23		220	8.59	11.08
24		240	6.87	9.07
25		260	5.50	7.43
26		280	4.40	6.08
27		300	3.52	4.98
28		320	2.81	4.08
29		340	2.25	3.34

Figure 17.7
Comparing the numerical and analytical results.

Step 8. Calculating the Error in the Numerical Solution. *Fractional error* can be computed as

$$\text{Fractional Error} = \frac{C_{A_{true}} - C_{A_{Euler}}}{C_{A_{true}}}. \tag{17.7}$$

A typical cell entry is

E13:   =(D13-C13)/D13

The results (partial listing) are shown in Figure 17.8. Note that the fractional error has been displayed in percent formatting; effectively, the fractional error has been increased by a factor of 100 and shown with a percent symbol.

When using a relatively large time step for the integration, the errors are clearly significant and getting worse (on a percentage basis) as the integration proceeds. In Figure 17.9, the percent error is plotted over the integration period.

To improve the accuracy of the result, the integration step size in cell C7 should be reduced. This will require more integration steps to cover the same span (i.e., 500 s), but adding the steps is accomplished simply by copying the last row of formulas down the worksheet as far as necessary. The graphs in Figure 17.10 and Figure 17.11 show the results when the step size is reduced to $\Delta t = 2$ seconds.

Figure 17.8
Computing the fractional
errors.

	A	B	C	D	E	F
10	Solution		Euler	Analytical		
11		Time (sec)	C_A (mg/mL)	C_A (mg/mL)	Frac. Error	
12		0	100.00	100.00	0%	
13		20	80.00	81.87	2%	
14		40	64.00	67.03	5%	
15		60	51.20	54.88	7%	
16		80	40.96	44.93	9%	
17		100	32.77	36.79	11%	
18		120	26.21	30.12	13%	
19		140	20.97	24.66	15%	
20		160	16.78	20.19	17%	
21		180	13.42	16.53	19%	
22		200	10.74	13.53	21%	

E12 ▾ f_x =(D12-C12)/D12

Figure 17.9
The errors in the numerical
solution.

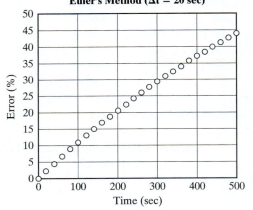

Figure 17.10
Comparing numerical and
analytical results with time
step = 2 seconds.

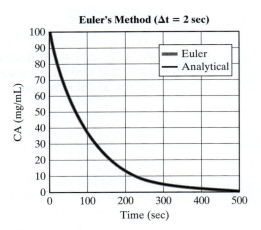

Note: In Figure 17.10, the way the graph is presented was changed because
there is so little difference between the two results.

Figure 17.11

Fractional errors with the smaller time step.

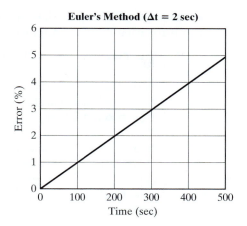

There is still some error, but the magnitudes of the errors have been greatly reduced by using the smaller integration step.

17.3 FOURTH-ORDER RUNGE–KUTTA METHOD

Runge–Kutta methods also use what is currently known about the variables (e.g., time and concentration) and the differential equation to predict the new values of the variable(s) at the end of the integration step. The difference is in how they compute the new values. The heart of the fourth-order Runge–Kutta method is a *weighted-average derivative estimate,* called dD_{avg} here (this nomenclature comes about by using D to stand for some arbitrary dependent variable value and dD to represent the derivative of that dependent variable):

$$dD_{avg} = \frac{1}{6}[dD_1 + 2\ dD_2 + 2\ dD_3 + dD_4]. \qquad (17.8)$$

Those four derivative values, dD_1 through dD_4, must be calculated from the differential equation and the currently known values of concentration (the dependent variable in our example). They could be computed directly in a worksheet, but the solution process gets messy. Instead, we will write a function to carry out one-integration step using Runge–Kutta's fourth-order technique, and we will call it **runge4**. But first we need another function that can be used to calculate the derivative value at any point. We'll call this function **dDdI,** because it will be used to calculate the derivative of a dependent variable (dD) with respect to an independent variable (dI). In the tank-washout problem (Example 17.1), the dependent variable is concentration, and the independent variable is time.

17.3.1 Evaluating the Derivative: Function dDdI

Since, in our example, C_A is the dependent variable, and t is the independent variable, the generic $\dfrac{dD}{dI}$ can be written as $\dfrac{dC_A}{dt}$ using the variables in Example 17.1.

Our differential equation is

$$\frac{dC_A}{dt} = \frac{1}{\tau}(C_{A_{inf}} - C_A). \qquad (17.9)$$

So, the left side is the derivative we need to evaluate, and it is equal to the algebra on the right side of the equation. We will use the algebra on the right side to evaluate the values of the derivative at any point in time. The **dDdI** function, written in VBA (Visual Basic for Applications) from Excel, looks as follows:

```
Public Function dDdI(Ival As Double, Dval As Double) As Double

    Dim Tau As Double
    Dim Cinf As Double
    Dim C As Double

    Tau = 100      'seconds
    Cinf = 0       'mg/ml

    C = Dval
    dDdI = (1/Tau) * (Cinf - C)

End Function
```

The independent and dependent variable values are passed into the function as parameters *Ival* and *Dval*, and the function returns a double-precision result through variable *dDdI*. The value passed into the function as Dval has been transferred to variable *C* just to make the last equation more recognizable as coming from the right side of our differential equation.

The independent variable passed into the function as Ival has not been used in this function. The independent variable is being passed to function **dDdI** for compatibility with function **runge4**. It is common practice to write Runge–Kutta functions that pass both the independent and dependent values to the derivative function in case the evaluation of the derivative for a particular problem requires both values.

17.3.2 How the Fourth-Order Runge–Kutta Method Works

The fourth-order Runge–Kutta method uses four estimates of the derivative to compute a weighted-average derivative that is used with the current value of concentration (in our example) to predict the next concentration value at the end of the time step. The four derivative estimates include the following:

First-Derivative Estimate

The **dDdI** function is used with the concentration value at the beginning of the integration step, $C_{A_{old}}$, to find the first-derivative estimate, called dD_1. This derivative estimate, shown as a slope in Figure 17.12, is used to project a concentration, C_1, halfway across the integration step.

Figure 17.12
First-derivative estimate.

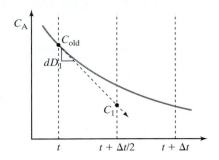

Mathematically, C_1 is computed as:

$$C_1 = C_{old} + dD_1 \cdot \frac{\Delta t}{2}. \tag{17.10}$$

Second-Derivative Estimate

The **dDdI** function is then used with the concentration value C_1 to calculate the second-derivative estimate, dD_2. This is illustrated in Figure 17.13.

Figure 17.13
Second-derivative estimate.

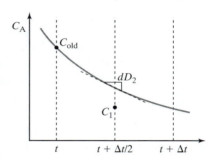

Then slope dD_2 is used back at time t to project another concentration estimate, C_2, halfway across the integration step. This is illustrated in Figure 17.14.

$$C_2 = C_{old} + dD_2 \cdot \frac{\Delta t}{2}. \tag{17.11}$$

Figure 17.14
Second concentration estimate.

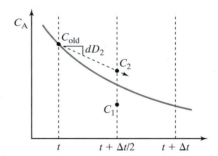

Third-Derivative Estimate

The **dDdI** function is then used with the concentration value C_2 to calculate the third-derivative estimate, dD_3, as illustrated in Figure 17.15.

Figure 17.15
The third-derivative estimate.

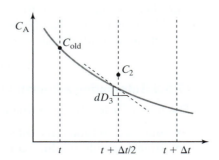

Shouldn't the second- and third-derivative values be the same? They are calculated for the same point in time, but they are based on different *estimates* of the concentration at $t + \Delta t/2$. If the estimates are equal, the derivatives calculated by using the estimates will be equal, but this is not the usual situation.

The third-derivative estimate, dD_3, is used with the original concentration to project the concentration at the end of the time step, as illustrated in Figure 17.16. This concentration estimate is called C_3.

$$C_3 = C_{\text{old}} + dD_3 \cdot \Delta t. \tag{17.12}$$

Figure 17.16
Estimating the third
concentration.

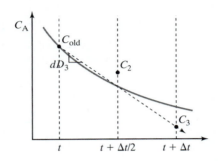

Fourth-Derivative Estimate

The **dDdI** function is used one last time with concentration value C_3 to calculate the fourth-derivative estimate, dD_4, as illustrated in Figure 17.17.

Figure 17.17
The fourth-derivative
estimate.

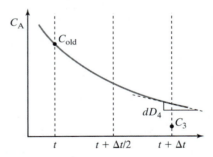

Once the four derivative estimates are available, a weighted-average derivative, dD_{avg}, is calculated:

$$dD_{\text{avg}} = \frac{1}{6}[dD_1 + 2\,dD_2 + 2\,dD_3 + dD_4]. \tag{17.13}$$

Note that the derivative estimates at the center of the time step are weighted more heavily than those at the beginning and end of the step.

The fourth-order Runge–Kutta method uses this weighted-average estimate of the derivative to predict the new concentration at the end of the integration step:

$$C_{\text{new}} = C_{\text{old}} + dD_{\text{avg}} \cdot \Delta t. \tag{17.14}$$

17.3.3 Automating a Fourth-Order Runge–Kutta: Function runge4()

The fourth-order Runge–Kutta method requires:

- four calls of function **dDdI**, to calculate derivative values dD_1 through dD_4, plus
- three dependent-value (concentration) calculations, then

- one weighted-average derivative calculation, and finally
- one calculation of the dependent variable value at the end the integration step.

These calculations can be done in the worksheet, but a function can be written to simplify the process:

```
Public Function runge4(Dt As Double, Ival As Double, Dval As
Double) As Double

    Dim D(3) As Double
    Dim dD(4) As Double
    Dim dDavg As Double

    dD(1) = dDdI(Ival, Dval)      'first-derivative estimate
    D(1) = Dval+dD(1)*Dt/2        'first concentration estimate (in
                                    this example)

    dD(2) = dDdI(Ival, D(1))      'second-derivative estimate
    D(2) = Dval+dD(2)*Dt/2        'second concentration estimate

    dD(3) = dDdI(Ival, D(2))      'third-derivative estimate
    D(3) = Dval+dD(3)*Dt          'third concentration estimate

    dD(4) = dDdI(Ival, D(3))      'fourth-derivative estimate

    dDavg = (1/6)*(dD(1)+2*dD(2)+2*dD(3)+dD(4)) 'Calculate
                                                  weighted-average
                                                  derivative value
    rk4=Dval+dDavg*Dt             'calculate and
                                    return next
                                    concentration
                                    value

End Function
```

If this function is written in VBA from Excel, it will be stored as part of the workbook and can be called from any worksheet in the workbook. Remember that with Excel 2007 and 2010, you must use macro-enabled workbooks to store VBA code.

Note: The program lines in the **dDdI** function depend on the equation that is being solved; the function must be modified for each differential equation you solve. The **runge4** function is independent of the equation being solved. This **runge4** function handles only one ODE, not simultaneous ODEs.

17.3.4 Implementing Fourth-Order Runge–Kutta in a Worksheet

The process of creating the worksheet begins, much as does the Euler method, by entering the parameters and initial conditions, as is illustrated in Figure 17.18.

The time at the end of the first integration step is computed by adding Δt to the previous time value, the zero stored in cell B12:

$$\text{B13:} \quad = \text{B12+\$C\$7}$$

We are purposely using a large time step again, this time to demonstrate how much more accurate the Runge–Kutta technique is, compared to Euler's method.

Figure 17.18
Preparing to integrate with
function **runge4**.

	A	B	C	D	E	F
1	Single ODE: Fourth-Order Runge-Kutta Method					
2						
3		Parameter	Value	Units		
4		V	10000	mL		
5		V_{dot}	100	mL/sec		
6		C_{A_inf}	0	mg/mL		
7		Δt	20	sec		
8		τ	100	sec		
9						
10	Solution		RK4	Analytical		
11		Time (sec)	C_A (mg/mL)	C_A (mg/mL)	Frac. Error	
12		0	100.00	100.00	0.00%	
13						

The concentration at the end of the first integration step is computed by using the **runge4** function in cell C13, as shown in Figure 17.19.

$$\text{C13:} \quad \text{=runge4}(\$C\$7, \text{B12}, \text{C12})$$

Figure 17.19
Calculating the results after
one-integration step.

C13		f_x	=runge4($C$7,B12,C12)			
	A	B	C	D	E	F
1	Single ODE: Fourth-Order Runge-Kutta Method					
2						
3		Parameter	Value	Units		
4		V	10000	mL		
5		V_{dot}	100	mL/sec		
6		C_{A_inf}	0	mg/mL		
7		Δt	20	sec		
8		τ	100	sec		
9						
10	Solution		RK4	Analytical		
11		Time (sec)	C_A (mg/mL)	C_A (mg/mL)	Frac. Error	
12		0	100.00	100.00	0.00%	
13		20	81.87	81.87	0.00%	
14						

The analytical solution and percent-error calculation are unchanged from the Euler solution presented earlier.

Once the formulas in cells B13 through E13 have been entered, simply copy those cells down the page to complete the solution. The results (at least the first 320 seconds) are shown in Figure 17.20.

Once the **runge4** function has been created, using the fourth-order Runge–Kutta technique is no more effort than using Euler's method, but the results are

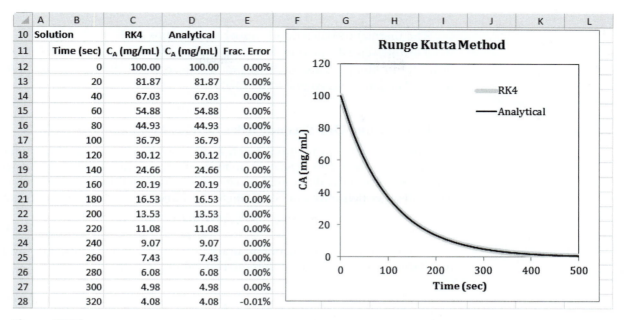

Figure 17.20
The completed integration.

much better. In this example, the solutions, even with a large time step of 20 seconds, are identical to two decimal places for the entire solution.

17.4 INTEGRATING TWO SIMULTANEOUS ODES BY USING THE RUNGE–KUTTA METHOD

Solving two *simultaneous ODEs* by using the fourth-order Runge–Kutta technique will be demonstrated here with the conduction–convection problem described in Example 17.2.

EXAMPLE 17.2

CONDUCTION WITH CONVECTION

A copper rod is placed between two containers: one contains boiling water and the other ice water. The rod is 1meter in length and is of diameter 4 cm (or 0.04 m). Ambient air at 25°C flows over the surface of the rod, causing the rod to exchange energy with the air stream. The heat-transfer coefficient for this convective heat transfer between the rod and the air is $h = 50 \text{ W/m}^2$ K. Calculate the temperature profile along the rod at steady state.

This steady-state conduction problem can be described mathematically as

$$\frac{d^2T}{dx^2} = \frac{4\,h}{D\,k}(T - T_{\text{amb}}) \quad \text{(second-order ODE)},$$

with boundary conditions

$$T = 100 \text{ C at } x = 0,$$
$$T = 0 \text{ C at } x = 1 \text{ m}.$$

(*continued*)

The thermal conductivity of copper is $k = 390$ W/m K.

The differential equation described in Example 17.2 is a second-order ODE. In order to use Runge–Kutta methods, we will need to rewrite the second-order differential equation as two first-order ODEs.

17.4.1 Rewriting One Second-Order ODE as Two First-Order ODEs

Because good solution methods (such as the fourth-order Runge–Kutta method) exist for first-order ODEs, it is common to rewrite higher-order differential equations as a series of first-order differential equations. This is easily done if you remember that a second-order derivative can be written as the derivative of a first-order derivative:

$$\frac{d^2 T}{dx^2} = \frac{d}{dx}\left[\frac{dT}{dx}\right]. \tag{17.15}$$

If the portion of the equation in brackets is simply given a new name, say, F (for first derivative), the original differential equation can be rewritten as

$$\frac{dF}{dx} = \frac{4\,h}{D\,k}(T - T_{\text{amb}}), \tag{17.16}$$

and the definition of F becomes the second differential equation:

$$\frac{dT}{dx} = F. \tag{17.17}$$

The original second-order differential equation has become two simultaneous first-order differential equations.

17.4.2 Boundary-Value Problems and the Shooting Method

The two first-order differential equations must be solved simultaneously. With the Runge–Kutta method used here, starting values of each dependent variable (T and F) must be known in order to carry out the integration. *Boundary-value problems* such as this one (Example 17.2) specify the value of a dependent variable at each end of the system (each boundary). We know the temperature at $x = 0$ and $x = 1$ m. We don't know anything about F at $x = 0$. For problems such as this, the *shooting method* is required.

The shooting method gets its name from artillery practice. The gunners aim the cannon, shoot, and see where the shell lands; then they adjust the angle on the cannon and try again. Eventually, they hit their target. The equivalent of "aiming the cannon" is setting the value of F. We will choose a value for F at $x = 0$, integrate the differential equation from $x = 0$ to $x = 1$ m, and see whether we get the right value of T (0°C) at $x = 1$ m. If we don't get the right boundary value, we choose a new F value at $x = 0$ and integrate again.

17.4.3 Modifying the runge4 Function for Two ODEs: Function runge4two

The changes required to function `runge4` are the following:

- To receive an integer code to indicate which equation is being integrated. (eqN = 1 for the dT/dx equation; eqN = 2 for the dF/dx equation).

- To receive two dependent variables (*T* and *F* in this example) that will be given generic names D1val and D2val.
- To call two different derivative functions, which we will call **dD1dI** and **dD2dI,** to calculate the derivative of each of the two differential equations.

Note: No programmer will consider this an elegant solution. It is a simple modification of function **runge4** that allows two ODEs to be integrated. There is a lot of room for improvement in this function, such as making it possible to handle more than two ODEs by passing the dependent variable values as a matrix. Many advanced math packages provide built-in Runge–Kutta integration routines.

The modified **runge4** function will be called **runge4two** and is listed here:

```
Public Function runge4two(eqN As Integer, Dt As Double, Ival
  AsDouble, _D1val As Double, D2val As Double) As Double

Dim D(3) As Double
Dim dD(4) As Double
Dim dDavg As Double

    Select Case eqN
        Case 1 'first differential equation
            dD(1)=dD1dI(Ival, D1val, D2val)
            D(1)=Dval+dD(1)*Dt/2
            dD(2)=dD1dI(Ival, D(1), D2val)
            D(2)=Dval+dD(2)*Dt/2
            dD(3)=dD1dI(Ival, D(2), D2val)
            D(3)=Dval+dD(3)*Dt
            dD(4)=dD1dI(Ival, D(3), D2val)
            dDavg=(1/6) * (dD(1)+2*dD(2)+2*dD(3)+dD(4))

            runge4two=D1val+dDavg*Dt

        Case 2 'second differential equation
            dD(1)=dD2dI(Ival, D1val, D2val)
            D(1)=Dval+dD(1)*Dt/2
            dD(2)=dD2dI(Ival, D1val, D(1))
            D(2)=Dval+dD(2)*Dt/2
            dD(3)=dD2dI(Ival, D1val, D(2))
            D(3)=Dval+dD(3)*Dt
            dD(4)=dD2dI(Ival, D1val, D(3))
            dDavg=(1/6) * (dD(1)+2*dD(2)+2*dD(3)+dD(4))

            runge4two=D2val+dDavg*Dt
        End Select
End Function
```

New Derivative Functions: dD1dI and dD2dI

The derivative functions are specific to the problem being solved. The functions required for this problem are listed here:

```
Public Function dD1dI(Ival As Double, D1val As Double, D2val As
    Double) As Double

    'used with equation dT/Dx=F
    '
    'D1val is T
    'D2val is F
    dD1dI=D2val

End Function

Public Function dD2dI(Ival As Double, D1val As Double, D2val As
    Double) As Double

    'used with equation dF/dx=(4 h)/(D k) * (T-Tamb)
    '
    'D1val is T
    'D2val is F
    Dim h As Double
    Dim D As Double
    Dim k As Double
    Dim T As Double
    Dim Tamb As Double
    h=50 'W/m2 K
    D=0.04 'm
    k=386 'W/m K
    Tamb=25 '°C
    T=D1val
    dD2dI=(4*h)/(D*k) * (T-Tamb)

End Function
```

Solving the ODEs

To allow the integration step size to be varied, Δx is specified as a parameter near the top of the worksheet (see Figure 17.21). A Δx value equal to one-tenth the length of the rod is a very large integration step. We use such a large step here to be able to easily show the entire solution on a page, but we will reduce the step size later to check the accuracy of the solution.

Figure 17.21
Setting up the worksheet for integration.

◢	A	B	C	D	E	F
1	Integrating Simultaneous ODEs with the Runge-Kutta Method					
2	Steady-State Conduction Along a Rod					
3						
4		Δx:	0.1	m		
5						
6		Position (m)	Temp. (°C)	F (°C/m)		
7		0.00	100.00	-100.00		
8						

Then the initial position ($x = 0$), temperature at $x = 0$ ($100°C$), and guess for F at $x = 0$ are also entered. In Figure 17.21, a guess value of -100 has been used for F. Here's why...

If the final temperature profile was a straight line from 100°C at $x = 0$ to 0°C at $x = 1$ m, the F value would be $-100°C/m$. The temperature profile will not be a straight line because of the energy transfer to the moving air, but 100°C/m is a reasonable (orderofmagnitude) first guess for F.

Next, the x position at the end of the first integration step is calculated as

B8:   =B7+$C$4

The temperature at the end of the first integration step is found, using **runge4two,** as

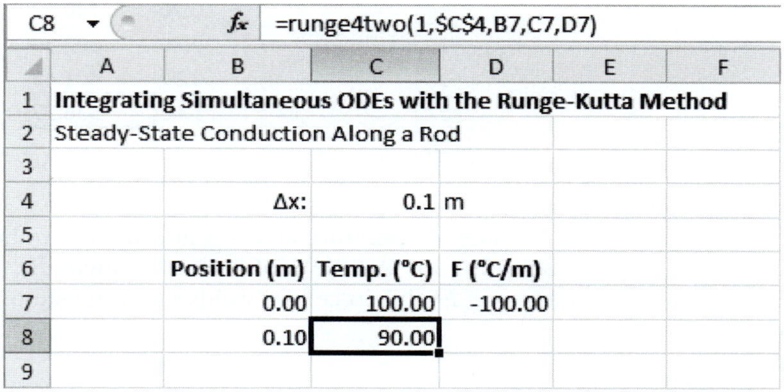

C8:   =runge4two(1,$C$4,B7,C7,D7)

In this equation, the "1" tells the function to integrate the first differential equation $(dT/dx = F)$, the $C$4 passes the integration step size ($\Delta x = 0.1$ m) to the function, and the remaining three arguments pass the current values of position, temperature, and F to the function. At this point, the worksheet looks like Figure 17.22.

Figure 17.22
Calculating T after one-integration step using function **runge4two**.

C8 ▾		f_x	=runge4two(1,$C$4,B7,C7,D7)		
	A	B	C	D	E	F
1	Integrating Simultaneous ODEs with the Runge-Kutta Method					
2	Steady-State Conduction Along a Rod					
3						
4		Δx:	0.1 m			
5						
6		Position (m)	Temp. (°C)	F (°C/m)		
7		0.00	100.00	-100.00		
8		0.10	90.00			
9						

The value of F at the end of the first integration step is calculated in similar fashion:

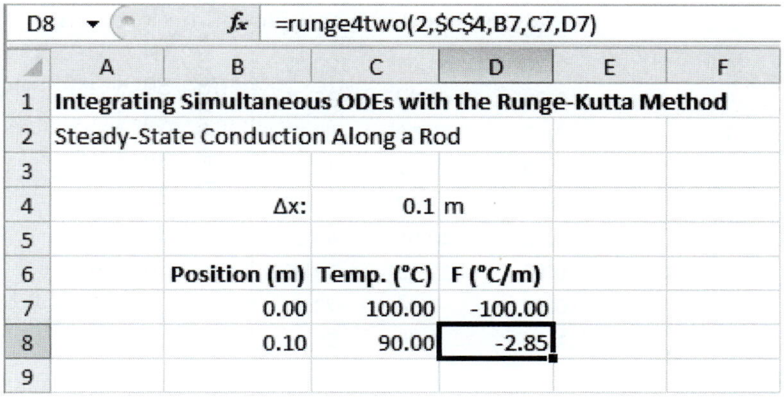

D8:   =runge4two(2,$C$4,B7,C7,D7)

Here the "2" tells **runge4two** to integrate the second differential equation (dF/dx). The result is shown in Figure 17.23.

Figure 17.23
Calculating F after one-integration step using function **runge4two**.

D8 ▾		f_x	=runge4two(2,$C$4,B7,C7,D7)		
	A	B	C	D	E	F
1	Integrating Simultaneous ODEs with the Runge-Kutta Method					
2	Steady-State Conduction Along a Rod					
3						
4		Δx:	0.1 m			
5						
6		Position (m)	Temp. (°C)	F (°C/m)		
7		0.00	100.00	-100.00		
8		0.10	90.00	-2.85		
9						

To complete the integration and solve for temperatures across the rod, simply copy the formulas in row 8 down the sheet to row 17, as shown in Figure 17.24.

Figure 17.24
The complete (but incorrect) integration, based on the guessed F value at x = 0.

	A	B	C	D	E	F
1		Integrating Simultaneous ODEs with the Runge-Kutta Method				
2		Steady-State Conduction Along a Rod				
3						
4			Δx:	0.1 m		
5						
6			Position (m)	Temp. (°C)	F (°C/m)	
7			0.00	100.00	-100.00	
8			0.10	90.00	-2.85	
9			0.20	89.72	81.35	
10			0.30	97.85	165.17	
11			0.40	114.37	259.54	
12			0.50	140.32	375.30	
13			0.60	177.85	524.68	
14			0.70	230.32	722.67	
15			0.80	302.59	988.63	
16			0.90	401.45	1348.20	
17			1.00	536.27	1835.83	
18						

The temperature at $x = 1$ m came out to be 536°C, not quite the 0°C we were shooting for. A little trial and error (actually, I used Excel's Goal Seek) with the starting F value in cell D7 yields the result shown in Figure 17.25.

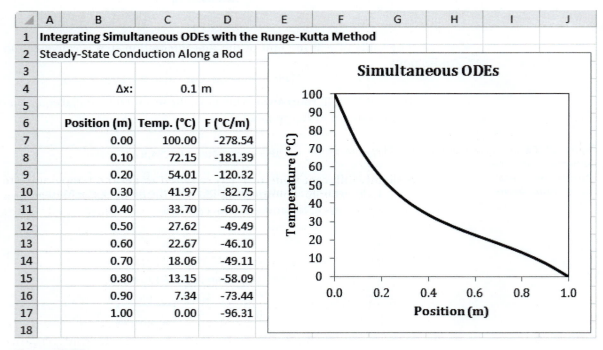

	A	B	C	D	E	F	G	H	I	J
1		Integrating Simultaneous ODEs with the Runge-Kutta Method								
2		Steady-State Conduction Along a Rod								
3										
4		Δx:	0.1 m							
5										
6		Position (m)	Temp. (°C)	F (°C/m)						
7		0.00	100.00	-278.54						
8		0.10	72.15	-181.39						
9		0.20	54.01	-120.32						
10		0.30	41.97	-82.75						
11		0.40	33.70	-60.76						
12		0.50	27.62	-49.49						
13		0.60	22.67	-46.10						
14		0.70	18.06	-49.11						
15		0.80	13.15	-58.09						
16		0.90	7.34	-73.44						
17		1.00	0.00	-96.31						
18										

Figure 17.25
The solution, with a step size of 0.1 m.

17.4.4 Checking for Accuracy

Numerical integration of differential equations always produces approximate results. The approximation gets better as the size of the integration step size goes to zero. It never hurts to reduce the size of the integration step to see whether the calculated results change significantly. If they do, the size of the step is affecting the results, and your integration step size might need to be made even smaller.

In Figure 17.26, the entire solution was recalculated (so that the two solutions could be compared) with a step size reduced by the factor 2 to $\Delta x = 0.5$. Twice as many integration steps were required, so the formulas were copied to more cells (row 7 to row 27).

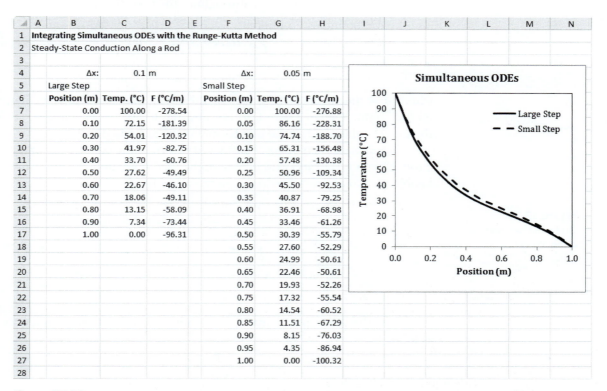

Figure 17.26

Comparing two solutions with different integration step sizes.

A slightly different value of F was required to get the temperature to 0°C at $x = 1$ m, so the solution definitely was affected by the size of the integration step. It would be a good idea to continue to make the step size smaller until the changes in the solution become insignificant.

17.5 IMPLICIT INTEGRATION METHODS

17.5.1 Using the Method of Lines to Get the General Finite-Difference Equation

The *method of lines* is used to convert a *PDE* into a series of ODEs, one per *grid point,* where the number of grid points is chosen as part of the solution process. The ODEs can thenbe solved for the temperatures at each grid point, either directly or simultaneously as a set of simultaneous linear equations.

EXAMPLE 17.3

PDE: UNSTEADY CONDUCTION

This example considers unsteady conduction along a metal rod. Mathematically, the problem can be summarized as

$$\frac{\partial T}{\partial t} = \alpha \frac{\partial^2 T}{\partial x^2} \qquad \text{partial differential equation (PDE)} \qquad (17.18)$$

subject to the initial condition

$$T = 35°C \qquad \text{throughout the rod at time zero} \qquad (17.19)$$

and the two boundary conditions

$$T = 100°C \qquad \text{at } x = 0,$$

and

$$T = 100°C \qquad \text{at } x = 1 \text{ m,} \qquad (17.20)$$

where α is the thermal diffusivity of the metal and has a prespecified value of $0.01 \text{ m}^2/\text{min}$. The goal is to find out how the temperature changes with time for the first 10 minutes after the end temperatures are raised to 100°C.

The example we have just considered describes the conduction of energy down a rod. Eleven (arbitrary) grid points will be used here, uniformly distributed over x between $x = 0$ and $x = 1$ m as shown in Figure 17.27.

Figure 17.27
Grid point locations along the rod.

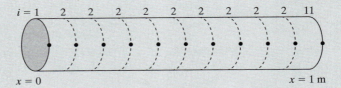

When the differential equation is written in terms of finite differences, it looks as follows:

$$\frac{T_{i_{\text{new}}} - T_{i_{\text{old}}}}{\Delta t} = \alpha \frac{T_{i-1} - 2T_i + T_{i+1}}{(\Delta x)^2} \qquad (i \text{ ranges from 1 to 11.)} \qquad (17.21)$$

Here, Δx is the distance between grid points, and Δt is the size of the integration step, which will be chosen as part of the solution process.

We need to rearrange these 11 equations to get one temperature by itself on the left side, set equal to an algebraic combination of other variables (T's, α, Δx). Depending on how the equations are rearranged, the temperatures on the right side can be the "old" temperatures (known at the beginning of each integration step) or the "new" temperatures (to be computed during each integration step). If the old temperature values are used on the right side, each of the 11 equations can be solved directly for the new temperatures—this is called an *explicit* solution technique. If the new temperatures are used on the right side, the 11 equations must be solved simultaneously for the 11 new temperature values—an *implicit* technique. We will use the implicit solution technique here:

$$\frac{T_{i_{\text{new}}} - T_{i_{\text{old}}}}{\Delta t} = \alpha \frac{T_{i-1_{\text{new}}} - 2T_{i_{\text{new}}} + T_{i+1_{\text{new}}}}{(\Delta x)^2} \qquad \text{general finite} - \text{difference equation.}$$

$$(17.22)$$

17.5.2 Eliminating Fictitious Points by Using Boundary Conditions

Whenever you use the method of lines to turn a PDE into a series of ODEs, you generate a problem on each end of the spatial domain. At $x = 0$ ($i = 1$), the equation becomes

$$\frac{T_{1_{new}} - T_{1_{old}}}{\Delta t} = \alpha \frac{\{T_{0_{new}}\} - 2\,T_{1_{new}} + T_{2_{new}}}{(\Delta x)^2}, \quad \text{when } i = 1, \qquad (17.23)$$

and the temperature at $i = 0$ is needed, but $i = 0$ doesn't exist; it is outside of the spatial domain of the problem. This is called a *fictitious point. Boundary conditions* are used to eliminate fictitious points before integration. In this case, the temperature at $x = 0$ ($i = 1$) is constant at 100°C throughout the integration. If the temperature at $i = 1$ never changes, then

$$T_{1_{new}} = T_{1_{old}} \quad \text{at } i = 1 \ (x = 0). \qquad (17.24)$$

This simpler equation can be used instead of the general finite-difference equation at $i = 1$.

Similarly, when $x = 1$ m ($i = 11$), the general finite-difference equation becomes

$$\frac{T_{11_{new}} - T_{11_{old}}}{\Delta t} = \alpha \frac{T_{10_{new}} - 2\,T_{11_{new}} + \{T_{12_{new}}\}}{(\Delta x)^2}, \quad \text{when } i = 11, \qquad (17.25)$$

and there is a fictitious point (T_{12}). Again, in this problem, the temperature at $x = 1$ m ($i = 11$) is constant at 100°C, so we can replace the general finite-difference equation evaluated at $i = 11$ with

$$T_{11_{new}} = T_{11_{old}} \quad \text{at } i = 11 (x = 1 \text{ m}). \qquad (17.26)$$

Summary of Equations to Be Solved
The 11 equations to be solved simultaneously are summarized in the following table:

Location	Grid Point	Equation
$x = 0$ m	$i = 1$	$T_{1_{new}} = T_{1_{old}}$
$x = 0.1$ m to 0.9 m	$i = 2$ to 10	$\dfrac{T_{i_{new}} - T_{i_{old}}}{\Delta t} = \alpha \dfrac{T_{i-1_{new}} - 2T_{i_{new}} + T_{i+1_{new}}}{(\Delta x)^2}$
$x = 1$ m	$i = 11$	$T_{11_{new}} = T_{11_{old}}$

In these equations, the T_{new} values will be the unknowns. Moving all of the unknown temperatures to the left side and the known (old) temperatures to the right side prepares the way for solving the equations via matrix methods:

Location	Grid Point	Equation
$x = 0$ m	$i = 1$	$1 T_{1_{new}} = T_{1_{old}}$
$x = 0.1$ m to 0.9 m	$i = 2$ to 10	$-\Phi T_{i-1_{new}} + (1 + 2\Phi) T_{i_{new}} - \Phi T_{i+1_{new}} = T_{i_{old}}$
$x = 1$ m	$i = 11$	$1 T_{11_{new}} = T_{11_{old}}$

Here

$$\Phi = \frac{\alpha\,\Delta t}{(\Delta x)^2}. \qquad (17.27)$$

These equations can be written in matrix form as a coefficient matrix C multiplying an unknown vector T_{new} set equal to a right-hand-side vector r, that is:

$$[C][T_{new}] = [r], \tag{17.28}$$

where

$$C = \begin{bmatrix} 1 & 0 & 0 & 0 & 0 & 0 & 0 & 0 & 0 & 0 & 0 \\ -\Phi & 1+2\Phi & -\Phi & 0 & 0 & 0 & 0 & 0 & 0 & 0 & 0 \\ 0 & -\Phi & 1+2\Phi & -\Phi & 0 & 0 & 0 & 0 & 0 & 0 & 0 \\ 0 & 0 & -\Phi & 1+2\Phi & -\Phi & 0 & 0 & 0 & 0 & 0 & 0 \\ 0 & 0 & 0 & -\Phi & 1+2\Phi & -\Phi & 0 & 0 & 0 & 0 & 0 \\ 0 & 0 & 0 & 0 & -\Phi & 1+2\Phi & -\Phi & 0 & 0 & 0 & 0 \\ 0 & 0 & 0 & 0 & 0 & -\Phi & 1+2\Phi & -\Phi & 0 & 0 & 0 \\ 0 & 0 & 0 & 0 & 0 & 0 & -\Phi & 1+2\Phi & -\Phi & 0 & 0 \\ 0 & 0 & 0 & 0 & 0 & 0 & 0 & -\Phi & 1+2\Phi & -\Phi & 0 \\ 0 & 0 & 0 & 0 & 0 & 0 & 0 & 0 & -\Phi & 1+2\Phi & -\Phi \\ 0 & 0 & 0 & 0 & 0 & 0 & 0 & 0 & 0 & 0 & 1 \end{bmatrix}$$

$$T_{new} = \begin{bmatrix} T_{1_{new}} \\ T_{2_{new}} \\ T_{3_{new}} \\ T_{4_{new}} \\ T_{5_{new}} \\ T_{6_{new}} \\ T_{7_{new}} \\ T_{8_{new}} \\ T_{9_{new}} \\ T_{10_{new}} \\ T_{11_{new}} \end{bmatrix}, \qquad r = \begin{bmatrix} T_{1_{old}} \\ T_{2_{old}} \\ T_{3_{old}} \\ T_{4_{old}} \\ T_{5_{old}} \\ T_{6_{old}} \\ T_{7_{old}} \\ T_{8_{old}} \\ T_{9_{old}} \\ T_{10_{old}} \\ T_{11_{old}} \end{bmatrix} \tag{17.29}$$

The initial right-hand-side vector contains the "old" temperatures that are known from the problem statement's initial condition. The problem doesn't actually start until the instant the end temperatures change to 100°C, so the end-grid-point initial temperatures are set at 100°C, while all interior temperatures are initially set at 35°C.

Solving the Matrix Problem Using Excel

As usual, the solution process begins with the entering of problem parameters into the worksheet, as shown in Figure 17.28.

The formulas entered in column C are

$$\begin{array}{lll} \text{C8:} & \text{=C6/(C7-1)} & \text{calculates } \Delta x \\ \text{C9:} & \text{=(C4*C5)/C8^2} & \text{calculates } \Phi \end{array} \tag{17.30}$$

Establishing the Coefficient Matrix

The coefficient matrix is mostly zeroes (a *sparse matrix*), with nonzero elements located on the diagonal and one element off the diagonal. This type of matrix is called a *tridiagonal matrix*. An 11×11 tridiagonal matrix can quickly be created in a worksheet by:

1. Creating a column of 11 zeroes.
2. Copying that column to create an 11×11 array of zeroes.
3. Enter the values and formulas on and near the diagonal.

The result is shown in Figure 17.29.

Figure 17.28
Entering parameter values for the integration.

	A	B	C	D	E	F	G	H
1	**Partial Differential Equation, Implicit Solution**							
2	Unsteady Conduction Along a Rod							
3								
4		α:	0.01	m²/min	property of the metal			
5		Δt:	1	min	chosen (smaller is better)			
6		L:	1	m	specified rod length			
7		N:	11	points	chosen (more makes Δx smaller)			
8		Δx:	0.1	m				
9		Φ:	1	no units				
10								

Figure 17.29
The coefficient matrix, C.

	A	B	C	D	E	F	G	H	I	J	K	L	M
1	**Partial Differential Equation, Implicit Solution**												
2	Unsteady Conduction Along a Rod												
3													
4		α:	0.01	m²/min		property of the metal							
5		Δt:	1	min		chosen (smaller is better)							
6		L:	1	m		specified rod length							
7		N:	11	points		chosen (more makes Δx smaller)							
8		Δx:	0.1	m									
9		Φ:	1	no units									
10													
11	[C]	1	0	0	0	0	0	0	0	0	0	0	
12		-1	3	-1	0	0	0	0	0	0	0	0	
13		0	-1	3	-1	0	0	0	0	0	0	0	
14		0	0	-1	3	-1	0	0	0	0	0	0	
15		0	0	0	-1	3	-1	0	0	0	0	0	
16		0	0	0	0	-1	3	-1	0	0	0	0	
17		0	0	0	0	0	-1	3	-1	0	0	0	
18		0	0	0	0	0	0	-1	3	-1	0	0	
19		0	0	0	0	0	0	0	-1	3	-1	0	
20		0	0	0	0	0	0	0	0	-1	3	-1	
21		0	0	0	0	0	0	0	0	0	0	1	
22													

- Every cell in the coefficient matrix in Figure 17.29 that displays a "3" contains the same formula:

$$\text{C12:} \quad \texttt{=1+2*\$C\$9} \qquad \text{diagonal elements}$$

So you can enter the formula once and paste it into all of the interior (non-corner) diagonal cells.

- All of the cells in the coefficient matrix in Figure 17.29 that display a "1" (except the corner cells) hold the value of $-\Phi$, pulled from the parameter list with this formula:

$$\text{B11:} \quad \texttt{=-\$C\$9}$$

Again, you can enter the formula once, copy it, and paste it into all of the other cells that need the formula.

- The ones in the top-left and bottom-right corner represent the coefficients on the left side of the first and last equations:

$$i = 1 \qquad 1T_{1_{new}} = T_{1_{old}},$$
$$i = 11 \qquad 1T_{11_{new}} = T_{11_{old}}.$$

Inverting the Coefficient Matrix

In solving the simultaneous equations by using matrix methods, the coefficient matrix must be inverted. This is done by using Excel's **MINVERSE** function.

This is an array function, so there are a couple of things to remember when entering the function:

- Select the cells that will receive the inverted matrix to set the size of the result matrix before entering the array function.
- After typing the formula into the top-left cell of the result matrix, press [Ctrl-Shift-Enter] to enter the formula into each of the cells in the result matrix.

While having to determine the size of the result is something of a nuisance, Excel's array functions do have advantages. The principal advantages are the ability to recalculate your matrix calculations automatically if any of the input data change, and the ability to copy matrix formulas by using relative addressing to repeat calculations. This will be extremely useful as we finish this example.

The coefficient matrix is inverted by using

{B22:L32}   =MINVERSE(B10:L20)   [Ctrl-Shift-Enter] to finish

The result is shown in Figure 17.30.

Figure 17.30

Inverting the coefficient matrix.

	A	B	C	D	E	F	G	H	I	J	K	L	M
10													
11	[C]	1	0	0	0	0	0	0	0	0	0	0	
12		-1	3	-1	0	0	0	0	0	0	0	0	
13		0	-1	3	-1	0	0	0	0	0	0	0	
14		0	0	-1	3	-1	0	0	0	0	0	0	
15		0	0	0	-1	3	-1	0	0	0	0	0	
16		0	0	0	0	-1	3	-1	0	0	0	0	
17		0	0	0	0	0	-1	3	-1	0	0	0	
18		0	0	0	0	0	0	-1	3	-1	0	0	
19		0	0	0	0	0	0	0	-1	3	-1	0	
20		0	0	0	0	0	0	0	0	-1	3	-1	
21		0	0	0	0	0	0	0	0	0	0	1	
22													
23	[C$_{inv}$]	1	0	0	0	0	0	0	0	0	0	0	
24		0.38	0.38	0.15	0.06	0.02	0.01	0	0	0	0	0	
25		0.15	0.15	0.44	0.17	0.06	0.02	0.01	0	0	0	0	
26		0.06	0.06	0.17	0.45	0.17	0.07	0.02	0.01	0	0	0	
27		0.02	0.02	0.06	0.17	0.45	0.17	0.07	0.02	0.01	0	0	
28		0.01	0.01	0.02	0.07	0.17	0.45	0.17	0.07	0.02	0.01	0.01	
29		0	0	0.01	0.02	0.07	0.17	0.45	0.17	0.06	0.02	0.02	
30		0	0	0	0.01	0.02	0.07	0.17	0.45	0.17	0.06	0.06	
31		0	0	0	0	0.01	0.02	0.06	0.17	0.44	0.15	0.15	
32		0	0	0	0	0	0.01	0.02	0.06	0.15	0.38	0.38	
33		0	0	0	0	0	0	0	0	0	0	1	
34													

The right-hand-side vector, *r*, contains the known (old) temperatures, which are the time-zero solution (*initial condition*) to the problem. They are entered on the worksheet as both the [*r*] matrix and the solution at time zero, as shown in Figure 17.31.

Figure 17.31
Setting the initial condition.

	A	B	C	D	E	F	G	H	I	J	K	L	M
34													
35	time:	0											
36													
37	[r]	100											
38		35											
39		35											
40		35											
41		35											
42		35											
43		35											
44		35											
45		35											
46		35											
47		100											
48													

As the integration proceeds, the computed temperature values will be displayed in columns C through L. To keep track of the times corresponding to the integration steps, the time is computed in row 35 by adding Δt to each previous time. The formula in cell C35 is:

C35:   =B35+$C$5   adds Δt from cell C5 to the previous time in cell B35

This calculation is shown in Figure 17.32.

Figure 17.32
Calculating the time after one-integration step.

| C35 | ▾ | | f_x | =B35+$C$5 | | | | | | | | |

	A	B	C	D	E	F	G	H	I	J	K	L	M
34													
35	time:	0	1										
36													
37	[r]	100											
38		35											
39		35											
40		35											
41		35											
42		35											
43		35											
44		35											
45		35											
46		35											
47		100											
48													

To solve for the temperatures at each grid point at the end of the first integration step, simply multiply the inverted coefficient matrix and the initial right-hand-side vector:

$$[C]^{-1}[r] \tag{17.31}$$

The multiplication is handled by Excel's **MMULT** array function. Because the new matrix result is temperatures at each grid point, the size of the result matrix will be the same as the size of [r]:

$$\{C37:C47\} \quad =\text{MMULT}(\$B\$23:\$L\$33, B36:B47)$$

The result is shown in Figure 17.33.

Figure 17.33
Calculating the temperatures after one-integration step.

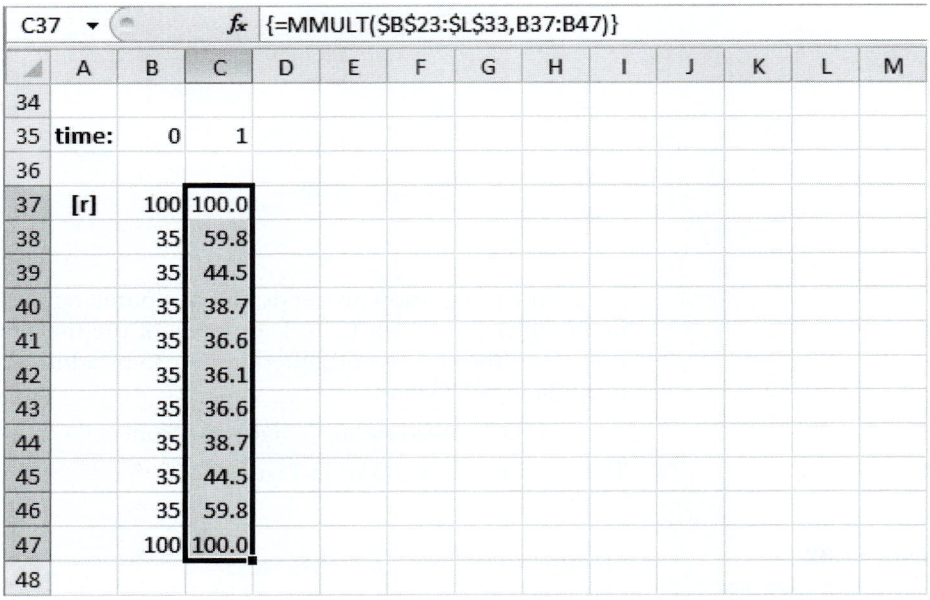

C37	▼		f_x	{=MMULT($B$23:$L$33,B37:B47)}								
◢	A	B	C	D	E	F	G	H	I	J	K	L	M
34												
35	time:	0	1									
36												
37	[r]	100	100.0									
38		35	59.8									
39		35	44.5									
40		35	38.7									
41		35	36.6									
42		35	36.1									
43		35	36.6									
44		35	38.7									
45		35	44.5									
46		35	59.8									
47		100	100.0									
48												

Note: Be sure to use an absolute reference (dollar signs) on the [C inv] matrix ($B$23:$L$33); it becomes important when the result is copied to integrate over additional time steps.

Completing the Solution

This is the step in which the power of Excel's array functions really becomes apparent. To calculate the results at additional times, simply copy cells C35:C46 to the right. The more copies you create, the more time steps you are integrating over. The temperatures at each time step are displayed in rows 37 through 47, and the elapsed time is displayed above each temperature result. Temperature profiles for the first 10 minutes are shown in Figure 17.34.

Figure 17.34
The completed integration for the first 10 minutes.

◢	A	B	C	D	E	F	G	H	I	J	K	L	M
34													
35	time:	0	1	2	3	4	5	6	7	8	9	10	
36													
37	[r]	100	100.0	100.0	100.0	100.0	100.0	100.0	100.0	100.0	100.0	100.0	
38		35	59.8	71.0	76.9	80.5	83.0	84.9	86.5	87.8	89.0	90.0	
39		35	44.5	53.1	59.6	64.6	68.6	71.8	74.5	76.9	79.1	81.0	
40		35	38.7	43.8	49.0	53.7	58.0	61.9	65.4	68.5	71.3	73.9	
41		35	36.6	39.6	43.4	47.6	51.8	55.8	59.6	63.1	66.4	69.4	
42		35	36.1	38.4	41.8	45.7	49.8	53.8	57.7	61.3	64.7	67.8	
43		35	36.6	39.6	43.4	47.6	51.8	55.8	59.6	63.1	66.4	69.4	
44		35	38.7	43.8	49.0	53.7	58.0	61.9	65.4	68.5	71.3	73.9	
45		35	44.5	53.1	59.6	64.6	68.6	71.8	74.5	76.9	79.1	81.0	
46		35	59.8	71.0	76.9	80.5	83.0	84.9	86.5	87.8	89.0	90.0	
47		100	100.0	100.0	100.0	100.0	100.0	100.0	100.0	100.0	100.0	100.0	
48													

KEY TERMS

Analytical solution
Boundary conditions
Boundary-value problems
Dependent variable
Differential equation
Euler's method
Explicit technique
Fictitious point
Fractional error
Grid point

Implicit technique
Independent variable
Initial condition
Integration
Integration step size
Method of lines
Ordinary differential equation (ODE)
Parameter
Partial differential equation (PDE)

Residence time
Runge–Kutta methods
Shooting method
Simultaneous ODEs
Sparse matrix
Time scale
Time step
Tridiagonal matrix
Weighted-average derivative estimate

SUMMARY

Euler's Method

The chief virtue of Euler's method is its simplicity; unfortunately, it can be fairly inaccurate unless you use a very small integration step size.

1. Rewrite the ODE using finite differences (converts calculus to algebra).
2. Solve for the new (unknown) value of the dependent variable.

3. Integrate as follows:

   a. Enter problem parameters into the worksheet.

   b. Enter the integration step.

   c. Enter the initial condition.

   d. Compute the new value of the independent variable using the integration step size.

   e. Compute the new value of the dependent variable using the algebra from Step 2.

   f. Copy the formulas used in Step e to finish the integration.

Fourth-Order Runge–Kutta Method

The fourth-order Runge–Kutta method requires that you write a function, but the function is pretty simple, and the Runge–Kutta method yields much better results with larger integration steps.

1. Rearrange the ODE to obtain only the derivative of the dependent variable on the left, with algebra on the right side of the equation.

2. Use the algebra on the right side of the equation to develop the **dDdI** function.

3. Integrate as follows:

   a. Enter problem parameters into the worksheet.

   b. Enter the integration step.

   c. Enter the initial condition.

   d. Compute the new value of the independent variable using the integration step size.

   e. Compute the new value of the dependent variable function **runge4**.

   f. Copy the formulas used in Step e to finish the integration.

Rewriting One Second-Order ODE as Two First-Order ODEs

1. Rewrite the second derivative as the derivative of a derivative

$$\frac{d^2 T}{dx^2} = \frac{d}{dx}\left[\frac{dT}{dx}\right].$$

2. Assign the derivative in brackets a variable name

   The two first-order ODEs are:

   - The original equation, now written as a first derivative of the new variable

   - The definition of the new variable (equal to the derivative of the original dependent variable)

Boundary-Value Problems and the Shooting Method

1. Guess the starting value for all unknown dependent variables.

2. Integrate across the solution domain.

3. Check to see if the integrated result agrees with the specified boundary values—if not, guess again.

You can often use Excel's Goal Seek to automate the iteration process.

PDEs

Use the method of lines to convert one PDE to a series of N ODEs, where N is the (chosen) number of grid points in the spatial domain.

Method of Lines

1. Use finite differences to convert the spatial derivatives (usually) to algebra. This generates one ODE for every grid point used with the finite differences.
2. The method of lines creates a fictitious point at each boundary; use boundary conditions to eliminate the fictitious points before integrating.

   PDEs can be integrated by using either explicit or implicit techniques. Implicit techniques, such as are illustrated in this chapter, tend to be more accurate for a given integration step size.

> **Note:** The fourth-order Runge–Kutta method does work for PDEs, but it is a pretty significant extrapolation from **runge4two,** the two-equation simultaneous ODE function used in this chapter, to a version for N simultaneous ODEs. Math packages such as Matlab and MathCad provide fourth-order Runge–Kutta functions for N simultaneous equations.

PROBLEMS

17.1 Concentration During a Wash-in

Example 17.1 in this chapter was a "washout" problem, in which a tank contained a specified concentration of some chemical species (called component A) in the example and no A in the influent flow. The concentration of A decreases with time as it is washed out of the tank.

Rework Example 17.1 as a "wash-in" problem. Assume that the initial concentration of A in the tank is zero, but that the influent contains A at a level of 100 mg/ml.

The analytical result still applies:

$$C_A = C_{A_{\text{inf}}} + (C_{A_{\text{inf}}} - C_{A_{\text{inf}}})\, e^{-t/\tau}. \tag{17.32}$$

Plot the concentration of A calculated by each method (Euler's method and analytical result) as a function of time.

17.2 Tank Temperature During a Wash-in

One evening, a few friends come over for a soak, but the hot tub has been turned off for days, so it has to be warmed up. As your friends turn on the hot water to warm up the tub, you want to know how long this is going to take and write out a quick energy balance on a well-mixed tank. You end up with a differential equation relating the temperature in the tank T to the temperature of the hot water entering the tank T_{in}, the volume of the tank V, and the volumetric flow rate of hot water $\dot{V}$:

$$\frac{dT}{dt} = \frac{\dot{V}}{V}(T_{\text{in}} - T). \tag{17.33}$$

If the initial temperature is 45°F (7.2°C) and you want to heat it to 100°F (37.8°C), you estimate that you will have plenty of time to solve this equation in Excel.

Suppose the hot water enters the tank at 120°F (54.4°C) at a rate of 30 liters per minute, and the tank contains 3000 liters of water.

Figure 17.35
Hot tub.

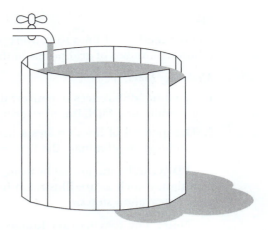

a. Integrate the differential equation, using either Euler's method or the fourth-order Runge–Kutta technique, to obtain tank-temperature values as a function of time.
b. Graph the temperature and time values.
c. Approximate how long it will take to heat up the tub to 100°F.
d. Compute how long it would take to heat the tub to 100°F if it was first drained and then refilled with hot water.

17.3 Steady-State Conduction

Example 17.2 used a heat-transfer coefficient of $h = 50$ W/m^{2}K, a value that might represent heat loss to a fast-moving air stream. Rework Example 17.2, but use the following heat-transfer coefficients (the h value is included in the user-written **dD2dI** function, which will have to be modified for this problem):

a. $h = 5$ W/m^{2}K (typical of energy transfer to slow-moving air)
b. $h = 300$ W/m^{2}K (typical of energy transfer to a moving liquid)

17.4 Unsteady Diffusion

A heat-transfer example was used in Example 17.3. The mass-transfer analog is a mass-diffusion problem, namely,

$$\frac{\partial C_A}{\partial t} = D_{AB}\frac{\partial^2 C_A}{\partial x^2}, \tag{17.34}$$

where

C_A is the concentration of A,
D_{AB} is the diffusivity of A diffusing through B—say, 0.00006 cm^2/sec,
t is time, and
x is position.

The units on C_A in this equation are not critical, because they are the same on both sides of the equation. However, D_{AB} is a constant only if the concentration of A is low.

One initial condition and two boundary conditions are needed to solve this problem:

$C_A = 0$ throughout the diffusion region ($0 \le x \le 10$ cm) at time zero;
$C_A = 100$ mg A/mL at $x = 0$ (all times);
$C_A = 0$ mg/mL at $x = 10$ cm (all times).

a. Use the method of lines to transform the PDE into a series of ODEs (one per grid point).

b. Use the boundary conditions to eliminate the fictitious points in the boundary ODEs.

c. Write out the coefficient matrix and right-hand-side vector required for the implicit method of solution.

d. Solve the systems of ODEs to find out how long it will take for the concentration at $x = 5$ cm to reach 20 mg/mL.

Index